Practical Statistical Mechanics

This book—the first of two companion volumes, together with a third dedicated entirely to exercises—is devoted to introducing the standard techniques of statistical mechanics; presenting a practical, operative, and attentive approach to the recurring structures that link different areas of statistical physics.

It strikes a balance between breadth and an accessible walk-through approach to calculations and concepts that students and researchers will encounter across the many corners of modern statistical mechanics. It is a broad, accessible starting point enabling readers to then move on to more focused and advanced treatments. It also touches on topics that have become increasingly relevant in recent years, including applications of machine learning to elements of quantum information and computing.

Chapters include short experimental boxes that connect theoretical concepts with laboratory demonstrations—from critical phenomena in fluids and magnets to frustrated spin systems. To support learning by doing, it also provides simple numerical codes that illustrate several topics throughout the text.

This will serve as an ideal textbook for undergraduate and graduate courses on Statistical Mechanics and Thermodynamics. It can also be used as supplementary reading for courses on Many-Body Physics.

Key Features:

- A one-stop compendium of major results of standard and modern Statistical Mechanics.
- Includes pedagogical boxes containing worked examples, historical perspectives, suggested further activities, and experimental notes and connections to real world systems.
- Each chapter is structured to serve as a two-hour lecture on the topic, with accompanying calculations.

Francesco Caravelli is a physicist whose research spans equilibrium and non-equilibrium statistical mechanics, dynamical and complex systems, and quantum technologies. He began his scientific career in quantum gravity at the University of Waterloo and the Perimeter Institute, later moving into complex systems during positions at the University of Oxford and University College London, followed by research in the private sector. Since joining Los Alamos National Laboratory in 2017, first as an Oppenheimer Fellow and later as a Staff Scientist in T4, he has investigated the dynamics of memristive and neuromorphic circuits, deriving exact results that link their memory properties to foundational principles of statistical physics. His recent work focuses on collective nonlinear dynamics, neuromorphic computing, low-power analog architectures, and emerging directions in quantum computing.

Diego A. R. Dalvit is a Senior Staff Member at the Theoretical Division of Los Alamos National Laboratory. He is a quantum optics theorist with expertise in Casimir physics, statistical mechanics, quantum sensing, and metamaterials. He received his Ph.D. in Physics from Universidad de Buenos Aires (Argentina) in 1998. He is a Fellow of the American Physical Society and Optica. He has authored more than 110 peer-reviewed papers, with more than 10,000 citations. He has also co-authored two physics textbooks, one a guide to the essence of Casimir physics, and one on problems on statistical mechanics.

Practical Statistical Mechanics

Thermodynamics and Statistical Mechanics, Volume 1

Francesco Caravelli and Diego A. R. Dalvit

CRC Press
Taylor & Francis Group
Boca Raton London New York

CRC Press is an imprint of the
Taylor & Francis Group, an informa business

Designed cover image: Shutterstock

First edition published 2027
by CRC Press
2385 NW Executive Center Drive, Suite 320, Boca Raton FL 33431

and by CRC Press
4 Park Square, Milton Park, Abingdon, Oxon, OX14 4RN

CRC Press is an imprint of Taylor & Francis Group, LLC

ISBN: 978-1-032-98204-5 (hbk)
ISBN: 978-1-032-99327-0 (pbk)
ISBN: 978-1-003-60352-8 (ebk)

DOI: 10.1201/9781003603528

Typeset in Latin Modern font
by KnowledgeWorks Global Ltd.

Publisher's note: This book has been prepared from camera-ready copy provided by the authors.

Dedication

Francesco Caravelli would like to dedicate this book to his father, mother, and sisters, and to his wife and son. In addition, a special mention to supportive friends and colleagues over the years.

Diego A. R. Dalvit dedicates this book to the memory of his grandparents.

Contents

Foreword

This Volume marks the beginning of a larger project: a three-part journey through statistical mechanics conceived between the two authors of this Volume, Francesco Caravelli and Diego A. R. Dalvit, and in its later stages with Alioscia Hamma and Juan Zanella. The present book—the first of two companion Volumes, together with a third dedicated entirely to exercises—is devoted to the standard techniques of statistical mechanics, presented with a direct spirit: practical, operative, and attentive to the recurring structures that link different areas of statistical physics. While the foundations remain those of standard books, we also touch on topics that have become increasingly relevant in recent years, from applications in machine learning to elements of quantum information and computing. The aim is to strike a balance between a broad coverage, together with a clear and detailed walk-through approach with calculations and concepts that students and researchers will encounter across many corners of modern statistical mechanics.

The second volume, written jointly by Caravelli, Zanella, and Dalvit, turns toward more advanced methods: exact solutions, non-equilibrium phenomena, disordered systems, quantum many-body systems, and the techniques that have grown around them. These subjects form the living frontier of statistical mechanics, where classical ideas meet quantum structure, and where new tools continue to emerge. The third volume, with the addition of coauthor Hamma, is deliberately and unapologetically practical: a collection of exercises that reflects our view that statistical mechanics is, for all its theoretical elegance, also a discipline grounded in computation, approximation, and hands-on technique. The choice to devote an entire volume to problems is a consequence of this philosophy. The third volume shares the spirit of a previous exercise book entitled *Problems on Statistical Mechanics* (Institute of Physics Publishing, Bristol, 1999), written by one of us (Dalvit) together with Jaime Frastai and Ian Lawrie. Some of the problems in Volume III have been taken from that previous book, and we are grateful to the other authors for their permission to reproduce them here.

These notes owe their existence to many colleagues, postdocs, students, and collaborators we encountered over the years. We are grateful to those with whom we have shared discussions, calculations, and ideas—too many to name fully, but whose influence touches these pages in ways both direct and indirect. Their questions and comments have often forced greater clarity; their enthusiasm has made the work worthwhile. We are particularly endebted to Lukasz Cincio, who serendipitously kickstarted this project.

Preface

In his *States of Matter*, David Goodstein famously reminds the reader that

> *Ludwig Boltzmann, who spent much of his life studying statistical mechanics, died in 1906, by his own hand. Paul Ehrenfest, carrying on the work, died similarly in 1933. Now it is our turn to study statistical mechanics. Perhaps it will be wise to approach the subject cautiously.*

With due respect to that warning, it is now ***your*** turn to embark on the same adventure.

This Volume is the first in a three-volume series. It represents the convergence of personal notes, lectures, and exercises assembled by Francesco Caravelli, Diego A. R. Dalvit, Alioscia Hamma, and Juan Zanella after many years of research and teaching in the field.

Statistical mechanics has evolved over more than 160 years since Maxwell's 1859 paper on the distribution of molecular velocities, and no finite collection can hope to capture every facet of such a vast subject. What we can offer, across thermodynamics and hard and soft condensed matter, is a coherent glimpse into its central ideas and methods. This also means that we faced many difficult decisions about what to include in these books: whether a topic deserved a full derivation, an exercise, a brief mention, or a dedicated box.

There are, of course, many excellent books on statistical mechanics, both classical and modern. Among the standard references, one may cite the texts by Fermi, Khinchin, Callen, Prigogine, Pathria and Beale, Huang, and Tolman, as well as more recent ones by Sethna, Gallavotti, Parisi, Kardar, and Mussardo, to mention only a fraction of what exists on statistical mechanics alone.

There may be many ways of solving a model, using a variety of techniques. Experts eventually learn all these approaches in their specific model of liking, but only after years of study and research. In these volumes, we aim to present many of these techniques side by side, and to include additional ones in the accompanying Volume III as exercises. As students, we often found ourselves scanning the literature in search of varied techniques before knowing which direction to pursue. These books are written with that exploratory stage in mind: a broad, accessible starting point before moving on to more focused and advanced treatments.

It is also worth stressing how little attention many textbooks devote to experimental aspects of statistical mechanics. Where possible, we have attempted to fill this gap by including short experimental boxes that connect theoretical concepts with laboratory demonstrations—from critical phenomena in fluids and magnets to frustrated spin systems.

To support learning by doing, we also provide simple numerical codes that illustrate several topics throughout the text. These are meant to serve not as polished software packages but as transparent, minimal examples that help the reader build intuition and experiment with the concepts introduced in the main chapters.

The word *Practical* in the title comes from a book that inspired one of us (Francesco Caravelli) for many years. This is *Practical Quantum Mechanics* by Flügge (Springer Berlin, Heidelberg, 1971), that explains quantum mechanics by solving problems. This is why in the original draft of these Volumes, we had solved exercises at the end of each chapter, while now they are in the aforementioned Volume III.

There are undoubtedly many books we do not know about, but we have made an effort to group the key references within each topical section. This process is, in fact, part of what convinced us to write these volumes.

We hope that this series will serve as a useful companion to the reader. The material has been developed and presented to the best of our understanding, to make this rich and beautiful subject accessible, engaging, and intellectually rewarding.

List of Figures

List of Tables

List of Symbols

Symbol	Meaning
k_B	Boltzmann constant
$\beta = 1/(k_B T)$	Inverse temperature
Z	Partition function
$\mathcal{G}$	Grand-canonical partition function
F	Helmholtz free energy
S	Entropy
$U = \langle H \rangle$	Internal energy
H	Hamiltonian / energy function
$\Omega(E)$	Density of states / micro-canonical multiplicity
ξ	Correlation length
χ	Susceptibility
C_V	Heat capacity
m	Magnetization / order parameter
$\langle \cdots \rangle$	Thermal expectation value
$\sigma_i = \pm 1$	Ising spin
$\vec{S}_i$	Classical spin vector
J_{ij}	Coupling between sites i and j
h_i, h	Local / uniform magnetic field
$\langle i, j \rangle$	Nearest-neighbor pair
A_{ij}	Adjacency matrix
d	Spatial dimension
a	Lattice spacing
$m(\mathbf{x})$	Coarse-grained order parameter
$\phi(\mathbf{x})$	Scalar field (LG theory)
$a(T), b(T)$	Landau expansion coefficients
t	Reduced temperature or LG mass
u	Quartic interaction strength
κ, K_1	Gradient stiffness
$\mathcal{F}[\phi]$	LG free-energy functional
$G(\mathbf{r}), G(q)$	Correlation function
$S(q)$	Structure factor
$\alpha, \beta, \gamma, \delta, \nu, \eta$	Critical exponents
D	Dimension (RG notation)
y_t, y_h	RG scaling dimensions
$\epsilon = 4 - D$	Expansion parameter
$\tilde{m}(\mathbf{q})$	Fourier transform of $m(\mathbf{x})$
q	Wavevector / momentum
$\int d^D q$	Momentum-space integral
$\hat{G}^{-1}(q) = q^2 + t$	Gaussian propagator denominator
K_ν	Modified Bessel function
b	RG rescaling factor

Λ	Momentum cutoff
K_0, K_1, K_2	LG couplings in RG
K'	Renormalized coupling
T_{ij}	Linearized RG matrix
K^*	RG fixed point
λ_i	RG eigenvalues
$\mathcal{R}$	RG transformation
$p(x)$	Probability distribution
$H(p)$	Shannon entropy
$D_{\mathrm{KL}}(p\|q)$	Kullback–Leibler divergence
$I(X:Y)$	Mutual information
$Q(x)$	Variational trial distribution
$F[Q]$	Variational free energy
$\ln Z$	Log-partition function
$b_i(x_i)$	Node belief (marginal)
$b_{ij}(x_i, x_j)$	Pair belief
ψ_i, ψ_{ij}	Local / pairwise potentials
F_{Bethe}	Bethe free energy
$m_{i\to j}$	BP message
$w_i \in \{0,1\}$	QUBO binary variable
M_{ij}, N_i	QUBO coefficients
C	Graph cut
X_C	Edge boundary of a cut
f_{ij}	Flow on edge
c_{ij}	Edge capacity
$G=(V,E)$	Graph
W_{uv}	Weight (TSP, cuts)
$\hat{H}(t)$	Time-dependent Hamiltonian
$\|\psi(t)\rangle$	Quantum state
Δ	Minimum spectral gap
$\Gamma(t), A(t)$	Annealing schedules
$\hat{\sigma}_i^{x,z}$	Pauli operators
$U(t)$	Time-evolution operator
$\theta_n(t)$	Dynamical phase
$\gamma_n(t)$	Berry phase
β_{eff}	Effective inverse temperature
r	Radial distance
N	Number of variables / spins
V	Volume or vertex set
E	Energy or edge set
d_i	Node degree
S_D	Surface of unit D-sphere

1

From the Theory of Fire to Classical Thermodynamics

1.1 Pre-Classical Thermodynamics: A Coarse Timeline of Events

Throughout history, fire has held an exceptionally significant role in numerous cosmological and physical systems developed over the ages [Mas70, Sas20]. Its importance to humankind is undeniable, and its presence has been consistently acknowledged and explored since ancient times. One pivotal theory that highlights this reverence for fire is the concept of the four elements—water, air, earth, and fire—which was first formulated by Empedocles around the fifth century BC. This theory endured for centuries, gaining acceptance until the eighteenth century.

In ancient Greek mythology, the fire was stolen from the Gods.

```
Prometheus stole fire and gave it to men. But when Zeus learned
of it, he ordered Hephaestus to nail his body to Mount Caucasus.
On it, Prometheus was nailed and kept bound for many years.
Every day an eagle swooped on him and devoured the lobes of
his liver, which grew by night.

    ---Apollodorus, The Library, book 1:7, second century BC
```

Thus, ancient Greeks had a fascination for the subject that, during the development of some of the most important philosophical advancements in history, also allowed the development of the sciences, and the sciences of fire, and in particular of hot and cold. Building upon Empedocles' groundwork, Aristotle expanded and refined the theory, introducing the idea of fundamental and contrary qualities: hot and cold, dry and moist. These qualities were believed to combine in non-contradictory ways to give rise to the four elements, with their respective proportions in various substances thought to determine their temperature.

The knowledge of dilation and thermal expansion in solids, liquids, and gases dates back to Classical Greece. Ingenious devices based on these principles were constructed by Philo of Byzantium (third-second century BC) and Heron of Alexandria (first century BC).

The historical journey through the theories of fire and temperature showcases humanity's fascination with these elemental aspects. From ancient philosophers to prominent scholars like Aristotle and Galen (a Greek doctor who worked in the Roman Empire), the quest to understand

DOI: 10.1201/9781003603528-1

and harness the power of fire has been an enduring pursuit that has shaped our understanding of the natural world. What was the theory of fire in ancient times is now the theory of thermodynamics, and its derivatives and extensions. The interest in fire and transformations was also considered to be magical at a certain point in history, even until more recent times. For instance, in the well-known book on magic by Arnold and Cahill [AC62], aside from normal "sleight-of-hand" tricks and so much more, there are whole sections on *Chemistry*, *Electricity*, *Galvanism*, and *Magnetism*.

The Aristotelian concept of heat faced challenges during the Scientific Revolution, leading to the emergence of two competing theories regarding heat. Followers of distant Greek atomism defended the materiality of fire, suggesting that it was composed of countless subtle, light, and mobile particles capable of penetrating ordinary matter to manifest their activity. On the other hand, a different theory explained fire through mechanical causes, like friction, without attributing it to direct particle action.

The lack of decisive rational or experimental evidence in favor of either theory made the choice difficult. Some, like Boyle in England, adopted a mixed theory that involved both fire particles and kinetic phenomena. Newton, on the other hand, maintained more ambiguous positions on the matter. Eventually, the corpuscular theory of heat gained prominence, driven by erroneous notions of calcination, the success of the phlogiston theory, and experimental confirmation of heat conservation in calorific processes.

The phlogiston theory was a scientific hypothesis proposed in the seventeenth century to explain the process of combustion and the burning of materials. According to this theory, substances that could burn contained a substance called *phlogiston*, which was released during combustion. This release of phlogiston was thought to be responsible for the observed changes during burning, such as the emission of heat and light.

The phlogiston theory attempted to explain various phenomena, including why substances like wood and coal decreased in weight after burning and why metals gained weight when they were calcined (heated in air). The theory suggested that when a material burned, it lost its phlogiston, leaving behind the *calx* (the ash or residue).

However, as scientific knowledge progressed and more experimental evidence accumulated, several problems and contradictions emerged with the phlogiston theory. One significant challenge was the discovery of oxygen by Antoine Lavoisier in the late eighteenth century. Lavoisier's experiments demonstrated that combustion was not the release of phlogiston, but the combination of substances with oxygen from the air.

Through meticulous experiments, Lavoisier showed that substances gained weight when they burned because they were combining with oxygen, not losing phlogiston. This directly contradicted the core idea of the phlogiston theory. Lavoisier's work led to the development of the concept of conservation of mass and the recognition of oxygen as an essential element in combustion processes.

As a result of these advancements, the phlogiston theory was gradually abandoned in favor of the oxygen theory of combustion, which paved the way for modern chemistry. This shift marked a fundamental transformation in scientific thinking, emphasizing the importance of empirical evidence and experimentation in shaping our understanding of natural phenomena.

This corpuscular theory held sway well into the nineteenth century, though the chemistry discoveries of the latter half of the eighteenth century brought continuous modifications to the theoretical interpretation of heat.

As time progressed, the concept of heat underwent further transformations. It shifted from being regarded as a corpuscular and ponderable matter to an indestructible, immaterial fluid known as caloric, as proposed by Lavoisier. However, the kinetic interpretation of heat persisted in the background, finding supporters throughout the eighteenth century such as Macquer, Davy, and Rumford, even as Laplace and others continued to defend the caloric theory.

The kinetic theory received notable support from the work of Rumford, who demonstrated through careful experiments that the generation of heat by friction seemed inexhaustible, implying that heat could not be a material substance. He argued that mechanical energy conversion was responsible for heat production. His research also showed that heat possesses no weight, reinforcing the idea that it is a form of motion rather than a distinct material entity.

Furthermore, the study of radiant heat provided additional evidence against the caloric theory. Numerous eighteenth-century experiments investigated the analogy between light and radiant heat, with most results pointing toward their shared nature. This growing body of data strengthened the mechanical interpretation of heat, further undermining the prevailing caloric view.

During the latter half of the eighteenth century, significant progress was made in understanding thermal phenomena, largely owing to the clarification of key concepts — notably, the distinction between temperature and quantity of heat, and the definitions of specific and latent heat.

Temperature is a physical concept that requires a measuring device, called a *thermometer*. *Thermometry*—the science of measuring temperature—originated from the basic human sensations of *hot* and *cold*, which can be experienced directly by touching or approaching an object. The results of such qualitative "measurements" might be described as *cold*, *warm*, *hot*, or *lukewarm*.

Early pioneers of thermometry lacked a precise physical understanding of what temperature represented, although they recognized that hot and cold corresponded to different physical states of matter. The ancient Greeks, for instance, knew that permanent magnets lose their magnetism when heated to very high temperatures. They also observed that the volume of a fluid or gas varies with the degree of hotness.

The history of thermometry is thus closely tied to the development of more refined temperature-measuring instruments. Galileo Galilei is often credited with inventing one of the earliest thermometers—or rather *thermoscopes*, devices that indicated changes in heat without providing numerical measurements. His design used a glass bulb containing air attached to a long tube immersed in water: changes in the air's volume caused the water level in the tube to rise or fall, providing a qualitative indication of temperature.

Later, Evangelista Torricelli's invention of the mercury barometer in 1643 laid the groundwork for understanding the relationship between temperature and atmospheric pressure, a crucial step toward the development of accurate thermometers.

Following this development, many other scholars became intrigued by building such instruments. They experimented with various substances, used different fixed temperatures as references (like candle flame, snow temperature, boiling water temperature, etc.), and proposed various arbitrary measurement scales.

Scales also allow us to check when two bodies are the same or different temperatures and, in particular, which of them is higher. However, although the measurement of temperature is important, as the development of the theory behind thermodynamics becomes more advanced, it also becomes extremely important to measure the condition of bodies with external conditions that are the same for every experiment. Anders Celsius and Jean-Pierre Christin made significant contributions to thermometry in the eighteenth century. Celsius introduced the *Celsius scale* in 1742, defining $0\,°\mathrm{C}$ as the boiling point and $100\,°\mathrm{C}$ as the freezing point of water (the reverse of today's convention). A year later, Christin proposed the *centigrade scale*, which used the modern orientation—$0\,°\mathrm{C}$ for the freezing point and $100\,°\mathrm{C}$ for the boiling point of water—and this version ultimately became the standard Celsius scale we use today.

Gabriel Fahrenheit, in the early eighteenth century, developed one of the first reliable alcohol thermometers and later the mercury thermometer, along with the *Fahrenheit temperature scale*, introduced in 1724. The Fahrenheit scale is still widely used today in the *United States*, as well as in the *Bahamas*, the *Cayman Islands*, and *Liberia* (with limited or mixed use reported in

Belize, *Palau*, the *Marshall Islands*, and the *Federated States of Micronesia*). The Celsius scale is used throughout most of the rest of the world.

Early thermometric measurements were often based on the melting or boiling points of easily available substances. For instance, butter was at one time used to gauge temperature by its melting point. Water later became the preferred reference, and by 1694 thermometers commonly used the reproducible freezing and boiling points of water as calibration marks. Temperature scales began to appear on thermometers around this same period.

A comparison of early temperature scales is shown in Fig. 1.2. In the eighteenth century, there was a growing belief that if heat were a tangible substance, it would possess a certain weight. Consequently, numerous attempts were made during this period to measure the weight of heat. Herman Boerhaave conducted experiments with iron but failed to detect any change in weight when heating a mass of the metal; see the first page of his book in Fig. 1.1 [Boe32]. However, other scientists of the eighteenth century reported different results. Georges-Louis Leclerc, Comte de Buffon, found that iron gained weight when heated, while John Roebuck and John Whitehurst reported the opposite—a decrease in weight upon heating. David Fordyce also conducted experiments with ice and liquid water, finding what he interpreted as evidence of a negative weight associated with heat.

Around 1732, Boerhaave proposed that the amount of heat contained in a body should be proportional to its volume, since temperature essentially measured the density of the caloric fluid. Prior to this, others had argued that the quantity of heat was directly proportional to the body's mass. However, experiments conducted by Fahrenheit involving mixtures of water and mercury challenged Boerhaave's proposition, leading to the first notable experimental discrepancy.

In 1739, George Martine further advanced the understanding of heat by observing that when water and mercury were heated together, the temperature of the mercury increased roughly twice as fast as that of the water. It was Joseph Black, professor of medicine and chemistry at the Universities of Glasgow and Edinburgh, who first clearly distinguished between two crucial concepts: *heat* and *temperature*. From the perspective of the caloric fluid theory, a thermometer measures the *density* of the fluid (temperature) rather than the total quantity of fluid (heat). In the kinetic theory viewpoint, temperature represents the *intensity* of the molecular motion that gives rise to heat. This understanding established that temperature is an *intensive variable*, independent of the mass of the substance, whereas heat is an *extensive property* that depends on the material's mass.

Black also introduced the concept of *latent heat* and performed pioneering experiments on heat absorption during melting and boiling processes. It was at this stage that thermometry and calorimetry began to develop as distinct branches of study. By measuring the equilibrium temperature of mixtures of different substances, Black deduced that the amount of heat contained in a body depends on its *heat capacity*, *mass*, and *temperature*. Heat capacity was understood as a constant property of each substance, independent of its temperature and of its interactions with other materials.

In 1831, the Swedish scientist Johan Carl Wilcke independently reached the same conclusion as Black, introducing the term *specific heat* instead of heat capacity. Specific heat is defined as the amount of heat required to raise the temperature of one unit mass of a homogeneous substance by $1\,^{\circ}\mathrm{C}$.

Additionally, important concepts such as the *heat of dissolution*, *heat of reaction*, and the *heat produced by living organisms* were also investigated. Together, these advances laid the groundwork for *calorimetry*—the quantitative study of thermal phenomena and processes.

The device used to measure heat is called a *calorimeter*. A calorimeter is an instrument designed to measure the heat absorbed or released during a chemical or physical process. It typically consists of two main parts: a container that holds the substance under study and a thermometer that records the corresponding temperature changes. When a chemical reaction or physical transformation takes place inside the calorimeter, the system either absorbs heat from its surroundings or releases heat into them, resulting in a measurable change in temperature within the calorimeter.

The calorimeter's container is usually insulated to minimize heat exchange with the surroundings, allowing for a more accurate measurement of the heat involved in the process. The thermometer is used to monitor the temperature change, as the heat absorbed or released is directly related to the temperature difference.

By measuring the temperature change and knowing the specific heat capacity of the substance under study, one can calculate the amount of heat involved in the process using the formula:

$$Q = mc\Delta T, \tag{1.1}$$

where Q is the heat absorbed or released, m is the mass of the substance, c is its specific heat capacity, and ΔT is the change in temperature.

Count Rumford (Benjamin Thompson), in the late eighteenth and early nineteenth centuries, investigated the relationship between heat and friction, challenging the prevailing caloric theory, which proposed the existence of an invisible heat fluid. The nineteenth century then saw the development of classical thermodynamics, with major contributions from *Sadi Carnot* and *Rudolf Clausius.* Carnot's work on the Carnot cycle laid the foundation for modern thermodynamics, as presented in his treatise *Réflexions sur la puissance motrice du feu et sur les machines propres à développer cette puissance.*

This shift forced the experimental community to focus on the external conditions of a gas or body, namely its *pressure* and *volume.* These conditions could be controlled by means of pistons or boundaries, such as the enclosing walls. Walls can be *adiabatic* or *diathermic*—idealized concepts introduced to describe different thermal behaviors. Adiabatic walls do not exchange any form of heat with the surroundings, whereas diathermic walls allow heat to be exchanged freely with the environment.

The first experiments aimed at understanding how heat was exchanged were carried out using water. In 1741, Georg Wolfgang Krafft, a physicist from the Saint Petersburg Academy of Sciences, discovered an arithmetical relationship connecting the final temperature (T) reached by a mixture of two masses of water with their initial temperatures (T_1, T_2) and masses (m_1, m_2). This relationship, described in the following section, became known as *Krafft's relationship.* His disciple and successor at the Academy, Georg Wilhelm Richmann, later refined Krafft's calculations by considering the influence of the experimental apparatus. The corrected expression is known as:

$$T = \frac{m_1 T_1 + m_2 T_2}{m_1 + m_2}. \tag{1.2}$$

Joseph Black further advanced this idea by replacing the masses with the corresponding heat capacities and generalizing *Richmann's formula*:

$$T = \frac{m_1 T_1 + m_2 T_2 + m_3 T_3 + \cdots}{m_1 + m_2 + m_3 + \cdots}. \tag{1.3}$$

Around 1769, Johan Carl Wilcke, a physicist at the Swedish Academy of Sciences, observed deviations from Richmann's law while conducting experiments. These discrepancies led him to the discovery of the *latent heat of fusion.* Similarly, Black had previously reached comparable conclusions, although his results had not yet been widely disseminated. Wilcke noticed that when he mixed two masses of liquid water, one at $T_1 = 0\,°\mathrm{C}$ and the other at temperature T_2, the final equilibrium temperature R followed Richmann's law. However, when he used ice instead of water at $0\,°\mathrm{C}$ in the same type of mixture, he obtained a different final temperature T, which no longer followed Richmann's law. This revealed a deficit of about $R - T = 72\,°\mathrm{C}$, which appeared to "disappear" or remain dormant while the ice melted. Black measured a similar deficit of approximately $78\,°\mathrm{C}$ by timing how long different mixtures of ice and water took to reach room temperature, taking into account the water equivalent of the containers used. The discovery of the *latent heat* associated with changes of state introduced a new calorimetric quantity and provided an explanation for natural phenomena such as the slow melting of snow in spring.

This improved understanding of *specific heat* made it possible to compare different substances, revealing that water possesses one of the highest specific heat values, whereas denser materials such as mercury exhibit much lower ones.

After grasping the distinction between heat and temperature, the path was paved for the interpretation and definition of other calorimetric quantities. As an early precursor to these definitions, the concept of *current heat power* for a body was introduced by Giovanni Marliani in the late fifteenth century, taking into account the body's volume and density. Subsequently, physicists at the Italian Academy of Cement explored the topic further and observed that equal masses of different liquids, when heated to the same temperature, melted unequal amounts of ice.

The differentiation between heat and temperature, along with the definitions of *specific heat* and *latent heat*, marked the emergence of *calorimetry* as a distinct scientific discipline. As the field advanced, other quantities such as the *heat of reaction* and the *heat of dissolution* were incorporated into its study. Toward the end of the eighteenth century, major research efforts in calorimetry were undertaken, including the one by Antoine-Laurent Lavoisier and Pierre-Simon Laplace [LL80].

The distinction between good conductors of heat, such as metals, and poor conductors, such as wood, had been known for a long time. However, it was not until the late eighteenth century that the first systematic experiments were conducted to understand the process of heat transfer from one body to another. In 1785, Jan Ingenhousz experimented to compare the thermal conductivities of various solids, particularly metal rods. He coated the rods with wax and then applied hot water to one end, observing that the wax melted at different rates along the rods, depending on their respective thermal conductivities. Ingenhousz, a botanist, was also the first to propose that plants perform photosynthesis.

Today we understand that thermal conductivity arises from the transfer of energy among neighboring molecules and, in the case of metals, from the motion of free electrons within the conducting medium. The rate of heat flow through a rod is directly proportional to its cross-sectional area A and to the temperature difference $T_2 - T_1$ between its ends, and inversely proportional to its length l. This empirical relationship is expressed as:

$$H = -k\frac{A}{l}(T_2 - T_1), \tag{1.4}$$

where H represents the heat flow rate and k is the thermal conductivity of the material [Ing85].

The negative sign in the equation arises from the fact that heat always flows from regions of higher temperature to regions of lower temperature, a fact that we also know to be connected to the second law of thermodynamics. His experiments were inspired by Benjamin Franklin, who wondered why silver teapots had wooden handles [Nat25].

Although there were qualitative experiments like Ingenhousz's, the theoretical and experimental investigation of heat propagation had to wait until the influential contributions of Fourier in the nineteenth century. His intervention marked a decisive turning point in the study of heat transfer.

1.2 Caloric Thermodynamics and Its Demise

With the invention of both the thermometer and the calorimeter, the quantitative exploration of thermodynamics could truly begin. One of the earliest problems examined concerned the mutual thermalization of two bodies at different temperatures, an investigation often carried out experimentally using water.

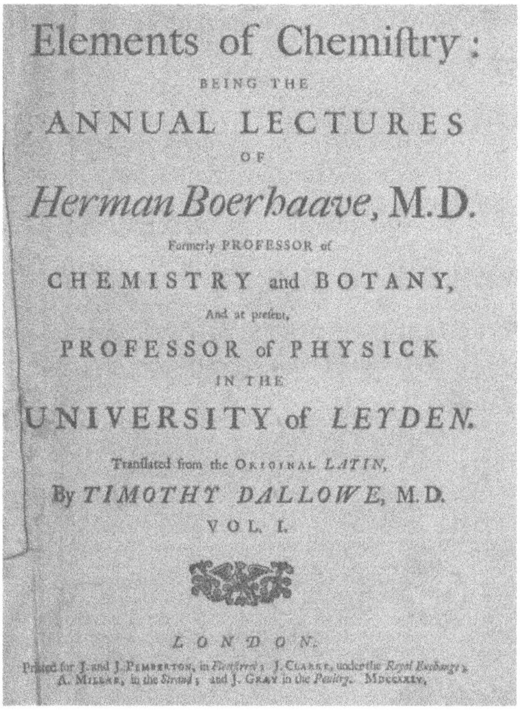
Elements of Chemiftry:
BEING THE
ANNUAL LECTURES
OF
Herman Boerhaave, M.D.
Formerly PROFESSOR of
CHEMISTRY and BOTANY,
And at prefent,
PROFESSOR of PHYSICK
IN THE
UNIVERSITY of LEYDEN.
Tranflated from the ORIGINAL LATIN,
By TIMOTHY DALLOWE, M.D.
VOL. I.
LONDON.
Printed for J. and J. PEMBERTON, in Fleetftreet; J. CLARKE, under the Royal Exchange; A. MILLAR, in the Strand; and J. GRAY in the Poultry. MDCCXXXV.

FIGURE 1.1 The first page of the book by Boerhaave on chemistry and botany.

The Richmann equation can be interpreted as an expression of the conservation of heat. If heat is defined as

$$\delta Q = m\,\delta t, \tag{1.5}$$

then the Richmann relation for two bodies can be obtained by assuming that

$$\delta Q_{\text{tot}} = \delta Q_1 + \delta Q_2 = 0, \tag{1.6}$$

that is, the heat gained by one body equals the heat lost by the other. This implies that if

$$\delta Q_1 = m_1(T - T_1), \tag{1.7}$$

$$\delta Q_2 = m_2(T - T_2), \tag{1.8}$$

the final equilibrium temperature is given by

$$T = \frac{m_1 T_1 + m_2 T_2}{m_1 + m_2}, \tag{1.9}$$

where m_i and T_i denote the masses and initial temperatures of the two bodies being mixed. For water, since the volume and mass change very little away from the freezing point, one could equivalently replace m_i with the corresponding volumes v_i. In principle, this suggested two possible definitions for the final temperature of mixing. Herman Boerhaave, however, had already pointed out that only one of these could be valid for substances in general. His reasoning later inspired Gabriel Fahrenheit to undertake further experiments, which were subsequently discussed by Joseph Black in his chemistry lectures at the University of Glasgow.

From these considerations, two hypotheses were formulated:

$$\delta Q = V\,\delta T, \tag{1.10}$$

$$\delta Q = M\,\delta T. \tag{1.11}$$

Black refined this picture by introducing the temperature difference $\delta T = T - T'$, yielding

$$\delta Q = mc\,\delta T, \tag{1.12}$$

where m is the mass and c a new quantity that he called the *affinity for heat*, now known as the *specific heat*. The product $C = mc$ defines the *heat capacity*, as understood today. This conceptual step marked the real beginning of theoretical thermodynamics and laid the foundation for the caloric theory of heat.

It was Pierre-Simon Laplace and Antoine-Laurent Lavoisier who first recognized that the heat capacity C is not a constant but a function of temperature, $C = C(T)$. At this stage, the caloric theory began to take on a form that resembles a more modern thermodynamic framework. The total caloric content of a body could then be expressed as

$$Q(T) = Q(T_0) + \int_{T_0}^{T} dt\, C(t), \tag{1.13}$$

where $Q(T_0)$ is the reference caloric content at temperature T_0. In this formulation, $C(t)$ was extended—thanks to Laplace and Lavoisier—to include the latent heat released or absorbed during phase transitions:

$$C(T) = C_0(T) + \sum_i L_i\, \delta(T - T_c^i), \tag{1.14}$$

where T_c^i are the characteristic temperatures at which the phase transitions occur. Equation (1.13) thus represents a formal caloric equation of state, capable of accounting for discontinuities associated with changes of phase.

These developments took place roughly a century after the introduction of the gas laws by Robert Boyle and Edme Mariotte [Boy62, Mar76].

Just as the conservation of mass and the invariance of chemical elements had become central principles in chemistry, the conservation of heat emerged as a guiding principle in the nascent theory of thermodynamics. Caloric theory, grounded in the notion of heat as a conserved fluid, gained considerable momentum and can be regarded as an early precursor of the modern principle of energy conservation. From an experimental standpoint, the caloric theory proved remarkably successful in explaining observed thermal phenomena. Its limitations, however, became evident in processes involving mechanical work. In most early experiments, the mechanical effects were negligible compared to the quantities of heat exchanged, so the inconsistencies of the theory were not yet apparent.

The crisis of the caloric theory was initiated by the work of Count Rumford (Benjamin Thompson), who demonstrated through his experiments on the boring of cannons that mechanical work could generate seemingly unlimited amounts of heat [Tho98, Mac96, Rol50]. Rumford's observations provided compelling evidence against the notion of heat as a conserved fluid and suggested, for the first time, a connection between heat, temperature, and molecular motion—what we would now describe as kinetic energy. Nevertheless, the full overthrow of the caloric theory would only come later, with the development of heat engines and the formal establishment of thermodynamics in the nineteenth century.

Thermodynamics is a branch of physics concerned with understanding how energy is transferred and transformed within matter. The early pioneers of the field—including Robert Boyle, Robert Hooke, and Denis Papin—conducted experiments to investigate the properties of gases and their behavior under different physical conditions. These studies laid the groundwork for the formulation of the gas laws, such as the *Boyle–Mariotte law* (circa 1660), which describes the inverse relationship between the pressure and volume of a gas when its temperature remains constant.

The Boyle–Mariotte law, also known as the *pressure–volume law*, is a fundamental principle in the study of gases. It was formulated independently by Boyle, an Irish natural philosopher, in the mid-seventeenth century, and by Mariotte, a French physicist and priest, a few years later. The law states that, at constant temperature, the product of the pressure (P) and the volume (V) of a given quantity of gas remains constant. Mathematically, it can be expressed as

$$P_1V_1 = P_2V_2, \tag{1.15}$$

where P_1 and V_1 are the initial pressure and volume of the gas, and P_2 and V_2 are their values after a change in pressure or volume.

The Boyle–Mariotte law can be understood by considering the behavior of gas molecules. When the pressure on a gas increases, the gas molecules are pushed closer together, leading to a decrease in volume. Conversely, if the pressure is decreased, the gas molecules have more space to move, resulting in an increase in volume. The Boyle–Mariotte law has several important implications. Firstly, it provides a quantitative relationship between the pressure and volume of a gas, allowing scientists and engineers to predict how gases will behave under different conditions. Secondly, it contributed to the development of the kinetic theory of gases, which describes the motion and behavior of gas molecules. Moreover, the law is essential in the understanding of gas behavior in various applications, such as in the design of pneumatic systems, the operation of engines, and the behavior of gases in the atmosphere. It has become a fundamental principle in the study of gases and continues to be a cornerstone of modern physics and chemistry. The theory of gases represented a crucial milestone in the development and understanding of engines. In 1824, Sadi Carnot published his seminal work on the motive power of heat, which brought a major advance in the understanding of heat engines and provided a theoretical framework for the study of energy transformations. His idealized *Carnot cycle*, involving the reversible expansion and compression of a gas, laid the foundation for the formulation of the second law of thermodynamics and introduced the concept of *entropy*.

A few decades later, James Prescott Joule conducted a series of experiments that led to the formulation of the first law of thermodynamics, or the *law of energy conservation*. This universal principle, applicable to all forms of energy, also deepened the understanding of gaseous behavior. Joule's experiments demonstrated the equivalence between mechanical work and heat and contributed to the development of the *kinetic theory of gases*, in which gases are described as collections of rapidly moving particles. He showed that the internal energy of a gas is directly related to the kinetic energy of its constituent molecules.

In parallel with these developments, Rudolf Clausius refined the theoretical foundations of thermodynamics by formalizing the second law and introducing the modern definition of entropy. His work not only clarified the principles governing energy transformations but also extended the kinetic theory of gases, relating entropy to the randomness and disorder associated with molecular motion.

The progress achieved in the study of gases during this period was instrumental in shaping the emerging field of *statistical mechanics*, which seeks to explain the macroscopic properties of matter through the statistical behavior of its microscopic constituents.

At the same time, the practical development of heat engines spurred engineers and scientists to deepen their understanding of gas behavior under real-world conditions. The contributions of pioneers such as Lord Kelvin, who introduced the concept of *absolute temperature*, further refined the theoretical description of gases and their thermodynamic properties.

The continuous interplay between theoretical advances in thermodynamics and practical innovations in engine design profoundly expanded the understanding of gases and their fundamental laws. These foundational principles continue to underpin modern engineering, physics, and our broader comprehension of the natural world. Chemistry, by contrast, has a long and diverse history that extends back to antiquity, as discussed earlier. However, it was during the eighteenth and nineteenth centuries that major advances were made in understanding chemical reactions and the composition of matter. The beginning of modern chemistry is often associated with the work of Antoine-Laurent Lavoisier, whose *Traité Élémentaire de Chimie* (1789) established the *law of conservation of mass* [Lav89]. Equally transformative was the contribution of John Dalton, who, in his *A New System of Chemical Philosophy* (1808), formulated the *atomic theory* [Dal08]. Together, these foundational principles revolutionized the way chemical transformations were understood and explained.

The convergence of thermodynamics and chemistry emerged as scientists began to investigate the relationship between heat and chemical change. A crucial development in this context was the study of the heat evolved or absorbed during chemical reactions. Early chemists observed that certain reactions absorbed heat (*endothermic*), while others released it (*exothermic*), leading to fundamental questions about the role of energy in chemical processes. Such investigations would not have been possible without the experimental methods of *calorimetry*, which provided the quantitative means to measure thermal effects in chemical reactions. In 1811, Amedeo Avogadro proposed what later became known as *Avogadro's hypothesis*. He stated that equal volumes of gases, under the same conditions of temperature and pressure, contain the same number of molecules. This simple yet groundbreaking idea established a crucial link between the observable macroscopic properties of gases and the microscopic characteristics of the particles composing them. It represented a fundamental step toward understanding the molecular nature of gases. One of the most significant consequences of Amedeo Avogadro's hypothesis was its impact on determining the *relative atomic masses* of the elements. By accurately measuring the densities and masses of gases, scientists were able to calculate the ratios of different elements involved in chemical reactions. Avogadro's insight greatly advanced the field of *stoichiometry*, which deals with the quantitative relationships governing chemical transformations. By revealing these atomic ratios, his work provided decisive experimental support for John Dalton's atomic theory, which held that elements consist of indivisible atoms combining in fixed proportions. Avogadro's hypothesis thus confirmed that chemical reactions follow simple whole-number ratios of combining gas volumes, lending strong empirical credibility to the atomic theory.

Moreover, Avogadro's ideas paved the way for the development of the *kinetic theory of gases* and, later, of *statistical mechanics*. Scientists now possessed a theoretical framework to explain the behavior of gases in terms of the motion and interactions of their constituent molecules. This represented a major step toward unifying the microscopic world of atoms and molecules with the macroscopic properties of gases.

It was through the pioneering work of Josiah Willard Gibbs, a nineteenth-century American mathematical physicist, that the connection between thermodynamics and chemistry was firmly established. Gibbs's seminal paper, *On the Equilibrium of Heterogeneous Substances* (1876–1878), laid the foundations of *chemical thermodynamics*. In this work, he introduced the concept of the *chemical potential* and developed a rigorous mathematical framework for analyzing chemical equilibrium in complex systems [Gib76].

It was Gibbs, after all, who emphasized that [Tis66]

> `Theoretical thermodynamics is anchored in observations and quantitative experiments pertaining to phenomena commonly denoted thermal.`

The close relationship between thermodynamics and chemistry continued to deepen as researchers explored phase transitions, reaction kinetics, and the thermodynamics of solutions. Concepts such as *entropy*, *free energy*, and *equilibrium constants* became indispensable tools for describing chemical systems, predicting reaction outcomes, and designing industrial processes.

This fruitful interplay extended into the twentieth century with the advent of *quantum mechanics*, which provided a more profound understanding of atomic and molecular behavior. Quantum mechanics and thermodynamics complemented one another, enabling scientists to uncover the microscopic foundations of thermodynamic properties and chemical phenomena. Indeed, the Planck constant h was first introduced in the theory of blackbody radiation and later

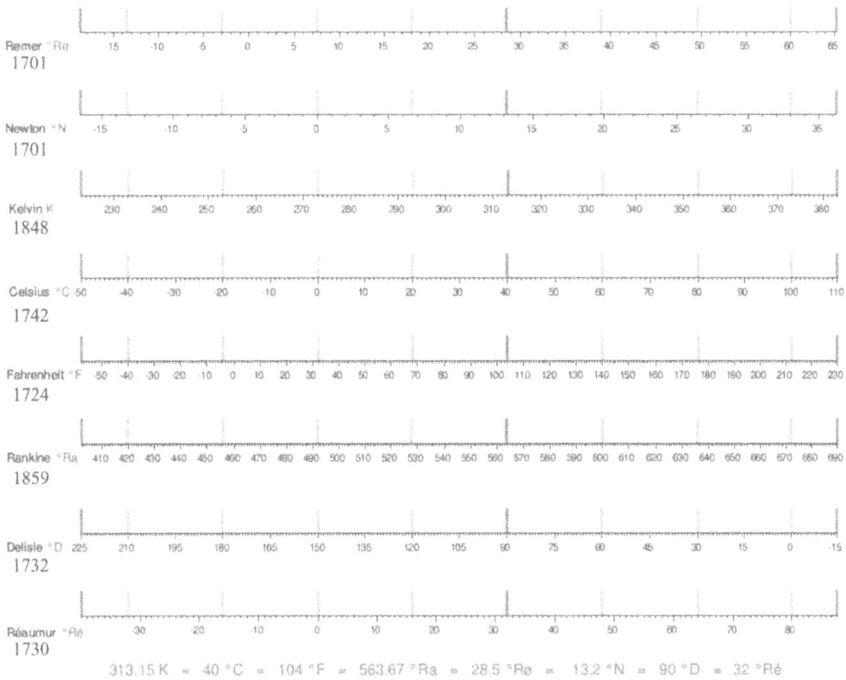

FIGURE 1.2 Temperature scales and their conversion by name and year of introduction. (Source: Wikipedia)

appeared in the statistical definition of the entropy of gases.

The first to introduce the notion of work through mechanical operations were Sadi Carnot and Lord Kelvin. Carnot's departure from the caloric theory that had dominated the previous century was based on the observation that temperature differences can produce mechanical effects in gases, such as variations in pressure. These pressure changes can in turn be harnessed to perform mechanical work, thus enabling the operation of an engine. To conceptualize this process, Carnot introduced the idea of *heat reservoirs.* In his model, an engine consists of two reservoirs at different temperatures, connected to a cylinder containing a (perfect) gas, which is itself linked to a mechanical device.

Carnot proposed that the operation of a heat engine is analogous to potential theory for water in a gravitational field. If a certain quantity x moves between two potentials V_1 and V_2, the work performed is $w = x(V_1 - V_2)$. Carnot assumed that it was *heat* that played the role of this quantity, since within the framework of the caloric theory no other conserved entity was recognized. Following Carnot, the efficiency of an engine is defined as the ratio

$$\eta = \frac{W}{Q'}, \tag{1.16}$$

where W denotes the mechanical work produced and Q' represents the quantity of caloric (or energy) transferred. Although Q' does not correspond exactly to what we now call heat, Carnot recognized that it must depend on the temperatures of the reservoirs.

Consider an ideal engine operating between two reservoirs at temperatures T_{h} (hot) and T_{c} (cold). In such a system, heat flows from the hot to the cold reservoir, and the engine is assumed to operate *reversibly*—that is, every mechanical or thermal process can be reversed without loss. The engine thus acts as a kind of "catalyst" between mechanical work and thermal exchange: a piston connected to a load mediates the conversion of heat transfer into mechanical motion.

Within this framework, Carnot conceived the idea of "caloric work," which he expressed in the form

$$W = Q'(T_{\rm h} - T_{\rm c}), \tag{1.17}$$

or, for small temperature differences,

$$\delta W = Q' \, \delta T. \tag{1.18}$$

Under the assumption of reversibility, Carnot went on to prove a fundamental result about engines operating between two thermal reservoirs, now known as *Carnot's principle.*

Carnot's principle is a cornerstone of thermodynamics, providing fundamental insight into the maximum efficiency attainable by a heat engine operating between two temperature reservoirs. Formulated by Carnot in 1824, this principle establishes a universal upper limit on the efficiency of any heat engine and serves as a benchmark for the performance of real machines. Carnot's ideas profoundly advanced our understanding of the laws governing heat and energy transfer.

Mathematically, Carnot's principle expresses the efficiency η of a reversible engine operating between two reservoirs at absolute temperatures $T_{\rm h}$ (hot) and $T_{\rm c}$ (cold). The maximum or *Carnot efficiency* is given by

$$\eta_{\rm Carnot} = 1 - \frac{T_{\rm c}}{T_{\rm h}}. \tag{1.19}$$

Consider a heat engine that absorbs heat $Q_{\rm h}$ from a hot reservoir, performs work $W_{\rm out}$, and rejects heat $Q_{\rm c}$ to a cold reservoir. According to the first law of thermodynamics, the energy balance is

$$Q_{\rm h} = W_{\rm out} + Q_{\rm c}. \tag{1.20}$$

The efficiency of the engine is then defined as the ratio of the useful work output to the heat absorbed:

$$\eta = \frac{W_{\rm out}}{Q_{\rm h}}. \tag{1.21}$$

For a reversible process, the ratio of heat exchanges is proportional to the temperatures of the reservoirs, $Q_{\rm c}/Q_{\rm h} = T_{\rm c}/T_{\rm h}$, which yields Carnot's efficiency expression above.

Carnot's principle thus sets a theoretical upper limit for all heat engines, offering a reference point for real systems and revealing that no engine can be more efficient than a reversible one operating between the same two temperatures. It also highlights the essential role of temperature differences in thermodynamic processes and the inevitable losses caused by irreversibilities.

Later, Lord Kelvin (William Thomson) reformulated Carnot's result independently of any specific temperature scale by introducing the *absolute Kelvin temperature.* This led to what is now known as the *Carnot–Kelvin principle*, valid provided that the quantity Q is properly defined in thermodynamic terms. Kelvin's elaboration refined Carnot's ideas and, for a time, extended the conceptual life of the caloric theory before it was ultimately replaced by the modern energy framework. Another difficulty of caloric thermodynamics was its dependence on the specific substance or gas under study. At the time, the principle of energy conservation was not yet established, and thus these ideas lacked a firm foundation in the physical laws of nature. For this reason, the quantity Q defined earlier was often misinterpreted.

The caloric theory began to show signs of collapse with the work of Robert Mayer [Can74] and James Prescott Joule, who independently introduced the principle of energy conservation. This principle marked a decisive turning point in nineteenth-century physics. Mayer (1814–1878), a German physician from Heilbronn, played a pivotal role in recognizing the equivalence between heat and mechanical work. While serving as a ship's doctor on a voyage to the Dutch East Indies in 1840, Mayer observed that the venous blood of sailors appeared lighter in color than expected in the tropical climate—an observation that led him to infer that less oxidation was required to maintain body temperature in warmer conditions. From this, Mayer deduced that heat and mechanical work are interchangeable forms of energy, anticipating the principle of conservation of energy. He published his results in 1842, slightly preceding Joule, whose more systematic and extensive experimental investigations, published in 1843 and continued over three decades, provided the quantitative foundation for the first law of thermodynamics.

It is worth recalling some early nineteenth-century experiments that explored the free expansion of gases. In Joseph Gay-Lussac's experiment on the free expansion of gases, a confined gas was allowed to expand into an evacuated container where no external forces acted on it [GL07]. This simple setup led to key insights into the behavior of gases. Gay-Lussac observed that when a gas expands freely into a larger volume, its temperature remains constant. This observation was consistent with the *ideal gas law*,

$$PV = nRT, \tag{1.22}$$

where P is pressure, V is volume, n is the number of moles of gas, R is the gas constant, and T is the absolute temperature. Gay-Lussac's findings helped clarify the relationship between pressure, volume, and temperature, and provided one of the first experimental indications that the internal energy of an ideal gas depends only on temperature. These experiments, along with his other contributions, had a lasting influence on the development of thermodynamics and the understanding of gas behavior.

At this stage, we return to Robert Mayer's considerations, which appeared to contradict Gay-Lussac's results [Can74]. Mayer reasoned as follows: when an ideal gas is heated at constant volume, the supplied heat increases its internal energy. If the gas is instead allowed to expand while being heated, additional heat is required to compensate for the work done by the gas. This reasoning implies that the heat capacity of a gas depends on its thermodynamic constraints—specifically, that the heat capacity at constant pressure, C_P, exceeds that at constant volume, C_V. Their ratio,

$$\gamma = \frac{C_P}{C_V}, \tag{1.23}$$

is approximately 1.6 for many diatomic gases. The apparent contradiction with Gay-Lussac's result is resolved by noting that Mayer's reasoning concerns an expansion against an external pressure (a piston), whereas Gay-Lussac's experiment described a free expansion into a vacuum.

It was through such insights that caloric thermodynamics began to fade, giving way to a more complete theory of energy conservation. From these studies, Mayer concluded that the work performed and the heat supplied are related by

$$W = JQ, \tag{1.24}$$

where Q is the quantity of heat and J is a proportionality constant known as the *mechanical equivalent of heat*. Later, James Prescott Joule confirmed and quantified this relationship through a series of systematic experiments [Jou43].

In modern notation, the *Mayer–Joule principle* is expressed as

$$\Delta U = Q - W, \tag{1.25}$$

where ΔU is the change in internal energy of the system, Q is the heat added to the system, and W is the work done by the system. In differential form, it reads

$$dU = \delta Q - \delta W, \tag{1.26}$$

where dU is the infinitesimal change in internal energy, δQ the infinitesimal heat added, and δW the infinitesimal work performed. This relation constitutes the *first law of thermodynamics*, which extends the principles of caloric theory into a more general and quantitative statement of energy conservation—accounting for heat, work, and internal energy as interchangeable forms of energy. Let us now consider a direct application of the first law of thermodynamics. Suppose a gas is enclosed within a piston–cylinder assembly and undergoes a reversible expansion against a constant external pressure P_{ext}. The infinitesimal work done by the gas is

$$\delta W = -P_{\text{ext}}\, dV, \tag{1.27}$$

and for a finite volume change, the total work becomes

$$W = -P_{\text{ext}}\, \Delta V, \tag{1.28}$$

where ΔV is the change in volume of the gas.

If a quantity of heat Q is added to the gas, the corresponding change in internal energy is given by

$$\Delta U = Q - W = Q + P_{\text{ext}}\, \Delta V. \tag{1.29}$$

The *Mayer–Joule principle*—a cornerstone of thermodynamics—finds application in diverse areas, including the analysis of heat engines, refrigeration cycles, and energy conversion processes. It provides a fundamental understanding of the relationship between heat and work, forming the basis of the *First Law of Thermodynamics*. As we shall see in the following chapter, the Mayer–Joule principle marks the conceptual boundary between pre-classical and classical thermodynamics.

Classical thermodynamics, in the form presented in modern textbooks, emerged from the effort to reconcile two seemingly contradictory principles. On one hand stood the *Mayer–Joule principle*, which expressed the equivalence between heat and work; on the other, *Carnot's principle*, which imposed limits on the conversion of heat into work.

It was Rudolf Clausius who resolved this apparent contradiction by reformulating the definitions of heat and work. The difficulty arose because, if one identifies Mayer's work in Eq. (1.24) with Carnot's work in Eq. (1.17), an inconsistency appears when Q' and Q are treated as equivalent quantities.

Clausius's key insight was that the constant J in Mayer's formulation represents a universal proportionality factor between heat and mechanical work. By choosing appropriate units such that $J = 1$, one can write $Q = W$, placing mechanical and thermal energy on the same energetic basis. Applying this idea to Carnot's expression for a reversible engine operating between two reservoirs and using the Mayer–Joule principle $W = Q_{\text{h}} - Q_{\text{c}}$, Clausius obtained

$$Q_{\text{h}} - Q_{\text{c}} = (T_{\text{h}} - T_{\text{c}})\, Q'. \tag{1.30}$$

Assuming that the reversible engine exchanges heat with the reservoirs at temperatures T_h and T_c, we may define the quantities $Q'_{\text{h}} = Q_{\text{h}}/T_{\text{h}}$ and $Q'_{\text{c}} = Q_{\text{c}}/T_{\text{c}}$, leading to the conservation of a new quantity,

$$Q' = \frac{Q}{T}. \tag{1.31}$$

Here Q represents the modern notion of heat, while Q' acquires a new interpretation. In Clausius's reformulation, this quantity corresponds to the conserved property known today as *entropy*, $S = Q/T$, for reversible processes.

Clausius further showed that for a Carnot engine the efficiency can be expressed as

$$\eta = \frac{W}{Q_{\text{h}}} = \frac{Q_{\text{h}} - Q_{\text{c}}}{Q_{\text{h}}} = 1 - \frac{T_{\text{c}}}{T_{\text{h}}}, \tag{1.32}$$

which is the modern expression for the efficiency of an ideal reversible engine.

It is at this historical juncture that thermodynamics began to crystallize into the classical framework we recognize today—one built upon the unified principles of energy conservation and entropy.

2

Thermodynamics

2.1 From Thermodynamics to Statistical Mechanics

Statistical mechanics is a branch of theoretical physics concerned with understanding the collective properties of systems composed of a large number of particles in *equilibrium*.

In the opening chapters of this book, we aim to bridge the gap between *thermodynamics* and statistical mechanics by addressing fundamental questions: *How do systems reach equilibrium? What happens when many particles collide and are confined within a given region of space?*

As discussed earlier, thermodynamics is a heuristic and phenomenological theory developed throughout the nineteenth century to describe the macroscopic behavior of gases, engines, and chemical reactions. Its central objective is to establish relationships among measurable quantities such as temperature, pressure, and volume.

The enduring importance of the *laws of thermodynamics* lies in their universality. They have survived at least three major scientific revolutions without modification: they remained valid through the rise of *relativity* and *electromagnetism*, which redefined the concepts of space, time, and energy; they endured the advent of *quantum mechanics*, which transformed our understanding of determinism and measurement; and they persisted even with the emergence of *chaos theory*, which challenged classical predictability. In this sense, the laws of thermodynamics are among the most resilient principles in all of physics.

As a starting point, we shall focus on the simplest systems—*ideal gases*—which are described by the *equation of state*

$$F(P,T,V) = PV - Nk_{\rm B}T = 0, \tag{2.1}$$

a relation that applies to dilute gases where the particle density satisfies $N/V \ll 1$. Above, $k_{\rm B} = 1.38 \times 10^{-23}\,{\rm J\,K^{-1}}$ is the Boltzmann constant and it represents a fundamental physical constant connecting macroscale physics to the microscale. In more complex systems, the equation of state can take on a more intricate form. For example, at higher densities, real gases require corrections to ideal behavior. One of the earliest and most important corrections was introduced through the *Van der Waals equation of state*,

$$(V-b)\left(P+\frac{a}{V^2}\right) = Nk_{\rm B}T, \tag{2.2}$$

DOI: 10.1201/9781003603528-2

where the constants a and b account for intermolecular attraction and the finite size of the molecules, respectively. Such modifications lead to a richer phenomenology, including the occurrence of *phase transitions*.

In the early development of thermodynamics, the microscopic or *kinetic* properties of gases were of secondary importance. The theory instead centered on the study of *thermodynamic processes*, transformations in which heat is exchanged or work is performed on a system, and attention is restricted to the initial and final equilibrium states. In the chapters that follow, we shall see how statistical mechanics provides the microscopic foundation for these macroscopic laws, revealing the bridge between individual particle dynamics and the collective behavior of matter.

The notion of equilibrium can be interpreted in many ways, but in thermodynamics, this implies a constant temperature across the gas. This implies that there cannot be part of the gas that is at different temperatures. At this stage, we are assuming that there exists a universal way of measuring temperature, e.g., a thermometer, but we will discuss thermometry in the next chapters.

In the sections that follow, we shall review the fundamental principles of *thermodynamics*. Subsequently—and particularly in the second volume of this series—we shall turn to the study of *kinetic theory*, developed primarily through the pioneering work of Ludwig Boltzmann. There, we shall derive, from the *Boltzmann transport equation*, the celebrated *Boltzmann–Maxwell distribution*.

This formulation marks the true beginning of *statistical mechanics*. The power of Boltzmann's arguments continues to resonate even today, extending beyond physics into the realm of *information theory*, which reveals a profound conceptual link between statistical physics and the microscopic role of information.

2.2 Modern Thermodynamics: Macroscopic Variables

Before we can fully address the connection between *thermodynamics* and *statistical mechanics*, it is useful to recall a few essential facts about how we can act on a gas [Fer36]. Let us consider a gas at a given *pressure* P, which represents the force exerted per unit area on the walls of its container.

2.2.1 Pressure

Is there a microscopic interpretation of pressure? Consider a gas confined within a piston (see also Fig. 2.1). Particles moving with an average velocity v collide elastically with the walls of the container. The number of particles striking a given area per unit time can be represented by the rate $I = nv$, where n denotes the number density of particles. Upon collision, each particle reverses its velocity from v to $-v$, producing a change in momentum of $2p = 2mv$. Hence, the pressure exerted on the wall is proportional to the rate of collisions and the momentum transfer per collision:

$$P = I\,\Delta p = 2nmv^2. \tag{2.3}$$

This simple argument captures the microscopic essence of pressure as arising from molecular impacts on the boundaries of the system.

2.2.2 Work versus Heat

Work and *heat* are the two fundamental mechanisms through which energy can be exchanged between a system and its surroundings. In mechanics, work corresponds to the energy transferred by macroscopic forces acting through a distance. As we have seen, it involves coherent, organized motion—such as the displacement of a piston or the change in the system's volume.

At the molecular level, work may be interpreted as the collective, directional motion of particles under the influence of an external force. For example, in a gas expanding against a piston, work is performed as the molecules collide with the piston, exerting a force that moves it outward. Conversely, when an external agent pushes the piston inward, the molecules are compressed, and energy is transferred to the gas.

As illustrated in Fig. 2.1, the infinitesimal work done on a gas through the motion of a piston is

$$\delta W = P\,dV = P\,A\,dh, \tag{2.4}$$

where A is the cross-sectional area of the piston and dh is its infinitesimal displacement. The total work performed in a transformation from an initial volume V_1 to a final volume V_2 is given by

$$W = \int_{V_1}^{V_2} P\,dV. \tag{2.5}$$

Thus, work involves the organized transfer of energy through macroscopic motion or volume change.

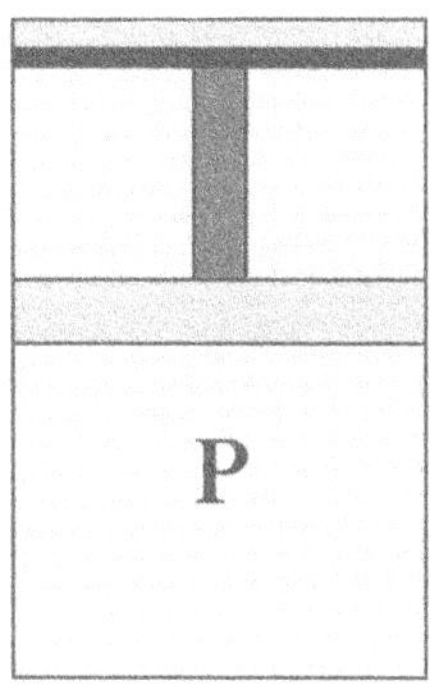

FIGURE 2.1 Work performed on a gas via a piston.

In contrast, heat is not a form of energy itself but rather an *energy transfer* associated with temperature differences between a system and its surroundings. At the microscopic level, *heat transfer* arises from the random thermal motion of particles. When two systems at different temperatures come into contact, molecules in the hotter system—having higher average kinetic energy—transfer energy to those in the colder system through collisions.

From a thermodynamic standpoint, infinitesimal heat transfer is related to entropy change through the expression

$$dQ = T\,dS, \tag{2.6}$$

where T is the absolute temperature and S is the *entropy*. This relation provides a natural bridge between the macroscopic laws of thermodynamics and the microscopic interpretation of energy exchange.

REMARK 2.1 Historically, thermodynamics used the symbol δ to denote *inexact differentials*—quantities such as heat δQ and work δW that depend on the path taken during a process. In modern thermodynamics, however, it is often more convenient to employ total differentials dQ and dW, emphasizing the mathematical structure of the theory and its connection with exact state variables such as energy and entropy. This shift in notation marks an important conceptual evolution from classical to modern thermodynamic formalisms.

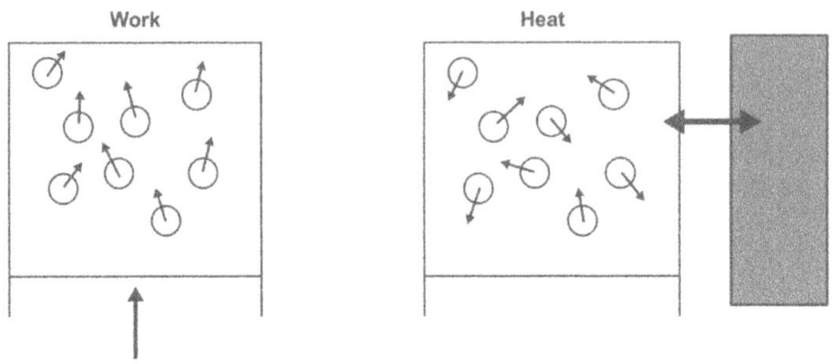

FIGURE 2.2 Work versus heat at a molecular level.

A useful graphical distinction between heat and work is illustrated in Fig. 2.2, highlighting the difference between organized and random modes of energy transfer.

2.2.3 Temperature

We often take the concept of *temperature* for granted. Before introducing the variable T in the equations that follow, it is worth pausing to reflect on what temperature means—at least from an operational standpoint. We assume that we can determine the temperature of an object by placing it in contact with a thermometer. After a short time, the two systems reach *thermal equilibrium*, and the reading on the thermometer provides the temperature of the object.

However, a fundamental question arises: How can we make two thermometers agree? Different thermometers may use different physical principles, and their readings might not vary linearly with temperature. To ensure consistency between measurements, we must use what is known as a *good thermometer*.

A thermometer is said to be "good" if its scale is based on the thermal expansion of an *ideal gas*, for which the expansion is linearly proportional to temperature. Paradoxically, this remains the only reliable way to verify whether a thermometer is consistent. An illustration of good and poor thermometers is shown in Fig. 2.4.

Another essential issue concerns the existence of an *absolute temperature scale* and the identification of a point corresponding to *absolute zero*. This problem, too, can be addressed using the properties of ideal gases. Different ideal gases satisfy equations of the form

$$\begin{aligned} PV &= a_1 T, \\ PV &= a_2 T, \\ &\vdots \\ PV &= a_n T. \end{aligned} \tag{2.7}$$

Despite the variation in constants a_i, all gases exhibit a unique temperature at which the product PV—a measure of the gas's energy—vanishes, corresponding to a vanishing volume as well. This common reference point defines the temperature known as *absolute zero*.

The modern *Kelvin scale* is constructed so that absolute zero lies at 0 K, while the triple point of water is defined to occur at 273.16 K. This definition preserves the same unit size as the Celsius scale but establishes an absolute reference independent of the properties of any particular substance.

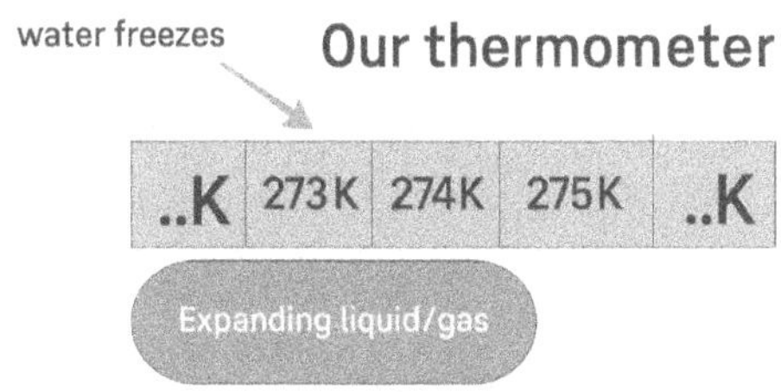

FIGURE 2.3 A thermometer with a Kelvin grading. The freezing point of water is around 273 Kelvin.

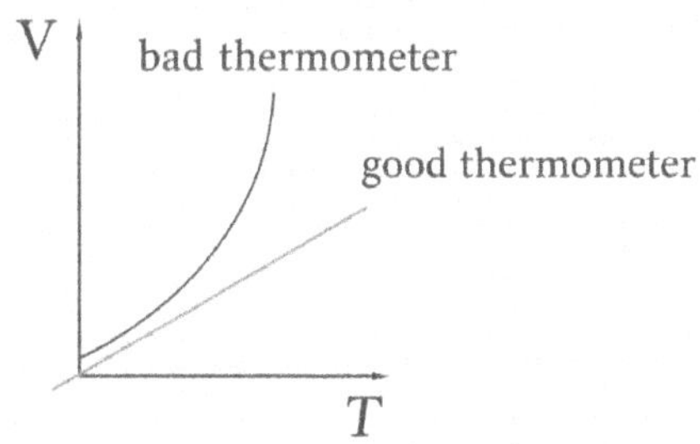

FIGURE 2.4 An example of a bad and good thermometer measuring the expansion of an ideal gas (at fixed pressure).

2.3 Thermodynamic Units and Molar Quantities

Before applying thermodynamic principles to chemical and physical systems, it is essential to establish the quantities and units commonly used in this context. Thermodynamics connects the microscopic world of molecules and atoms with macroscopic observables such as temperature, pressure, and energy. This connection is mediated through the concepts of the *mole*, *Avogadro's number*, and the universal *gas constant*.

On the microscopic level, energy and temperature are linked through Boltzmann's constant,

$$k_{\mathrm{B}} = 1.380649 \times 10^{-23}\ \mathrm{J\,K^{-1}}. \tag{2.8}$$

Boltzmann's constant sets the scale relating the energy of a single particle to the macroscopic temperature of a system. In many derivations, it is useful to define

$$\beta = (k_B T)^{-1}, \tag{2.9}$$

which often appears in statistical mechanics and probability distributions. To give a sense of scale, at the boiling point of water ($T = 373$ K) we have $\beta = 1.94 \times 10^{20}\ \mathrm{J^{-1}}$, while at the freezing point ($T = 273$ K), $\beta = 2.65 \times 10^{20}\ \mathrm{J^{-1}}$.

In macroscopic thermodynamics, we usually deal not with individual molecules but with *moles* of substance. One mole contains exactly

$$N_{\mathrm{A}} = 6.02214076 \times 10^{23} \tag{2.10}$$

elementary entities (atoms, molecules, or ions) and represents the bridge between microscopic and laboratory-scale quantities. The mole is defined as the number of atoms in 12 g of pure carbon-12 [Giu15]. If N is the number of microscopic particles and n the number of moles, the two are related by $N = nN_A$.

The Gas Constant and the Equation of State

The *universal gas constant* R connects microscopic and macroscopic descriptions of thermodynamic systems:

$$nR = Nk_B. \tag{2.11}$$

Here n is the number of moles and N the total number of molecules in the system. In SI units, the value of the gas constant is

$$R = 8.314462618 \text{ J mol}^{-1}\text{ K}^{-1}. \tag{2.12}$$

Equation (2.11) demonstrates the equivalence between the microscopic and macroscopic viewpoints: k_B relates energy to temperature per particle, while R does so per mole.

Energy Units and Molar Volume

Energy in thermodynamics is measured in *joules* (J), the SI unit of energy. One joule is defined as the work done by a force of one newton acting over a distance of one meter:

$$1 \text{ J} = 1 \text{ N m} = 1 \text{ kg m}^2\text{ s}^{-2}. \tag{2.13}$$

In thermal terms, one joule corresponds to the amount of heat required to raise the temperature of approximately 0.239 g of water by 1°C. In older systems, energy was often expressed in *calories*, where

$$1 \text{ cal} = 4.184 \text{ J}. \tag{2.14}$$

The joule is the standard unit in the first law of thermodynamics and is used consistently for both mechanical and thermal energy.

The *molar volume*, V_m, is defined as the volume occupied by one mole of a substance:

$$V_m = \frac{V}{n}. \tag{2.15}$$

It is typically expressed in liters per mole (L mol^{-1}), where 1 L $= 10^{-3}$ m^3. For ideal gases at standard temperature and pressure ($T = 273.15$ K, $P = 1$ atm), the molar volume is

$$V_m = \frac{RT}{P} \approx 22.4 \text{ L mol}^{-1}. \tag{2.16}$$

This quantity is essential for connecting the macroscopic equation of state to experimental observations and chemical stoichiometry.

2.3.1 Zeroth, First, and Second Laws of Thermodynamics

Before turning to the formal statements of the laws of thermodynamics, it is useful to recall that the macroscopic behavior of a gas or liquid can be described phenomenologically by an *equation of state* (EOS) of the form

$$F(P, V, T) = 0, \tag{2.17}$$

where P denotes the pressure, V the volume, and T the temperature of the system. We assume that F is a differentiable function, which implies that these variables cannot in general be independent.

To illustrate this mathematically, let us generalize the discussion. Suppose that we have N measurable physical quantities $z_1, \ldots, z_{n+m}$ that satisfy a system of n constraint equations,

$$f_r(z_1, \ldots, z_{n+m}) = 0, \qquad r = 1, \ldots, n. \tag{2.18}$$

We can represent this as a mapping $f : \mathbb{R}^{n+m} \to \mathbb{R}^n$, where the existence of constraints ($m \geq 1$) implies that some variables are functionally dependent on others. If we partition the variables into two groups, $x \in \mathbb{R}^n$ and $y \in \mathbb{R}^m$, then according to the *implicit function theorem* (Cauchy, Dini), there exists a differentiable function $g : \mathbb{R}^n \to \mathbb{R}^m$ such that

$$f(x, g(x)) = 0. \tag{2.19}$$

This result guarantees that certain thermodynamic quantities may be expressed as functions of others, consistent with the structure of Eq. (2.17).

In thermodynamics, this reasoning leads directly to the conclusion that an equation of state can be written in equivalent but not independent forms:

- $F(P, V, T(P, V)) = 0$, where P and V are chosen as independent variables,
- $F(P, V(P, T), T) = 0$, where P and T are chosen as independent variables, or
- $F(P(V, T), V, T) = 0$, where V and T are chosen as independent variables.

For an *ideal gas*, these relations take particularly simple forms:

- $P = \dfrac{Nk_{\mathrm{B}}T}{V}$,
- $V = \dfrac{Nk_{\mathrm{B}}T}{P}$,
- $T = \dfrac{PV}{Nk_{\mathrm{B}}}$,

where N is the number of particles and k_{B} is the Boltzmann constant.

REMARK 2.2 The formalism introduced above highlights a key feature of thermodynamics: macroscopic variables such as P, V, and T are not independent but constrained by equations of state. The *zeroth, first, and second laws of thermodynamics* provide the physical principles that underlie these constraints—defining the conditions for thermal equilibrium, energy conservation, and the direction of spontaneous processes. It is through these laws that the abstract relationships among variables acquire their empirical and conceptual meaning.

Zeroth Law

The *zeroth law of thermodynamics* establishes the fundamental notion of *thermal equilibrium.* It states that temperature is a transitive property that allows systems to be compared and measured consistently.

Consider the situation illustrated in Fig. 2.5, where three systems are brought into thermal contact. If the three systems initially have different temperatures, they will exchange energy until equilibrium is reached. According to the zeroth law, if system 1 is in thermal equilibrium with system 2, and system 2 is in thermal equilibrium with system 3, then system 1 must also be in equilibrium with system 3. Consequently, their temperatures must be equal:

$$T_1 = T_2 = T_3. \tag{2.20}$$

In this sense, the zeroth law is not an absolute statement about temperature but rather a relational one—it defines temperature as the property that determines whether two systems are in thermal equilibrium. From a practical perspective, the zeroth law states the *existence of temperature* as a macroscopic quantity *in equilibrium.*

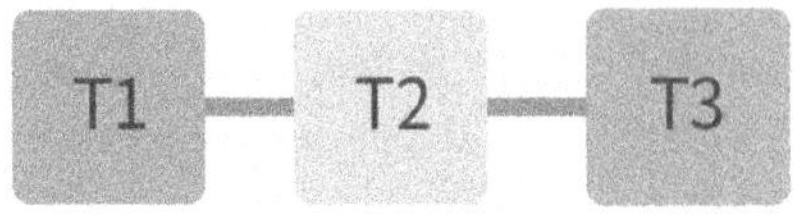

FIGURE 2.5 Three systems at different temperatures in contact.

First Law

The *first law of thermodynamics* expresses the conservation of energy for macroscopic systems. It connects three fundamental quantities associated with any thermodynamic process:

- *Work* (W)—the energy exchanged through macroscopic, organized motion;
- *Heat* (Q)—the energy exchanged due to random microscopic motion; and
- *Internal energy* (U)—the total energy associated with the internal state of the system.

Having already introduced the concept of work, let us now consider heat. Heat represents the transfer of energy between a system and its surroundings that does not result in mechanical work. Microscopically, it corresponds to the energy stored in the random motion of the particles composing the gas or liquid.

The internal energy U can be understood as the energy associated with the internal configuration of the system. In the microscopic description of a gas, this energy corresponds to the *Hamiltonian* H of the system, written in terms of particle positions and momenta $\{x_i, p_i\}$, so that $U = H(x,p)$. However, classical thermodynamics abstracts away from this microscopic detail: it treats U as a macroscopic state function depending only on measurable quantities such as P, V, and T. For convenience, we use the symbol U in thermodynamics, while in *statistical mechanics* it is customary to denote energy by E. In the present context, we shall not distinguish between the two.

The first law of thermodynamics asserts the existence of a conserved potential, the *internal energy* U, such that infinitesimal changes in the state of a system satisfy

$$dU = \delta Q - \delta W. \tag{2.21}$$

This relation expresses that the change in internal energy equals the heat absorbed minus the work performed by the system. For a *reversible process*, the exchange of heat and work can be determined unambiguously from the state variables of the system. In general, however, the distinction between differentials is essential.

The symbol dU represents an *exact differential*. This means that U is a *state function*—its value depends only on the thermodynamic state of the system and not on the particular path or process taken to reach it. If the system evolves between two states A and B, the integral of dU over any process connecting them yields the same result:

$$\int_A^B dU = U(B) - U(A). \tag{2.22}$$

In other words, U has a single-valued potential surface in the thermodynamic space spanned by its natural variables, such as entropy S and volume V, or equivalently temperature T and pressure P.

By contrast, the symbols δQ and δW denote *inexact differentials*. They describe process-dependent quantities: the amount of heat or work exchanged *depends on the specific path* taken in the thermodynamic space. Thus, the integrals

$$\int_A^B \delta Q \quad \text{and} \quad \int_A^B \delta W \tag{2.23}$$

cannot be evaluated solely from the initial and final states but require detailed knowledge of the process connecting them.

In physical terms, this distinction reflects the difference between energy stored within a system (the internal energy U) and energy exchanged with the surroundings (heat Q and work W). The first law therefore, establishes that while energy can change its form, the total amount is conserved.

We shall return to the notion of reversibility shortly. For now, we restrict our discussion to *ideal gases*, for which the state of the system can be represented in either the pressure–volume (P–V) or pressure–temperature (P–T) planes. In these diagrams, reversible processes appear as well-defined curves connecting equilibrium states, and the integral of dU along any path depends only on the endpoints:

$$\oint dU = 0, \tag{2.24}$$

confirming that U is a true thermodynamic potential. By contrast, the cyclic integrals of δQ and δW are not zero individually but must satisfy the conservation condition

$$\oint (\delta Q - \delta W) = 0. \tag{2.25}$$

This is the precise mathematical statement of the first law: in any closed cycle, the net change in internal energy vanishes, and the total heat absorbed equals the total work done by the system.

The notion of *process* is extremely important in thermodynamics. Various thermodynamic processes—such as isothermal, adiabatic, and isobaric transformations—can then be represented as distinct paths on these diagrams, as illustrated in Fig. 2.6.

REMARK 2.3 The first law provides the foundation upon which all of classical thermodynamics is built. It defines energy as a measurable, conserved quantity while distinguishing between its macroscopic (*work*) and microscopic (*heat*) modes of transfer. This distinction will later acquire a deeper interpretation within statistical mechanics, where the conservation of energy arises naturally from the microscopic dynamics of particles.

The First Law and Sign Conventions

The first law of thermodynamics expresses the conservation of energy: there exists a thermodynamic potential, the *internal energy* U, such that infinitesimal changes in the system's state satisfy

$$dU = \delta Q - \delta W. \tag{2.26}$$

Here:

- δQ is the infinitesimal *heat absorbed by the system.*
- δW is the infinitesimal *work done by the system on its surroundings.*

With this sign convention, $dU > 0$ when the system absorbs heat or when work is done

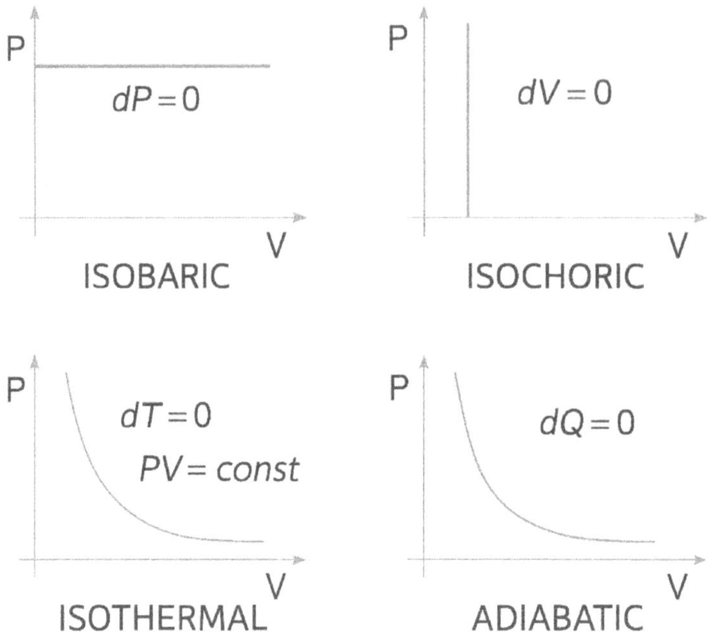

FIGURE 2.6 Isobaric, isochoric, isothermal, and adiabatic processes represented in the P–V plane.

on the system (i.e., $\delta W < 0$) and combinations thereof. Equation (2.21) is standard in physics, particularly in thermodynamics of gases and heat engines, where expansion work corresponds to energy leaving the system.

Alternative Convention

In chemistry and engineering, it is often convenient to write instead

$$dU = \delta Q + \delta W, \tag{2.27}$$

where now δW denotes the work done on the system by its surroundings. The two conventions are mathematically equivalent; they differ only in the interpretation of the sign of W. This is the one that can be easily remembered, typically.

Interpretation and Usage

- The $dU = \delta Q - \delta W$ form emphasizes energy *leaving* the system—useful for engines and expansions where the system performs work.
- The $dU = \delta Q + \delta W$ form emphasizes energy *entering* the system—appropriate for compressions, chemical reactions, and engineering analyses.

Both conventions are consistent provided they are applied uniformly. The choice depends only on whether we regard the system as an *agent* (doing work) or a *recipient* (receiving work).

Summary of Conventions

Convention	**Expression**	**Meaning of $W > 0$**
Physics / Thermodynamics	$dU = \delta Q - \delta W$	Work done *by* the system
Chemistry / Engineering	$dU = \delta Q + \delta W$	Work done *on* the system

REMARK 2.4 In either form, the essence of the first law remains unchanged: the internal energy U is a state function that changes only through exchanges of heat and work with the surroundings. The sign convention simply reflects a choice of perspective.

Thermodynamic Processes in Ideal Gases

For an *ideal gas*, the analysis of thermodynamic processes can be greatly simplified. This simplification originates from the celebrated *Joule experiment*, in which it was established that the total *internal energy* of a gas depends solely on its temperature. Mathematically, this implies that

$$U = C_V T, \tag{2.28}$$

where C_V denotes the *heat capacity at constant volume*.

Similarly, when heat is exchanged with the system at constant pressure, the infinitesimal heat transfer can be expressed as

$$\delta Q = C\,dT, \tag{2.29}$$

where C is the *specific heat*, representing the amount of heat required to raise the temperature of the system by one kelvin.

For an ideal gas, the *first law of thermodynamics* can thus be written as

$$\delta Q = C_V\,dT + P\,dV. \tag{2.30}$$

Adiabatic Processes

In an *adiabatic process*, there is no exchange of heat between the system and its surroundings, so $\delta Q = 0$. Applying the first law yields

$$0 = C_V\,dT + P\,dV. \tag{2.31}$$

Using the ideal gas law $PV = Nk_BT$, we can rewrite this as

$$0 = C_V\,\frac{dT}{T} + Nk_B\,\frac{dV}{V}. \tag{2.32}$$

Integrating between two states (T_1, V_1) and (T_2, V_2) gives

$$\ln\left(\frac{T_2}{T_1}\right) = -\frac{Nk_B}{C_V}\ln\left(\frac{V_2}{V_1}\right), \tag{2.33}$$

or equivalently,

$$TV^{\frac{Nk_B}{C_V}} = \text{constant}. \tag{2.34}$$

It is common to introduce the adiabatic index

$$\gamma = \frac{C_P}{C_V}, \tag{2.35}$$

for which the ideal gas law implies the familiar adiabatic relation

$$PV^{\gamma} = \text{constant}. \tag{2.36}$$

Isothermal Processes

In an *isothermal process*, the temperature of the system remains constant, $dT = 0$. Assuming the process proceeds slowly enough for the system to remain in thermal equilibrium with a large reservoir, we have

$$\delta Q = P\,dV. \tag{2.37}$$

For an ideal gas, $P = Nk_{\mathrm{B}}T/V$, so integrating between two volumes V_1 and V_2 gives

$$Q(V_2) - Q(V_1) = Nk_{\mathrm{B}}T \ln\left(\frac{V_2}{V_1}\right). \tag{2.38}$$

This expression shows that, for an isothermal expansion of an ideal gas, the heat absorbed by the system is entirely converted into work performed on the surroundings.

REMARK 2.5 The analysis of these processes illustrates how the first law links microscopic quantities such as temperature and molecular motion to macroscopic observables like pressure, volume, and heat transfer. The results derived here also provide the foundation for understanding more complex cycles, such as the Carnot and Otto engines, which we shall encounter later in this book.

2.3.2 The Heat Theorem and the Existence of Temperature as an Integrating Factor

The first law, Eq. (2.21), distinguishes between energy exchanges through heat and work:

$$dU = \delta Q - \delta W. \tag{2.39}$$

Neither δQ nor δW is an exact differential; they depend on the path taken between two states. However, empirical experience, going back to Carnot and Clausius, shows that for a *reversible quasi-static process*, the ratio $\delta Q/T$ behaves as an *exact differential.* This remarkable fact is summarized in what is historically called the *heat theorem* [Cam24, Cla65].

For a quasi-static infinitesimal transformation,

$$\frac{\delta Q}{T} = dS, \tag{2.40}$$

where T is the temperature at which the infinitesimal heat δQ is exchanged. The function S is the *entropy* of the system, and is a state function.

Although δQ is not itself an exact differential, the existence of the *integrating factor* $1/T$ makes the combination $\delta Q/T$ exact. This implies that in the space of thermodynamic variables $(E, \lambda_1, \lambda_2, \dots)$, where λ_i are external parameters (volume, magnetic field, etc.), the differential form

$$\omega = \frac{dE + \sum_i X_i\, d\lambda_i}{T} \tag{2.41}$$

is exact, i.e.,

$$\omega = dS, \tag{2.42}$$

so that

$$\left(\frac{\partial S}{\partial E}\right)_{\lambda_i} = \frac{1}{T}, \tag{2.43}$$

$$\left(\frac{\partial S}{\partial \lambda_i}\right)_{E} = \frac{X_i}{T}. \tag{2.44}$$

Equations (2.43)–(2.44) express the fact that the vector field $(1, X_1, X_2, \ldots)/T$ in the space of extensive coordinates $(E, \lambda_1, \lambda_2, \ldots)$ is conservative, implying the existence of a potential function $S(E, \lambda_1, \lambda_2, \ldots)$. This is a highly nontrivial property: not every inexact differential admits an integrating factor. The discovery that $1/T$ is such a factor constitutes one of the deepest insights of nineteenth-century thermodynamics [Car09, Gib02]. In geometric language, the heat theorem states that the differential form $\delta Q/T$ is *exact* for reversible transformations. Temperature thus appears as an integrating factor making the energy–work form $\delta Q = dE + \sum_i X_i \, d\lambda_i$ integrable. Entropy is therefore the *potential function* associated with this exact form, just as the electric potential is associated with a conservative electrostatic field.

The heat theorem can also be written in integral form:

$$\int_A^B \frac{\delta Q_{\mathrm{rev}}}{T} = S(B) - S(A), \tag{2.45}$$

i.e., the integral of $\delta Q/T$ between two equilibrium states depends only on the endpoints and not on the path. This defines entropy as a true state function, completing the bridge from empirical thermodynamics to its mechanical foundations. The heat theorem is a foundational result at the basis of the second law of thermodynamics.

For a modern pedagogical discussion, see Campisi [Cam24]. For historical sources, see Clausius's original memoir on the heat theorem [Cla65], Carathéodory's axiomatic formulation [Car09], and Gibbs's *Elementary Principles in Statistical Mechanics* [Gib02].

2.3.3 Differential Relations Between Thermodynamic Quantities

One of the key implications of the equation of state, Eq. (2.17), is that the *internal energy* U, and hence the *first law of thermodynamics*, can be expressed as a function of different pairs of thermodynamic variables. Depending on which quantities are held fixed, we can obtain various differential forms that connect measurable parameters such as pressure, volume, and temperature.

U(V, P)

Let us first assume that the internal energy can be expressed as a function of volume and pressure, $U = U(V, P)$. The total differential of U is then

$$dU = \left(\frac{\partial U}{\partial V}\right)_P dV + \left(\frac{\partial U}{\partial P}\right)_V dP. \tag{2.46}$$

Since U is a state function, it possesses an exact differential. This implies that its mixed second derivatives must be equal, i.e.,

$$\left(\frac{\partial}{\partial P}\left(\frac{\partial U}{\partial V}\right)_P\right)_V = \left(\frac{\partial}{\partial V}\left(\frac{\partial U}{\partial P}\right)_V\right)_P. \tag{2.47}$$

This mathematical condition ensures that U is well-defined independently of the path taken during any thermodynamic transformation.

U(V, T)

Next, consider U as a function of volume and temperature, $U = U(V, T)$. From the first law of thermodynamics, $dU = \delta Q - \delta W$, and using $\delta W = P\,dV$, we can write

$$dU = \left(\frac{\partial U}{\partial V}\right)_T dV + \left(\frac{\partial U}{\partial T}\right)_V dT. \tag{2.48}$$

Therefore,

$$\left[\left(\frac{\partial U}{\partial V}\right)_T + P\right] dV + \left(\frac{\partial U}{\partial T}\right)_V dT = \delta Q. \tag{2.49}$$

From previous results, we know that

$$\left(\frac{\partial Q}{\partial T}\right)_V = C_V, \tag{2.50}$$

where C_V is the *heat capacity at constant volume.* This form makes explicit the dependence of the internal energy on both temperature and volume.

U(P, T)

Finally, let us express the internal energy as a function of pressure and temperature, $U = U(P,T)$. Its total differential is

$$dU = \left(\frac{\partial U}{\partial P}\right)_T dP + \left(\frac{\partial U}{\partial T}\right)_P dT = \delta Q - \delta W. \tag{2.51}$$

Rearranging terms gives

$$\delta Q = \left(\frac{\partial U}{\partial P}\right)_T dP + \left(\frac{\partial U}{\partial T}\right)_P dT + \delta W. \tag{2.52}$$

Since $\delta W = P\,dV$, and noting that the volume itself can be expressed as a function of T and P, i.e.,

$$dV = \left(\frac{\partial V}{\partial T}\right)_P dT + \left(\frac{\partial V}{\partial P}\right)_T dP, \tag{2.53}$$

we obtain

$$\delta Q = \left[\left(\frac{\partial U}{\partial P}\right)_T + P\left(\frac{\partial V}{\partial P}\right)_T\right] dP + \left[\left(\frac{\partial U}{\partial T}\right)_P + P\left(\frac{\partial V}{\partial T}\right)_P\right] dT. \tag{2.54}$$

2.3.4 Relation Between C_P and C_V

From these relations, we can now define the heat capacities at constant pressure and volume, respectively:

$$C_P = \left(\frac{\partial Q}{\partial T}\right)_P = \left(\frac{\partial U}{\partial T}\right)_P + P\left(\frac{\partial V}{\partial T}\right)_P, \tag{2.55}$$

$$C_V = \left(\frac{\partial Q}{\partial T}\right)_V = \left(\frac{\partial U}{\partial T}\right)_V. \tag{2.56}$$

For an ideal gas, the internal energy depends only on temperature, so that C_V is identical to the constant obtained from $U(T)$:

$$C_V = \left(\frac{\partial U}{\partial T}\right)_V = C_v. \tag{2.57}$$

Using the ideal gas relation $PV = Nk_\mathrm{B}T$, it follows that

$$C_P - C_V = Nk_\mathrm{B}. \tag{2.58}$$

Typical values are $C_V = 3Nk_\mathrm{B}/2$ for monoatomic gases and $C_V = 5Nk_\mathrm{B}/2$ for diatomic gases.

REMARK 2.6 The differential relations derived above form the mathematical backbone of thermodynamics. They illustrate how state variables are interconnected and how measurable quantities such as heat capacities encode the response of a system to infinitesimal changes in its environment. For ideal gases, these relationships acquire particularly elegant forms, while in real systems they reveal the complexity and richness of intermolecular interactions.

3

Cycles and Thermodynamic Entropy

3.1 Second Law of Thermodynamics

In the previous chapter, we examined the *zeroth* and *first laws of thermodynamics*. We now turn to the *second law of thermodynamics*, which governs the direction of thermodynamic processes

DOI: 10.1201/9781003603528-3

and introduces the concept of *entropy*.

For clarity in what follows, we adopt the convention that $\Delta(\,\cdot\,)$ denotes the finite change of a thermodynamic quantity during a process, while $d(\,\cdot\,)$ represents an infinitesimal differential. For instance, ΔP refers to the change in pressure between two states, whereas

$$dP(T,V) = \frac{\partial P}{\partial T}dT + \frac{\partial P}{\partial V}dV \tag{3.1}$$

expresses how pressure varies infinitesimally with temperature and volume.

In practical terms, the second law of thermodynamics asserts that energy cannot be converted into work without limitation. More precisely, it places a fundamental constraint on the direction of energy transfer and on the efficiency of any cyclic process.

> *Let us assume that two systems are in equilibrium at a temperature T_0. It is not possible to perform a process between these two systems that results in the extraction of net work.* — **Lord Kelvin**

FIGURE 3.1 The *Nautilus* submarine exploited the temperature difference between its upper and lower sections to produce work, which was then converted into electricity (this is a model from the 1954 Disney film adaptation).

A vivid way to recall this principle is through a literary example. In Jules Verne's *Twenty Thousand Leagues Under the Sea*, Captain Nemo explains that the *Nautilus* generated electricity by exploiting the temperature difference between the warmer water near the ocean surface and the colder water in the depths (Fig. 3.1). The thermoelectric device on board converted this temperature difference into usable energy—an elegant illustration of the second law in action.

An equivalent but complementary formulation of the second law was given by Rudolf Clausius, who emphasized the unidirectional nature of heat transfer:

> *Let us assume that two systems are at different temperatures, T_1 and T_2. Heat cannot flow spontaneously from a body at the lower temperature to one at the higher temperature.* — **Clausius**

Both Kelvin's and Clausius's statements capture the same underlying principle: while energy is conserved (as expressed by the first law), the second law introduces a fundamental asymmetry in nature, a preferred direction for thermodynamic processes.

REMARK 3.1 The second law marks a profound conceptual shift in physics. It introduces irreversibility as a fundamental feature of natural processes, distinguishing the possible from the impossible. In the following sections, we shall see how this law leads naturally to the definition of entropy and to the quantitative formulation of thermodynamic cycles.

The equivalence between the different formulations of the *second law of thermodynamics* is not immediately evident. When thermodynamics was first developed, these statements were largely motivated by experimental observations rather than by theoretical reasoning. It was through the work of Sadi Carnot, and in particular through his idealized *Carnot engine*, that the equivalence between these formulations could be established. Carnot's reasoning, based on the thought experiment known as the *Carnot cycle*, demonstrated that all these statements are consistent with one another and that they express a single fundamental principle of nature.

As we shall see in the next sections, there is a third formulation of the second law, introduced by Carnot himself. We shall state it here for completeness and return to it once the concept of the Carnot cycle has been developed in detail:

> *A Carnot engine operating between two given temperatures has the greatest possible efficiency of any heat engine operating between those two temperatures.* — **Carnot**

This statement implies that no real or hypothetical machine operating between two thermal reservoirs can exceed the efficiency of a reversible Carnot engine:

$$\eta \leq \eta_{\mathrm{C}}, \tag{3.2}$$

where η denotes the efficiency of any engine and η_{C} that of the ideal Carnot engine.

An alternative and historically influential formulation of the second law was introduced by *Rudolf Clausius* in 1865. It is expressed in terms of a new state function—*entropy*—which he introduced to give a quantitative expression to the directionality of thermodynamic processes. Clausius stated that:

> *The only possible processes are those in which the total entropy of the combined system and its surroundings does not decrease.* — **Clausius**

Mathematically, for two heat reservoirs exchanging infinitesimal amounts of energy,

$$\Delta S_{\mathrm{total}} = \Delta S_1 + \Delta S_2 \geq 0. \tag{3.3}$$

Equality holds for reversible processes ($\Delta S = 0$), while strict inequality applies to irreversible ones ($\Delta S > 0$). This statement is known as the *Clausius inequality*. For an infinitesimal reversible process, Clausius further noted that there exists a function S such that

$$dS = \frac{\delta Q_{\mathrm{rev}}}{T}, \tag{3.4}$$

where T is the absolute temperature at which the heat δQ_{rev} is exchanged. Equation (3.4) is the local, differential form of the second law, and constitutes what Clausius himself called the *heat theorem* [Cla65, Cam24]. The Clausius inequality is therefore the *integral* statement of the heat theorem:

$$\oint \frac{\delta Q}{T} \leq 0, \tag{3.5}$$

where the equality holds for reversible (quasi-static) cycles, and the inequality for irreversible ones. For any process between two equilibrium states A and B,

$$\int_A^B \frac{\delta Q_{\mathrm{rev}}}{T} = S(B) - S(A), \tag{3.6}$$

showing that entropy S is a *state function*—it depends only on the initial and final equilibrium states, not on the path connecting them.

Physical and Geometric Interpretation

The Clausius formulation emphasizes the irreversibility of real processes: entropy can only increase globally. The heat theorem, on the other hand, identifies temperature as the *integrating factor* that renders the inexact differential δQ exact. This realization led to the definition of entropy as a potential function whose differential governs all reversible heat exchanges.

In modern geometric terms, the second law implies that in the space of equilibrium states $(E, V, \dots)$, the one-form $\delta Q/T$ is *closed* and hence exact. The existence of a scalar function $S(E, V, \dots)$ satisfying

$$dS = \frac{\delta Q_{\text{rev}}}{T} \tag{3.7}$$

ensures the integrability of thermodynamics, as first rigorously established by Constantin Carathéodory [Car09] and later discussed in statistical terms by Josiah Gibbs [Gib02].

REMARK 3.2 Clausius's empirical discovery that $\delta Q/T$ behaves as an exact differential was one of the most profound insights of nineteenth-century physics. It transformed thermodynamics from a phenomenological description of heat engines into a theory founded on the existence of an integrating factor—temperature—and a corresponding potential—entropy.

A further equivalent formulation, known as the *Kelvin–Planck statement*, emphasizes the impossibility of converting all the heat from a single reservoir into mechanical work:

> *It is impossible to construct a device that operates in a cycle and produces no effect other than the transfer of heat from a single thermal reservoir into mechanical work.*
> — **Kelvin–Planck**

This statement asserts that no engine can operate using only one heat reservoir; in order to produce work, heat must flow between two bodies at different temperatures.

REMARK 3.3 The various formulations of the second law—those of Kelvin, Clausius, Carnot, and Planck—are conceptually equivalent, though they emphasize different aspects of irreversibility and efficiency. Carnot's approach highlights the upper bound on performance, Clausius's formulation introduces the state function of entropy, and the Kelvin–Planck statement captures the physical impossibility of perpetual motion machines of the second kind. Together, these formulations provide a unified and self-consistent framework for the second law, establishing the limits of energy conversion and the meaning of irreversibility in thermodynamic systems [Cal85].

3.2 Cycles and Thermodynamic Entropy

One important consequence of the fact that the energy function U is an exact differential is that, for any reversible closed cycle performed in the P–V plane, the total change in internal energy must vanish. In other words, after the system returns to its initial state, all state functions—including U—recover their original values.

Let us define a *cycle* as a sequence of quasistatic transformations:

$$(P_1, V_1) \xrightarrow{\text{Process 1}} (P_2, V_2) \xrightarrow{\text{Process 2}} (P_3, V_3) \xrightarrow{\text{Process 3}} \cdots \xrightarrow{\text{Process } k} (P_1, V_1). \tag{3.8}$$

The total change in internal energy over such a closed path is

$$\sum_{t=1}^{k-1} \int_{(P_t,V_t)}^{(P_{t+1},V_{t+1})} dU + \int_{(P_k,V_k)}^{(P_1,V_1)} dU = \oint dU. \tag{3.9}$$

Since U is a state function, each integral $\int dU$ depends only on the initial and final points, and therefore

$$\oint dU = 0. \tag{3.10}$$

This condition holds for any reversible cycle, a property that will soon play a central role in the analysis of heat engines. The notion of a *reversible process* will be discussed shortly, but for now, we can proceed to the most famous example of a reversible thermodynamic cycle.

The Carnot Cycle

Few concepts in thermodynamics are as fundamental as the *Carnot cycle*, a theoretical construct that reshaped our understanding of heat engines and established the foundation of the second law of thermodynamics. Introduced by the French physicist Sadi Carnot in 1824 in his treatise *Reflections on the Motive Power of Fire*, the Carnot cycle demonstrated that the efficiency of any heat engine operating between two temperatures depends solely on those temperatures and not on the nature of the working substance.

The Carnot engine operates between two temperature reservoirs, T_1 (cold) and T_2 (hot), and consists of four reversible stages: two isothermal and two adiabatic transformations. These are illustrated in Fig. 3.2 and can be described as follows:

- $A \to B$ *(Isothermal Expansion)* — The system is placed in contact with the hot reservoir at temperature T_2. The gas expands slowly from V_A to V_B, performing work on a piston while absorbing heat Q_2 from the reservoir.
- $B \to C$ *(Adiabatic Expansion)* — The system is thermally isolated and continues to expand from V_B to V_C. No heat is exchanged ($\delta Q = 0$), and the gas cools from T_2 to T_1.
- $C \to D$ *(Isothermal Compression)* — The system is placed in contact with the cold reservoir at temperature T_1. The gas is compressed from V_C to V_D, releasing heat Q_1 to the cold reservoir.
- $D \to A$ *(Adiabatic Compression)* — The system is again isolated and compressed adiabatically from V_D back to V_A, restoring the initial temperature T_2.

The purpose of the Carnot cycle is to convert heat into mechanical work through the controlled, reversible interaction between the system and the two heat reservoirs. It is crucial to note that the two reservoirs are never connected simultaneously: the working substance exchanges heat with only one reservoir at a time.

During the isothermal stages, heat exchange occurs according to the first law of thermodynamics. Since the temperature remains constant ($dT = 0$), we have

$$0 = C_V dT = \delta Q - \delta W = \delta Q - P\,dV. \tag{3.11}$$

Hence, the heat absorbed during the isothermal expansion $A \to B$ is

$$Q_2 = \int_{V_A}^{V_B} N k_B T_2 \frac{dV}{V} = N k_B T_2 \ln\left(\frac{V_B}{V_A}\right). \tag{3.12}$$

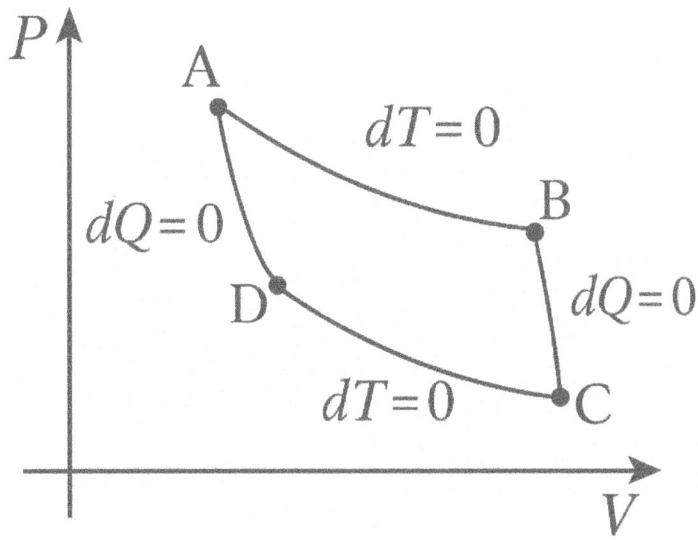

FIGURE 3.2 The Carnot cycle, which is composed of: (AB) an isothermal transformation $PV = Nk_BT_2$; (BC) an adiabatic transformation; (CD) an isothermal transformation $PV = Nk_BT_1$; and (DA) an adiabatic transformation. Diagrams of this type in the PV axis are called Clapeyron diagrams.

Similarly, during the isothermal compression $C \to D$, the system releases heat

$$Q_1 = \int_{V_C}^{V_D} Nk_BT_1 \frac{dV}{V} = Nk_BT_1 \ln\left(\frac{V_D}{V_C}\right). \tag{3.13}$$

In the adiabatic branches $B \to C$ and $D \to A$, by definition $\delta Q = 0$.

During the expansion $A \to B$, where $V_B > V_A$, the system absorbs heat Q_2 from the hot reservoir. During the compression $C \to D$, the system releases heat Q_1 to the cold reservoir. The net work performed by the gas over one complete cycle is therefore

$$\Delta W = Q_2 - Q_1. \tag{3.14}$$

The *efficiency* of the cycle, defined as the ratio of useful work output to the total heat absorbed from the hot reservoir, is given by

$$\eta = \frac{\Delta W}{Q_2} = 1 - \frac{Q_1}{Q_2}. \tag{3.15}$$

This expression represents the maximum possible efficiency for any engine operating between the two temperatures T_1 and T_2, regardless of the working substance.

REMARK 3.4 The Carnot cycle exemplifies the ideal reversible process and establishes an upper bound on the efficiency of all real heat engines. Its beauty lies in its universality: the efficiency depends only on the temperatures of the two reservoirs, not on the material properties of the working substance. In the next section, we shall demonstrate that this efficiency can be expressed purely in terms of temperature ratios, leading directly to the definition of entropy.

Let us now examine the relations among the volumes involved in the Carnot cycle, which are not independent. During the first isothermal expansion ($A \to B$), the heat *absorbed* by the system is

$$Q_2 = Nk_BT_2 \ln\left(\frac{V_B}{V_A}\right). \tag{3.16}$$

During the third branch ($C \to D$), corresponding to isothermal compression, the heat *released* to the cold reservoir is

$$Q_1 = Nk_BT_1 \ln\left(\frac{V_C}{V_D}\right). \tag{3.17}$$

Since $V_C > V_D$, the logarithm is positive, and thus $Q_1 > 0$. The efficiency of the Carnot engine is therefore

$$\eta_C = 1 - \frac{Q_1}{Q_2} = 1 - \frac{T_1 \ln\left(\frac{V_C}{V_D}\right)}{T_2 \ln\left(\frac{V_B}{V_A}\right)}. \tag{3.18}$$

However, the four volumes V_A, V_B, V_C, and V_D are not independent. From the adiabatic relations for an ideal gas,

$$T_2V_B^\gamma = T_1V_C^\gamma, \qquad T_1V_D^\gamma = T_2V_A^\gamma, \tag{3.19}$$

where $\gamma = C_P/C_V$ is the adiabatic index, we obtain

$$\frac{V_B}{V_A} = \frac{V_C}{V_D}. \tag{3.20}$$

Substituting this relation into the expression for the efficiency yields the celebrated result:

$$\eta_C = 1 - \frac{T_1}{T_2}. \tag{3.21}$$

This shows that the efficiency of a reversible Carnot engine depends solely on the ratio of the temperatures of the two heat reservoirs and not on any specific property of the working substance.

We are now in a position to demonstrate the equivalence between the Clausius and Kelvin formulations of the second law of thermodynamics.

- *Clausius* $\Rightarrow$ *Kelvin*: When $T_1 = T_2$, the efficiency $\eta_C = 0$, and the process produces no net work on the piston. This is consistent with Kelvin's statement that it is impossible to extract work from two systems at the same temperature; hence, the Kelvin statement follows from Clausius's.
- *Kelvin* $\Rightarrow$ *Clausius*: Conversely, if we were to assume $T_1 > T_2$, this would imply $\Delta W < 0$, meaning the system would perform negative work—a violation of the Kelvin statement. Therefore, the Clausius formulation must also hold.

REMARK 3.5 The Carnot cycle not only defines the maximum theoretical efficiency of any heat engine but also provides the logical bridge between the different formulations of the second law. Through the relations among heat, work, and temperature, Carnot demonstrated that thermodynamic efficiency depends only on temperature ratios, unifying the insights of Clausius and Kelvin into a single, consistent framework.

According to Carnot's formulation of the second law, the efficiency η_C represents the *maximum possible efficiency* attainable by any heat engine operating between two heat reservoirs at different temperatures. No real engine can surpass this value without violating the fundamental principles of thermodynamics.

To appreciate the meaning of this result, let us consider a hypothetical scenario. Suppose, contrary to Carnot's conclusion, that we could construct a heat engine more efficient than a Carnot engine operating between the same two temperatures. By coupling this hypothetical engine to a real Carnot engine operating in reverse (that is, as a refrigerator or heat pump), we could design a composite system that extracts heat from a single reservoir and converts it entirely into work. Such a machine would contradict the *Kelvin–Planck statement* of the second law, which asserts that no device operating in a cycle can produce work while exchanging heat with only one reservoir.

Alternatively, imagine attempting to build an engine that operates with a single bath, effectively treating two identical reservoirs at the same temperature as distinct sources and sinks. In this case, no temperature difference would exist to drive heat flow, and therefore no work could be produced. This again reinforces the principle that thermal energy can be converted into mechanical work only when a temperature gradient is present.

REMARK 3.6 The Carnot theorem thus provides a profound insight: the efficiency of a reversible Carnot engine sets an absolute, universal upper bound for all heat engines. Any claim of higher efficiency would imply the construction of a perpetual-motion machine of the second kind, an impossibility under the laws of thermodynamics. This reasoning connects the abstract concept of reversibility with the physical impossibility of single-reservoir engines, cementing the link between the Carnot, Clausius, and Kelvin formulations of the second law.

There is yet another way to demonstrate the equivalence between the Clausius and Kelvin formulations of the second law [Atk10]. Consider two heat engines, A and B. Engine A is connected only to a hot reservoir, while engine B is connected to both hot and cold reservoirs. Engine A is used to drive engine B. By reasoning *ad absurdum*, suppose that engine A can convert all the heat it receives from the hot reservoir entirely into work. This work could then drive engine B in reverse, causing it to transfer heat from the cold to the hot reservoir. Thus, if it were possible to convert heat into work at no cost, heat could spontaneously flow from a lower to a higher temperature source—contradicting Clausius's formulation of the second law. Therefore, Clausius's statement implies Kelvin's.

Now consider the reverse argument. Suppose a thermal engine operates between a hot and a cold reservoir, and imagine that Clausius's second law were false—so that heat could be moved from the cold to the hot reservoir without external work. In that case, it would be possible to construct a device that converts all absorbed heat into mechanical work without any expenditure—a perpetual-motion machine of the second kind. This outcome directly violates Kelvin's formulation, and therefore Kelvin implies Clausius. Hence, both statements are logically equivalent and consistent with the Carnot theorem.

A similar reasoning applies to the *Kelvin–Planck statement*. Assume, for contradiction, that this statement is false. Then one could expel heat into the same reservoir by connecting the hot and cold sides to the same bath, which would violate Kelvin's version of the second law by allowing heat to flow spontaneously from higher to lower temperature reservoirs without producing entropy. Such a scenario would imply the existence of a Carnot engine operating with perfect efficiency, in contradiction with Carnot's own result. Thus, the Kelvin and Kelvin–Planck formulations are also equivalent.

REMARK 3.7 The chain of reasoning presented above establishes that the formulations of Clausius, Kelvin, and Kelvin–Planck are mutually consistent and express the same fundamental limitation: no process can convert heat entirely into work without a temperature difference, and no spontaneous transfer of heat can occur from a colder to a hotter body. Together, these statements form the logical framework that underpins the second law of thermodynamics.

Thermodynamic Entropy

Having established the operation and efficiency of the Carnot engine, we can now use it to introduce one of the most profound concepts in thermodynamics: *entropy.* Following Clausius's reasoning, we imagine repeating the Carnot cycle multiple times between a central reservoir at temperature T_0 and a set of other reservoirs at lower temperatures $T_1, T_2, \ldots, T_n < T_0$. The purpose of this construction is to identify a new differential quantity,

$$dS = \frac{\delta Q}{T}, \tag{3.22}$$

which we call the *change in entropy,* and to show that for all natural processes

$$dS \geq 0. \tag{3.23}$$

The word "entropy" itself, introduced by Rudolf Clausius, derives from the Greek *tropé,* meaning "transformation," reflecting its role as a measure of change or irreversibility in a thermodynamic system.

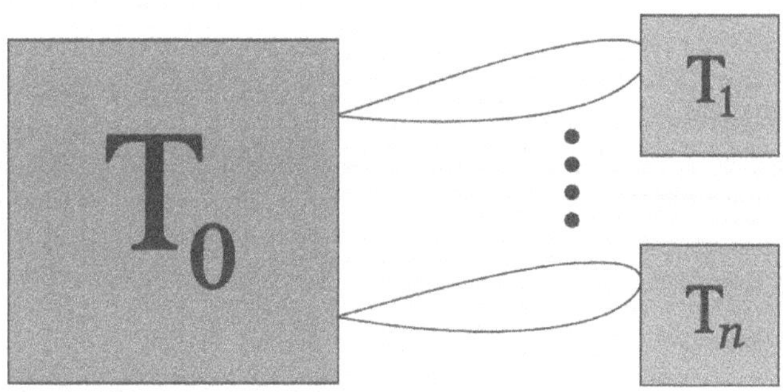

FIGURE 3.3 Multiple Carnot cycles connecting a central reservoir at temperature T_0 to several reservoirs at lower temperatures $T_1, \ldots, T_n$. We assume $T_0 > T_i$.

Let us consider n such Carnot cycles. Denote by Q_i^0 the quantity of heat released by the central bath at temperature T_0 in the i-th cycle, and by Q_i the heat absorbed by the reservoir at temperature T_i. Because each cycle is a reversible Carnot cycle, we know that

$$\frac{Q_i^0}{Q_i} = \frac{T_0}{T_i}. \tag{3.24}$$

The total heat released by the central bath is then

$$Q_0 = \sum_{i=1}^{n} Q_i^0 = \sum_{i=1}^{n} \frac{T_0}{T_i} Q_i. \tag{3.25}$$

Since $T_0 > T_i$ for all i, and because the second law forbids the net absorption of heat by the hotter reservoir, Q_0 must be negative. This leads to the inequality

$$\sum_{i=1}^{n} \frac{Q_i}{T_i} \leq 0. \tag{3.26}$$

This relation implies the existence of a new quantity, which Clausius defined as the *entropy* S, such that

$$\oint \frac{\delta Q}{T} \leq 0. \tag{3.27}$$

For reversible cycles, the equality sign holds, and we can define entropy differentials through

$$dS = \frac{\delta Q_{\text{rev}}}{T}. \tag{3.28}$$

For any arbitrary (possibly irreversible) process, we then have

$$\int dS = -\int \frac{\delta Q}{T} \geq 0, \tag{3.29}$$

which follows because any transformation in the P–V plane can be decomposed into a series of infinitesimal Carnot cycles.

In the case of a reversible cycle, we know that the total change in entropy vanishes:

$$\oint \frac{\delta Q_{\text{rev}}}{T} = 0. \tag{3.30}$$

This condition implies that entropy is an *exact differential*, allowing us to write

$$\int_A^B \frac{\delta Q_{\text{rev}}}{T} = S(B) - S(A). \tag{3.31}$$

For irreversible processes, however, the inequality applies:

$$\int_A^B \frac{\delta Q}{T} \leq S(B) - S(A), \tag{3.32}$$

which leads to

$$S(B) \geq S(A). \tag{3.33}$$

Hence, entropy can only increase (or, in the ideal reversible case, remain constant) in an isolated system.

The *second law of thermodynamics* can thus be restated in Clausius's definitive form:

The entropy of a closed system can never decrease.

REMARK 3.8 This formulation provides a profound reinterpretation of the second law: while energy is conserved, entropy measures the degree of irreversibility in natural processes. Entropy increase is the quantitative signature of the arrow of time, indicating that real processes evolve toward equilibrium, where entropy reaches its maximum possible value.

We can now combine the *first law of thermodynamics* with the definition of entropy to obtain one of the fundamental relations of thermodynamics. Starting from

$$dU = \delta Q - \delta W, \tag{3.34}$$

and substituting $\delta Q = T\,dS$, we have

$$dU = T\,dS - \delta W. \tag{3.35}$$

Since the infinitesimal work is given by $\delta W = P\,dV$, this becomes

$$dU = T\,dS - P\,dV, \tag{3.36}$$

which represents the *fundamental thermodynamic relation* for a system of fixed particle number. This equation expresses the internal energy as a function of the natural variables S and V. This compact expression, $dU = T\,dS - P\,dV$, is sometimes called the *Gibbs relation*, after J. Willard Gibbs, who formalized thermodynamics using differential geometry.

For an *ideal gas*, we can relate entropy to temperature and volume by combining Eq. (3.36) with the ideal gas law $PV = Nk_{\mathrm{B}}T$. Using $dU = C_V\,dT$, we find

$$dS = C_V\,\frac{dT}{T} + Nk_{\mathrm{B}}\,\frac{dV}{V}. \tag{3.37}$$

Integrating between two states A and B gives

$$S(B) - S(A) = \int_A^B dS = C_V \ln\left(\frac{T_B}{T_A}\right) + Nk_{\mathrm{B}} \ln\left(\frac{V_B}{V_A}\right). \tag{3.38}$$

We can now identify the entropy changes for common thermodynamic processes:

- *Isothermal process:* $\Delta S = Nk_{\mathrm{B}} \ln\left(\frac{V_B}{V_A}\right)$, since $T_B = T_A$.
- *Adiabatic process:* $\Delta S = 0$, as $\delta Q = 0$.
- *Isochoric process:* $\Delta S = C_V \ln\left(\frac{T_B}{T_A}\right)$, since $V_B = V_A$.
- *Adiabatic* $\Delta S = 0$, since $dQ = 0$.

For an ideal gas, both C_V and C_P are approximately constant and independent of temperature—a result derived later from kinetic theory.

REMARK 3.9 The entropy changes discussed above can be used to understand various types of efficiency during cycles.

3.2.1 Entropy and the Carnot Cycle

Let us now return to the Carnot cycle and examine its entropy balance. The process is *reversible* precisely because the total entropy change over one complete cycle vanishes. Entropy changes occur only during the two isothermal branches, since $\delta Q = 0$ for the adiabatic stages. Therefore,

$$\oint_{\text{Carnot}} dS = Nk_{\mathrm{B}} \ln\left(\frac{V_A V_C}{V_B V_D}\right). \tag{3.39}$$

From the adiabatic relations, we know that $V_A V_D = V_B V_C$, which implies

$$\Delta S_{\text{Carnot}} = 0. \tag{3.40}$$

Thus, for the Carnot cycle, the total change in entropy over a complete cycle is zero, confirming that the process is reversible.

Historically, Clausius first recognized that in a Carnot cycle the sum $\sum_i Q_i/T_i = 0$, introducing entropy as a new state function. Boltzmann later provided its microscopic interpretation in terms of probability and disorder.

REMARK 3.10 The equality $\Delta S_{\text{Carnot}} = 0$ provides a rigorous mathematical criterion for reversibility: a process is reversible if and only if the total entropy of the system and its surroundings remains constant. This insight bridges the macroscopic formulation of thermodynamics and the microscopic statistical interpretation of entropy, linking Carnot's idealized cycles to the irreversible processes that govern real physical systems.

Adiabatic Processes

For clarity and future reference, we summarize here several physical mechanisms that can give rise to an *adiabatic process.* The concept, as formalized by Max Planck, provides an operational foundation for understanding transformations in which no heat is exchanged with the surroundings. Planck's systematic discussion of adiabatic processes appeared in his *Treatise on Thermodynamics* (1897).

Planck's definition can be stated as follows:

> *Consider a system that changes from a state A to a state B in the P–V diagram through interaction with a device—mechanical or electrical—such that the amount of work performed can be measured, for instance, by the displacement of a weight in a gravitational field. At the end of the transformation, the mechanical device has returned to its original state, while the weight may not have.*

This definition highlights that the distinction between heat and work is not absolute but depends on the nature of the interaction between the system and its surroundings. In an adiabatic process, the system is thermally isolated so that $\delta Q = 0$, and any energy exchange with the environment occurs solely in the form of work.

The term *adiabatic* comes from the Greek *adiabatos*, meaning "not passable," emphasizing the absence of heat flow across the system's boundary. Several examples of adiabatic processes are of practical and theoretical interest:

- *Frictional or mechanical work:* energy transfer through rubbing or stirring.
- *Free expansion or compression:* changes in volume without heat exchange, whether involving external weights or not.
- *Natural internal processes:* transformations occurring within an isolated compound, such as chemical reactions or nuclear interactions, where energy redistribution is internal.
- *Electrical interactions:* work performed by or on electrical devices within a thermally insulated container.
- *Combinations:* any process involving multiple mechanisms of work while maintaining thermal isolation.

In general, the property of being "adiabatic" is not intrinsic to the process itself but rather to the *boundary conditions* of the system—specifically, the thermal insulation of the container that prevents heat exchange with the surroundings.

REMARK 3.11 Adiabatic processes are central to thermodynamics because they allow the conversion of internal energy entirely into work (or vice versa) without heat exchange. They provide the conceptual bridge between purely mechanical descriptions of energy transfer and the thermodynamic distinction between heat and work, which lies at the heart of the first law.

3.3 Thermodynamic Potentials

The study of *thermodynamic potentials* provides powerful tools for describing equilibrium and stability in physical systems. Among these, the *Massieu potentials*, introduced by the French physicist François Massieu in 1869, offer a distinctive formulation in which entropy is treated as an independent variable [Mas69, Bal17]. This entropy-based representation provides an alternative to the traditional energy-centered framework of thermodynamics and reveals deep relationships among thermodynamic variables.

Historically, Massieu's work represented a conceptual shift. Massieu's approach inspired later developments by Gibbs, who extended the formal structure of thermodynamic potentials to include the free energies and the grand potential.

Instead of using the internal energy U or enthalpy H as the central thermodynamic potential, he proposed considering entropy S as a fundamental independent variable alongside other extensive quantities such as volume V and particle number N. This innovation provided a new perspective on equilibrium and stability, particularly for systems where entropy changes are more natural to measure or calculate than energy.

By adopting entropy as an independent variable, Massieu formulated a family of potentials—now known as the *Massieu functions*—that re-express the thermodynamic relations in a concise and symmetric form. For a system characterized by a set of extensive variables $X_1, X_2, \ldots$, the general form of a Massieu potential Φ_S is written as

$$\Phi_S = f(S, X_1, X_2, \ldots), \tag{3.41}$$

where S is entropy and the X_i denote other extensive quantities such as volume, particle number, or magnetization, depending on the system considered.

For a single-component system, the differential form of the fundamental relation becomes

$$d\Phi_S = \left(\frac{\partial \Phi_S}{\partial S}\right)_{X_1, X_2, \ldots} dS + \sum_i \left(\frac{\partial \Phi_S}{\partial X_i}\right)_{S, X_{j \neq i}} dX_i. \tag{3.42}$$

Since temperature is defined through the relation

$$\frac{1}{T} = \left(\frac{\partial S}{\partial U}\right)_{X_i}, \tag{3.43}$$

we can equivalently express

$$d\Phi_S = \frac{dS}{T} + \sum_i Y_i \, dX_i, \tag{3.44}$$

where the conjugate quantities Y_i correspond to generalized forces (e.g., pressure, chemical potential, etc.) associated with the extensive variables X_i.

The introduction of the Massieu functions thus extends the thermodynamic formalism by emphasizing entropy as the organizing principle of equilibrium. Although the more familiar energy-based potentials (Helmholtz, Gibbs, and grand potentials) eventually became standard, Massieu's approach revealed the deep duality between energy and entropy representations and anticipated modern statistical mechanics, where entropy plays a central role in defining ensembles and equilibrium conditions.

REMARK 3.12 Massieu's framework provides a unified viewpoint from which these traditional thermodynamic potentials can be derived. By treating entropy as a central variable, his formulation reveals that all equilibrium conditions can be expressed either in terms of energy minimization or entropy maximization, depending on which variables are held fixed. This dual perspective offers valuable insight into systems operating under non-standard conditions, or in contexts where entropy variations are the natural quantities of interest.

Among the most familiar examples of Massieu's potentials are the *Helmholtz* and *Gibbs free energies*, which can be regarded as particular cases of entropy-based thermodynamic functions. These potentials re-express the thermodynamic equilibrium conditions under constraints of temperature, pressure, and volume.

In the sections that follow, we shall first discuss these potentials in the case where the number of particles N is fixed, and later generalize the formulation to include systems in which particle exchange or chemical reactions occur.

3.4 Thermodynamic Potentials: Enthalpy and Free Energies

Thermodynamic potentials provide compact ways of describing equilibrium and energy exchange under specific experimental conditions. Each potential corresponds to a particular set of natural variables, allowing us to express the thermodynamic state of a system and its spontaneous behavior in differential form.

Enthalpy $H = U + PV$

The *enthalpy*, whose name originates from the Greek *enthalpein* ("to warm"), is defined as

$$H = U + PV, \tag{3.45}$$

where U is the internal energy, P the pressure, and V the volume of the system. Differentiating this expression gives

$$dH = dU + P\,dV + V\,dP. \tag{3.46}$$

Substituting $dU = T\,dS - P\,dV$ from the first law (for a fixed number of particles) yields

$$dH = T\,dS + V\,dP. \tag{3.47}$$

Hence, enthalpy naturally depends on the variables (S, P), so we may write $H = H(S, P)$. The corresponding thermodynamic derivatives are then

$$\begin{aligned} T &= \left(\frac{\partial H}{\partial S}\right)_P, \\ V &= \left(\frac{\partial H}{\partial P}\right)_S. \end{aligned} \tag{3.48}$$

These are the first of the celebrated *Maxwell relations*.

For an *isobaric process*, $dP = 0$, and the enthalpy change equals the heat absorbed:

$$\Delta H = \Delta Q = T\Delta S. \tag{3.49}$$

Thus, under constant pressure and reversible conditions, the change in enthalpy directly measures the heat transferred.

The concept of enthalpy is particularly useful in chemical thermodynamics and engineering, where many processes occur at constant pressure. For instance, the heat of reaction at constant P equals the enthalpy change.

REMARK 3.13 Later in this Volume, we will discuss statistical mechanics starting from the notion of Hamiltonian. Historically, both Hamiltonians and enthalpy are described using the letter H; although they have different meanings, they both have the dimensionality of energy. It should be clear from the context that we are discussing different quantities, but it is important to stress that one has a thermodynamic and macroscopic meaning, i.e., the enthalpy, and one represents a microscopic description, i.e., the Hamiltonian.

Helmholtz Free Energy $F = U - TS$

Another important potential is the *Helmholtz free energy*, denoted by F, and defined as

$$F = U - TS. \tag{3.50}$$

This potential is most relevant for systems maintained at constant volume and temperature, such as closed systems undergoing reversible transformations. In this case, the Helmholtz free energy measures the maximum work that can be extracted from a system other than the $P\,dV$ work, and a process is spontaneous when $\Delta F < 0$.

As we see later, the Helmholtz free energy is often interpreted as the Legendre transform of the internal energy U with respect to the variable S, replacing entropy by temperature as an independent parameter.

Differentiating, we find

$$dF = dU - T\,dS - S\,dT. \tag{3.51}$$

Using $dU = T\,dS - P\,dV$, we obtain

$$dF = -S\,dT - P\,dV. \tag{3.52}$$

Thus, the Helmholtz free energy depends on the natural variables (T, V), and we can write $F = F(T, V)$. The corresponding relations are

$$\begin{aligned} S &= -\left(\frac{\partial F}{\partial T}\right)_V, \\ P &= -\left(\frac{\partial F}{\partial V}\right)_T. \end{aligned} \tag{3.53}$$

These constitute the second pair of Maxwell relations.

For an *isothermal process* ($dT = 0$), we can express the finite change in F as

$$\Delta F \leq -\Delta W, \tag{3.54}$$

with equality for reversible processes. Hence, F decreases in spontaneous processes and remains constant at equilibrium, where it attains its minimum value.

The Helmholtz potential is particularly relevant for systems held at constant temperature and volume, such as canonical ensembles in statistical mechanics. Its minimum defines the condition for thermodynamic equilibrium.

Gibbs Free Energy $G = U - TS + PV$

The *Gibbs free energy*, denoted by G, is defined as

$$G = U - TS + PV, \tag{3.55}$$

where U is the internal energy, T the absolute temperature, S the entropy, P the pressure, and V the volume of the system.

The Gibbs free energy is particularly useful in *chemical thermodynamics*, since it determines the spontaneity of processes occurring at constant temperature and pressure. A process is spontaneous if and only if $\Delta G < 0$ under these conditions.

Gibbs introduced the concept of free energy in the 1870s while developing the geometrical and analytical foundations of chemical thermodynamics. His work unified mechanics, chemistry, and thermodynamics in a single formalism. The *Gibbs free energy*, G, combines the definitions of F and H:

$$G = F + PV = H - TS = U - TS + PV. \tag{3.56}$$

Differentiating, we find

$$dG = dU - T\,dS - S\,dT + P\,dV + V\,dP. \tag{3.57}$$

Using $dU = T\,dS - P\,dV$, this simplifies to

$$dG = -S\,dT + V\,dP. \tag{3.58}$$

The Gibbs free energy thus depends on the natural variables (T, P), so we can write $G = G(T, P)$. The corresponding thermodynamic derivatives are

$$\begin{aligned} S &= -\left(\frac{\partial G}{\partial T}\right)_P, \\ V &= \left(\frac{\partial G}{\partial P}\right)_T. \end{aligned} \tag{3.59}$$

These constitute the final pair of Maxwell relations.

Finally, for *isothermal and isobaric* processes, we have

$$\Delta G \leq -\Delta W, \tag{3.60}$$

with equality for reversible transformations. Therefore, G serves as the criterion of equilibrium for systems held at fixed temperature and pressure. At constant T and P, a process is called *spontaneous* when $\Delta G < 0$, and equilibrium corresponds to the minimum of G.

REMARK 3.14 The four thermodynamic potentials—U, H, F, and G—together with their differential forms, provide a complete framework for describing equilibrium under different constraints. The symmetry among their derivatives, encapsulated by the *Maxwell relations*, reflects the deep mathematical structure underlying thermodynamics and anticipates the geometric formulation later developed by Gibbs and Carathéodory.

Legendre Transform and the Relationship Between Potentials

The thermodynamic potentials introduced above are not independent; rather, they are related by a remarkable symmetry. This symmetry can be visualized, as in Fig. 3.4, by placing the variables P, T, V, and S at the corners of a square, with each side representing one of the four fundamental potentials as a function of the adjacent variables. Two potentials that share a common variable are connected by "swapping" the pair of variables by which they differ. This is not coincidental—the potentials are connected through a mathematical operation known as the *Legendre transform*.

In this framework, the variables (V, S) are *extensive*, while (P, T) are *intensive*. The internal energy $U(S, V)$ is unique among the thermodynamic potentials because it depends solely on

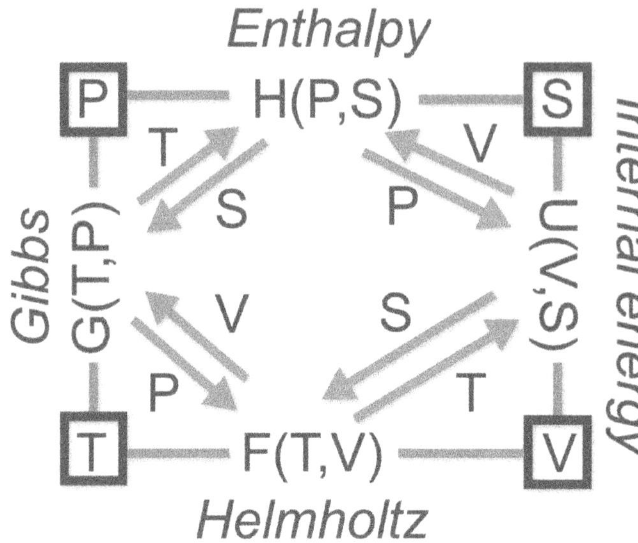

FIGURE 3.4 Diagram showing the relationships among the thermodynamic potentials. Each potential is obtained from its neighbor through a Legendre transform with respect to one conjugate pair of variables.

extensive variables. The other potentials arise through Legendre transforms of U with respect to one or more conjugate pairs of variables. If these potentials are convex in their respective natural variables, they are equivalent representations of the same thermodynamic information.

To understand the Legendre transform, consider a real function $f(x)$ that is smooth and strictly convex, i.e., $f''(x) > 0$. The *Legendre transform* of f is defined as the function

$$G(s) = s\,x(s) - f(x(s)), \tag{3.61}$$

where the variable s is related to x by

$$s = f'(x). \tag{3.62}$$

The transformation exchanges the dependence on x for one on its conjugate variable s. If $f(x)$ is convex, $f'(x)$ is strictly monotonic, ensuring a one-to-one mapping between x and s. This means that the transformed function $G(s)$ carries the same information as $f(x)$, but expressed in terms of the conjugate variable.

An equivalent and often more general definition can be written as

$$G(s) = \sup_x L(x, s) = \sup_x [x\,s - f(x)], \tag{3.63}$$

where $\sup_x$ denotes the supremum over all x. From this, it immediately follows that

$$G(s) \geq s\,x - f(x), \tag{3.64}$$

a result known as *Fenchel's inequality*. Equality holds when $s = f'(x)$, i.e., when x and s are conjugate variables. A graphical representation is shown in Fig. 3.5, where the tangent line of slope s touches $f(x)$ at a single point, and its intercept defines $G(s)$.

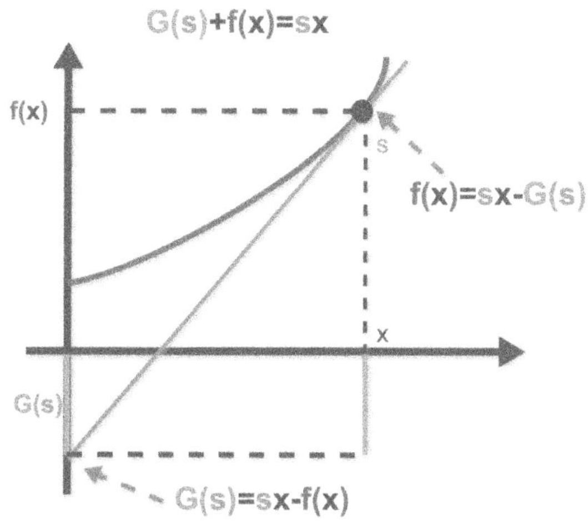

FIGURE 3.5 Graphical representation of the Legendre transform relation $G(s) + f(x) = s\,x$. The tangent line to $f(x)$ of slope s defines the corresponding value of $G(s)$.

Differentiating with respect to s and x, we obtain the mutual relations

$$\frac{dG}{ds} = x, \quad \frac{df}{dx} = s, \tag{3.65}$$

$$\frac{d^2G}{ds^2} = \frac{dx}{ds}, \quad \frac{d^2f}{dx^2} = \frac{ds}{dx}. \tag{3.66}$$

It then follows immediately that

$$\frac{d^2G}{ds^2}\frac{d^2f}{dx^2} = \frac{dx}{ds}\frac{ds}{dx} = 1. \tag{3.67}$$

This relationship confirms that convexity is preserved under the Legendre transform: if $f(x)$ is convex ($f''(x) > 0$), then so is $G(s)$.

The Legendre transform provides a systematic way of interchanging natural variables in thermodynamics: transforming from $U(S, V)$ to $H(S, P)$ replaces the variable V by its conjugate P, while transforming from $U(S, V)$ to $F(T, V)$ replaces S by its conjugate T. Each transformation swaps one pair of conjugate variables and yields a new potential whose natural variables are suited to the physical constraints of the system.

REMARK 3.15 The Legendre transform lies at the heart of thermodynamics, mechanics, and field theory alike. In classical mechanics, it connects the Lagrangian $L(q, \dot{q})$ and the Hamiltonian $H(q, p)$; in thermodynamics, it relates energy-based and entropy-based descriptions of equilibrium. Both interpretations rely on the same geometric principle: transforming between equivalent but complementary representations of the same physical information.

Legendre Transforms and Thermodynamic Potentials

The same mathematical symmetry connects the various thermodynamic potentials. To formalize this, let us introduce a concise notation. Given a function $F(x_1, \ldots, x_n)$, its Legendre transform with respect to the variable x_m is denoted by

$$F'(\ldots, s, \ldots) = \mathcal{L}_{s,x_m} F, \tag{3.68}$$

where $\mathcal{L}_{s,x_m}$ denotes the Legendre transform operation replacing x_m by its conjugate s.

Applying this operation successively to the internal energy $U(S,V)$ yields the familiar hierarchy of thermodynamic potentials:

$$\begin{aligned}
H(P,S) &= \mathcal{L}_{V,P}\, U(V,S), && (3.69)\\
F(T,V) &= \mathcal{L}_{S,T}\, U(V,S), && (3.70)\\
G(T,P) &= \mathcal{L}_{P,V}\, F(T,V), && (3.71)\\
H(P,S) &= \mathcal{L}_{T,S}\, G(T,P). && (3.72)
\end{aligned}$$

These relationships encapsulate the logical symmetry among the four fundamental thermodynamic potentials. They also lead to a set of reciprocal relations:

$$\begin{aligned}
H(P,S) + U(V,S) &= PV, && (3.73)\\
F(T,V) + U(V,S) &= TS, && (3.74)\\
G(T,P) + F(T,V) &= PV, && (3.75)\\
H(P,S) + G(T,P) &= TS. && (3.76)
\end{aligned}$$

From these, we can see that

$$\begin{aligned}
H(P,S) + U(V,S) &= G(T,P) + F(T,V), && (3.77)\\
H(P,S) + G(T,P) &= F(T,V) + U(V,S), && (3.78)
\end{aligned}$$

which demonstrate the deep internal consistency of the thermodynamic formalism: each potential encodes the same physical information but expressed through different natural variables.

REMARK 3.16 The network of Legendre-transformed thermodynamic potentials illustrates the geometric structure underlying equilibrium thermodynamics. Each potential represents a projection of the same multidimensional surface, $U(S,V)$, onto different coordinate planes in the space of extensive and intensive variables. This geometric interpretation, first explored by Gibbs and later generalized by Carathéodory, lies at the foundation of modern thermodynamic theory.

3.5 General Systems with Varying Particle Number

As we have discussed, thermodynamic potentials are *state functions*. Together with the equations of state, they describe the behavior of a system in terms of its macroscopic, measurable, and natural variables. So far, we have focused on a simple system—an ideal gas—whose natural variables are pressure and volume. In that case, we wrote the differential form of the internal energy as

$$dU = T\,dS - P\,dV, \qquad (3.79)$$

where (T,S) and (P,V) form conjugate pairs of thermodynamic variables.

For more general systems, such as those involving particle exchange, electromagnetic interactions, or chemical reactions, the first law must include additional terms accounting for all possible work contributions. The most general differential form of the internal energy can be written as [Cal85, Atk10]:

$$dU = T\,dS + \sum_{i=1}^{v} F_i\,dq_i + \sum_{j=1}^{s} \mu_j\,dN_j, \qquad (3.80)$$

where:

- $T\,dS$ represents the thermal contribution,
- the second term $\sum_i F_i\,dq_i$ includes generalized work terms (mechanical, electrical, magnetic, etc.), and
- $\sum_j \mu_j\,dN_j$ accounts for changes in the energy due to variation in the number of particles of each species N_j, where μ_j is the *chemical potential.*

The first term in Eq. (3.80) is invariant across all systems, but the other terms depend on the specific physical processes taking place. For a simple gas, we can identify the conjugate pair $(F, q) = (-P, V)$, which reduces Eq. (3.80) to the familiar form of Eq. (3.36). Pressure and volume are thus conjugate variables: one extensive (V) and one intensive (P).

In thermodynamics, conjugate variables appear as measurable pairs: one intensive quantity (such as pressure) represents a generalized force, while the corresponding extensive variable (such as volume) represents a generalized displacement. Their product has the dimension of energy.

In general, we may write the first law in compact notation as

$$dU = \sum_i \mathcal{F}_i\,dq_i, \tag{3.81}$$

where the set $\{\mathcal{F}_i, q_i\}$ includes all conjugate pairs—such as (T, S), $(-P, V)$, and (μ_j, N_j)—that fully characterize the thermodynamic system. Each conjugate pair describes how the system responds to small changes in its macroscopic variables.

The identification of conjugate variables follows from the observation that they are linked through measurable responses. For example, in a gas, increasing the volume V by moving a piston while keeping other variables constant results in a measurable change in pressure P. This pressure change represents the system's linear response to the external action of changing volume.

Chemical Potential and Particle Exchange

Let us now consider a thought experiment in which we add a small number of particles to a thermodynamic system. When particles interact—such as in gases, liquids, or solutions—the internal energy changes not only because of the particles' kinetic energy but also due to their interactions with existing particles. Adding a particle can either increase or decrease the system's energy depending on whether new interactions are attractive or repulsive.

For instance, in a liquid solution, adding solute particles disrupts existing solvent–solvent interactions and forms new solute–solvent interactions. The energy required to overcome or create these interactions determines the change in internal energy per particle added, which is precisely the *chemical potential* μ.

The concept of chemical potential was first formalized by J. Willard Gibbs in 1875. It quantifies the change in a system's internal energy when a particle is added while keeping entropy and volume constant: $\mu = \left(\frac{\partial U}{\partial N}\right)_{S,V}$. The equality of chemical potentials across subsystems defines equilibrium in open systems.

The inclusion of $\mu_j\,dN_j$ in Eq. (3.80) therefore generalizes the first law to systems in which both energy and matter can be exchanged with the surroundings. This extension is crucial in chemical thermodynamics, statistical mechanics, and in the study of phase equilibrium.

REMARK 3.17 The chemical potential unifies mechanical, thermal, and compositional equilibrium within a single framework. While temperature equalization governs heat flow and pressure equalization governs mechanical balance, the equality of chemical potentials determines diffusive and chemical equilibrium. This completes the general statement of the first law for open systems.

3.6 Systems Subject to External Fields and Generalized Conjugate Variables

In systems subject to external fields or additional constraints, new forms of work appear beyond the mechanical $P\,dV$ term. If the constituent particles carry charge or experience long-range forces, energy changes may include contributions from interactions with gravitational, electric, or magnetic fields.

For example, in a gravitational field, the internal energy changes with height due to the work required to lift or lower particles against gravity. In electrochemical systems such as batteries or fuel cells, the energy includes an electrostatic term arising from the motion of charged particles across a potential difference. This separation of charges establishes an electric field that performs work on the particles as they move through the system.

The generalization of the first law to include field contributions dates back to Gibbs (1875) and has since been extended to electromagnetism, elasticity, and even relativistic systems.

Energy Changes in Open Systems

When a system exchanges both energy and matter with its surroundings, the infinitesimal change in internal energy can be written as

$$dU = T\,dS - P\,dV + \mu\,dN, \tag{3.82}$$

where μ is the *chemical potential*. For processes involving the addition or removal of particles,

$$\Delta U = \mu\,\Delta N, \tag{3.83}$$

so that μ represents the change in internal energy per particle added at fixed entropy and volume:

$$\mu = \left(\frac{\partial U}{\partial N}\right)_{S,V}. \tag{3.84}$$

The equality of chemical potentials across subsystems defines diffusive equilibrium, just as the equality of temperature defines thermal equilibrium and the equality of pressure defines mechanical equilibrium.

From this expression, the entropy differential can be written as

$$dS = \frac{dU}{T} + \frac{P}{T}\,dV - \frac{\mu}{T}\,dN, \tag{3.85}$$

showing explicitly that entropy depends on energy, volume, and the number of particles. Consequently, the internal energy can be regarded as a function of the three extensive variables:

$$U = U(S, V, N). \tag{3.86}$$

Legendre Transformations and Generalized Potentials

By performing Legendre transformations on $U(S, V, N)$, we can derive new thermodynamic potentials suited to different experimental conditions. The most common cases are:

- The *Enthalpy* $H(S, P, N) = U + PV$, whose differential is

$$dH = T\,dS + V\,dP + \mu\,dN. \tag{3.87}$$

- The *Helmholtz free energy* $F(T, V, N) = U - TS$, with differential

$$dF = -S\,dT - P\,dV + \mu\,dN. \tag{3.88}$$

- The *Gibbs free energy* $G(T, P, N) = U - TS + PV$, whose differential is

$$dG = -S\,dT + V\,dP + \mu\,dN. \tag{3.89}$$

These generalized potentials are useful because they identify the natural variables associated with specific physical constraints: $U(S, V, N)$ for isolated systems, $F(T, V, N)$ for isothermal processes, $H(S, P, N)$ for isobaric systems, and $G(T, P, N)$ for processes at constant temperature and pressure.

Generalized Conjugate Variables

The language of conjugate variables provides a unifying perspective on the diversity of thermodynamic systems. Each pair consists of an *intensive variable*—a generalized force—and its conjugate *extensive variable*—a generalized displacement. Their product always has dimensions of energy.

Some of the most common examples include:

- *Pressure (P) and Volume (V)*: the mechanical pair governing compressional work. Expanding or compressing a gas changes its internal energy through the work $-P\,dV$.
- *Temperature (T) and Entropy (S)*: the thermal pair connecting energy transfer to disorder. Heat exchange occurs through the product $T\,dS$, representing reversible energy transfer at temperature T.
- *Chemical Potential (μ) and Particle Number (N)*: the compositional pair describing energy changes due to variations in particle number. When two systems are in diffusive equilibrium, their chemical potentials are equal.
- *Electric Potential (Φ) and Charge (Q)*: the electrostatic pair relevant to batteries, capacitors, and electrolytes. Work is done by or against the electric field as charged particles move across a potential difference, with energy change $\Phi\,dQ$.
- *Electric Field (E) and Polarization (P)*: the pair characterizing dielectric materials. The field E aligns molecular dipoles, increasing the polarization P and storing electrostatic energy through $E\,dP$.
- *Magnetic Field (B) and Magnetization (M)*: the magnetic pair relevant to ferromagnetic or paramagnetic materials. The work term $B\,dM$ quantifies energy exchange between an applied field and the internal magnetic moments.
- *Stress (τ) and Strain (ϵ)*: the mechanical pair associated with the deformation of solids. Work is done on the material as it stretches or compresses, contributing an energy term $\tau\,d\epsilon$.

Each of these pairs contributes a term $\mathcal{F}_i\,dq_i$ to the total energy differential, leading to the general expression

$$dU = \sum_i \mathcal{F}_i\,dq_i, \tag{3.90}$$

which encompasses all forms of work and energy exchange in classical thermodynamics.

REMARK 3.18 The notion of conjugate variables is one of the unifying ideas of modern thermodynamics. It allows gravitational, electrical, magnetic, mechanical, and chemical systems to be described within a single mathematical framework. Each pair reflects a measurable cause-and-effect relationship between an external field and the system's response, revealing the profound symmetry that underlies all thermodynamic processes.

3.7 Extensive and Intensive Quantities, and the Grand Potential

The second law of thermodynamics retains its universal form,

$$\Delta S \geq 0, \tag{3.91}$$

for an isolated system, with equality at equilibrium where $dS = 0$. Thus, for any reversible process under the conditions $U = \text{const}$, $V = \text{const}$, and $N = \text{const}$, the entropy increases and reaches its maximum value at stationary equilibrium. For irreversible processes, the corresponding inequalities read

$$T\,dS \geq \delta Q = dU + P\,dV - \mu\,dN, \qquad \delta W = -P\,dV. \tag{3.92}$$

For an isolated system, equilibrium requires that no exchange of energy, volume, or particles occurs:

$$\delta Q = 0, \tag{3.93}$$

$$dV = 0, \tag{3.94}$$

$$dN = 0, \tag{3.95}$$

which implies $dU = 0$ and therefore a stationary state.

Extensive versus Intensive Variables

In thermodynamics, all physical quantities can be classified as either *extensive* or *intensive*. Extensive quantities scale with the size of the system, while intensive quantities remain independent of it.

For instance, the fundamental potentials

$$U(S,V,N), \quad H(S,P,N), \quad F(T,V,N), \quad G(T,P,N), \tag{3.96}$$

as well as S, V, and N, are all *extensive*: if the system is scaled by a factor c, these quantities also scale by c. In contrast, temperature, pressure, and chemical potential,

$$T,\ P,\ \mu, \tag{3.97}$$

are *intensive*, remaining constant under such scaling.

This distinction traces back to Gibbs, who emphasized that equilibrium between two systems requires the equality of their intensive variables: temperature, pressure, and chemical potential.

The Grand Potential (Landau Free Energy)

A thermodynamic potential of particular importance in the study of open systems is the *grand potential*, denoted by Ω. It is sometimes referred to as the *Landau free energy*. The grand potential is defined by

$$\Omega = U - TS - \mu N = F - \mu N. \tag{3.98}$$

Differentiating, we obtain

$$d\Omega = dU - T\,dS - S\,dT - \mu\,dN - N\,d\mu. \tag{3.99}$$

Substituting $dU = T\,dS - P\,dV + \mu\,dN$ gives

$$d\Omega = -P\,dV - S\,dT - N\,d\mu. \tag{3.100}$$

From this differential form, we immediately read off the partial derivatives:

$$\left(\frac{\partial\Omega}{\partial V}\right)_{T,\mu} = -P, \tag{3.101}$$

$$\left(\frac{\partial\Omega}{\partial T}\right)_{V,\mu} = -S, \tag{3.102}$$

$$\left(\frac{\partial\Omega}{\partial \mu}\right)_{V,T} = -N. \tag{3.103}$$

At thermodynamic equilibrium, the grand potential is minimized with respect to its natural variables (T, V, μ):

$$\left(\frac{\partial\Omega}{\partial T}\right)_{V,\mu} = \left(\frac{\partial\Omega}{\partial V}\right)_{T,\mu} = \left(\frac{\partial\Omega}{\partial \mu}\right)_{V,T} = 0. \tag{3.104}$$

The grand potential is particularly important in the grand canonical ensemble of statistical mechanics, where T, V, and μ are fixed, and Ω plays the role of the thermodynamic potential governing fluctuations in energy and particle number.

Extensivity and Pressure

Since both T and μ are intensive, the grand potential Ω must be *extensive*, scaling linearly with the system volume V. We can thus write

$$\Omega = V\,\omega(\mu, T), \tag{3.105}$$

where $\omega(\mu, T)$ is an intensive quantity, the *grand potential density*. Differentiating with respect to V gives

$$\omega = \left(\frac{\partial\Omega}{\partial V}\right)_{T,\mu} = -P, \tag{3.106}$$

which leads to the important identity

$$\Omega = -PV. \tag{3.107}$$

Thus, the grand potential is directly proportional to the negative of the pressure times the system volume. In this sense, the grand potential measures the mechanical work that can be extracted from an open system at fixed temperature and chemical potential.

REMARK 3.19 The grand potential unifies thermal, mechanical, and chemical equilibrium in a single framework. It is the natural thermodynamic potential for open systems, controlling processes where both energy and matter can be exchanged with the surroundings. In statistical mechanics, Ω corresponds to the logarithm of the grand partition function, linking macroscopic observables such as pressure, entropy, and particle number to microscopic states. This bridge between thermodynamics and microscopic theory will be the focus of the following chapters.

The Euler Identity

Having discussed the role of extensivity in thermodynamics, we can now derive one of its most important consequences: the *Euler identity.* This identity expresses how the internal energy and other thermodynamic quantities are related through the property of homogeneity of extensive variables.

This result is a direct application of a general theorem by Leonhard Euler (1707–1783), the Swiss mathematician who formalized the properties of homogeneous functions long before the birth of thermodynamics.

A function $f(x_1, x_2, \ldots, x_n)$ is said to be *homogeneous of degree one* if scaling all its arguments by a constant factor c scales the function itself by the same factor:

$$f(cx_1, cx_2, \ldots, cx_n) = c\, f(x_1, x_2, \ldots, x_n). \tag{3.108}$$

In thermodynamics, this property corresponds to the *extensivity* of macroscopic quantities. For example, doubling the size of a system (and therefore its entropy S, volume V, and number of particles N) doubles its internal energy U:

$$U(cS, cV, cN) = c\, U(S, V, N). \tag{3.109}$$

Similarly, the entropy is extensive in its natural variables (U, V, N):

$$S(cU, cV, cN) = c\, S(U, V, N). \tag{3.110}$$

Differentiating both sides with respect to c and then setting $c = 1$ yields Euler's theorem for homogeneous functions:

$$S = \left(\frac{\partial S}{\partial U}\right)_{V,N} U + \left(\frac{\partial S}{\partial V}\right)_{U,N} V + \left(\frac{\partial S}{\partial N}\right)_{U,V} N. \tag{3.111}$$

Using the standard thermodynamic relations

$$\frac{1}{T} = \left(\frac{\partial S}{\partial U}\right)_{V,N}, \qquad \frac{P}{T} = \left(\frac{\partial S}{\partial V}\right)_{U,N}, \qquad -\frac{\mu}{T} = \left(\frac{\partial S}{\partial N}\right)_{U,V}, \tag{3.112}$$

we can rewrite the equation above as

$$S = \frac{U}{T} + \frac{PV}{T} - \frac{\mu N}{T}. \tag{3.113}$$

Multiplying both sides by T gives the familiar form of the *Euler identity*:

$$U = TS - PV + \mu N. \tag{3.114}$$

This result expresses the internal energy U as a sum of its contributions from entropy, volume, and particle number, each weighted by its corresponding intensive variable (temperature, pressure, and chemical potential, respectively). The Euler identity thus reflects the intrinsic scaling symmetry of equilibrium thermodynamics.

REMARK 3.20 The Euler identity is more than a mathematical curiosity: it reveals how extensivity and equilibrium constrain the relationships between thermodynamic quantities. It also provides the starting point for the *Gibbs–Duhem relation*, which connects the differentials of intensive variables (T, P, μ) and ensures the internal consistency of thermodynamic systems. We shall derive this relation in the next section.

Extensivity of the Gibbs Free Energy

As discussed in previous sections, thermodynamic potentials behave as extensive state functions: when the system size changes, extensive quantities scale linearly with it. For the Gibbs free energy, this means that doubling the amount of substance doubles the total Gibbs energy:

$$G(T, P, cN) = c\,G(T, P, N). \tag{3.115}$$

If the system contains several chemical species, labeled by indices $j = 1, \ldots, k$, this property generalizes to

$$G(T, P, cN_1, \ldots, cN_k) = c\,G(T, P, N_1, \ldots, N_k). \tag{3.116}$$

Differentiating both sides with respect to c and evaluating at $c = 1$, we obtain

$$G(T, P, N_1, \ldots, N_k) = \sum_{j=1}^{k} \left(\frac{\partial G}{\partial N_j} \right)_{T,P,N_{r \neq j}} N_j. \tag{3.117}$$

Recognizing that the chemical potential of species j is defined as

$$\mu_j = \left(\frac{\partial G}{\partial N_j} \right)_{T,P,N_{r \neq j}}, \tag{3.118}$$

we obtain the important relation

$$G(T, P, N_1, \ldots, N_k) = \sum_{j=1}^{k} \mu_j N_j. \tag{3.119}$$

Equation (3.119) expresses the extensivity of the Gibbs free energy and is sometimes referred to as the *Gibbs–Duhem identity*. It states that the total Gibbs energy of a mixture equals the sum of the individual chemical potentials multiplied by their respective particle numbers. For a single-component system, this reduces to

$$g = \frac{G(T, P, N)}{N} = \mu, \tag{3.120}$$

showing that the Gibbs free energy per particle equals the chemical potential.

Since $G = U - TS + PV = \mu N$, we can rearrange to recover the familiar expression for the internal energy,

$$U = TS - PV + \mu N, \tag{3.121}$$

which is consistent with the Euler identity derived earlier.

This identity forms the starting point for the Gibbs–Duhem relation, which we shall discuss shortly. It expresses the differential constraint among intensive variables T, P, and μ.

Thermodynamic Response Functions

In real materials, measurable quantities often describe how the system responds to external perturbations such as changes in temperature or pressure. These *response functions* provide a bridge between thermodynamic theory and experimental observation. Three particularly important examples are the following:

Thermal Expansion Coefficient

The *coefficient of thermal expansion* α quantifies how the volume of a material changes with temperature at constant pressure:

$$\alpha = \frac{1}{V}\left(\frac{\partial V}{\partial T}\right)_P. \tag{3.122}$$

A positive α means that the material expands upon heating—typical for gases, liquids, and most solids.

Isothermal Compressibility

The *isothermal compressibility* K_T measures the relative change in volume with pressure at constant temperature:

$$K_T = -\frac{1}{V}\left(\frac{\partial V}{\partial P}\right)_T. \tag{3.123}$$

It reflects how compressible a substance is; gases have large K_T, while solids have very small values.

Adiabatic Compressibility

The *adiabatic compressibility* K_S describes the same response but under adiabatic conditions (no heat exchange):

$$K_S = -\frac{1}{V}\left(\frac{\partial V}{\partial P}\right)_S. \tag{3.124}$$

Because adiabatic processes restrict heat exchange, K_S is generally smaller than K_T.

The ratio $\gamma = K_T/K_S$ is the adiabatic index, equal to the ratio of specific heats C_P/C_V for ideal gases. It plays a key role in acoustics and fluid dynamics.

These coefficients are experimentally measurable and fundamental to understanding the mechanical and thermal stability of materials. They characterize how substances expand, contract, and resist compression under various thermodynamic conditions.

REMARK 3.21 The extensivity of the Gibbs free energy and the associated response functions together define how macroscopic systems behave under changes in temperature, pressure, and composition. They provide the link between the abstract framework of thermodynamic potentials and the concrete, measurable properties of real materials—serving as a bridge between theoretical and experimental thermodynamics.

Examples of Thermodynamic Relations

We provide two examples of entropy calculations.

Example 1: Entropy of an Ideal Gas

For an ideal gas, the entropy can be expressed as

$$S(P, V) = S_0 + nC_V \ln\frac{P}{P_0} + nC_P \ln\frac{V}{V_0}. \tag{3.125}$$

Since $PV = Nk_BT$ and $U = (3/2)Nk_BT$, we have $PV = (3/2)U$, which implies

$$\frac{P}{P_0} = \frac{U}{U_0}\frac{V_0}{V}. \tag{3.126}$$

Substituting this into the expression for the entropy yields

$$S(P(U), V) = S_0 + nC_V \ln\frac{U}{U_0} + n(C_P - C_V)\ln\frac{V}{V_0}. \tag{3.127}$$

This gives the entropy in terms of U and V.

Example 2: Bounds Between Free Energies

The thermodynamic potentials satisfy

$$H = U + PV, \qquad F = U - TS, \qquad G = U - TS + PV = H - TS = F + PV. \tag{3.128}$$

From these relations,

$$G = H - TS. \tag{3.129}$$

Since $S \geq 0$ and $T \geq 0$, it follows that $G \leq H$. Moreover,

$$G = F + PV, \tag{3.130}$$

and because $P \geq 0$ and $V \geq 0$, we have $G \geq F$. Combining both inequalities leads to the result

$$F \leq G \leq H. \tag{3.131}$$

which provides a hierarchy between potentials.

3.8 The Third Law of Thermodynamics

Up to this point, we have discussed the *first* and *second laws of thermodynamics*, which describe the conservation and transformation of energy and the direction of natural processes. We now turn to the *third law of thermodynamics*, which concerns the behavior of matter as it approaches absolute zero temperature.

The origin of the third law can be traced to the work of the German chemist Walther Nernst in the early twentieth century. Based on his studies of heat capacities and the thermodynamic properties of substances at very low temperatures, Nernst formulated a general statement about the vanishing of entropy changes in such limits [Ner66]. His work, later refined through discussions with Einstein and Planck, gave rise to the modern formulation of the third law. Although its status as a "fundamental" thermodynamic law has been debated, its implications for both chemistry and physics are profound.

Nernst's Heat Theorem

Nernst's observation, known today as the *Nernst heat theorem*, can be stated as follows:

> *Any system undergoing a process between equilibrium states as a result of external influences—such as changes in pressure or volume—experiences a change in entropy that tends to zero as the temperature approaches absolute zero.*

In other words,

$$\lim_{T \to 0} \Delta S = 0. \tag{3.132}$$

This means that all isothermal processes between equilibrium states become indistinguishable in terms of entropy changes as the system approaches absolute zero. The entropy of any pure crystalline substance thus tends toward a constant value, often taken to be zero.

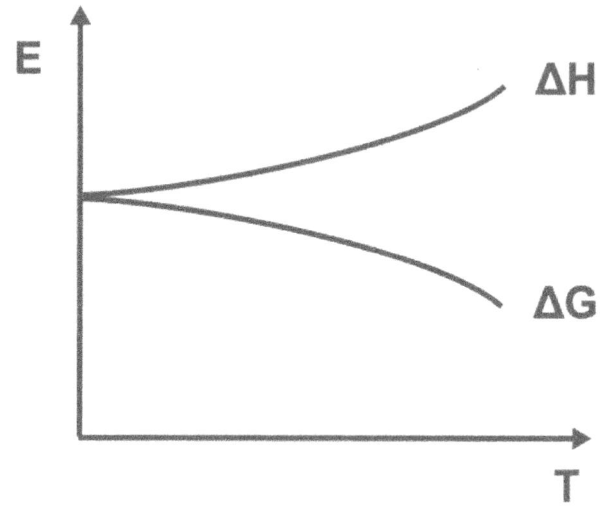

FIGURE 3.6 Schematic representation of Nernst's observation: as temperature decreases, the enthalpy H and Gibbs free energy G converge at $T = 0$, implying $\Delta S \to 0$.

Nernst observed that for many chemical reactions, the change in enthalpy ΔH increases with temperature, while the change in Gibbs free energy ΔG decreases. At very low temperatures, the two functions converge toward the same value, as illustrated in Fig. 3.6. We recall that

$$\left(\frac{\partial H}{\partial T}\right)_P = \Delta S, \qquad \left(\frac{\partial G}{\partial T}\right)_P = -\Delta S, \tag{3.133}$$

and that these quantities are related through

$$\Delta H - \Delta G = T\,\Delta S. \tag{3.134}$$

At $T = 0$, $H = G$, and the slopes of the two curves vanish if and only if $\Delta S = 0$. This captures the essence of Nernst's theorem: as temperature approaches absolute zero, entropy differences between all equilibrium states tend to vanish.

REMARK 3.22 The third law establishes the thermodynamic unattainability of absolute zero: no finite sequence of processes can reduce a system's temperature to exactly zero. It also sets the foundation for defining absolute entropies, which are essential for low-temperature physics, cryogenics, and quantum thermodynamics.

3.8.1 Formulations of the Third Law

Nernst's original formulation of the third law focuses on the vanishing of entropy differences between equilibrium states as temperature approaches absolute zero. It can be stated as follows:

> **Nernst's Original Formulation:** *At zero temperature, for a finite system with entropy S independent of any macroscopic external parameter x, we have*
>
> $$\lim_{T\to 0}\left[S(T, x_1) - S(T, x_2)\right] = 0. \tag{3.135}$$

This expresses that at $T \to 0$, the entropy of all equilibrium states tends to the same value, regardless of external parameters. In differential form, it can be written as

$$\lim_{T\to 0} \left(\frac{\partial S}{\partial x}\right)_T = 0, \tag{3.136}$$

meaning that at very low temperatures, entropy ceases to depend on macroscopic parameters such as pressure, volume, or magnetic field.

In many chemistry texts, this observation is rephrased as a statement about chemical reactions:

> **Nernst's Statement (Chemical Form):** *At temperatures near absolute zero, any reaction occurring in a system at internal equilibrium proceeds without producing any measurable change in entropy.*

In physics, however, the third law is most often presented in the version due to *Max Planck*, who strengthened Nernst's principle by proposing a universal entropy limit:

> **Planck's Formulation:** *For all systems in internal equilibrium, the entropy approaches a universal constant as the temperature tends to absolute zero; for a pure crystalline substance, this constant can be taken as zero.*

$$\lim_{T\to 0} S(T,x) = S_0, \qquad S_0 = 0 \text{ for a perfect crystal.} \tag{3.137}$$

Planck's formulation implies that the entropy of a perfect crystal at absolute zero is zero because the system occupies a single microscopic state. However, this universality is not always valid. Albert Einstein pointed out that if residual disorder remains—due to atomic configurations, magnetic orientations, or other macroscopic parameters—then this disorder contributes a finite entropy even at $T = 0$.

Einstein's critique anticipated the discovery of *residual entropy* in systems such as ice, glasses, and disordered crystals—phenomena later explained through quantum statistical mechanics.

Einstein therefore proposed a more general statement:

> **Einstein's Formulation:** *As the temperature approaches absolute zero, the entropy of a system approaches a constant value that may depend on macroscopic parameters:*

$$\lim_{T\to 0} S(T,x) = S_0(x). \tag{3.138}$$

At the time, Einstein was studying the heat capacity of solids. He noted that the classical expression for entropy at constant volume,

$$dS = \frac{C_V}{T}\, dT, \tag{3.139}$$

becomes divergent as $T \to 0$, indicating a fundamental problem with classical theory. This paradox was resolved by quantum mechanics, which predicts that $C_V \to 0$ as $T \to 0$, making $S(T)$ finite and consistent with the third law.

To respond to Einstein's critique, Nernst later introduced an alternative formulation emphasizing the *operational* meaning of the third law:

> **Nernst's Unattainability Principle:** *The absolute zero of temperature cannot be reached by any thermodynamic process in a finite number of steps.*

This principle, sometimes called the *unattainability formulation* of the third law, highlights the practical impossibility of achieving $T = 0$. It complements the entropy-based formulations by expressing the third law as a limitation on attainable states of matter rather than on entropy itself.

REMARK 3.23 The different formulations of the third law—by Nernst, Planck, and Einstein—reflect its evolution from an empirical observation about entropy changes to a deeper principle linking thermodynamics, statistical mechanics, and quantum theory. Nernst's and Planck's versions emphasize the vanishing of entropy differences, Einstein's acknowledges residual disorder, and the unattainability principle expresses the operational limits of cooling. Together, they capture the essence of the third law: as temperature approaches absolute zero, entropy variations vanish, the approach to equilibrium slows dramatically, and absolute zero remains an unreachable ideal.

Equivalence Between the Nernst Heat Theorem and the Unattainability Principle

The *Nernst unattainability principle* and the *Nernst heat theorem* are, in fact, equivalent formulations of the third law of thermodynamics. Both describe the same physical limitation—that entropy changes vanish and that absolute zero cannot be reached by any finite process—but they do so from complementary perspectives: one thermodynamic, the other operational.

From the Heat Theorem to the Unattainability Principle

Let us first assume that the Nernst heat theorem holds. Suppose, by contradiction, that there exists a finite process capable of cooling a system to absolute zero in a finite number of steps. According to the heat theorem, as $T \to 0$, the entropy change between any two equilibrium states must tend to zero:

$$\lim_{T \to 0} \Delta S = 0. \tag{3.140}$$

However, if a finite process could reach $T = 0$, the system would still possess a finite residual entropy at absolute zero. This would violate the heat theorem, which requires $\Delta S \to 0$ as $T \to 0$. Hence, the assumption of a finite-step process must be false, and the unattainability principle follows: absolute zero cannot be reached through any finite sequence of thermodynamic transformations.

From the Unattainability Principle to the Heat Theorem

Conversely, consider the contrapositive argument. If the entropy S of a system does not tend to a constant value as $T \to 0$—that is, if there remains a finite entropy difference between states—then it would, in principle, be possible to exploit this difference to extract work and further reduce the temperature. This would imply that absolute zero could be achieved in a finite number of steps, in direct contradiction to the unattainability principle. Therefore, if the unattainability principle holds, the Nernst heat theorem must also hold. The two statements are thus logically equivalent.

The formal equivalence between the Nernst heat theorem and the unattainability principle was clarified in the mid-twentieth century through rigorous thermodynamic analyses by Planck and Francis Simon.

Modern Interpretation

In the modern statistical-mechanical framework, the third law acquires a deeper microscopic meaning. At very low temperatures, the entropy of a system is given by the Boltzmann relation,

$$S = k_B \ln \Omega_0, \tag{3.141}$$

where Ω_0 is the degeneracy of the ground state of the Hamiltonian. Hence, the validity of the Nernst heat theorem depends on whether the ground-state degeneracy Ω_0 remains constant as external parameters (such as pressure, volume, or magnetic field) are varied. If the degeneracy changes with these parameters, residual entropy may persist even at $T = 0$. Residual entropy is observed, for example, in disordered crystals, spin ices, and glasses, where multiple ground-state configurations exist even at zero temperature.

Thus, in modern terms, the third law does not demand that $S(0) = 0$ universally, but rather that entropy becomes independent of all macroscopic parameters at $T = 0$. Whether or not this condition is satisfied depends on the microscopic structure and degeneracy of the system's ground state.

REMARK 3.24 The equivalence of the heat theorem and the unattainability principle reveals that the third law is both a statement about thermodynamic reversibility and a reflection of quantum mechanical constraints. It tells us that perfect order—the complete elimination of entropy—cannot be achieved through any finite process, and that the behavior of systems near absolute zero is governed by the discrete nature of their microscopic energy spectra.

Alternative Formulations and Modern Interpretation

In summary, the third law of thermodynamics states that as the temperature of a system approaches absolute zero (0 K), its entropy approaches a minimum value. For a perfect crystal, whose ground state is unique and completely ordered, this minimum entropy is zero:

$$\lim_{T \to 0} S(T) = 0. \tag{3.142}$$

This is the familiar formulation derived from Nernst's heat theorem and Planck's extension of it. It expresses the idea that, at absolute zero, a system in perfect order has no residual randomness and thus no entropy.

Historically, Nernst's heat theorem already contained this idea, though stated it more operationally: the entropy change of any process between equilibrium states tends to zero as $T \to 0$. Planck later strengthened it by asserting that the entropy itself must vanish for perfect crystals, thus defining an absolute zero of entropy.

A further refinement was offered by Francis Simon in the early twentieth century, who clarified the third law's physical content by emphasizing the independence of distinct internal contributions to entropy. His formulation can be stated as follows:

> **Simon's Statement:** *As the temperature approaches absolute zero, the contribution to the entropy of a system from each aspect of the system that is in internal thermodynamic equilibrium diminishes, ultimately tending toward zero.*

Simon's interpretation recognizes that different degrees of freedom—translational, rotational, vibrational, electronic, magnetic—"freeze out" at different rates as temperature decreases [Sim31]. Each component of the system gradually ceases to contribute to the total entropy as its excitations become energetically inaccessible. Therefore, as $T \to 0$, the entropy approaches a constant value,

$$\lim_{T \to 0} S(T) = S_0, \tag{3.143}$$

where $S_0 = 0$ only if the system has a unique ground state. For systems with residual degeneracy—such as disordered crystals, glasses, or spin ices—$S_0 > 0$. We will examine these counterexamples later in this book.

Residual entropy was first observed experimentally in ice by William Giauque and Ashley Stout (1936) [GS36], providing a striking example of a system whose macroscopic disorder persists even as $T \to 0$. Also, see the discussion in [Pau35] on the residual entropy of ice, which we will discuss later and in Volume II.

REMARK 3.25 The third law, in its various formulations, from Nernst's operational principle to Simon's refinement, marks the boundary where thermodynamics meets quantum mechanics. It asserts that entropy cannot decrease below a limiting value as temperature approaches absolute zero, encapsulating both the impossibility of reaching $T = 0$ and the microscopic ordering of matter at low temperatures. In this sense, the third law defines the ultimate reference point for all thermodynamic quantities.

4

Work–Entropy Relationships

4.1 Entropy as Non-Available Energy

In many introductory texts on thermodynamics, entropy is loosely described as a measure of "non-available energy"—that is, the portion of a system's energy that cannot be used to perform mechanical work. It is also frequently interpreted as a measure of *disorder* or as the quantity defining the *arrow of time.*

E. T. Jaynes argued that entropy does not define an intrinsic arrow of time in equilibrium thermodynamics, but rather quantifies information and uncertainty.

While such interpretations capture certain aspects of the concept, they can be misleading unless one distinguishes between two related but distinct ideas: *entropy* itself and *entropy production.*

Entropy, in its thermodynamic sense, is a state function—a variable that characterizes the equilibrium condition of a system. Entropy production, by contrast, measures the extent to which a process deviates from reversibility. In Clausius's formulation, the infinitesimal change in entropy is given by

$$dS = \frac{\delta Q}{T}, \tag{4.1}$$

DOI: 10.1201/9781003603528-4

where δQ is the infinitesimal amount of heat absorbed by the system at temperature T. The *second law of thermodynamics*, in Clausius's differential form, then states that

$$dS \geq 0, \tag{4.2}$$

with equality holding for reversible transformations. For detailed discussions of entropy production and the second law, see Zemansky [Zem97], Tolman and Fine [TF48], and Marcella [Mar92].

The quantity of *lost work*, or the portion of energy rendered unavailable for performing mechanical work during an irreversible process, is proportional to the increase in entropy:

$$W_{\text{lost}} = T\,\Delta S. \tag{4.3}$$

In a reversible process, $\Delta S = 0$, and all absorbed heat can, in principle, be converted into work. In any real (irreversible) process, $\Delta S > 0$, and some energy is irrevocably dissipated as heat.

Entropy Change in a Thermodynamic Process

For any finite thermodynamic transformation, the total entropy change is the sum of the entropy changes of the system and its surroundings:

$$\Delta S = \Delta S_{\text{sys}} + \Delta S_{\text{env}}. \tag{4.4}$$

This relation ensures that, although the system's entropy may decrease locally, the total entropy of the system plus environment never decreases.

Example: Isothermal Expansion of an Ideal Gas

Consider an ideal gas undergoing an isothermal expansion at fixed temperature T, described by the equation of state

$$PV = Nk_BT.$$

Since the temperature is held constant, heat must flow into the system from a thermal reservoir to compensate for the work performed during expansion. From the first law,

$$\delta Q = P\,dV,$$

and integrating from V_1 to V_2 yields

$$Q = Nk_BT\ln\left(\frac{V_2}{V_1}\right).$$

The corresponding entropy change of the gas is therefore

$$\Delta S_{\text{sys}} = Nk_B\ln\left(\frac{V_2}{V_1}\right).$$

However, the heat Q absorbed by the gas must come from a reservoir at temperature $T_1 > T_2 = T$. The entropy change of the environment is

$$\Delta S_{\text{env}} = -\frac{Q}{T_1}.$$

Hence the total entropy change is

$$\Delta S = Q\left(\frac{1}{T_2} - \frac{1}{T_1}\right) \geq 0,$$

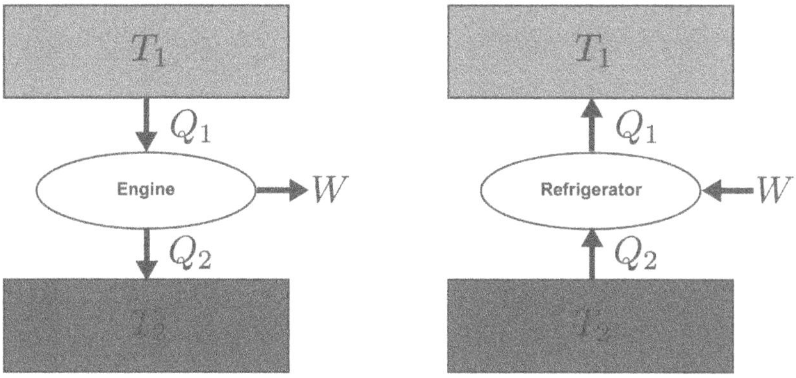

FIGURE 4.1 Schematics of the flows of heat in a thermal engine, extracting work (left); schematics of the flows of heat in a refrigerator (right). We can see that work has to be done in order for the system to cool.

in agreement with Clausius's formulation of the second law. Equality holds only when $T_1 = T_2$, corresponding to a reversible isothermal expansion.

This simple example also illustrates a general principle: when two thermal reservoirs at different temperatures are brought into contact, heat flows spontaneously from the hotter to the colder body, increasing the total entropy even in the absence of mechanical work. Thus, the elementary process of *heat transfer* is inherently irreversible.

REMARK 4.1 Entropy quantifies the energy that cannot be used to perform mechanical work and provides a universal measure of irreversibility. Every real process in nature—from heat conduction to diffusion and chemical reactions—involves an increase in total entropy, marking the natural direction of time and defining the efficiency limits of all energy conversions.

4.2 Work–Entropy Relationship for a Heat Engine

To understand the relationship between entropy and useful work, let us consider a simple heat engine operating between two heat reservoirs. The engine absorbs heat Q_2 from a high-temperature reservoir at temperature T_2 and rejects heat Q_1 to a low-temperature reservoir at $T_1 < T_2$. A portion of the absorbed heat is converted into mechanical work during each cycle, while the remainder is expelled into the cold reservoir.

According to the *first law of thermodynamics*, the work performed by the engine in one cycle is the difference between the heat absorbed and the heat expelled:

$$W = Q_2 - Q_1. \tag{4.5}$$

The total entropy change of the two reservoirs during this cycle is given by

$$\Delta S = -\frac{Q_1}{T_1} + \frac{Q_2}{T_2}. \tag{4.6}$$

Combining Eqs. (4.5) and (4.6), we can express the work in terms of entropy and temperature differences:

$$W = \frac{Q_2(T_2 - T_1)}{T_2} - T_1\,\Delta S. \tag{4.7}$$

This general expression for the work output of a heat engine, valid for both reversible and irreversible processes, was discussed in great detail by Tolman and Fine [TF48].

Reversible and Irreversible Cases

For a reversible engine, no entropy is produced during the cycle ($\Delta S = 0$), and Eq. (4.7) simplifies to

$$W_{\text{rev}} = Q_2\left(1 - \frac{T_1}{T_2}\right), \tag{4.8}$$

which defines the *maximum possible work* that can be extracted from the system. The factor

$$\eta = 1 - \frac{T_1}{T_2} \tag{4.9}$$

represents the *Carnot efficiency*—the upper bound on efficiency for any engine operating between two temperature reservoirs.

For irreversible processes, entropy is produced ($\Delta S > 0$), and Eq. (4.7) becomes

$$W = W_{\text{rev}} - T_1\,\Delta S. \tag{4.10}$$

This shows that any production of entropy reduces the amount of useful work obtainable from a given heat flow. The term

$$W_{\text{lost}} = T_1\,\Delta S \tag{4.11}$$

represents the *lost work* or the portion of energy rendered unavailable due to irreversibility.

Since $\Delta S \geq 0$ for all real (irreversible) processes, the efficiency of any practical engine must always be less than that of the reversible Carnot engine. This also leads to the Clausius form of the second law:

$$\Delta S = -\frac{Q_1}{T_1} + \frac{Q_2}{T_2} = \Delta S_1 + \Delta S_2 \geq 0, \tag{4.12}$$

with equality holding for reversible cycles.

This formulation unifies the concepts of energy conservation and entropy production: while the first law limits what can be done (energy conservation), the second law dictates how efficiently it can be done (irreversibility).

REMARK 4.2 The work–entropy relationship provides a direct measure of the cost of irreversibility in energy conversion. It quantifies how the generation of entropy—through friction, turbulence, or heat conduction—reduces the maximum extractable work from a process. In this sense, entropy truly represents the fraction of energy that can no longer be converted into useful work.

Historical Note: Work–Entropy Relationships

Although modern textbooks derive the relationship between work output and entropy change in a straightforward manner, its historical development involves several lesser-known milestones.

One early precursor is found in the work of Gibbs (1873) [Gib73], who introduced the geometric representation of thermodynamic potentials. In Gibbs's construction, the maximum obtainable work corresponds to areas under specific surfaces in the (S, T) and (V, p) planes, implicitly linking mechanical work with the curvature of the entropy surface.

Another overlooked contribution comes from Wilhelm Ostwald (1890) [Ost90], whose "energetic" program sought to express all physical processes in terms of energy transformations. Although Ostwald rejected atomism, his inequality-based treatment of irreversible processes anticipates later entropy-production terms in the work–entropy relation.

A surprisingly modern formulation appears in Max Planck's *Vorlesungen über Thermodynamik* (1897) [Pla97]. Planck derived a general expression relating useful work to the entropy generated during a cycle, effectively providing a precursor to the irreversible work formulas that would be formalized decades later.

These foundational but often forgotten threads were later unified and clarified by Richard C. Tolman and Paul C. Fine (1948) [TF48], who provided one of the first explicit expressions for the work output of a general (not necessarily reversible) heat engine in terms of entropy and temperature differences, closely resembling the modern form used today.

Perfect Engines and the Second Law

We can now understand why the work lost in a heat engine can be expressed as $W_{\mathrm{lost}} = T\,\Delta S$. In a real engine, entropy production arises from irreversibilities such as friction, turbulence, finite temperature differences during heat exchange, or other dissipative effects.

Returning to the picture of a heat flow from a hot reservoir to a cold one through an engine, we recall that $W = Q_2 - Q_1$. Rearranging, the heat rejected to the cold reservoir can be written as

$$Q_2 = Q_1 - W = Q_1(1-\eta) + T_2\,\Delta S = Q_1\frac{T_2}{T_1} + W_{\mathrm{lost}}. \tag{4.13}$$

Here, $Q_2^{\mathrm{min}} = Q_1(T_2/T_1)$ represents the *minimum* amount of energy that must be expelled to the cold reservoir to maintain reversibility. If the engine is irreversible ($\Delta S > 0$), some additional heat is rejected, corresponding to work that could not be converted—the *lost work*. This process is schematically illustrated in Fig. 4.1 (left).

This discussion parallels the early treatment by Richard C. Tolman and Paul C. Fine [TF48], who emphasized the role of entropy production in reducing the available work in real engines.

Now let us imagine, by contradiction, a *perfect engine*, one that converts all absorbed heat into work ($W = Q_1$), with no heat ejected into the cold reservoir. From Eq. (4.7), such an engine would satisfy

$$-\frac{T_2}{T_1}Q_1 - T_2\,\Delta S = 0, \tag{4.14}$$

which gives

$$\Delta S = -\frac{Q_1}{T_1} < 0. \tag{4.15}$$

This result directly violates the Kelvin–Planck statement of the second law, which forbids the complete conversion of heat into work in a cyclic process. Hence, a perfect (100%) efficient engine is impossible: any real engine must produce entropy, and its efficiency must therefore be less than that of the reversible Carnot engine.

Thermal Engine Efficiency

The efficiency of a heat engine is defined as the ratio of the useful work output to the heat absorbed from the high-temperature reservoir:

$$\eta = \frac{W}{Q_1} = \frac{Q_1 - Q_2}{Q_1} = 1 - \frac{Q_2}{Q_1}. \tag{4.16}$$

We can rewrite this in terms of entropy and irreversibility using the work–entropy relationship:

$$\eta = \frac{W_{\text{rev}} - T_2\,\Delta S}{Q_1} = \eta_{\text{rev}} - \frac{T_2\,\Delta S}{Q_1}. \tag{4.17}$$

Since for a reversible engine $W_{\text{rev}} = Q_1(1 - T_2/T_1)$, we identify

$$\eta_{\text{rev}} = 1 - \frac{T_2}{T_1}, \tag{4.18}$$

and the efficiency for any real engine becomes

$$\eta = \eta_{\text{rev}} - \frac{T_2}{Q_1}\,\Delta S \leq \eta_{\text{rev}}. \tag{4.19}$$

The second term, proportional to ΔS, represents the reduction in efficiency due to entropy production. Since $\Delta S > 0$ for all irreversible processes, the efficiency of any real engine is necessarily lower than the Carnot efficiency.

REMARK 4.3 The impossibility of a perfect engine and the reduction of efficiency by entropy production are two faces of the same principle—the second law of thermodynamics. They express the universal limitation that governs all energy conversions: while energy is conserved, not all of it can be converted into useful work. Irreversibility, embodied by entropy production, defines the boundary between idealized and real processes and determines the efficiency limits of every heat engine.

Refrigerators and Heat Pumps

Let us now consider the reverse process of a heat engine—a *refrigerator* or *heat pump*—illustrated schematically in Fig. 4.1 (right). In this case, heat is made to flow from a low-temperature reservoir to a high-temperature one by performing external work on the system.

For a refrigeration cycle, energy conservation gives

$$-W = -(Q_1 - Q_2), \tag{4.20}$$

or equivalently,

$$Q_1 = Q_2 + W, \tag{4.21}$$

where Q_2 is the heat absorbed from the cold reservoir at temperature T_2, and Q_1 is the heat rejected to the hot reservoir at temperature $T_1 > T_2$.

The total entropy change during the process is

$$\Delta S = \frac{Q_1}{T_1} - \frac{Q_2}{T_2}. \tag{4.22}$$

Combining the above equations yields

$$W = Q_2\frac{T_1 - T_2}{T_2} + T_1\,\Delta S. \tag{4.23}$$

The first term on the right-hand side corresponds to the reversible work, while the second represents the extra work that must be supplied because of entropy production. For a reversible refrigerator ($\Delta S = 0$), we have

$$W_{\text{rev}} = Q_2\frac{T_1 - T_2}{T_2}. \tag{4.24}$$

In the general (irreversible) case,

$$W = W_{\text{rev}} + T_1\,\Delta S. \tag{4.25}$$

The term $T_1\Delta S$ thus represents the additional work input required to overcome irreversibilities, such as friction, heat leaks, or finite temperature gradients.

In principle, a *perfect refrigerator* would operate with $W = 0$, moving heat from cold to hot spontaneously. However, such a process would imply

$$\Delta S = -\frac{W_{\text{rev}}}{T_1} < 0, \tag{4.26}$$

which violates the *Clausius statement* of the second law: it is impossible for heat to flow from a colder to a hotter body without external work. Hence, just as there can be no perfect engine, there can also be no perfect refrigerator.

Non-Cyclic Processes with a Single Heat Reservoir

As a final example, consider a system in contact with a single thermal reservoir at temperature T. The system exchanges heat Q with the reservoir and performs work W during a non-cyclic process. From the first law of thermodynamics,

$$W = Q - \Delta U_{\text{sys}}, \tag{4.27}$$

where ΔU_{sys} is the change in internal energy of the system.

The total entropy change for the combined system and reservoir is

$$\Delta S_{\text{tot}} = \Delta S_{\text{sys}} - \frac{Q}{T}. \tag{4.28}$$

Combining the two expressions, we obtain

$$W = T(\Delta S_{\text{sys}} - \Delta S_{\text{tot}}) - \Delta U_{\text{sys}}. \tag{4.29}$$

If the process is reversible ($\Delta S_{\text{tot}} = 0$), this reduces to

$$W_{\text{rev}} = T\,\Delta S_{\text{sys}} - \Delta U_{\text{sys}}. \tag{4.30}$$

Therefore, the general work–entropy relationship for a single-reservoir system is

$$W = W_{\text{rev}} - T\,\Delta S_{\text{tot}}. \tag{4.31}$$

The available work thus decreases linearly with the entropy produced during the process, with the proportionality factor given by the temperature of the reservoir. This does not contradict the Kelvin–Planck statement, because the process is not cyclic: energy is extracted once, and the system's state changes permanently.

REMARK 4.4 The symmetry between heat engines and refrigerators illustrates the two complementary manifestations of the second law. Engines convert heat into work but inevitably lose efficiency due to entropy production; refrigerators consume work to transfer heat against its natural flow and require extra work input for the same reason. Even non-cyclic single-reservoir processes obey this universal rule: entropy production limits the amount of useful energy transformation possible in nature.

4.3 Thermodynamic Cycles in Standard Processes

In the previous sections, we have introduced the relationship between work and entropy for cyclic transformations. We now examine several *elementary thermodynamic processes*—adiabatic, isothermal, isochoric, and isobaric—which serve as the fundamental components of many thermodynamic cycles. Each of these processes represents a limiting case in which one or more thermodynamic variables are held constant, allowing us to understand the interplay between work, heat, and entropy.

These processes can be analyzed from two equivalent perspectives:

- the traditional *P–V diagram*, which emphasizes mechanical work and volume changes, and
- the *T–S diagram*, which provides insight into heat exchange and entropy variation.

In either view, the results are consistent and interconnected through the first law of thermodynamics:

$$\Delta U = Q - W, \tag{4.32}$$

where ΔU is the change in internal energy, Q is the heat exchanged with the surroundings, and W is the work performed *by the system.*

Sign Conventions and Efficiencies

Before proceeding, we recall the sign convention typically adopted in physics:

- Work done *by the system* (expansion) is *negative*, i.e., energy leaves the system.
- Work done *on the system* (compression) is *positive*, i.e., energy enters the system.

Thus, in an expansion, $W < 0$ and $Q > 0$ if heat is absorbed, while in a compression, $W > 0$ and $Q < 0$. The efficiency of a cycle can then be expressed as

$$\eta = \frac{\text{Net work done by the engine}}{\text{Heat absorbed by the engine}} = \frac{\text{Net work done on the engine}}{\text{Heat released by the engine}}, \tag{4.33}$$

and since not all absorbed heat is converted into work, $\eta < 1$ for all real cycles.

In the following examples, we make frequent use of the heat capacities

$$C_P = \frac{\gamma N k_B}{\gamma - 1}, \qquad C_V = \frac{N k_B}{\gamma - 1}, \tag{4.34}$$

where $\gamma = C_P/C_V$ is the adiabatic index. These relationships will simplify many of our derivations.

1. Adiabatic Process ($Q = 0$)

In an adiabatic transformation, the system is thermally insulated, so that $\Delta Q = 0$ and consequently $\Delta S = 0$. From the first law, this implies $\Delta U = -W$: any work done by the system reduces its internal energy.

For an ideal gas, the adiabatic condition is given by

$$PV^\gamma = \text{constant}, \tag{4.35}$$

or, equivalently,

$$TV^{\gamma-1} = \text{constant}. \tag{4.36}$$

Between two states A and B, we can write

$$P_A V_A^\gamma = P_B V_B^\gamma. \tag{4.37}$$

The work done during an adiabatic process is obtained by integrating

$$W_{AB} = -\int_A^B P\,dV = -P_A V_A^\gamma \int_A^B \frac{dV}{V^\gamma}. \tag{4.38}$$

Performing the integral yields

$$W_{AB} = -\frac{P_A V_A^\gamma}{\gamma - 1}\left(\frac{1}{V_B^{\gamma-1}} - \frac{1}{V_A^{\gamma-1}}\right) = -\frac{P_B V_B^\gamma}{\gamma - 1}\left(\frac{1}{V_B^{\gamma-1}} - \frac{1}{V_A^{\gamma-1}}\right). \tag{4.39}$$

Because $\Delta Q = 0$, the internal energy change is $\Delta U = -W_{AB}$, and the process is isentropic.

2. Isochoric Process ($V = \text{constant}$)

In an isochoric process, the volume remains fixed ($\Delta V = 0$), so no mechanical work is done: $\Delta W = 0$. Hence, from the first law, $\Delta Q = \Delta U$. For an ideal gas,

$$\Delta U = \int_A^B C_V\,dT = C_V(T_B - T_A), \tag{4.40}$$

and the entropy change is given by

$$\Delta S = \int_A^B \frac{\delta Q}{T} = C_V \ln\left(\frac{T_B}{T_A}\right). \tag{4.41}$$

Entropy increases if the gas is heated ($T_B > T_A$), reflecting the greater number of accessible microstates.

3. Isobaric Process ($P = \text{constant}$)

In an isobaric process, pressure is held constant ($\Delta P = 0$), so the work performed is simply

$$W_{AB} = -P(V_B - V_A). \tag{4.42}$$

The internal energy change is

$$\Delta U_{AB} = \int_A^B C_V\,dT = C_V(T_B - T_A). \tag{4.43}$$

Using the ideal gas law, $PV = Nk_\mathrm{B}T$, we find

$$\Delta U_{AB} = \frac{C_V}{Nk_\mathrm{B}} P(V_B - V_A) = \frac{1}{\gamma - 1} P(V_B - V_A). \tag{4.44}$$

The total heat exchanged is then

$$\Delta Q_{AB} = \Delta U_{AB} - \Delta W_{AB} = \frac{\gamma}{\gamma - 1} P(V_B - V_A), \tag{4.45}$$

consistent with the definition of $C_P = \gamma C_V$.

4. Isothermal Process (T = constant)

In an isothermal process, the temperature remains constant, so $\Delta U = 0$. Thus, $\Delta Q = -\Delta W$: all heat absorbed is converted into work.

For an ideal gas, $PV = Nk_{\mathrm{B}}T$, so

$$W_{AB} = -\int_A^B P\,dV = -\int_A^B \frac{Nk_{\mathrm{B}}T}{V}\,dV = -Nk_{\mathrm{B}}T \ln\left(\frac{V_B}{V_A}\right). \tag{4.46}$$

Hence,

$$\Delta Q_{AB} = -\Delta W_{AB} = Nk_{\mathrm{B}}T \ln\left(\frac{V_B}{V_A}\right), \tag{4.47}$$

and the corresponding entropy change is

$$\Delta S_{AB} = \frac{\Delta Q_{AB}}{T} = Nk_{\mathrm{B}} \ln\left(\frac{V_B}{V_A}\right). \tag{4.48}$$

Entropy increases when the system expands and decreases when it is compressed.

REMARK 4.5 The four basic thermodynamic processes described above form the building blocks of all standard thermodynamic cycles, such as the Carnot, Otto, and Diesel cycles. Each process embodies a distinct form of the work–entropy relationship: adiabatic transformations conserve entropy, isothermal ones convert heat entirely into work, and isochoric and isobaric processes involve partial conversion limited by heat capacity. In the next sections, we will combine these elementary processes to construct complete thermodynamic cycles and examine their efficiencies.

4.4 Thermodynamic Cycles in the T-S Diagram

To bring together the results of the previous section, let us now analyze a complete thermodynamic cycle, in particular the *Carnot cycle.* This cycle represents the idealized reversible engine composed of two isothermal and two adiabatic processes. When represented on a T–S diagram (see Fig. 4.2), the Carnot cycle forms a rectangle: the isothermal expansions and compressions appear as horizontal lines (constant temperature, heat exchange), while the adiabatic segments appear as vertical lines (constant entropy, no heat exchange).

The total work and heat exchanged over one full cycle can be expressed as the sum of the contributions from the four processes:

$$\Delta W_{\mathrm{tot}} = \Delta W_{AB} + \Delta W_{BC} + \Delta W_{CD} + \Delta W_{DA}, \tag{4.49}$$

$$\Delta Q_{\mathrm{tot}} = \Delta Q_{AB} + \Delta Q_{BC} + \Delta Q_{CD} + \Delta Q_{DA}. \tag{4.50}$$

In the Carnot cycle:

- AB and CD are *isothermal* processes, so $\Delta U_{AB} = \Delta U_{CD} = 0$.
- BC and DA are *adiabatic* processes, so $\Delta Q_{BC} = \Delta Q_{DA} = \Delta S_{BC} = \Delta S_{DA} = 0$.

We assign $T_2 = T_{AB}$ and $T_1 = T_{CD}$, with $T_2 > T_1$, corresponding to the high- and low-temperature isotherms, respectively.

Before calculating the efficiency of the Carnot engine, it is convenient to summarize the thermodynamic relations for the four elementary processes. These are reported in Table 4.1, where the sign convention corresponds to the work done by the system on the environment and the heat absorbed by the system.

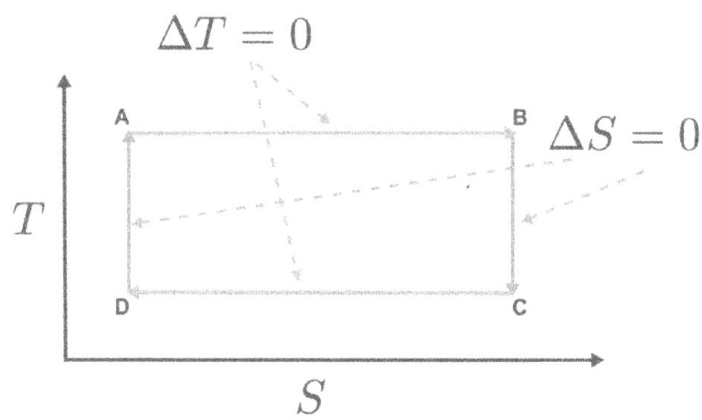

FIGURE 4.2 The Carnot cycle represented in the T–S diagram. Horizontal lines (AB, CD) correspond to isothermal processes; vertical lines (BC, DA) to adiabatic processes. The enclosed area represents the net work done during the cycle.

TABLE 4.1 Thermodynamic quantities for n moles of an ideal gas, from point A to point B on a P–V diagram. The sign convention corresponds to work done by the system and heat absorbed.

Process	W	Q	ΔS	ΔU
Isothermal	$W = nRT_A ln(V_B/V_A)$	$nRT_A \ln \frac{V_B}{V_A}$	$nR \ln \frac{V_B}{V_A}$	0
Adiabatic	$W = -nC_V(T_B - T_A)$	0	0	$nC_V(T_B - T_A)$
Isochoric	0	$nC_V(T_B - T_A)$	$nC_V \ln \frac{T_B}{T_A}$	$nC_V(T_B - T_A)$
Isobaric	$P_A(V_B - V_A)$	$nC_P(T_B - T_A)$	$nC_P \ln \frac{T_B}{T_A}$	$nC_V(T_B - T_A)$

Analysis of the Carnot Engine

Let us now apply these relations to the four steps of the Carnot cycle. During the isothermal expansion AB at temperature T_2, heat Q_2 is absorbed from the hot reservoir, and the system does positive work on the surroundings. During the isothermal compression CD at temperature T_1, heat Q_1 is released to the cold reservoir, and the system does negative work. The adiabatic branches BC and DA connect these isotherms without heat exchange ($Q = 0$, $\Delta S = 0$).

The efficiency of the Carnot engine can be interpreted as the fraction of absorbed heat that is converted into useful work. From the isothermal processes we have

$$Q_{AB} = nRT_2 \ln\left(\frac{V_B}{V_A}\right), \tag{4.51}$$

$$Q_{CD} = nRT_1 \ln\left(\frac{V_D}{V_C}\right), \tag{4.52}$$

where Q_{AB} is positive (heat absorbed) and Q_{CD} is negative (heat released).

Using the adiabatic relations,

$$T_A V_A^{\gamma-1} = T_B V_B^{\gamma-1}, \tag{4.53}$$

$$T_B = T_C, \tag{4.54}$$

$$T_C V_C^{\gamma-1} = T_D V_D^{\gamma-1}, \tag{4.55}$$

$$T_D = T_A, \tag{4.56}$$

we find that

$$\frac{V_C}{V_D} = \frac{V_B}{V_A}. \tag{4.57}$$

TABLE 4.2 Work, heat, entropy, and energy in the Carnot cycle. These are the work done by the engine and the heat released. Note that in the CD process, heat is released. Above, T_2 is the hot reservoir and T_1 is the cold reservoir.

Process	Work done	Heat absorbed	Entropy	Energy
AB (isot)	$Nk_B T_2 \log \frac{V_B}{V_A}$	$Nk_B T_2 \log \frac{V_B}{V_A}$	$Nk_B \log \frac{V_B}{V_A}$	0
BC (adia)	$C_V(T_2 - T_1)$	0	0	$-C_V(T_2 - T_1)$
CD (isot)	$Nk_B T_1 \log \frac{V_D}{V_C}$	$-Nk_B T_1 \log \frac{V_D}{V_C}$	$-Nk_B \log \frac{V_D}{V_C}$	0
DA (adia)	$C_V(T_1 - T_2)$	0	0	$C_V(T_1 - T_2)$
Total	$Nk_B\left(T_2 \log \frac{V_B}{V_A} + T_1 \log \frac{V_D}{V_C}\right)$	$Nk_B\left(T_2 \log \frac{V_B}{V_A} - T_1 \log \frac{V_D}{V_C}\right)$	$Nk_B \log \frac{V_B V_D}{V_A V_C}$	0

Substituting this into the expression for the efficiency yields

$$\eta = 1 - \frac{T_1}{T_2}. \tag{4.58}$$

This simple but profound result shows that the efficiency of any reversible heat engine operating between two temperatures depends only on those temperatures and not on the working substance or the details of the process.

Entropy Change in the Carnot Cycle

During the Carnot cycle, the total change in entropy of the system and the reservoirs is

$$\Delta S_{\text{tot}} = -\frac{Q_2}{T_2} + \frac{Q_1}{T_1}. \tag{4.59}$$

Since the Carnot engine operates reversibly, no entropy is produced, and thus

$$\Delta S_{\text{tot}} = 0. \tag{4.60}$$

Equivalently, from the efficiency relation

$$\eta = \frac{W}{Q_2} = 1 - \frac{Q_1}{Q_2}, \tag{4.61}$$

we find that

$$Q_1 = Q_2 \frac{T_1}{T_2}, \tag{4.62}$$

which again gives

$$\frac{Q_1}{T_1} = \frac{Q_2}{T_2}. \tag{4.63}$$

Hence, the total entropy change is zero and the process is perfectly reversible.

REMARK 4.6 In the T–S diagram, the area enclosed by the Carnot cycle represents the total work performed during one cycle. Because $\Delta S_{\text{tot}} = 0$, all heat transfers occur reversibly, and the engine operates at the maximum possible efficiency for given T_1 and T_2. Any deviation from reversibility—friction, finite temperature gradients, or non-equilibrium effects—would tilt the sides of the rectangle, reducing the enclosed area and therefore the available work.

The Stirling Engine

Let us now consider another thermodynamic cycle: the *Stirling engine*. This cycle, shown schematically in Fig. 4.7(a), consists of two isothermal and two isochoric processes: an isothermal expansion ($A \to B$), an isochoric cooling ($B \to C$), an isothermal compression ($C \to D$), and an isochoric heating ($D \to A$).

During the isothermal expansion $A \to B$ at temperature T_2, the gas absorbs heat from the hot reservoir and performs work:

$$W_{AB} = -Nk_{\mathrm{B}}T_2 \ln\left(\frac{V_B}{V_A}\right). \tag{4.64}$$

During the isochoric cooling $B \to C$, no work is done ($W_{BC} = 0$), but heat is released:

$$\Delta Q_{BC} = -C_V(T_1 - T_2) < 0. \tag{4.65}$$

In the isothermal compression $C \to D$ at temperature T_1, the gas releases heat to the cold reservoir and work is done on the system:

$$W_{CD} = -Nk_{\mathrm{B}}T_1 \ln\left(\frac{V_C}{V_D}\right). \tag{4.66}$$

Finally, during the isochoric heating $D \to A$, no work is performed ($W_{DA} = 0$), but heat is absorbed:

$$\Delta Q_{DA} = C_V(T_2 - T_1) > 0. \tag{4.67}$$

Net Work and Efficiency

The total work done over one cycle is therefore

$$W = W_{AB} + W_{BC} + W_{CD} + W_{DA} = Nk_{\mathrm{B}}(T_2 - T_1)\ln\left(\frac{V_B}{V_A}\right), \tag{4.68}$$

while the total heat absorbed by the engine (from the hot reservoir and during isochoric heating) is

$$Q_{\mathrm{in}} = Q_{AB} + Q_{DA} = Nk_{\mathrm{B}}T_2 \ln\left(\frac{V_B}{V_A}\right) + C_V(T_2 - T_1). \tag{4.69}$$

The efficiency of the Stirling engine is then given by

$$\begin{aligned} \eta_{\mathrm{Stirling}} &= \frac{W}{Q_{\mathrm{in}}} = \frac{Nk_{\mathrm{B}}(T_2 - T_1)\ln(\frac{V_B}{V_A})}{Nk_{\mathrm{B}}T_2\ln(\frac{V_B}{V_A}) + C_V(T_2 - T_1)} \\ &= \frac{T_2 - T_1}{T_2 + (\gamma - 1)\frac{(T_2 - T_1)}{\ln(V_B/V_A)}}, \end{aligned} \tag{4.70}$$

where we used the relation $C_V = Nk_{\mathrm{B}}/(\gamma - 1)$. Since $T_2 > T_1$, $\gamma > 1$, and $V_B > V_A$, the denominator exceeds T_2, implying that

$$\eta_{\mathrm{Stirling}} < \eta_{\mathrm{Carnot}}, \tag{4.71}$$

for the same temperature limits T_1 and T_2. Thus, the Stirling engine—though reversible in its idealized form—cannot exceed the efficiency of the Carnot engine.

Suggested Experiment: Observing the Mechanical Efficiency of a Stirling Engine

A simple and engaging experiment can be performed to connect the abstract ideas of thermodynamic work and energy transformation to something visibly mechanical. Following the design proposed in [YQ21], one may use a small β-type Stirling engine with a transparent chamber to observe the conversion of heat into rotational motion.

Place the engine on a cup of boiling water so that the bottom plate reaches a temperature $T_H \approx 373$ K, while the top plate remains near room temperature $T_C \approx 293$ K. Within a minute, the flywheel begins to spin, driven solely by the temperature difference across the plates. As the water cools, the wheel gradually slows, its frequency faithfully tracking the decreasing temperature gradient.

The rotation frequency f of the flywheel can be measured with a simple light gate and oscilloscope, while thermocouples measure the hot and cold plate temperatures. The physical parameters of a typical inexpensive engine are $M = 6.5 \times 10^{-2}$ kg, $r = 4.3 \times 10^{-2}$ m, $n = 3.5 \times 10^{-3}$ mol, and a compression ratio $C_R \approx 1.3$.

Mechanical Efficiency Derivation

The mechanical efficiency is defined as the ratio of the useful work done by the wheel to the thermodynamic work performed by the gas:

$$\eta_{\text{mech}} = \frac{W_{\text{wheel}}}{W_{\text{gas}}}. \tag{4.72}$$

The rotational kinetic energy of the wheel, $W_{\text{wheel}} = I\omega^2/2$ with $I = Mr^2/2$ and $\omega = 2\pi f$, gives

$$W_{\text{wheel}} = Mr^2\pi^2 f^2. \tag{4.73}$$

The work done by the gas in an ideal Stirling cycle is

$$W_{\text{gas}} = nR(T_H - T_C)\ln C_R. \tag{4.74}$$

Hence,

$$\eta_{\text{mech}} = \frac{Mr^2\pi^2 f^2}{nR\,(T_H - T_C)\ln C_R}. \tag{4.75}$$

Using the measured quantities, Yeadon and Quinn found $\eta_{\text{mech}} \approx 10\%$, with a relative uncertainty of about 20%. Despite its simplicity, this tabletop demonstration offers a vivid and quantitative illustration of how heat can be transformed into useful mechanical work.

Carnot Theorem

To see why this is the case, recall Eq. (4.19), which relates the efficiency to entropy production:

$$\eta = \eta_{\text{rev}} - \frac{T_2}{Q_1}\Delta S. \tag{4.76}$$

The maximum efficiency is achieved only when $\Delta S = 0$, corresponding to a completely reversible process. Since $\eta_{\text{rev}} = \eta_{\text{Carnot}}$, we obtain directly:

> **Carnot's Theorem:** *No thermal engine operating between two heat reservoirs can be more efficient than a reversible Carnot engine operating between the same temperatures.*

This theorem establishes the absolute upper bound for thermal efficiency and is independent of the working substance or the particular form of the engine.

Entropy in the Stirling Cycle

We can also express the entropy production in the Stirling engine using the work–entropy relationship:

$$\begin{aligned}\Delta S_{\text{Stirling}} &= (\eta_{\text{rev}} - \eta_{\text{Stirling}})\frac{Q_2}{T_2} \\ &= \frac{(\gamma-1)\left(1-\frac{T_1}{T_2}\right)^2}{(\gamma-1)(T_2-T_1)+T_2\ln(V_B/V_A)}\frac{Q_2}{T_2} \\ &= \frac{(\gamma-1)\,\eta_{\text{rev}}^2}{(\gamma-1)\,\eta_{\text{rev}}+\ln(V_B/V_A)}\frac{Q_2}{T_2}. \end{aligned} \tag{4.77}$$

This result explicitly shows how the entropy produced in the cycle quantifies its deviation from reversibility. When $\Delta S_{\text{Stirling}} = 0$, the cycle reduces to the ideal Carnot case.

REMARK 4.7 The Stirling engine demonstrates how the combination of isothermal and isochoric processes can approach, but never exceed, the Carnot limit. Its lower efficiency arises from entropy production during the isochoric steps, where heat transfer occurs through finite temperature differences. In this sense, the Stirling engine illustrates the universality of Carnot's theorem and the quantitative power of the work–entropy relationship.

4.5 Perpetual Motion in History

From the earliest days of alchemy and natural philosophy, humanity has been fascinated by the dream of creating machines capable of producing work indefinitely without any external energy input—the so-called *perpetual motion machines.* This idea predates modern physics and has reappeared in countless guises throughout history, from the wheels and fountains of medieval inventors to the intricate mechanical contraptions of the Renaissance.

The fascination persisted well into the nineteenth century, even as the *laws of thermodynamics* were being formulated. Each class of perpetual motion machine can be identified by the thermodynamic principle it violates:

- A *perpetual motion machine of the first kind* violates the *first law of thermodynamics* by creating energy from nothing, thereby breaking energy conservation.
- A *perpetual motion machine of the second kind* violates the *second law of thermodynamics* by spontaneously converting heat entirely into work without a temperature gradient or by making heat flow from cold to hot without external input.

Despite centuries of ingenious attempts, such machines have never succeeded. Nevertheless, the quest for perpetual motion has had a curious legacy: it has driven many of the conceptual clarifications that eventually shaped classical thermodynamics itself. See Arthur W. J. G. Ord-Hume, *Perpetual Motion: The History of an Obsession*, St. Martin's Press, New York (1977).

Beyond science, the idea of perpetual motion has captivated artists and thinkers alike, serving as a metaphor for the tension between imagination and physical law. It has inspired numerous artistic interpretations, notably in the works of M. C. Escher and Salvador Dalí. Escher's famous lithograph *Waterfall*, shown in Fig. 4.3, depicts an impossible machine in which water appears to flow uphill in a closed loop, endlessly turning a mill wheel—a visual paradox that playfully defies the second law of thermodynamics.

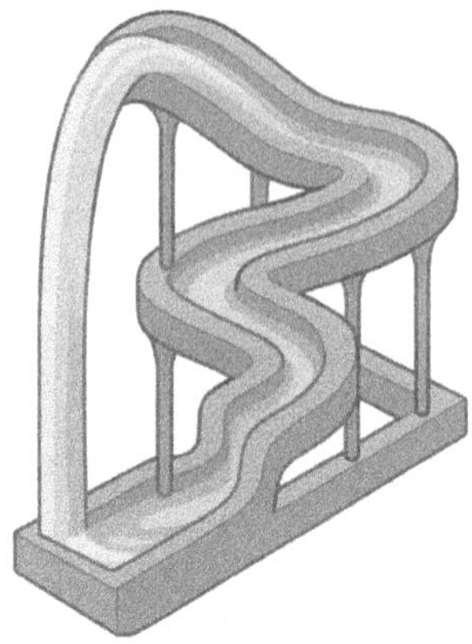

FIGURE 4.3 Stylized version of M. C. Escher's *Waterfall* (1961), a visual representation of an impossible perpetual motion machine. The water appears to rise continuously, and attaching a mill at the waterfall would, humorously, imply a violation of the second law of thermodynamics.

4.6 Real Engines and Perpetual Motion of the First Kind

The components of a real heat engine vary depending on its type, but all share the same fundamental principles governed by the laws of thermodynamics. Work is typically performed either mechanically, via a *piston* in a reciprocating engine, or through a *turbine* that rotates as high-pressure gas expands and flows through its blades. In many modern power plants, the turbine is connected to an electrical generator, converting the mechanical work into electrical energy. During one complete thermodynamic cycle, two key processes always occur: *heat injection* and *heat rejection.*

Heat Injection (Boiler Stage)

Heat injection is usually achieved in a *boiler chamber*, a closed vessel in which water or another working fluid is heated. The boiler transfers heat from a fuel source—such as coal, natural gas, oil, or biomass—to the fluid, typically through combustion in a separate chamber. Hot gases produced during combustion flow past the boiler's heat exchanger, transferring energy to the water and generating steam or high-temperature vapor. This vapor is stored under pressure and used to drive a piston or turbine, converting the internal energy of the vapor into mechanical work.

Heat Rejection (Condenser Stage)

Heat rejection occurs in the *condenser*, where the working fluid releases its heat to a cooler medium, such as water or air, and condenses back into a liquid state. The condenser receives high-temperature vapor from the turbine or compressor, which then comes into contact with cooler surfaces or coils. As the vapor cools, it releases its *latent heat of vaporization* and transitions into a liquid. This released heat is transferred to the cooling medium, raising its temperature. The liquid is then returned to the boiler to restart the cycle, completing a closed-loop system.

Condensers are not unique to power plants: they are also found in refrigeration and air conditioning systems, where the working fluid absorbs heat in an evaporator and rejects it through the condenser, ensuring continuous heat transfer across a temperature gradient.

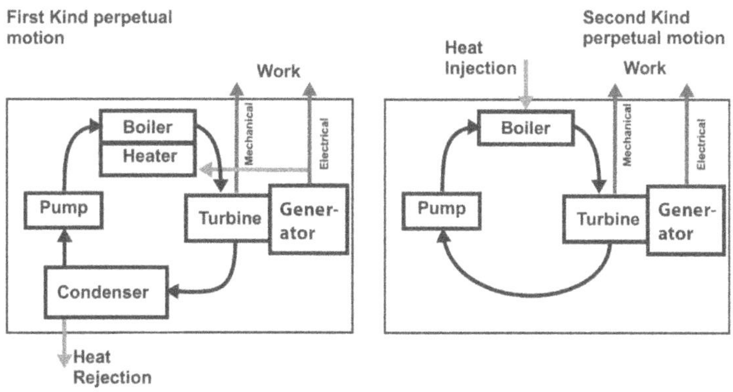

FIGURE 4.4 Schematics of engines which claim perpetual motion of the first and second kind. The perpetual motion of the first kind has no heat injection, only heat rejection of some sort (for instance, through friction) and work performed. Typically, the system has a (hidden) internal source of energy which, once depleted, leads to a stop of the operation of the engine. Perpetual motion of the second kind typically has a single reservoir performing heat injection, but no heat rejection. This implies that all energy is used somehow, leading to a violation of the second law and the limits of the efficiency of engines.

Real Engines and Thermodynamic Boundaries

Every real engine obeys the first and second laws of thermodynamics. The *first law* enforces energy conservation: the total energy input equals the sum of the work output and the energy rejected as heat. The *second law* restricts the direction of energy transfer, ensuring that some portion of energy is always lost as waste heat.

Perpetual motion machines, in contrast, appear to violate these boundaries. Their impossibility can be seen by identifying where, in the heat injection–rejection–work loop, the supposed violation occurs (see Fig. 4.4). Machines that claim to produce continuous work without an external energy source violate the first law of thermodynamics.

4.6.1 Examples of Perpetual Motion of the First Kind

A *perpetual motion machine of the first kind* violates the first law of thermodynamics by purporting to create energy from nothing. One of the most famous historical examples is the *perpetual watermill*, often attributed to the seventeenth-century English physician and mystic Robert Fludd (1574–1637).

The device is typically depicted as a water wheel that lifts water to a higher level, allowing it to flow back down and turn the wheel again in a continuous cycle. In principle, such a mechanism would require no additional energy once started, apparently achieving perpetual motion. However, it violates the first law of thermodynamics because it assumes that gravitational potential energy can be continuously replenished without work input.

In reality, to raise the water, the wheel must perform work against gravity, and friction and viscous losses dissipate energy as heat. The potential energy of the water must satisfy

$$U = Q - W, \tag{4.78}$$

where Q is the heat input and W the mechanical work done. Maintaining motion indefinitely would require a constant energy source to compensate for these losses. In ornamental devices based on the same principle (see Fig. 4.5, right), a concealed motor or battery provides the missing energy, keeping the illusion of perpetual motion alive.

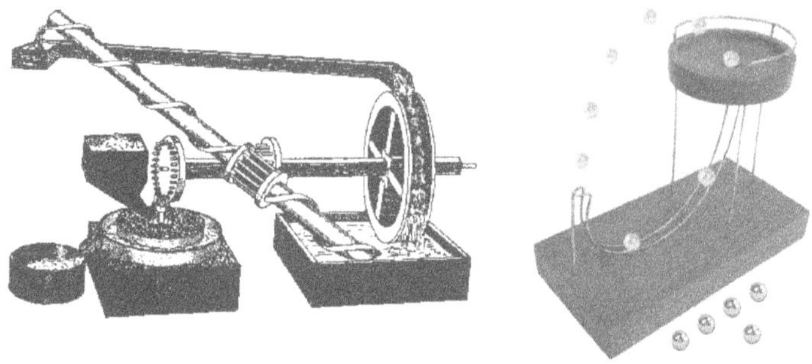

FIGURE 4.5 Illustrations of Robert Fludd's perpetual watermill. (left) historical schematic showing a water wheel lifting water to the top of a channel, creating an apparent perpetual flow. (right) modern reproductions often function as decorative devices powered by hidden batteries or pumps.

REMARK 4.8 Perpetual motion machines of the first kind serve as historical reminders of the importance of the first law of thermodynamics. They illustrate that even seemingly simple mechanical cycles—such as a falling and rising fluid—cannot sustain motion without continuous energy input. The recognition of this impossibility marked a critical step in the transition from speculative natural philosophy to modern energy science.

Examples of Perpetual Motion of the Second Kind

A *perpetual motion machine of the second kind* is one that violates the *second law of thermodynamics*. Unlike machines of the first kind, which create energy from nothing, this type of device appears to convert thermal energy from a single reservoir entirely into mechanical work—something the Kelvin–Planck statement of the second law explicitly forbids.

In practice, two key red flags reveal a second-law violation:

1. The engine claims to operate at an efficiency greater than that of the *Carnot efficiency*, despite using two heat reservoirs.
2. The device claims to operate using only a single thermal reservoir, continuously converting heat into work without any external temperature gradient.

One of the most famous historical examples of such a proposed device is *Gamgee's ammonia motor*, also known as the *Zeromotor*. It is often cited as a classic example of a misunderstanding of the second law. For a historical overview of ammonia-based engines, see [Sel24].

Gamgee's Ammonia Motor

The device was proposed in the late nineteenth century by the English chemist and physician Sampson Gamgee. Gamgee claimed that his "Zeromotor" could operate by absorbing heat directly from the ambient environment and converting it into mechanical work without any external energy source.

The working fluid in the engine was *ammonia*, chosen for its favorable thermodynamic properties: a relatively low boiling point (-33 °C) and high vapor pressure under ambient conditions. At room temperature, ammonia vaporizes, creating a vapor pressure of several atmospheres, which can be used to drive a piston or turbine. As the vapor expands, it cools and condenses, supposedly completing the cycle by returning to the liquid reservoir.

At first glance, the design seems plausible, as it involves a familiar sequence of evaporation, expansion, and condensation. However, the critical flaw lies in the assumption that condensation

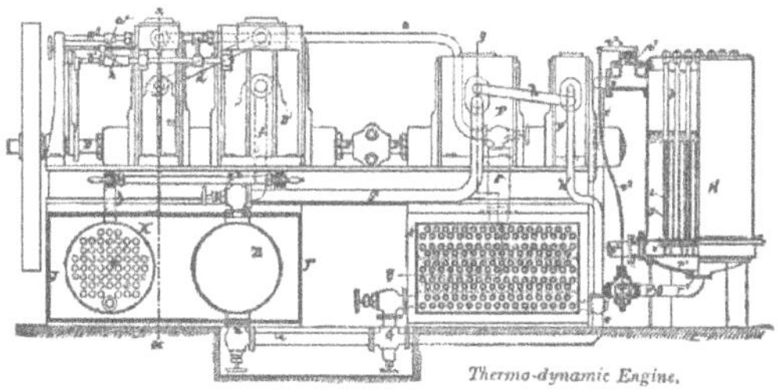

FIGURE 4.6 A picture of the Zeromotor device scheme. The engine was claimed to operate by vaporizing and condensing ammonia in a closed loop, supposedly extracting work continuously from the surrounding air. *The Practical Dictionary of Mechanics* by Edward H. Knight, 1884 [Kni84].

can occur spontaneously at the same ambient temperature that drives vaporization. For the vapor to condense, it must be cooled below its saturation temperature, which requires refrigeration—and refrigeration itself requires work input.

In other words, the engine's operation demands that heat flow from a colder body (the condenser) to a hotter one (the ambient environment) without external work, directly violating the *Clausius statement* of the second law. Furthermore, if the device could indeed absorb heat from a single reservoir and convert it entirely into work, it would also violate the *Kelvin–Planck statement.*

Thus, Gamgee's Zeromotor is an example of a proposed perpetual motion machine of the second kind: it relies on the spontaneous flow of heat in the wrong direction and assumes that ambient thermal energy can be harvested indefinitely without compensation.

REMARK 4.9 Devices like the ammonia Zeromotor highlight the subtlety of the second law. Unlike the first law, which simply forbids the creation of energy, the second law forbids the *direction* of certain processes—in this case, the spontaneous conversion of ambient heat into work. The persistence of such proposals throughout history underscores the intuitive difficulty of grasping entropy and irreversibility, and the ingenuity of those who tried to circumvent them.

4.6.2 "Perpetual" Motion Engines that Actually Work

Although true perpetual motion is impossible, certain devices appear to operate continuously for long periods without an obvious energy supply. In reality, such systems exploit subtle energy gradients in their environment and therefore do not violate the laws of thermodynamics. One famous example of this type of self-sustaining "perpetual motion" engine is the *drinking bird*, a classic thermodynamic toy.

The Drinking Bird

The drinking bird is a small heat engine that converts a temperature difference into periodic mechanical motion. It was patented by Miles V. Sullivan in 1945 [Sul46]. The device consists of two glass bulbs connected by a narrow glass tube, partially filled with a volatile liquid such as *dichloromethane*. The lower bulb acts as the reservoir or "boiler," while the upper bulb and head are covered with felt that can absorb water. The system operates cyclically as follows.

Initially, the bird is vertical with its head wet. Evaporation of water from the felt cools the head because of the latent heat of vaporization. This temperature decrease causes some dichloromethane vapor in the head to condense, lowering the pressure inside the upper bulb. The vapor pressure in the warmer lower bulb (the base) now exceeds that in the cooler head. This pressure difference forces liquid up through the neck tube into the head. As liquid rises, the bird becomes top-heavy and tips forward. When the bird tips, a bubble of warm vapor travels upward from the base, displacing the liquid back into the lower bulb and restoring pressure equilibrium between the two bulbs. The liquid's weight in the base causes the bird to swing back to its upright position. Once upright, the bottom bulb is warmed by ambient air—typically a few degrees warmer than the evaporatively cooled head. If a glass of water is positioned so that the bird's beak dips into it, the head remains wet, allowing the evaporation–condensation cycle to repeat indefinitely. The process continues as long as there is a temperature difference between the head and the base.

Thermodynamic Interpretation

From a thermodynamic perspective, the drinking bird is a miniature *heat engine.* The warm lower bulb acts as the heat source, while the cooled upper bulb serves as the heat sink. The working fluid—dichloromethane vapor and liquid—undergoes repeated vaporization, expansion, condensation, and compression. The system thus converts a small portion of thermal energy from the ambient environment into mechanical work (the oscillatory tipping motion), while the remainder is dissipated as heat.

The device is not a perpetual motion machine because its operation relies on a continuous energy input—the heat supplied by the temperature difference between its head and base. When this temperature gradient disappears, the motion ceases. The "drinking bird" cleverly demonstrates the principles of energy conversion, entropy production, and the necessity of a heat source and sink in every real engine.

REMARK 4.10 The drinking bird is an interesting machine that is the embodiment of the second law of thermodynamics in action. Although it appears to run indefinitely, it merely exploits a small and constant thermal gradient with the surrounding environment. Its periodic motion elegantly illustrates how nature's simplest systems can operate as true heat engines—not by violating thermodynamic laws, but by using them to their fullest extent.

4.6.3 The Rainmaker

Among the more imaginative engineering proposals inspired by thermodynamics is the so-called *rainmaker*, proposed by the meteorologist and physicist J. P. Peixoto in 1960 and later discussed in a 1973 article in *Scientific American* [PK73]. The concept was based on well-known atmospheric principles: adiabatic expansion, cooling, and condensation of moist air under the influence of gravity.

Thermodynamic Basis

Peixoto's idea relies on the *barometric formula* and the *adiabatic cooling* of gases. As a parcel of air ascends under gravity, the ambient pressure decreases with altitude, causing the air to expand and cool. For an ideal gas undergoing adiabatic expansion, the temperature and pressure at two altitudes are related by

$$T_2 = T_1 \left(\frac{P_2}{P_1} \right)^{\frac{\gamma-1}{\gamma}}, \tag{4.79}$$

where T_1 and P_1 are the temperature and pressure at the lower altitude, T_2 and P_2 at a higher altitude, and γ is the adiabatic index.

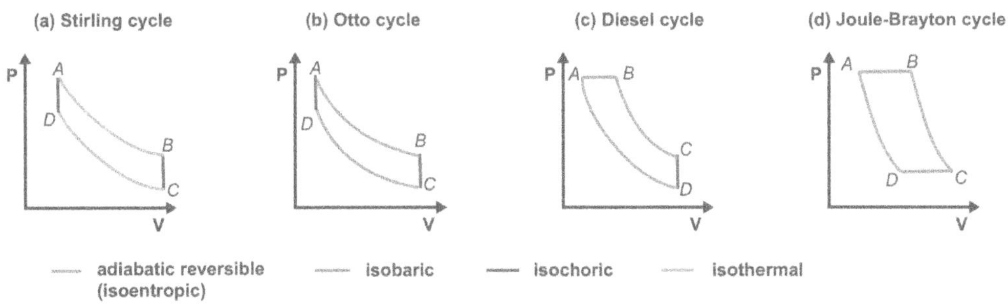

FIGURE 4.7 (a) Stirling cycle, composed of isothermal and isochoric compressions and expansions. (b) Otto cycle, composed of isochorics and adiabatic compressions and expansions. (c) Diesel cycle, which is similar to the Otto cycle, but an isochoric becomes an isobaric. (d) Joule–Brayton cycle, composed of isobaric and adiabatic compressions and expansions.

To include the effect of gravity, the variation of pressure with height h is given by the *barometric formula*

$$P_2 = P_1\, e^{-\frac{Mgh}{RT_1}}, \tag{4.80}$$

where M is the molar mass of air, g the gravitational acceleration, R the universal gas constant, and T_1 the local temperature near the surface.

Combining Eqs. (4.79) and (4.80) shows that an air parcel cools rapidly as it rises, leading to condensation if its temperature drops below the dew point. This is the same principle that drives cloud formation and precipitation in the atmosphere.

Peixoto's Proposal

Peixoto observed that near coastal deserts, especially in parts of the Middle East, a thin layer of warm, moist air exists near the surface, capped by a temperature inversion. He proposed constructing a tall, lightweight vertical tube (essentially a *man-made chimney*) to trigger a controlled updraft of this moist air.

If a small initial impulse (for instance, mechanical pumping) could push some of the warm, humid air upward through the tube, adiabatic expansion would cause it to cool as it rises. The cooling would lead to condensation and precipitation of water droplets inside or around the structure, forming a continuous flow of fresh water that could be collected near the base.

In principle, once initiated, the process could sustain itself: warm, moist air entering at the bottom would drive cooler, drier air out from the top, creating a convection cycle powered by the natural temperature gradient between the surface and higher altitudes. Peixoto estimated that a vertical structure on the order of 3 km in height could, in theory, maintain such circulation, effectively functioning as a "thermodynamic fountain" generating rain.

Although the design was never built, the concept is thermodynamically sound in principle and highlights the same mechanisms that drive natural atmospheric circulation. The "rainmaker" serves as a striking example of how the laws of thermodynamics govern not only engines and machines but also large-scale processes in the environment.

REMARK 4.11 Peixoto's rainmaker is another fascinating attempt to harness atmospheric thermodynamics for human benefit. While the practical challenges—such as structural feasibility, stability, and control—make its realization unlikely, the concept illustrates how adiabatic cooling and latent heat release can, in principle, be engineered to mimic nature's own water cycle.

5

Applications of Thermodynamics

Thermodynamics is among the most universal theories of physics, applicable to an extraordinary range of systems—from engines and biological organisms to planetary atmospheres and stars. Classical thermodynamics, however, may break down when systems are far from equilibrium or when quantum phenomena dominate. Nevertheless, the same fundamental principles continue to apply: both *heat* and *work* are mechanisms through which energy is transferred between a system and its surroundings.

5.1 Generalized Form of the First Law

When external potentials are present, the *first law of thermodynamics* must be extended to include all forms of energy exchange. The most general infinitesimal form can be written as

$$\begin{aligned} dU &= \delta Q - \delta W \\ &+ \delta E_{\text{el}} + \delta E_{\text{magnetic}} + \delta E_{\text{chemical}} + \delta E_{\text{gravitational}} + \delta E_{\text{elastic}} + \delta E_{\text{other}}, \end{aligned} \tag{5.1}$$

where each term represents a specific mode of energy transfer.

In this expression, the symbol dU denotes an *exact differential*, because the internal energy U is a *state function*. Its change depends only on the initial and final states of the system, not on the particular process or path connecting them. Thus, for any process between states A and B,

$$\int_A^B dU = U(B) - U(A). \tag{5.2}$$

By contrast, δQ and δW are *inexact differentials*: the heat and work exchanged during a transformation depend on the path followed in thermodynamic space. Hence, dU appears with

DOI: 10.1201/9781003603528-5

a regular differential, while all other process-dependent contributions are written with the symbol δ.

In Eq. (5.1), each term represents a distinct physical mechanism by which a system can exchange energy with its surroundings. We discuss their meanings and typical realizations below.

The quantity dU is the infinitesimal change in the system's *internal energy*. It includes the sum of all microscopic kinetic and potential energies associated with molecular motion and interactions. In gases, U encompasses translational, rotational, and vibrational degrees of freedom; in solids and liquids, it includes lattice vibrations, bond energies, and electronic excitations. Because U is a *state function*, its differential is exact: the change in U depends only on the initial and final states, not on the path taken between them.

The term δQ denotes the infinitesimal amount of energy exchanged as *heat* due to a temperature difference between the system and its surroundings. Heat represents disordered microscopic energy transfer through molecular collisions, conduction, convection, or radiation. For example, when a metal rod is heated at one end, energy diffuses along its length by conduction. In thermodynamic calculations, δQ is often expressed as $\delta Q = C\,dT$ for processes at constant pressure or volume, where C is the appropriate heat capacity.

The contribution δW corresponds to *mechanical work*, the ordered transfer of energy associated with macroscopic forces acting through distances. For a compressible fluid, $\delta W = P\,dV$, where P is the pressure and V the volume, describing the work done by a gas expanding in a piston. Other forms of mechanical work include the stretching of a membrane or compression of a spring, where $\delta W = F\,dx$ with F the applied force and x the displacement.

The term δE_{el} accounts for *electrical work*. It becomes relevant whenever electric charges move through a potential difference. For a particle of charge q moved through a voltage ΔV, the work done is $\delta E_{\mathrm{el}} = q\,\Delta V$. In electrochemical cells or batteries, the collective electrical work is expressed as $\delta E_{\mathrm{el}} = \Phi\,dQ$, where Φ is the electric potential and dQ the differential charge. This form of work powers devices ranging from electric motors to portable electronics.

Magnetic systems involve another form of energy exchange, represented by $\delta E_{\mathrm{magnetic}}$. When a magnetic material is placed in a varying magnetic field, work is performed to align its magnetic moments. The infinitesimal magnetic work is given by $\delta E_{\mathrm{magnetic}} = \vec{h} \cdot d\vec{M}$, where $\vec{h}$ is the magnetic field and $\vec{M}$ the magnetization. This term is fundamental in magnetic refrigeration, data storage, and the operation of electric generators, where magnetic fields convert mechanical energy into electrical energy.

Chemical work, denoted $\delta E_{\mathrm{chemical}}$, describes energy changes due to variations in chemical composition. It is written as $\delta E_{\mathrm{chemical}} = \sum_i \mu_i\,dN_i$, where μ_i is the chemical potential and N_i the number of particles of species i. This form of energy exchange governs chemical reactions, phase transformations, and the operation of fuel cells. For example, in a hydrogen fuel cell, chemical potential differences drive the flow of electrons, producing both electrical work and heat.

The term $\delta E_{\mathrm{gravitational}}$ corresponds to *gravitational potential energy*. When a system's mass distribution changes height in a gravitational field, work is done according to $\delta E_{\mathrm{gravitational}} = m\,g\,dh$, where m is the mass, g the gravitational acceleration, and dh the change in height. This contribution is significant in geophysical and atmospheric systems—such as rising air parcels that cool adiabatically—or in hydroelectric plants, where gravitational potential is converted into mechanical and electrical energy.

Elastic or deformation work, $\delta E_{\mathrm{elastic}}$, arises when the shape or size of a solid changes under stress. For a linear elastic material, $\delta E_{\mathrm{elastic}} = \sigma\,d\varepsilon$, where σ is the stress and ε the strain. This form of energy is central in materials science and engineering: stretching a steel spring, deforming a rock under tectonic stress, or compressing a polymer all involve elastic energy storage.

Finally, $\delta E_{\mathrm{other}}$ includes additional or specialized contributions not covered by the categories above. Examples include surface work ($\gamma\,dA$, where γ is the surface tension and A the surface area), radiation pressure in optical systems, and spin or polarization work in quantum and nanoscale devices. Though often small, such effects can dominate in high-precision measurements, microelectromechanical systems (MEMS), or astrophysical environments.

Each of these terms represents a specific mode of energy exchange permitted by the first law of thermodynamics, which is at its root a form of *energy accounting.* Together, they ensure that all possible forms of work and heat transfer—mechanical, electrical, magnetic, chemical, gravitational, and elastic—are encompassed in the universal statement of energy conservation.

While *heat* is associated with temperature differences between a system and its surroundings, *work* represents any energy transfer that is not due to temperature gradients. The most common example is *mechanical work*, such as that performed by a piston in a gas engine. If a system with pressure P undergoes a volume change dV, the infinitesimal mechanical work is

$$\delta W = P\,dV, \tag{5.3}$$

and integrating between two states gives

$$W = \int_A^B P\,dV. \tag{5.4}$$

This follows directly from the classical definition of work, $w = F\,\Delta x$, since pressure is the force per unit area acting over a displacement of the system's boundary.

Electrical Work as an Important Example

If the system carries an electrical charge or is placed in an external electric field, energy can also be transferred through electrical work. For a particle of charge q moving through a potential difference Δv, the infinitesimal electrical work is

$$\delta E_{\text{el}} = q\,\Delta v. \tag{5.5}$$

In macroscopic systems, the corresponding work is often expressed in terms of the electric field $\vec{E}$ and displacement $d\vec{r}$ as $\delta E_{\text{el}} = q\vec{E}\cdot d\vec{r}$, or, in continuum form, $\delta E_{\text{el}} = \int \vec{E}\cdot d\vec{P}$, where $\vec{P}$ is the polarization. The inclusion of this term in Eq. (5.1) ensures that all types of energy exchange—mechanical, thermal, electrical, magnetic, or chemical—are accounted for under a unified thermodynamic framework.

REMARK 5.1 The use of dU in Eq. (5.1) highlights the central role of internal energy as a true thermodynamic potential: it is a state property that tracks the cumulative result of all possible energy exchanges. Heat and work, by contrast, are process-dependent quantities whose separate values depend on the details of the path, but whose combined effect is fully captured in U. This distinction underlies much of what follows in this chapter, where we explore specific applications of the first law to mechanical, electrical, magnetic, and chemical systems.

5.2 Phase Changes and Latent Heat

One of the most familiar applications of thermodynamics concerns *phase transitions*, processes that occur every day in nature and technology: the melting of ice, the evaporation of water, or the condensation of steam. During such transformations, a system absorbs or releases energy while its temperature remains constant. This energy exchange is known as *latent heat.* It represents the amount of energy required to change the phase of a substance without changing its temperature [Zem97, PD61, Fer36].

Definition of Latent Heat

Latent heat can be defined as the amount of heat absorbed or released during a phase change per unit mass (or per mole):

$$L_m = \frac{Q}{m}, \tag{5.6}$$

where Q is the total heat exchanged and m the mass of the substance. It is conventional to consider phase transitions at constant pressure, as is typical in laboratory and environmental conditions.

The latent heat is closely related to the change in *enthalpy*, defined by $H = U + PV$. For a system at constant pressure, the heat transferred equals the change in enthalpy:

$$\delta Q_P = dH. \tag{5.7}$$

Hence, the latent heat associated with a specific phase change is simply the enthalpy difference between the two phases:

$$L_{\text{evaporation}} = H_{\text{vap}} - H_{\text{liq}}, \tag{5.8}$$

$$L_{\text{fusion}} = H_{\text{liq}} - H_{\text{sol}}, \tag{5.9}$$

$$L_{\text{sublimation}} = H_{\text{vap}} - H_{\text{sol}}. \tag{5.10}$$

From these relations, it immediately follows that

$$L_{\text{sublimation}} = L_{\text{fusion}} + L_{\text{evaporation}}. \tag{5.11}$$

Thus, the latent heat of sublimation equals the sum of the heats of fusion and evaporation, since it corresponds to a direct transition from solid to vapor.

Entropy and Phase Transitions

Entropy provides a natural framework for understanding latent heat. From Clausius's definition,

$$dS = \frac{\delta Q_{\text{rev}}}{T}, \tag{5.12}$$

the change in entropy associated with a phase transition at constant temperature T is

$$\Delta S = \frac{L}{T}. \tag{5.13}$$

This expression shows that a phase transition is accompanied by a discontinuous change in entropy, reflecting the change in molecular organization between phases. For example, the entropy of vapor is greater than that of liquid because gas molecules have access to a much larger number of microstates.

Boltzmann provided a microscopic interpretation of entropy in terms of the number of accessible configurations of the system:

$$S = k_{\text{B}} \ln W, \tag{5.14}$$

where W is the number of microstates consistent with the macroscopic state. Thus, an increase in entropy corresponds to an increase in the number of possible molecular arrangements.

From the differential form $dS = \frac{dQ}{T}$, we can also express temperature as

$$T = \frac{dQ}{dS}. \tag{5.15}$$

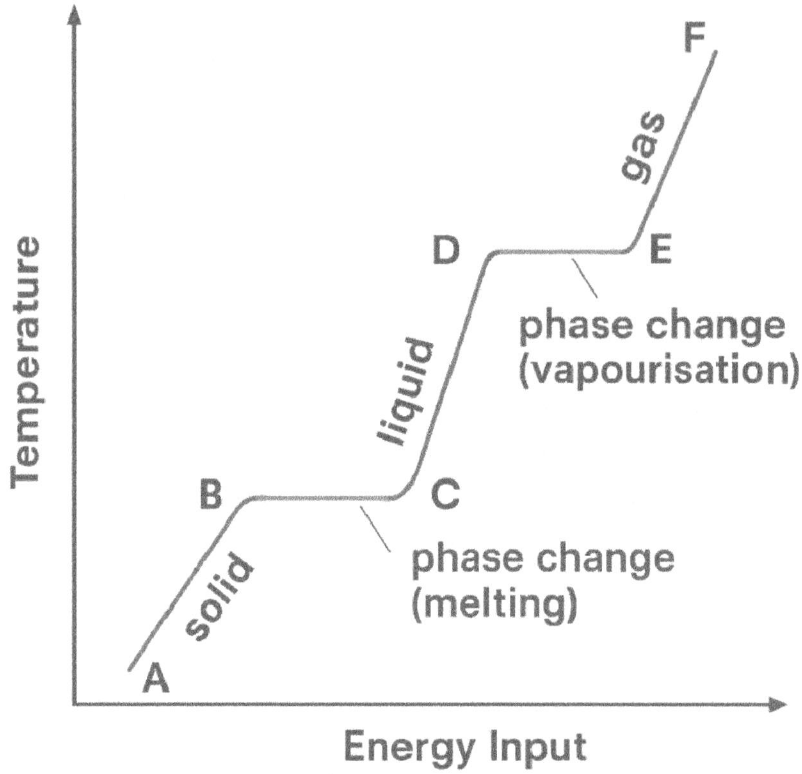

FIGURE 5.1 Energy required for melting and vaporization of a system. At characteristic transition temperatures, heat is absorbed without a change in temperature, corresponding to the latent heat of fusion or vaporization.

During a phase change, heat is absorbed ($dQ > 0$) but the temperature remains constant, implying that dS increases while T is fixed. This behavior characterizes first-order phase transitions, where the system reorganizes structurally while maintaining thermal equilibrium with its surroundings.

REMARK 5.2 At the temperature of a phase transition, the heat capacity of the system effectively diverges: an infinitesimal change in temperature corresponds to a finite energy transfer. This divergence reflects the system's structural rearrangement and the large change in entropy associated with the transition. Phase transitions thus provide one of the most direct manifestations of the interplay between energy, entropy, and the microscopic organization of matter.

Evaporation and Condensation

Evaporation is the process by which molecules at the surface of a liquid acquire sufficient kinetic energy to overcome intermolecular attractions and escape into the vapor phase. It occurs spontaneously at all temperatures but becomes significant near the boiling point, when the vapor pressure of the liquid equals the external pressure. Evaporation is an *endothermic* process—it absorbs heat from the surroundings—because the escaping molecules remove internal energy from the liquid. The opposite process, *condensation*, releases heat as vapor molecules lose kinetic energy and rejoin the liquid.

Entropy and Latent Heat

During evaporation at constant temperature and pressure, the entropy change ΔS is given by

$$\Delta S = \frac{L}{T}, \tag{5.16}$$

where L is the latent heat of evaporation and T the temperature at which the process occurs. The increase in entropy reflects the greater molecular disorder of the vapor compared to the liquid: the molecules in a gas have far more accessible microstates and move freely throughout the volume.

Typical values of the latent heat of evaporation are remarkably large:

- Water: $L = 2260$ kJ/kg,
- Ethanol (C_2H_5OH): $L = 841$ kJ/kg,
- Acetone (C_3H_6O): $L = 509$ kJ/kg.

These large values illustrate the strength of the intermolecular forces that must be overcome during vaporization—especially hydrogen bonding in the case of water.

The Clausius–Clapeyron Relation

The equilibrium between liquid and vapor phases can be described quantitatively by the *Clausius–Clapeyron equation*, which relates the rate of change of vapor pressure P with temperature T:

$$\frac{dP}{dT} = \frac{L}{T\,\Delta V}, \tag{5.17}$$

where $\Delta V = V_{\text{gas}} - V_{\text{liq}}$ is the change in specific volume during evaporation. Since $V_{\text{gas}} \gg V_{\text{liq}}$, the equation can often be approximated by

$$\frac{dP}{dT} \approx \frac{LP}{RT^2}, \tag{5.18}$$

which integrates to

$$\ln P = -\frac{L}{RT} + \text{const.} \tag{5.19}$$

This form shows that the vapor pressure increases exponentially with temperature—a relationship verified experimentally for most liquids.

Microscopic Interpretation

From the microscopic viewpoint, evaporation corresponds to the transition of molecules from a strongly interacting condensed phase to a weakly interacting gaseous phase. The latent heat represents the average energy per unit mass required to overcome intermolecular attractions and increase molecular freedom. The associated entropy change measures the increase in configurational possibilities for the system:

$$\Delta S = k_{\text{B}} \ln \frac{W_{\text{vapor}}}{W_{\text{liquid}}}, \tag{5.20}$$

where W_{vapor} and W_{liquid} denote the numbers of accessible microstates in each phase.

REMARK 5.3 Evaporation and condensation are central to natural and technological processes: they regulate Earth's climate through the water cycle, drive meteorological phenomena, and underlie the operation of refrigeration and distillation systems. In all cases, the balance between latent heat and entropy production embodies the deep connection between microscopic molecular dynamics and macroscopic thermodynamic laws.

Solidification and Liquefaction (Fusion)

Solidification (or freezing) is the process by which a liquid transforms into a solid as its temperature drops below the freezing point. During this transformation, the particles in the liquid lose kinetic energy and organize into a more ordered, often crystalline, arrangement characteristic of the solid phase. Solidification is an *exothermic* process: it releases heat to the surroundings as intermolecular bonds are strengthened and potential energy decreases.

The reverse process, *liquefaction* or *fusion* (melting), occurs when a solid absorbs heat and transitions into the liquid phase. This process is *endothermic*, since it requires breaking the ordered structure of the solid to allow the particles to move more freely.

The entropy change during solidification (or melting) is given by the Clausius–Clapeyron relation:

$$\Delta S = -\frac{L}{T}, \tag{5.21}$$

where L is the latent heat of solidification. The negative sign indicates that entropy decreases as the system transitions from a disordered (liquid) to an ordered (solid) state. For melting, the sign is reversed because entropy increases.

Representative values of latent heat for solidification include:

- Water: $L = 334$ kJ/kg,
- Aluminum: $L = 397$ kJ/kg,
- Lead: $L = 23$ kJ/kg.

These values show the wide range of energy associated with different substances: metals with strong atomic bonding require much more energy to melt than substances with weaker intermolecular interactions.

Sublimation and Deposition

Sublimation is the direct transition of a substance from the solid to the vapor phase without passing through the liquid state. This occurs when the vapor pressure of the solid exceeds the external (usually atmospheric) pressure at a given temperature. Sublimation is an *endothermic* process: heat must be absorbed for the molecules to escape the solid lattice and enter the gaseous phase.

The reverse process, *deposition*, occurs when vapor molecules lose energy and form a solid directly, bypassing the liquid state. Deposition is *exothermic*, as it releases the latent heat of sublimation back into the surroundings.

The entropy change during sublimation is positive and can be written as

$$\Delta S = \frac{L}{T}, \tag{5.22}$$

where L is the latent heat of sublimation. The increase in entropy corresponds to the transition from the highly ordered solid structure to the more disordered gaseous phase.

Typical latent heats of sublimation are:

- Carbon dioxide (dry ice): 593 kJ/kg;
- Iodine: 246 kJ/kg;
- Camphor: 341 kJ/kg.

These values emphasize that sublimation generally requires more energy than either fusion or vaporization, since it involves overcoming both solid–liquid and liquid–gas cohesive forces simultaneously.

REMARK 5.4 Evaporation, melting, and sublimation all follow the same thermodynamic principle: a phase transition occurs when the Gibbs free energies of two phases become equal. The latent heat quantifies the energy required to rearrange intermolecular bonds, while the corresponding entropy change measures the difference in microscopic order between the phases. In each case, the direction of the process—and whether it releases or absorbs heat—is determined by the sign of the entropy change.

Uses of Latent Heat

Latent heat plays a vital role in many natural and technological processes, as it enables large amounts of energy to be stored or released during phase transitions without significant changes in temperature. This property makes it an efficient mechanism for temperature regulation and energy management in both engineered systems and the environment.

Evaporative Cooling

One of the most familiar uses of latent heat is in *evaporative cooling*, where a liquid evaporates and absorbs latent heat from its surroundings, thereby lowering the local temperature. This phenomenon underlies the operation of *evaporative coolers* or *swamp coolers*. In these devices, water evaporates into the air stream, extracting heat from the air and reducing its temperature. The same principle explains the cooling effect of perspiration: when sweat evaporates from the skin, it removes latent heat from the body, maintaining thermal balance.

Refrigeration and Air Conditioning

Modern *refrigeration and air conditioning systems* also rely on latent heat through the controlled evaporation and condensation of a refrigerant fluid. The refrigerant circulates through a closed cycle of compression, condensation, expansion, and evaporation:

- During the *evaporation phase*, the refrigerant absorbs heat from the surrounding air, cooling the interior of the system.
- During *condensation*, the refrigerant releases this absorbed heat to the outside environment.

Because the refrigerant undergoes phase changes at well-defined pressures and temperatures, it enables precise thermal control with high energy efficiency.

Phase Change Materials

Phase change materials (PCMs) are substances engineered to store and release large amounts of energy in the form of latent heat during melting and solidification. They are increasingly used in *passive cooling systems*, such as building insulation, smart textiles, and electronic devices. In PCM-enhanced walls or coatings, the material absorbs excess heat during the day as it melts and releases it at night as it solidifies, maintaining a more stable indoor temperature. This principle is also applied in temperature-regulating clothing and portable medical coolers.

Thermal Energy Storage

Latent heat is also utilized in large-scale *thermal energy storage* systems, which store excess thermal energy for later use. During periods of low energy demand (or when electricity is inexpensive), surplus energy is used to melt a substance with a high latent heat capacity. When energy demand increases, the stored heat is released as the substance solidifies, providing thermal regulation or cooling without additional energy input. Such systems are particularly valuable for balancing renewable energy supply and demand, as they allow intermittent solar or wind power to be stored in thermal form.

REMARK 5.5 The practical applications of latent heat—from the cooling of our bodies and homes to the storage of renewable energy—demonstrate the profound connection between phase transitions and energy transfer. In all these cases, the underlying principle is the same: the reversible exchange of heat through structural transformations of matter. These processes embody the first and second laws of thermodynamics, converting microscopic molecular motion into macroscopic thermal regulation.

Calculations Involving Latent Heat

The latent heat of a substance quantifies the energy absorbed or released during a phase change at constant temperature and pressure. It is defined by

$$Q = m\,L, \tag{5.23}$$

where Q is the total heat exchanged, m is the mass of the substance, and L is the specific latent heat corresponding to the transition.

In the following examples, we illustrate how to compute the energy involved in several common phase changes.

Examples: Latent Heat and Phase Transitions

Example 1: Evaporation of Water

Consider the evaporation of water at its boiling point, 100 °C, under atmospheric pressure. The latent heat of evaporation for water at this temperature is approximately

$$L_{\text{evap}} = 2260 \text{ kJ/kg}.$$

If 1 kg of liquid water completely evaporates at constant temperature, the total heat absorbed is

$$Q = m\,L_{\text{evap}} = 1 \text{ kg} \times 2260 \text{ kJ/kg} = 2260 \text{ kJ}.$$

Thus, 2260 kJ of heat energy must be absorbed to convert 1 kg of water at 100 °C into steam at the same temperature. Because the process occurs isothermally, this energy goes entirely into breaking intermolecular hydrogen bonds, not into raising the temperature.

Example 2: Sublimation of Dry Ice

Now consider the sublimation of solid carbon dioxide (*dry ice*) at atmospheric pressure, where the sublimation temperature is approximately −78.5 °C. The latent heat of sublimation for dry ice is about

$$L_{\text{sub}} = 571 \text{ kJ/kg}.$$

For 1 kg of dry ice sublimating to gaseous carbon dioxide, the heat absorbed is

$$Q = m\,L_{\text{sub}} = 1 \text{ kg} \times 571 \text{ kJ/kg} = 571 \text{ kJ}.$$

Thus, 571 kJ of energy are absorbed from the surroundings, which leads to strong cooling. This explains the intense cold sensation associated with handling dry ice: the sublimation process consumes a large amount of heat.

Example 3: Solidification of Water

Finally, consider the solidification (freezing) of water. The latent heat of solidification is approximately

$$L_{\text{solid}} = 334\ \text{J/g} = 334\ \text{kJ/kg}.$$

If 100 g of water (0.1 kg) freeze at 0 °C, the heat released to the surroundings is

$$Q = m\, L_{\text{solid}} = 100\ \text{g} \times 334\ \text{J/g} = 33{,}400\ \text{J}.$$

Thus, 33.4 kJ of heat energy are released as 100 g of water freeze. This heat warms the surroundings slightly while maintaining the ice–water mixture at 0 °C.

These examples illustrate how latent heat governs the energetic balance of phase transitions. Large amounts of energy are absorbed or released during melting, freezing, vaporization, or sublimation without any temperature change. Such processes are central to climate regulation, refrigeration cycles, and thermal energy storage, where latent heat acts as a natural buffer between different phases.

5.3 Thermodynamics and Chemical Reactions

Thermodynamics provides a fundamental framework for understanding the energetics of chemical reactions. Chemical reactions involve the rearrangement of atomic or molecular bonds, transforming one or more substances (*reactants*) into others (*products*). While chemistry is often experimental and quantitative, the driving forces that determine whether a reaction occurs—and under what conditions—are entirely thermodynamic in nature [Gib73, Eck67, BLB+18, PD61].

A typical chemical reaction may be written as

$$\frac{1}{2}\,\text{O}_2(g) + \text{H}_2(g) \to \text{H}_2\text{O}(l), \tag{5.24}$$

representing the formation of liquid water from hydrogen and oxygen gases. The arrow $\to$ indicates that, under standard conditions, the reaction proceeds *spontaneously* from left to right. However, spontaneity depends strongly on temperature, pressure, and the physical states of the reactants and products.

It is therefore customary to indicate the phase of each species explicitly:

$$\begin{aligned} X(g) &: \text{ gas phase,} \\ X(l) &: \text{ liquid phase,} \\ X(s) &: \text{ solid phase.} \end{aligned}$$

Each phase corresponds to a local minimum of free energy in the relevant thermodynamic space (typically T and P). For example,

$$\text{H}_2\text{O}(l) \to \text{H}_2\text{O}(s), \tag{5.25}$$

is spontaneous only below 0 °C at atmospheric pressure, illustrating that spontaneity depends on both temperature and pressure.

Enthalpy and Energy Exchange

The *enthalpy*, H, represents the total heat content of a system at constant pressure and is defined by

$$H = U + PV. \tag{5.26}$$

For a chemical reaction, the change in enthalpy is given by

$$\Delta H = \Delta U + P\Delta V, \tag{5.27}$$

where ΔU is the change in internal energy and $P\Delta V$ represents the work done by expansion or compression. In practice, ΔH measures the amount of heat absorbed or released by a system at constant pressure:

- If $\Delta H < 0$, the process is *exothermic* (heat is released).
- If $\Delta H > 0$, the process is *endothermic* (heat is absorbed).

For instance, combustion reactions are exothermic, whereas the melting of ice is endothermic.

Gibbs Free Energy and Spontaneity

To determine whether a chemical reaction can occur spontaneously, we must consider not only the heat exchange but also the change in entropy. At constant pressure and temperature, the relevant potential is the *Gibbs free energy*, defined as

$$G = H - TS. \tag{5.28}$$

The change in Gibbs free energy during a process is then

$$\Delta G = \Delta H - T\Delta S. \tag{5.29}$$

The Gibbs free energy represents the maximum amount of non-expansion work obtainable from a process at constant T and P. A process is:

- *Spontaneous* if $\Delta G < 0$,
- *At equilibrium* (static) if $\Delta G = 0$,
- *Non-spontaneous* if $\Delta G > 0$.

This criterion follows directly from the second law of thermodynamics: for a spontaneous process, the total entropy of the universe increases ($\Delta S_{\text{univ}} > 0$), which, under constant T and P, is equivalent to $\Delta G < 0$.

Temperature Dependence and Reaction Direction

The sign of ΔG depends on both ΔH and ΔS. Reactions that release heat ($\Delta H < 0$) and increase entropy ($\Delta S > 0$) are spontaneous at all temperatures. If $\Delta H > 0$ and $\Delta S < 0$, the reaction is never spontaneous. However, for reactions where the two terms compete, temperature can determine the direction of spontaneity:

$$\Delta G = \Delta H - T\Delta S. \tag{5.30}$$

At high T, the entropy term dominates; at low T, enthalpy dominates. For example, the freezing of water,

$$\mathrm{H_2O}(l) \rightarrow \mathrm{H_2O}(s), \tag{5.31}$$

is spontaneous at low temperatures (where $\Delta H < 0$ outweighs $T\Delta S$), but melting becomes spontaneous at higher temperatures.

Thermodynamics versus Kinetics

It is important to note that thermodynamics predicts whether a process is *possible*, not how *fast* it occurs. A reaction may be thermodynamically favorable ($\Delta G < 0$) but still proceed very slowly if it must overcome a high *activation energy*. The reaction

$$2H_2(g) + O_2(g) \to 2H_2O(l), \tag{5.32}$$

for example, is strongly exothermic ($\Delta H < 0$) and spontaneous ($\Delta G < 0$), yet it proceeds negligibly without a spark or catalyst because the activation barrier is large. Thermodynamics determines whether a reaction *can* happen; kinetics determines whether it *will* happen in a reasonable time.

REMARK 5.6 The Gibbs free energy unifies the first and second laws of thermodynamics into a single predictive framework for chemical reactions. It identifies the balance between energy release and entropy increase that dictates spontaneity. From combustion to photosynthesis, every chemical process follows this same principle, first articulated by J. Willard Gibbs in the 1870s: nature spontaneously evolves toward states of lower free energy at constant temperature and pressure.

Example: Fusion (Melting) of Ice at Atmospheric Pressure

Worked Example: Fusion of Ice

To illustrate how enthalpy, volume changes, and heat flow are related in a first-order phase transition, consider the melting of ice at atmospheric pressure:

$$H_2O(s) \longrightarrow H_2O(l),$$

occurring at the equilibrium temperature $T = 0\ °C$ and pressure $P = 1$ atm. The molar enthalpy of fusion under these conditions is

$$\Delta H_{fus} = 6.01\ \text{kJ mol}^{-1}.$$

The corresponding molar volumes are

$$V_{ice} = 0.0197\ \text{L mol}^{-1}, \qquad V_{liq} = 0.0180\ \text{L mol}^{-1}.$$

Because the melting occurs at constant pressure, the heat absorbed per mole is equal to the enthalpy change:

$$Q = \Delta H_{fus} = 6.01\ \text{kJ/mol}.$$

Thus, melting one mole of ice (approximately 18 g) requires

$$6.01\ \text{kJ}$$

of heat energy. This energy does not raise the temperature: instead, it is entirely used to break and rearrange hydrogen-bond networks as the crystalline solid transforms into the more liquid phase.

REMARK 5.7 The careful use of consistent units—joules for energy, moles for substance, and kelvins for temperature—is essential when applying the first law of thermodynamics. It ensures that energy balances are meaningful across different processes, whether in mechanical work, heat transfer, or chemical reactions. By linking microscopic quantities (k_B, individual particles) to macroscopic ones (R, moles), thermodynamics provides a unified description of energy transformations from atoms to engines.

5.4 Free Energy of Mixing and the Method of Intercepts

When two or more substances are mixed, the resulting system may exhibit changes in energy and entropy due to molecular interactions between the components. The thermodynamic quantity that governs the stability of the mixture is the *Gibbs free energy of mixing*, denoted ΔG_{mix}. It represents the change in Gibbs free energy when initially separate components are combined to form a homogeneous mixture at constant temperature and pressure:

$$\Delta G_{\text{mix}} = \Delta H_{\text{mix}} - T\Delta S_{\text{mix}}. \tag{5.33}$$

Here, ΔH_{mix} is the enthalpy of mixing and ΔS_{mix} the entropy of mixing.

Enthalpy and Entropy of Mixing

The *enthalpy of mixing* ΔH_{mix} measures the energetic effects of intermolecular interactions between different species. When the interactions between unlike molecules are primarily attractive, the enthalpy of mixing is negative ($\Delta H_{\text{mix}} < 0$), indicating that energy is released and the process is energetically favorable. For example, ethanol and water mix exothermically because hydrogen bonding between their molecules lowers the total potential energy of the mixture. In contrast, when interactions are mainly repulsive, as in oil and water, $\Delta H_{\text{mix}} > 0$, and mixing is unfavorable unless entropy compensates.

The *entropy of mixing* ΔS_{mix} quantifies the increase in disorder upon mixing. Even if the energy of interaction does not change significantly, the number of possible molecular arrangements grows enormously when two species are mixed, leading to $\Delta S_{\text{mix}} > 0$. For an ideal binary mixture with mole fractions x_1 and x_2, the entropy of mixing is given by

$$\Delta S_{\text{mix}} = -nR\left(x_1 \ln x_1 + x_2 \ln x_2\right), \tag{5.34}$$

where n is the total number of moles and R the gas constant.

Spontaneity and Stability of Mixing

The sign of ΔG_{mix} in Eq. (5.33) determines whether mixing occurs spontaneously:

- If $\Delta G_{\text{mix}} < 0$, mixing is spontaneous and the system tends to form a stable solution.
- If $\Delta G_{\text{mix}} > 0$, the components separate spontaneously and form distinct phases.

Temperature plays a crucial role in this balance. At high temperatures, the $T\Delta S_{\text{mix}}$ term dominates and favors mixing due to entropy. At low temperatures, the enthalpic term ΔH_{mix} may dominate, leading to phase separation. This competition explains why some alloys or polymer blends mix only above a critical temperature, known as the *upper critical solution temperature* (UCST).

The free energy of mixing is therefore a fundamental indicator of the stability of solutions, alloys, and multicomponent systems. It also provides the thermodynamic foundation for processes such as dissolution, diffusion, and chemical equilibrium.

Gibbs's Method of Intercepts

The *method of intercepts*, introduced by J. Willard Gibbs in his 1876–1878 memoirs on heterogeneous systems [Gib78], is a graphical technique used to determine the equilibrium compositions of coexisting phases in multicomponent systems. It is particularly useful for visualizing phase equilibria in binary mixtures.

In a two-component system at fixed temperature and pressure, the total Gibbs free energy G can be expressed as a function of the composition variable n, representing the number of moles (or mole fraction) of one component. Plotting $G(n)$ reveals how the total free energy varies with composition.

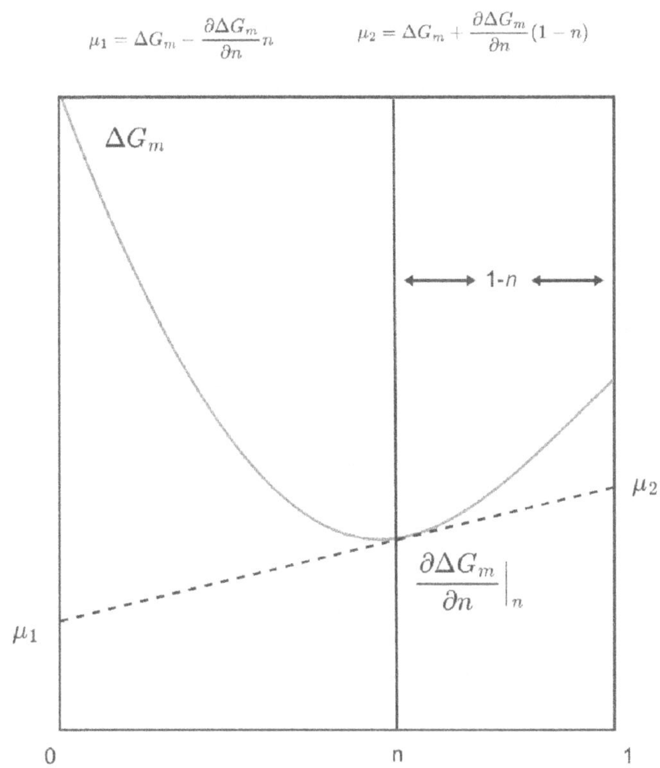

FIGURE 5.2 Graphical construction of the *method of intercepts* (after Gibbs). The Gibbs free energy G is plotted as a function of composition n at fixed T and P. The common tangent to the curve at points A and B identifies the chemical potentials of the two coexisting phases, given by the slope $\mu = (\partial G/\partial n)_{T,P}$. The points A and B represent the equilibrium compositions of the two phases. The tie line connecting them corresponds to the coexistence region.

The *chemical potential* of each component is given by the slope of the $G(n)$ curve:

$$\mu = \left(\frac{\partial G}{\partial n}\right)_{T,P}. \tag{5.35}$$

At equilibrium between two phases, the chemical potentials of each species must be equal in both phases. Graphically, this condition is represented by a common tangent to the $G(n)$ curve, as shown in Fig. 5.2. The points of tangency (A and B) correspond to the equilibrium compositions of the two coexisting phases. The horizontal projection between these points defines the *tie line*, whose endpoints give the relative phase compositions.

This graphical method allows one to identify regions of single-phase stability and two-phase coexistence directly from the curvature of the Gibbs free energy. The construction provides an intuitive understanding of how free energy minimization determines equilibrium in multicomponent systems and forms the basis for modern phase diagram analysis.

REMARK 5.8 The concept of the Gibbs free energy of mixing and Gibbs's method of intercepts together establish the thermodynamic foundation of solution theory and phase equilibria. They demonstrate that all spontaneous mixing and phase separation processes can be understood as the system's tendency to minimize its Gibbs free energy under the constraints of constant temperature and pressure.

Chemical Reactions and the Extent of Reaction

Chemical reactions offer a particularly clear illustration of how thermodynamics governs the transformation of matter. Macroscopically, the composition of a reacting system evolves continuously from an initial state toward equilibrium. This viewpoint was formalized by Théophile de Donder and later extended by Ilya Prigogine, giving rise to the modern theory of *irreversible thermodynamics*, in which reaction progress, chemical affinity, and entropy production are deeply interconnected [DR36, Pri68]. We wish to anticipate certain features of irreversible thermodynamics that we will discuss in detail in Volume II in order to show the extensions of thermodynamic to dynamical processes.

A general chemical reaction is written as

$$\sum_i \nu_i A_i = 0,$$

where ν_i are the *stoichiometric coefficients* (negative for reactants, positive for products). Stoichiometry ensures conservation of atoms and mass, but by itself does not specify reaction direction or rate.

To quantify how far a reaction has progressed, de Donder introduced the *degree of advancement* (or *extent of reaction*) ξ. For a closed system,

$$dn_i = \nu_i \, d\xi,$$

so that all species change consistently as the reaction proceeds. Given initial amounts $n_{i,0}$,

$$\xi = \frac{n_i - n_{i,0}}{\nu_i},$$

which is independent of the species chosen and therefore provides a single, global parameter describing reaction progress.

De Donder defined the *affinity* A as the thermodynamic force driving the reaction. At constant temperature, volume, and chemical potentials,

$$A = -\left(\frac{\partial \Omega}{\partial \xi}\right)_{T,V,\mu}, \tag{5.36}$$

where Ω is the grand potential. The infinitesimal work associated with the reaction is $A\,d\xi$, and at equilibrium $\partial\Omega/\partial\xi = 0$, so $A = 0$.

Because $dn_i = \nu_i \, d\xi$, one may also express the affinity in terms of chemical potentials:

$$A = -\sum_i \nu_i \mu_i.$$

Thus:

- $A > 0$: the reaction proceeds forward,
- $A < 0$: it proceeds backward,
- $A = 0$: the system is at equilibrium.

In a reacting system, the total entropy differential is decomposed into an exchange term (d_eS) and an internal term (d_iS):

$$dS = d_eS + d_iS.$$

For an isolated system ($d_eS = 0$), the second law requires $d_iS \geq 0$. Using the expression for A, de Donder obtained the fundamental relation

$$d_iS = \frac{A}{T}\, d\xi, \tag{5.37}$$

showing that entropy is produced whenever the reaction advances in the direction dictated by its affinity. Entropy production ceases only at equilibrium, where $A = 0$ and $d\xi = 0$.

This formulation provides a unified thermodynamic description of chemical change: reaction progress is governed by a generalized force (affinity), and the second law ensures that the system evolves spontaneously toward the state where that force vanishes.

Example: Extent of Reaction and Mass Conservation

Consider the Haber–Bosch process,

$$N_2 + 3\,H_2 \longrightarrow 2\,NH_3.$$

From the stoichiometric coefficients $\nu_{N_2} = -1$, $\nu_{H_2} = -3$, and $\nu_{NH_3} = 2$, the extent of reaction ξ satisfies

$$\frac{dn_{N_2}}{-1} = \frac{dn_{H_2}}{-3} = \frac{dn_{NH_3}}{2} = d\xi.$$

Thus, all mole-number changes are proportional to a single variable $d\xi$, which uniformly parametrizes the progress of the reaction.

Because mass is conserved in chemical reactions, summing over all species yields

$$dm = \sum_i \nu_i M_i\, d\xi = 0,$$

where M_i denotes the molar mass of species i. This implies the stoichiometric constraint

$$\sum_i \nu_i M_i = 0,$$

the macroscopic expression of Lavoisier's principle.

As a concrete illustration, consider methane combustion:

$$CH_4 + 2\,O_2 \longrightarrow CO_2 + 2\,H_2O.$$

With molar masses CH_4: 16 g/mol, O_2: 32 g/mol, CO_2: 44 g/mol, H_2O: 18 g/mol, the total mass of reactants,

$$16 + 2 \times 32 = 80 \text{ g/mol},$$

equals the mass of the products,

$$44 + 2 \times 18 = 80 \text{ g/mol}.$$

This verifies explicitly that mass is conserved.

Affinity and Chemical Potentials

Once the extent of reaction ξ is introduced, the *chemical affinity* A provides the thermodynamic force that drives the reaction. For a reaction

$$\sum_i \nu_i A_i = 0,$$

de Donder defined the affinity as

$$A = -\sum_i \nu_i \mu_i,$$

where μ_i are the chemical potentials and ν_i the stoichiometric coefficients (positive for products, negative for reactants). If $A > 0$, the reaction proceeds in the forward direction; if $A < 0$, in the reverse direction; and at equilibrium $A = 0$.

At constant temperature and pressure, the affinity is simply related to the Gibbs free energy change of the reaction:

$$A = -\Delta_r G.$$

Irreversible thermodynamics expresses the internal entropy production of a reacting system as

$$\frac{d_i S}{dt} = \frac{A}{T}\frac{d\xi}{dt} \geq 0,$$

making the direction of reaction directly thermodynamic. Thus a spontaneous reaction corresponds to $A\,d\xi > 0$, and entropy production vanishes only when the system reaches equilibrium.

For instance, for the reversible reaction

$$\mathrm{X} + \mathrm{Y} \rightleftharpoons 2\,\mathrm{Z},$$

the stoichiometric coefficients are $\nu_X = \nu_Y = -1$ and $\nu_Z = 2$. The affinity is therefore

$$A = \mu_X + \mu_Y - 2\mu_Z,$$

and the entropy production rate is

$$\frac{d_i S}{dt} = \frac{1}{T}(\mu_X + \mu_Y - 2\mu_Z)\frac{d\xi}{dt}.$$

At equilibrium, $A = 0$ and the entropy production rate vanishes.

If several independent reactions labeled by r occur simultaneously, each with extent ξ_r and affinity A_r, the total entropy production is additive:

$$\frac{d_i S}{dt} = \sum_r \frac{A_r}{T}\frac{d\xi_r}{dt} \geq 0.$$

This general expression forms the basis of the de Donder–Prigogine theory of irreversible thermodynamics.

REMARK 5.9 The de Donder–Prigogine formalism unifies stoichiometry, equilibrium thermodynamics, and kinetics by treating the extent of reaction as a generalized coordinate and the affinity as its conjugate thermodynamic force. Chemical reactions thus appear as dissipative macroscopic processes governed by the same conservation and entropy production principles that underlie heat conduction, diffusion, and other irreversible phenomena.

5.5 The Grand Potential and Open Systems

In the thermodynamic potentials discussed so far, the number of particles N was considered fixed. However, many real systems—such as chemical reactions, adsorption processes, or phase transitions—are *open systems* that exchange both energy and matter with their surroundings. To describe such systems, it is useful to introduce a potential whose natural variables include the *chemical potential* μ, rather than the total particle number N.

The chemical potential is a fundamental thermodynamic quantity that measures how the internal energy of a system changes when the number of particles varies. Mathematically, it is defined as

$$\mu = \left(\frac{\partial U}{\partial N}\right)_{S,V}, \tag{5.38}$$

that is, the change in internal energy when one particle is added to the system while keeping entropy and volume constant. Physically, μ represents the *energy cost or gain* of adding an additional particle. In a multicomponent system, each species has its own chemical potential μ_i, which determines its tendency to move, react, or change phase.

Intuitively, the chemical potential plays the same role for matter as temperature does for heat: just as heat flows from regions of high temperature to low temperature, particles flow from regions of high chemical potential to low chemical potential. Equilibrium between two phases or two subsystems is achieved when their chemical potentials are equal,

$$\mu_1 = \mu_2, \tag{5.39}$$

meaning that there is no net flow of particles between them. This condition governs not only diffusion and phase coexistence but also chemical equilibrium and electrochemical processes.

With this physical picture in mind, we now introduce the *grand potential*, the thermodynamic function that naturally describes open systems in terms of the variables (T, V, μ).

The *grand potential* Ω is defined for systems at constant temperature T, volume V, and chemical potential μ as

$$\Omega = \langle E\rangle - T\langle S\rangle - \mu\langle N\rangle, \tag{5.40}$$

where $\langle E\rangle$ is the average energy, $\langle S\rangle$ the entropy, and $\langle N\rangle$ the average number of particles. At equilibrium, the grand potential is minimized for given (T, V, μ):

$$\delta\Omega = 0. \tag{5.41}$$

This condition defines equilibrium in open systems, just as minimization of the Gibbs free energy G does at fixed T and P for closed systems.

Physically, Ω measures the amount of useful work a system can perform when both heat and particles may be exchanged with a reservoir. In statistical mechanics, Ω corresponds to the logarithm of the grand partition function and thus provides the thermodynamic link between microscopic particle statistics and macroscopic observables.

The grand potential can be obtained by a Legendre transformation of the internal energy $U(S, V, N)$:

$$\Omega = U - TS - \mu N. \tag{5.42}$$

Differentiating, we find

$$d\Omega = -S\,dT - P\,dV - N\,d\mu. \tag{5.43}$$

Hence, the grand potential is naturally expressed as a function of (T, V, μ), and its derivatives yield key measurable quantities:

$$S = -\left(\frac{\partial\Omega}{\partial T}\right)_{V,\mu}, \quad P = -\left(\frac{\partial\Omega}{\partial V}\right)_{T,\mu}, \quad N = -\left(\frac{\partial\Omega}{\partial \mu}\right)_{T,V}.$$

All other thermodynamic properties of open systems can be derived from Ω. In particular, the internal energy follows from

$$E = \Omega + TS + \mu N. \tag{5.44}$$

Applications and Examples

The grand potential plays a central role in describing matter under conditions of particle exchange. Important examples include:

- *Phase coexistence.* In a liquid–gas system at equilibrium, the equality of grand potentials ensures equal chemical potentials of the two phases:

$$\Omega_{\text{liquid}} = \Omega_{\text{gas}} \quad \Longleftrightarrow \quad \mu_{\text{liquid}} = \mu_{\text{gas}}. \tag{5.45}$$

- *Chemical equilibrium.* For a general reaction $A+B \rightleftharpoons C+D$, the equilibrium condition is

$$\Delta\Omega = \Omega_{\text{products}} - \Omega_{\text{reactants}} = 0, \tag{5.46}$$

  which expresses the balance of chemical potentials among all species.
- *Adsorption and electrochemical systems.* In surface reactions or electrode processes, Ω governs the exchange of particles (atoms, ions, or electrons) between the interface and its environment.

Because Ω naturally includes the chemical potential, it provides the most convenient potential for treating systems in which composition can change. At constant temperature and chemical potential, the pressure is related to the grand potential by

$$P = -\left(\frac{\partial\Omega}{\partial V}\right)_{T,\mu}, \tag{5.47}$$

and the total grand potential is proportional to the system's negative pressure times its volume:

$$\Omega = -PV. \tag{5.48}$$

Ellingham Diagrams

An *Ellingham diagram* is a graphical representation of the Gibbs free energy change $\Delta_r G$ for a chemical reaction as a function of temperature. It provides a powerful visual tool for analyzing the thermodynamic feasibility of oxidation–reduction (redox) reactions and metal extraction processes. First introduced by H. J. T. Ellingham in 1944 [Ell44], such diagrams are widely used in metallurgy and materials science to predict which reactions will proceed spontaneously at a given temperature.

For a general reaction,

$$aA + bB \to cC + dD, \tag{5.49}$$

the Gibbs free energy change is given by

$$\Delta_r G = \Delta_r H - T\Delta_r S, \tag{5.50}$$

where $\Delta_r H$ and $\Delta_r S$ are the reaction enthalpy and entropy, respectively. If both quantities are approximately constant over the temperature range of interest, $\Delta_r G$ varies linearly with temperature, producing a straight line on the diagram. The slope of each line is $-\Delta_r S$, and the intercept at $T = 0$ is $\Delta_r H$. Reactions with large negative slopes therefore have large positive entropies of reaction.

In an Ellingham diagram, the most thermodynamically favorable reaction at a given temperature is the one with the lowest Gibbs free energy (or equivalently, the most negative line).

The intersection points between lines indicate temperatures where two reactions have equal $\Delta_r G$, meaning that both reactions are equally favorable. Above this temperature, one reaction becomes spontaneous, while below it, the other dominates.

For oxidation or reduction reactions, the diagram shows which reducing agent can successfully reduce a metal oxide. For instance, in the reduction of metal oxides by carbon,

$$\mathrm{C} + \frac{1}{2}\mathrm{O_2} \to \mathrm{CO}, \quad \text{or} \quad \mathrm{C} + \mathrm{O_2} \to \mathrm{CO_2}, \tag{5.51}$$

the intersection between the CO/CO_2 line and the M/MO line (where M is a metal) marks the minimum temperature at which carbon can reduce the oxide to metallic form. This principle underlies the thermodynamics of processes such as iron smelting in blast furnaces.

The Ellingham diagram is essentially a plot of the negative of the reaction affinity $A = -\Delta_r G$ versus temperature. A reaction proceeds spontaneously when $A > 0$, or equivalently when $\Delta_r G < 0$. Hence, the diagram can also be viewed as a graphical representation of the temperature dependence of chemical affinities.

REMARK 5.10 Ellingham diagrams transform the abstract relationship between Gibbs free energy, entropy, and temperature into an intuitive graphical tool. They allow rapid prediction of the conditions under which chemical or metallurgical reactions become thermodynamically feasible, linking the microscopic notion of affinity with practical, industrial applications.

5.6 Thermodynamics of Magnets

Magnetism has been known since antiquity. The ancient Greeks described naturally magnetized stones—lodestones—found in Magnesia, which gave rise to the term *magnet.* In China, similar materials led to the invention of the magnetic compass. However, a quantitative theory of magnetism only emerged in the nineteenth and twentieth centuries, when electromagnetism and thermodynamics were unified. Pioneering work by Pierre Weiss [Wei07] and later Lev Landau [Lan37] established the foundations of magnetic thermodynamics and phase transitions.

Magnetic Work and the First Law

We consider a magnetic body of total magnetic moment $\vec{M}$ immersed in an external magnetic field $\vec{h}$ and in contact with a thermal reservoir at temperature T. If the system is thermally isolated and the magnetic field is changed externally, the only contribution to the change in internal energy arises from magnetic work:

$$dU = \delta W = \mu_0 \, \vec{h} \cdot d\vec{M}, \tag{5.52}$$

where μ_0 is the magnetic permeability of free space. The sign convention here implies that work is done *on* the system when $\vec{M}$ tends to align with $\vec{h}$.

When the system can also exchange heat with the environment, the first law reads

$$dU = \delta Q + \mu_0 \, \vec{h} \cdot d\vec{M}, \tag{5.53}$$

where δQ is the infinitesimal heat absorbed from the bath. For reversible processes, $\delta Q = T\,dS$, and thus

$$dU = T\,dS + \mu_0 \, \vec{h} \cdot d\vec{M}. \tag{5.54}$$

This expression is the magnetic analog of the mechanical first law, with $\vec{h}$ playing the role of a generalized force and $\vec{M}$ that of a generalized displacement.

Adiabatic Demagnetization: A Landmark Experiment in Magnetic Thermodynamics

Adiabatic demagnetization was one of the great experimental achievements of early low–temperature physics. Proposed independently by Peter Debye in 1926 and demonstrated experimentally by William Francis Giauque in the early 1930s, the method provided the first practical route to temperatures below 1 K, and it directly showcased magnetic work.

A paramagnetic salt (such as cerium magnesium nitrate) is first placed in a strong magnetic field while in thermal contact with a bath. The field aligns the magnetic moments, reducing the magnetic entropy. Once the sample reaches equilibrium, it is thermally isolated. If the applied magnetic field is then reduced *adiabatically*, the spins are allowed to randomize, increasing magnetic entropy. Because the process is adiabatic, this increase in entropy must be compensated by a decrease in the lattice temperature, producing strong cooling.

In the magnetic first law $dU = T\,dS + \mu_0 H\,dM$ the term $\mu_0 H\,dM$ represents the magnetic work performed as spins reorient. During ideal adiabatic demagnetization, $dS = 0$ and the change in M associated with decreasing H leads to a reduction in internal energy that manifests as a drop in temperature. This experiment powerfully demonstrates the coupling between magnetic degrees of freedom and thermal energy.

Debye's and Giauque's theoretical proposal and Giauque's and McDougall experimental realization established magnetic cooling as a central tool of cryogenic physics [Deb26, GM33]. The method is still used today in dilution refrigerators and nuclear demagnetization stages to reach microkelvin temperatures.

Helmholtz and Gibbs Free Energies

For systems at fixed temperature, it is useful to introduce the Helmholtz free energy,

$$F = U - TS, \tag{5.55}$$

whose differential follows from Eq. (5.54) as

$$dF = -S\,dT + \mu_0\,\vec{h}\cdot d\vec{M}. \tag{5.56}$$

At constant temperature, the system evolves to minimize F. This means that the magnetic moment aligns in a way that reduces the total free energy, corresponding to an increase in entropy at constant internal energy.

In many experimental situations, it is more convenient to fix the external magnetic field $\vec{h}$ rather than the magnetization $\vec{M}$. We therefore perform a Legendre transformation to define the Gibbs free energy,

$$G = F - \mu_0\,\vec{h}\cdot\vec{M} = U - TS - \mu_0\,\vec{h}\cdot\vec{M}. \tag{5.57}$$

Its differential form is

$$dG = -S\,dT - \mu_0\,\vec{M}\cdot d\vec{h}. \tag{5.58}$$

At fixed T and $\vec{h}$, equilibrium corresponds to the minimum of G. The equation of state follows directly:

$$\left(\frac{\partial G}{\partial \vec{h}}\right)_T = -\mu_0\,\vec{M}. \tag{5.59}$$

This relation defines the magnetization as the thermodynamic conjugate variable to the magnetic field.

Stability and Phase Transitions

A stable equilibrium requires that small deviations in $\vec{M}$ increase the free energy. At constant temperature,

$$\left(\frac{\partial^2 F}{\partial \vec{M}^2}\right)_T > 0. \tag{5.60}$$

Violation of this condition signals an instability, as occurs in ferromagnets near their critical (Curie) temperature, where the system develops spontaneous magnetization. Landau's theory of phase transitions models this behavior by expanding the free energy as a power series in $\mathbf{M}$, capturing the emergence of multiple minima corresponding to different magnetic domains.

In systems with metastable states, the free energy landscape $G^*(\vec{M}, \vec{h})$ may exhibit several local minima separated by barriers. Thermal activation or external fields can drive transitions between these states, leading to hysteresis phenomena characteristic of ferromagnetic materials.

Thermodynamic Potentials and Magnetic Equilibrium

At constant temperature and magnetic field, equilibrium requires that

$$\frac{\delta G^*}{\delta \vec{M}} = 0. \tag{5.61}$$

This extremum condition expresses the balance between the external magnetic field and the internal magnetic interactions that tend to align or oppose it. It provides the macroscopic thermodynamic basis for magnetization curves and magnetic susceptibility measurements.

The thermodynamics of magnetism reached maturity in the early twentieth century through the works of Weiss, who introduced the concept of the *molecular field* to explain spontaneous magnetization, and Landau, who developed the general theory of phase transitions and critical phenomena. These thermodynamic descriptions laid the groundwork for modern statistical mechanics of spin systems. While classical thermodynamics captures equilibrium behavior, non-equilibrium effects such as hysteresis, Barkhausen noise, and domain-wall motion require a dynamical treatment using the Landau–Lifshitz–Gilbert equation and related models. We will return to these topics in later chapters, especially when discussing Ising and spin models.

REMARK 5.11 Magnetism exemplifies the deep unity between microscopic order and macroscopic thermodynamics. The alignment of magnetic moments under external fields, the formation of domains, and the onset of spontaneous magnetization all arise from the same variational principles that govern equilibrium in gases, liquids, and solids.

5.7 Hurricanes

In the previous chapter, we mentioned that the *rainmaker* proposed by Peixoto was thermodynamically analogous to a natural process: the hurricane. The idea that hurricanes can be modeled as *heat engines* operating between the warm ocean surface and the cold upper atmosphere was first formulated by Kerry A. Emanuel in the 1980s [Ema86, Ema03]. Although the full theory of tropical cyclones is extremely complex and involves fluid dynamics, moisture transport, and turbulence, the thermodynamic approach captures their essential energy balance.

A hurricane (or tropical cyclone) is a large-scale atmospheric vortex that forms over warm tropical waters. Depending on the region, it is called a *hurricane* (North Atlantic and Eastern Pacific) or a *typhoon* (Western Pacific). Formation requires sea surface temperatures above about 26 °C, which is why global warming increases the likelihood of their occurrence.

Thermodynamic Model of a Hurricane

A mature hurricane can be idealized as a thermodynamic cycle composed of moist air, water vapor, and droplets in approximate equilibrium. While true equilibrium is never achieved in such a vast, dynamic system, a small radial–vertical section of the storm can be approximated as a heat engine.

The process begins with the evaporation of warm seawater, which injects both moisture and entropy into the boundary layer of the atmosphere. The moist air rises, expands, and cools adiabatically according to the barometric law. As it ascends, water vapor condenses into clouds and releases latent heat, sustaining the upward motion. In the upper troposphere, heat is radiated to space, and the cooled air descends, completing the cycle. The overall process can be compared to a Carnot engine operating between a hot reservoir (the ocean surface) and a cold reservoir (the upper atmosphere) [Ema06].

Typical temperatures are $T_h \approx 300$ K at the sea surface ($\sim$27 °C) and $T_c \approx 200$ K at the top of the storm (-73 °C). The idealized hurricane cycle proceeds through the following stages:

- *A–B (Heat input):* Warm seawater evaporates at roughly constant temperature, adding moisture and heat to the boundary layer.
- *B–C (Adiabatic expansion):* The moist air rises and expands adiabatically along the eyewall, cooling as it ascends.
- *C–D (Heat rejection):* In the upper atmosphere, heat is radiated away, approximately at constant temperature, through infrared emission to space.
- *D–A (Adiabatic compression):* The dry air descends, compresses, and warms adiabatically, completing the cycle.

These four steps correspond closely to those of a Carnot cycle, though only approximately realized in nature. Nevertheless, the analogy provides a useful means to estimate the efficiency and total energy involved.

We recall that in the Carnot picture, the maximum efficiency of energy conversion is

$$\eta_c = 1 - \frac{T_c}{T_h}. \tag{5.62}$$

With $T_h \simeq 300$ K and $T_c \simeq 200$ K, we find $\eta_c \approx 0.33$. This efficiency represents the fraction of absorbed heat that can, in principle, be converted into mechanical work (kinetic energy of the wind). Assuming that the total kinetic energy of the hurricane is derived from the absorbed heat Q_{in}, we can write

$$K = \eta_c \, Q_{\text{in}}. \tag{5.63}$$

While this is an idealization, it gives a useful first-order estimate of the enormous energy involved in tropical cyclones.

The Carnot model provides an upper bound on the efficiency of hurricanes but neglects many realistic effects, such as turbulent dissipation, moisture asymmetries, and non-equilibrium processes. Several authors have discussed the limitations of representing hurricanes as perfect Carnot engines [MGL08, MSC22]. In reality, entropy production within the system, frictional losses, and complex heat–moisture coupling reduce the actual efficiency well below the theoretical limit.

Nevertheless, this idealization elegantly demonstrates how natural phenomena (even complex systems like hurricanes) can be analyzed, although in an approximate manner, through the lens of thermodynamics. The hurricane is thus a striking example of how nature approaches the principles of the heat engine on a planetary scale. We will approach these principles to even a larger scale in the next section.

REMARK 5.12 Emanuel's theory transformed the understanding of tropical cyclones by showing that their intensity can be predicted using simple thermodynamic principles. The hurricane, seen as a heat engine driven by temperature gradients between the sea and the sky, beautifully illustrates how energy conversion, entropy production, and atmospheric dynamics are unified by the laws of thermodynamics.

5.8 Stars

Polytropic Transformations

In addition to adiabatic processes, which are characterized by

$$dQ = 0, \tag{5.64}$$

stars and other self-gravitating gaseous systems can undergo more general transformations known as *polytropic processes.* The concept of polytropic transformations was first introduced by Gustav Zeuner [Zeu06] and later applied to astrophysical systems by Robert Emden in his theory of stellar structure. These processes generalize adiabatic transformations by allowing a nonzero but controlled heat exchange with the environment.

Let us consider a gas obeying

$$\frac{dU}{dT} = C_V, \tag{5.65}$$

so that, from the first law of thermodynamics,

$$dQ = dU + dW = C_V\, dT + P\, dV. \tag{5.66}$$

For an adiabatic transformation, $dQ = 0$. For a *polytropic transformation,* we instead assume that the heat exchanged per unit temperature change is constant:

$$\frac{dQ}{dT} = c, \tag{5.67}$$

where c is a constant independent of T and V (distinct from C_P or C_V). Substituting Eq. (5.67) into the first law gives

$$c\, dT = C_V\, dT + P\, dV, \tag{5.68}$$

or equivalently,

$$(c - C_V)\, dT = P\, dV. \tag{5.69}$$

Equation of State for a Polytropic Process

For an ideal gas, $P = Nk_B T/V$, and substituting this relation into Eq. (5.69) yields

$$(c - C_V)\, \frac{dT}{T} = Nk_B\, \frac{dV}{V}. \tag{5.70}$$

Integrating between two states (T_1, V_1) and (T_2, V_2), we obtain

$$(c - C_V)\, \ln\left(\frac{T_2}{T_1}\right) = Nk_B\, \ln\left(\frac{V_2}{V_1}\right). \tag{5.71}$$

Using $Nk_B = C_P - C_V$, this becomes

$$(c - C_V)\, \ln\left(\frac{T_2}{T_1}\right) = (C_P - C_V)\, \ln\left(\frac{V_2}{V_1}\right), \tag{5.72}$$

which reduces to the adiabatic relation when $c = 0$. We can thus define a *modified adiabatic exponent*:

$$\gamma' = \frac{C_P - c}{C_V - c}. \tag{5.73}$$

By analogy with adiabatic processes, the polytropic equations of state can then be written as

$$P V^{\gamma'} = \text{const}, \qquad P^{1-\gamma'} T^{\gamma'} = \text{const}, \qquad T V^{\gamma'-1} = \text{const}. \tag{5.74}$$

These relations show that polytropic transformations behave like adiabatic ones, but with an effective exponent γ' that depends on the parameter c controlling the heat exchange.

Entropy in a Polytropic Transformation

Unlike an adiabatic process, a polytropic transformation involves a finite change in entropy. From the definition $dS = dQ/T$ and Eq. (5.67), we have

$$dS = c\,\frac{dT}{T}. \tag{5.75}$$

Alternatively, using the relation between γ' and $\gamma = C_P/C_V$, this can be written as

$$dS = C_V\,\frac{\gamma' - \gamma}{\gamma' - 1}\,\frac{dT}{T}. \tag{5.76}$$

For an adiabatic transformation, $\gamma' = \gamma$ and $dS = 0$, as expected. Integrating, the entropy change between two temperatures T_1 and T_2 is

$$S(T_2) - S(T_1) = c\,\ln\left(\frac{T_2}{T_1}\right). \tag{5.77}$$

Polytropic transformations encompass a continuous range of processes between adiabatic and isothermal limits. When $c = 0$, the process is purely adiabatic; when $c = C_P$, it becomes isothermal. Intermediate cases describe systems where heat is exchanged with the environment at a rate proportional to the temperature change.

In stellar interiors, polytropic processes are used to model the balance between pressure, gravity, and temperature gradients. Assuming that the pressure and density are related by a *polytropic equation of state*,

$$P = K\,\rho^{1+1/n}, \tag{5.78}$$

with n the polytropic index and K a constant, leads to the *Lane–Emden equation*, which describes the radial structure of stars in hydrostatic equilibrium. This connection between thermodynamics and astrophysics will be discussed in the following sections.

REMARK 5.13 Polytropic transformations generalize adiabatic behavior by including controlled heat exchange. Their introduction by Zeuner and application by Emden to stellar physics illustrate how thermodynamic principles extend naturally from laboratory gases to the structure and evolution of stars.

Hydrostatic Equilibrium

A star exists in a delicate balance between two competing effects: the inward pull of gravity and the outward push of pressure. When these two forces exactly balance, the star is said to be in *hydrostatic equilibrium*. This condition provides the mechanical counterpart to the thermodynamic relations described in the previous section.

The balance of forces in a spherically symmetric star is expressed by

$$\frac{dP(r)}{dr} = -\rho(r)\frac{GM(r)}{r^2}, \tag{5.79}$$

where $P(r)$ is the local pressure, $\rho(r)$ the mass density, and

$$M(r) = \int_0^r 4\pi r'^2 \rho(r')\, dr' \tag{5.80}$$

is the mass enclosed within radius r. Equation (5.79) ensures that every thin spherical shell of the star is in mechanical equilibrium.

If the thermodynamic state of the stellar material is described by a *polytropic equation of state*,

$$P = K\rho^{1+1/n}, \tag{5.81}$$

where K is a constant and n the polytropic index, then Eqs. (5.79) and (5.81) can be combined to yield the *Lane–Emden equation* for the dimensionless function $\theta(\xi)$:

$$\frac{1}{\xi^2}\frac{d}{d\xi}\left(\xi^2\frac{d\theta}{d\xi}\right) = -\theta^n, \tag{5.82}$$

where ξ is a dimensionless radial coordinate proportional to r. This equation, first derived by Emden [Emd07] and later refined by Eddington and Chandrasekhar, describes the internal structure of self-gravitating polytropic spheres.

The Lane–Emden equation provides analytical or numerical models for different types of stars:

- $n = 0$ corresponds to a star of uniform density.
- $n = 1.5$ describes fully convective stars.
- $n = 3$ models massive, radiation-dominated stars.

Each case yields a distinct radial density and pressure profile, illustrating how thermodynamics and gravity cooperate to maintain stellar equilibrium.

REMARK 5.14 The thermodynamics of hydrostatic equilibrium link microscopic equations of state with the macroscopic stability of stars. Chandrasekhar's seminal work on stellar equilibrium and degeneracy [Cha39] extended this framework to white dwarfs, showing that beyond a certain mass limit, pressure can no longer balance gravity, leading to gravitational collapse. Thus, hydrostatic equilibrium represents not only a condition of balance but also a gateway to understanding stellar evolution and instability.

5.9 Entropy and Callen's Reformulation of Thermodynamics

In the 1960s, Herbert B. Callen proposed a remarkably clear and general reformulation of classical thermodynamics [Cal60, Cal85]. His approach is often referred to as *thermostatics*, to emphasize that equilibrium thermodynamics concerns static states rather than time-dependent processes. Callen's formulation is conceptually parallel to statistical mechanics and provides a rigorous foundation for extending thermodynamics to more complex or non-equilibrium systems.

Traditional thermodynamics is built around two complementary but sometimes awkwardly presented principles: the *maximum entropy principle* and the *minimum energy principle*. Callen's reformulation unifies these into a precise set of postulates that treat energy and entropy symmetrically and place the concept of equilibrium at the center of thermodynamics. In this view, all equilibrium states can be obtained from the maximization of the entropy function or, equivalently, the minimization of a suitable thermodynamic potential under the given constraints.

Thermodynamics deals with the macroscopic properties of matter, especially those related to temperature, energy, and mechanical work. The systems of interest are typically large enough that surface and quantum effects can be neglected, and are characterized by a small number of macroscopic variables. Callen defines such systems as *simple systems*: homogeneous, isotropic, uncharged, chemically inert, and not subject to external fields. These assumptions make it possible to focus on the fundamental relationships among energy, entropy, and the extensive variables of the system.

The practical motivation for Callen's framework is that it provides a logically consistent foundation for all of thermodynamics without recourse to microscopic models. Every measurable thermodynamic property of a system can be derived from a single scalar function—the *entropy*—and its derivatives with respect to the appropriate extensive parameters.

Callen's Postulates

Callen's formulation of equilibrium thermodynamics rests on four postulates:

1. *Existence of equilibrium states.* The equilibrium states of a simple macroscopic system are completely specified by a set of extensive parameters such as the internal energy U, volume V, and the mole numbers N_k of its chemical components.
2. *Existence of entropy.* There exists a function, called the *entropy* $S(U, V, \{N_k\})$, which is defined for all equilibrium states of the system. The entropy is an extensive, continuous, and differentiable function of its arguments. In the absence of internal constraints, the equilibrium values of the extensive parameters are those that *maximize* the total entropy of the system.
3. *Additivity and monotonicity.* The entropy of a composite system is the sum of the entropies of its subsystems. It is a monotonically increasing function of the internal energy U, ensuring that temperature—defined by $T^{-1} = \partial S/\partial U$—is always positive in equilibrium.
4. *Definition of zero entropy.* There exists a state of the system, corresponding to the third law of thermodynamics, in which

$$\left(\frac{\partial U}{\partial S}\right)_{V,N} = 0. \tag{5.83}$$

   This defines the reference point at which the entropy vanishes.

These four postulates are sufficient to derive all of equilibrium thermodynamics. The second postulate provides the fundamental criterion for equilibrium: the system evolves spontaneously toward the state of maximal entropy, consistent with its external constraints.

Extensive and Intensive Variables

A thermodynamic system is described by its extensive parameters—those that scale with the size of the system—such as U, V, and N_k. Corresponding to these are the intensive parameters—temperature, pressure, and chemical potential—defined as partial derivatives of U or S:

$$\frac{1}{T} = \left(\frac{\partial S}{\partial U}\right)_{V,N}, \qquad \frac{P}{T} = \left(\frac{\partial S}{\partial V}\right)_{U,N}, \qquad -\frac{\mu_k}{T} = \left(\frac{\partial S}{\partial N_k}\right)_{U,V}. \tag{5.84}$$

The signs and ratios ensure that energy, work, and matter flow from high to low values of the respective intensive quantities until equilibrium is reached.

Constraints and Simple Systems

A system can exchange heat, work, or matter with its surroundings depending on the nature of its enclosing *walls.* A wall that restricts a given extensive parameter is called *adiabatic* or *impermeable* for that variable; one that allows free exchange is called *diathermal* or *permeable.* If energy, volume, and composition are all restricted, the system is said to be *closed.* When several simple systems are combined into a composite system, the constraints that prevent exchanges between subsystems are called *internal constraints.* If some of these restrictions are removed, the system evolves to a new equilibrium by redistributing its extensive quantities so as to maximize the total entropy.

The Energy and Entropy Representations

For an infinitesimal quasi-static process in a simple system with constant mole numbers, the heat absorbed is

$$\delta Q = dU + P\,dV, \tag{5.85}$$

and the mechanical work is $\delta W = -P\,dV$. From the second postulate, the heat exchanged in a reversible process is $\delta Q = T\,dS$, so that

$$dU = T\,dS - P\,dV. \tag{5.86}$$

This is the differential form of the first law for a simple, closed system. It defines the *energy representation* of thermodynamics, where $U(S, V, N)$ is the fundamental relation. Alternatively, one may use the *entropy representation* $S(U, V, N)$, from which all thermodynamic quantities and equations of state can be derived by differentiation.

REMARK 5.15 Callen's reformulation elevates entropy to the central organizing principle of thermodynamics. It replaces the somewhat heuristic traditional laws with a set of logically complete postulates and places equilibrium as a condition of entropy maximization under constraints. This modern, axiomatic viewpoint not only clarifies classical thermodynamics but also provides a natural gateway to statistical mechanics and the study of non-equilibrium systems.

Equivalence of the Energy and Entropy Representations

Postulate 2 of Callen's formulation asserts that equilibrium corresponds to the *maximum entropy* compatible with external constraints. Since the entropy is extensive and additive, for a composite system composed of subsystems labeled by α, we may write

$$S = \sum_{\alpha} S_\alpha(U_\alpha, V_\alpha, \{N_{k,\alpha}\}), \tag{5.87}$$

where α indexes each subsystem, U_α is its energy, V_α its volume, and $N_{k,\alpha}$ the mole number of component k.

Because the entropy must be a monotonically increasing function of energy, we require

$$\left(\frac{\partial S}{\partial U}\right)_{V,N} > 0. \tag{5.88}$$

This guarantees that $S(U, V, N)$ can be inverted to obtain $U(S, V, N)$. Thus, the two formulations—entropy maximization and energy minimization—are completely equivalent descriptions of equilibrium thermodynamics.

In the *energy representation*, the total differential of the internal energy is

$$dU = \left(\frac{\partial U}{\partial S}\right)_{V,N} dS + \left(\frac{\partial U}{\partial V}\right)_{S,N} dV + \sum_j \left(\frac{\partial U}{\partial N_j}\right)_{S,V} dN_j. \tag{5.89}$$

Defining

$$T = \left(\frac{\partial U}{\partial S}\right)_{V,N}, \qquad P = -\left(\frac{\partial U}{\partial V}\right)_{S,N}, \qquad \mu_j = \left(\frac{\partial U}{\partial N_j}\right)_{S,V}, \tag{5.90}$$

we obtain the familiar form of the First Law:

$$dU = T\,dS - P\,dV + \sum_j \mu_j\,dN_j. \tag{5.91}$$

For quasi-static processes with constant composition, $dQ = T\,dS$, so the heat absorbed increases the entropy. Temperature and pressure are thus functions of the system's extensive variables.

The Entropy Representation

In the *entropy representation*, the fundamental differential is obtained by inverting Eq. (5.91):

$$\begin{aligned} dS &= \left(\frac{\partial S}{\partial U}\right)_{V,N} dU + \left(\frac{\partial S}{\partial V}\right)_{U,N} dV + \sum_j \left(\frac{\partial S}{\partial N_j}\right)_{U,V} dN_j \\ &= \frac{1}{T}\left(dU + P\,dV - \sum_j \mu_j\,dN_j\right). \end{aligned} \tag{5.92}$$

This shows that in either representation, the complete thermodynamic information is encoded in a single fundamental relation—either $U(S, V, N)$ or $S(U, V, N)$.

Equilibrium Between Subsystems

Consider a closed composite system composed of two subsystems, 1 and 2, separated by internal walls. The total entropy is

$$S = S_1(U_1, V_1, N_1) + S_2(U_2, V_2, N_2), \tag{5.93}$$

with $U_1 + U_2 = U_{\text{tot}}$ constant. Differentiating, we obtain

$$dS = \frac{dU_1}{T_1} + \frac{dU_2}{T_2}. \tag{5.94}$$

Because $dU_1 = -dU_2$, the condition for maximum entropy ($dS = 0$) yields

$$T_1 = T_2. \tag{5.95}$$

Thus, equilibrium between subsystems implies equal temperatures. Repeating the argument for mechanical exchange yields equality of pressures, and for chemical exchange, equality of chemical potentials:

$$\mu_1 = \mu_2. \tag{5.96}$$

Hence, thermal, mechanical, and chemical equilibrium emerge naturally from the entropy maximization principle.

When $\mu_1 > \mu_2$, material spontaneously flows from subsystem 1 to subsystem 2; similarly, heat flows from high to low temperature, and expansion occurs from high to low pressure. All these processes increase total entropy until equilibrium is reached.

Energy Minimization Principle

By an entirely analogous reasoning, the same equilibrium conditions can be derived from the *energy minimization principle.* For a closed composite system,

$$U = U_1(S_1, V_1, N_1) + U_2(S_2, V_2, N_2), \tag{5.97}$$

and at fixed S_{tot}, V_{tot}, and N_{tot}, the total differential is

$$dU = T_1\, dS_1 + T_2\, dS_2. \tag{5.98}$$

At equilibrium, $dU = 0$ implies $T_1 = T_2$, again recovering the same conditions derived from entropy maximization. Thus, entropy maximization and energy minimization are equivalent, depending on whether the independent variables are extensive or intensive.

Legendre Transformations and Thermodynamic Potentials

Formally, thermodynamic potentials are obtained by performing Legendre transforms on the fundamental relation $\eta = \eta(M_1, \ldots, M_k)$, where the M_k are extensive variables and their conjugate intensive variables are

$$P_k = \frac{\partial \eta}{\partial M_k}. \tag{5.99}$$

The Legendre transform introduces new potentials of the form

$$\psi = \eta - \sum_k P_k M_k, \tag{5.100}$$

whose differential is

$$d\psi = -\sum_k M_k\, dP_k. \tag{5.101}$$

In thermodynamics, these potentials correspond to physically meaningful functions:

- *Helmholtz free energy:* $F = U - TS$, with $dF = -S\,dT - P\,dV + \sum_k \mu_k\, dN_k$. At constant T, V, and N_k, equilibrium minimizes F.
- *Enthalpy:* $H = U + PV$, with $dH = T\,dS + V\,dP + \sum_k \mu_k\, dN_k$. Equilibrium at constant S, P, and N_k minimizes H.
- *Gibbs free energy:* $G = U - TS + PV$, with $dG = -S\,dT + V\,dP + \sum_k \mu_k\, dN_k$. Equilibrium at constant T, P, and N_k minimizes G.

The Legendre transformations exchange pairs of conjugate variables (extensive $\leftrightarrow$ intensive) and introduce the minus sign that connects the dual principles of entropy maximization and energy minimization.

REMARK 5.16 The equivalence between the entropy and energy formulations expresses the deep duality underlying thermodynamics. Every equilibrium state can be found either by maximizing entropy at fixed energy or by minimizing the appropriate energy function at fixed constraints. This dual structure, made explicit in Callen's formulation, forms the mathematical backbone of both equilibrium and non-equilibrium thermodynamics.

5.10 Beyond Equilibrium: Irreversible Processes and Prigogine's Formulation of the Second Law

Classical thermodynamics, as developed in the nineteenth century, primarily concerns systems in or near equilibrium, and the quasi-static processes connecting them. In this equilibrium context—sometimes more accurately called *thermostatics*, as emphasized by Callen—the second law is expressed through the equality

$$dS = \frac{\delta Q}{T}. \tag{5.102}$$

However, most physical processes are *irreversible.* Heat flows across gradients, diffusion equalizes concentrations, and chemical reactions proceed toward equilibrium—all occurring in finite time and away from reversible paths. In such processes, the total entropy increases, but the traditional formalism provides no quantitative expression for how irreversibility arises or how entropy is *produced* inside the system.

Entropy Production and the Modern Second Law

Prigogine's central contribution was to extend the second law to systems arbitrarily far from equilibrium by explicitly separating *entropy exchange* with the environment from *entropy production* within the system itself [Pri68]. Following Clausius, we write

$$dS \geq \frac{\delta Q}{T}, \tag{5.103}$$

but introduce the decomposition

$$dS = d_e S + d_i S, \tag{5.104}$$

where

- $d_e S$ is the entropy exchanged with the environment (via heat or matter flows),
- $d_i S$ is the *entropy produced internally* by irreversible processes.

The second law now takes the local and quantitative form

$$d_i S \geq 0, \tag{5.105}$$

with equality only for reversible processes. This is the modern, generalized expression of the second law of thermodynamics.

As we will show in detail in the second volume of this series, irreversible processes can be expressed as *fluxes* J_k driven by conjugate *thermodynamic forces* F_k. The instantaneous rate of entropy production is then

$$\frac{d_i S}{dt} = \sum_k F_k J_k \geq 0, \tag{5.106}$$

where each term in the sum corresponds to a specific irreversible mechanism:

- *Heat conduction:* $F = \nabla(1/T), \quad J \to \vec{q}$ (heat flux);
- *Matter diffusion:* $F = -\nabla(\mu/T), \quad J \to \vec{J}_N$ (particle flux);
- *Chemical reactions:* $F = A/T, \quad J \to d\xi/dt$, where A is the affinity and ξ the extent of reaction.

Each product $F_k J_k$ represents a rate of irreversible dissipation and contributes positively to entropy production. Equation (5.106) thus embodies the essence of the modern second law: *entropy is produced whenever generalized fluxes act against their conjugate forces.*

This decomposition naturally distinguishes three types of systems, depending on whether energy or matter can cross their boundaries:

Isolated systems

In this case, no energy or matter exchange occurs:

$$d_e S = 0, \quad \Rightarrow \quad dS = d_i S \geq 0. \tag{5.107}$$

Entropy increases solely due to internal irreversible processes.

Closed systems

Energy may be exchanged as heat, but not matter:

$$d_e S = \frac{\delta Q}{T}, \qquad d_i S \geq 0. \tag{5.108}$$

The total entropy change has two contributions: entropy flow from the environment and entropy produced internally.

Open systems

Both energy and matter can be exchanged with the surroundings:

$$d_e S = \frac{dU + P\,dV}{T} + (d_e S)_{\text{matter}}, \qquad d_i S \geq 0, \tag{5.109}$$

where $(d_e S)_{\text{matter}}$ represents the entropy carried by inflowing or outflowing matter. Here again, $d_i S$ captures only the *internal* production of entropy due to irreversible processes such as chemical reactions, viscosity, or conduction.

Local Entropy Balance

Consider two subsystems, 1 and 2, that can exchange energy, volume, or matter across a common boundary. The total entropy change is

$$dS = dS_1 + dS_2 = (d_e S_1 + d_i S_1) + (d_e S_2 + d_i S_2). \tag{5.110}$$

Since the exchange terms are equal and opposite,

$$d_e S_1 = -\,d_e S_2, \qquad \Rightarrow \qquad d_e S_1 + d_e S_2 = 0. \tag{5.111}$$

Therefore, the total entropy change reduces to

$$dS = d_i S_1 + d_i S_2 = d_i S \geq 0, \tag{5.112}$$

demonstrating that entropy production is *strictly local.* No process in one subsystem can reduce the entropy of another; irreversible processes always contribute non-negatively to the total entropy of the composite system.

Examples include:

- Heat flowing from a hot body to a colder one increases entropy in both;
- Diffusion across a concentration gradient increases entropy on both sides;
- Chemical reactions or phase transitions within each subsystem generate positive entropy locally.

REMARK 5.17 Prigogine's reformulation of thermodynamics transforms the second law from a global inequality into a local, quantitative law of evolution. It identifies the sources of irreversibility and establishes entropy production as a measurable quantity, bridging the gap between equilibrium thermodynamics and real processes. This framework laid the foundation for modern non-equilibrium thermodynamics and for the study of *dissipative structures*—organized patterns emerging far from equilibrium.

Local Irreversibility and Entropy Production

Example. Consider two gas containers, one at high pressure and the other at low pressure, separated by a permeable membrane. When the valve is opened, particles diffuse from the high-pressure side to the low-pressure side. During this process, the entropy of each subsystem increases:

$$d_i S_1 = \int \frac{\delta Q_1}{T_1} + (\text{diffusion term}),$$
$$d_i S_2 = \int \frac{\delta Q_2}{T_2} + (\text{diffusion term}).$$

Each subsystem satisfies $d_i S_i \geq 0$, and therefore the total entropy production

$$d_i S = d_i S_1 + d_i S_2 > 0. \tag{5.113}$$

The increase of entropy in both subsystems reflects the local and additive nature of irreversibility.

The cornerstone of Prigogine's non-equilibrium formulation is the *hypothesis of local equilibrium.* Although a system as a whole may be far from equilibrium, each infinitesimal element is assumed to be in local thermodynamic equilibrium. This allows macroscopic fields such as entropy, temperature, and chemical potential to be defined locally:

$$S = \int s(\vec{r}, t)\, d^3 r, \tag{5.114}$$

where $s(\vec{r}, t)$ is the local entropy density. The local balance equation for entropy reads

$$\frac{\partial s}{\partial t} + \nabla \cdot \vec{J}_s = \sigma(\vec{r}, t), \tag{5.115}$$

where $\mathbf{J}_s$ is the entropy flux and $\sigma(\vec{r}, t) \geq 0$ is the local entropy production rate. This equation encapsulates the fact that entropy may flow across regions but can only be *produced*, never destroyed, within any local volume element.

A key feature of Prigogine's theory is that entropy production is non-negative in every subsystem:

$$d_i S = d_i S_1 + d_i S_2 \geq 0, \qquad \text{with each } d_i S_i \geq 0. \tag{5.116}$$

Irreversibility is therefore *local and additive*: entropy can only increase locally, and a decrease in one region cannot be compensated by an increase elsewhere.

For instance, in a thermoelectric system with spatially varying temperature and chemical potential, each infinitesimal region produces entropy through Joule heating, the Seebeck effect, and Peltier heat flow. Integrating the local entropy production density $\sigma(\mathbf{r})$ over the device volume gives the total irreversible entropy production, consistent with the second law.

From Local Irreversibility to Modern Non-Equilibrium Thermodynamics

Prigogine's reformulation transforms thermodynamics into a general theory of irreversible processes. Entropy production, expressed as the scalar product of thermodynamic forces and fluxes, provides the foundation for the dynamical description of non-equilibrium systems. This approach has enormous predictive power across physics, chemistry, and biology.

In the following volume, we will develop these ideas further, including:

- The derivation of the *Onsager reciprocal relations* [Ons31a, Ons31b];
- Applications to heat conduction, diffusion, and viscous flow;
- The emergence of self-organization and *dissipative structures* in far-from-equilibrium systems.

The modern second law, in this formulation, is not merely a restriction on possible processes—it becomes a *dynamical principle* governing how systems evolve in time and how, paradoxically, order can emerge from irreversible dissipation.

Extension: Stochastic Thermodynamics

In this context, entropy production is defined along individual stochastic trajectories, and fluctuation theorems quantify the probability of transient entropy-reducing events. These ideas will be discussed in detail in Volume II of this series. Stochastic thermodynamics is a framework, based on a variety of results derived in the late 90s the 1990s elucidating the role of thermodynamics in systems out of equilibrium and for a small number of particles. For example, two cornerstone results of stochastic thermodynamics, specifically the *Jarzynski equality* [Jar97] and the *Crooks fluctuation theorem* [Cro99], establish precise relationships between microscopic work fluctuations and macroscopic free-energy differences. The Jarzynski equality states that

$$\left\langle e^{-W/k_{\mathrm{B}}T} \right\rangle = e^{-\Delta F/k_{\mathrm{B}}T}, \tag{5.117}$$

while the Crooks relation connects the work distributions of the forward and reverse processes through

$$\frac{P_{\mathrm{F}}(W)}{P_{\mathrm{R}}(-W)} = e^{(W-\Delta F)/k_{\mathrm{B}}T}. \tag{5.118}$$

Together, these expressions show that even far from equilibrium, fluctuating microscopic trajectories encode the equilibrium free-energy difference ΔF exactly. These results show that the macroscopic second law emerges statistically from microscopic dynamics. Nonetheless, entropy production remains the fundamental measure of irreversibility, connecting thermodynamics from the cosmic to the molecular scale.

REMARK 5.18 Prigogine's insight, that entropy is produced locally and irreversibly, has proven to be one of the most profound in modern physics. It reveals the second law as a universal principle describing the directionality of time, the emergence of complexity, and the evolution of structure in open systems. Far from being a static constraint, the second law becomes the driving principle behind the dynamic, creative aspect of nature.

Experimental Tests of the Second Law Beyond Equilibrium

While the systematic development of irreversible thermodynamics and stochastic thermodynamics—together with the historical motivations behind each framework—will be presented in detail in Volume II, it is useful to anticipate their significance already here. Before the emergence of fluctuation theorems in the late 1990s, it was widely assumed (and reflected in most textbooks published before the 1990s) that the second law expressed an *absolute* prohibition rather than a fundamentally statistical statement. A statistical interpretation of the second law predates modern fluctuation theorems and originates in the work of Ludwig Boltzmann. In his *Lectures on Gas Theory* and related papers, Boltzmann repeatedly emphasized that the second law is not an inviolable dynamical principle but a probabilistic statement valid only for systems with an enormous number of degrees of freedom. He noted that entropy-decreasing fluctuations are not impossible, only exceedingly improbable when $N \gg 1$, and that the apparent irreversibility of macroscopic processes results from the overwhelming predominance of microstates corresponding to higher-entropy configurations [Bol64, Bol95]. This perspective, that the second law is true "for all practical purposes" in large systems, anticipated the modern view established quantitatively a century later through stochastic thermodynamics.

Irreversible thermodynamics formalizes macroscopic entropy production near equilibrium, whereas modern *stochastic thermodynamics* extends these concepts to microscopic regimes where thermal fluctuations become observable and the statistical character of the second law becomes explicit. In this setting, entropy production is defined along individual stochastic trajectories, and fluctuation theorems quantify the probability of temporary, entropy-reducing events—rare microscopic "violations" of the second law. Over the past two decades, several landmark experiments have directly tested these predictions. Single-molecule pulling experiments on RNA [LDS+02, CRJ+05], optical-trap measurements on colloidal particles [WSM+02], and even microscale stochastic heat engines [BB12] have verified both the *Jarzynski equality* [Jar97] and the *Crooks fluctuation theorem* [Cro99]. These experiments demonstrate that transient negative entropy production is not only possible but quantitatively predicted by the theory.

Such results show that the second law is not an *absolute*, inviolable constraint at the microscopic level. Instead, it emerges as a *statistical law*, and irreversible behavior appears only after averaging over many fluctuating realizations. Thus, the macroscopic second law and Prigogine's thermodynamic framework arise statistically from underlying time-symmetric microscopic dynamics.

REMARK 5.19 Prigogine's insight—that entropy is produced locally and irreversibly—finds a natural extension in this stochastic language. It reveals the second law as a statistical arrow of time: valid overwhelmingly, but not absolutely, and grounded in the probability structure of microscopic fluctuations. Seen in this light, irreversibility becomes a unifying principle spanning molecular machines, biological systems, and macroscopic thermodynamics.

Despite its apparent simplicity and the familiarity of its laws, thermodynamics admits many logically distinct formulations. At its core lie concepts—equilibrium, entropy, irreversibility, and adiabatic accessibility—that are operational in nature and can be interpreted in different mathematical ways. Some approaches emphasize the geometric structure of equilibrium states, others focus on algebraic relations between processes, while more recent treatments highlight order-theoretic or information-theoretic aspects. The existence of multiple axiomatizations reflects the attempt to clarify which assumptions are essential, which are empirical, and which follow from logical consistency alone.

Axiomatizations of Thermodynamics

Axiomatic thermodynamics seeks to derive the structure of the theory—and in particular the existence and uniqueness of entropy—from a minimal set of physically motivated postulates. The objective is to make the second law a logical consequence of basic relations between equilibrium states, rather than an empirical principle. Two broad families of approaches have emerged:

- *Dynamic frameworks*, such as those of Onsager and Prigogine [Ons31a, Pri47], which connect thermodynamic behavior to microscopic dynamics and treat entropy production as a measure of irreversibility.
- *Static frameworks*, beginning with Carathéodory [Car09] and culminating in Lieb and Yngvason [LY99], which take the set of equilibrium states and their relations of accessibility as primitive, independent of time or microscopic detail.

Carathéodory (1909): Geometry and Inaccessibility

Carathéodory [Car09] formulated the second law in differential-geometric language. He postulated that in the neighborhood of any equilibrium state there exist states that are *adiabatically inaccessible*. This condition imposes an integrability constraint on the Pfaffian form representing heat exchange,

$$\delta Q = dE + p\,dV + \sum_i \mu_i\,dN_i, \tag{5.119}$$

implying the existence of a function $S(E, V, N_i)$ and an integrating factor $1/T$ such that $\delta Q = T\,dS$. Entropy S and temperature T thus emerge not as empirical quantities but as *mathematical consequences* of the inaccessibility postulate.

Carathéodory's approach reinterprets the second law as a theorem on the *integrability of differential forms* on the manifold of equilibrium states. Each equilibrium state corresponds to a point in this manifold, and adiabatic processes correspond to curves along which $\delta Q = 0$. The statement that some neighboring states are adiabatically inaccessible is equivalent to the non-integrability of $\delta Q = 0$, which ensures the existence of a potential function S. This geometric viewpoint established the logical necessity of entropy but remained local and formal: it assumes differentiability and infinitesimal transformations, with limited operational interpretation. Nevertheless, it inspired later efforts to make the axioms algebraic and global.

Giles (1964): Algebraic Processes and Quasi-Entropy

Giles [GS64] eliminated Carathéodory's geometric structure entirely and replaced it by a purely algebraic system of *states* and *processes*. A *state* represents an equilibrium configuration of a system, while a *process* is an ordered pair of states (a, b), denoted symbolically as $a \to b$. Two primitive operations are defined:

$$a + b \quad \text{(composition of systems)}, \qquad a \to b \quad \text{(feasible transformation)}. \tag{5.120}$$

These obey a small set of axioms: associativity and commutativity of $+$, transitivity of $\to$, and compatibility under composition, i.e., $a + c \to b + c$ implies $a \to b$. Within this structure, Giles proved the existence of a real-valued function $S(a)$—the *quasi-entropy*—and a family of additive conserved quantities $Q(a)$—the *contents*—such that

$$a \to b \quad \text{if and only if} \quad S(a) \leq S(b), \quad Q(a) = Q(b). \tag{5.121}$$

The quasi-entropy S is extensive and additive, and its monotonicity encodes the second law. This theorem shows that entropy can be deduced purely from algebraic relations between feasible processes, without assuming continuity, differentiability, or microscopic structure.

In contrast to Carathéodory, who derived the *local* existence of entropy through integrability, Giles's framework establishes its *global* existence as the unique monotone function compatible with the order structure of thermodynamic processes. His formulation thus turns thermodynamics into an order theory of physical transformations, anticipating later developments in both information theory and quantum resource frameworks.

Falk and Jung (1959–1970s): Order and Comparability

Falk and Jung [FJ59] simplified Giles's structure by assuming that adiabatic accessibility defines a *total preorder*: for any two equilibrium states a and b, either $a \to b$ or $b \to a$.

This completeness postulate leads directly to the existence of a scalar function S such that

$$a \to b \iff S(a) \leq S(b), \tag{5.122}$$

with S additive and extensive under composition and scaling. Their work recasts Giles's algebraic system into a simpler order-theoretic form, in which entropy becomes the numerical representation of an accessibility order. They thus bridge the gap between Carathéodory's differential geometry (local structure) and Giles's algebra (global process algebra).

Lieb and Yngvason (1990s): Axiomatic approach

Elliot Lieb and Yakub Yngvason [LY99] synthesized these traditions into a modern axiomatization with explicit operational meaning. They take the relation $a \prec b$—"b is adiabatically accessible from a"—as a primitive physical notion and introduce axioms capturing empirical properties of macroscopic systems: composition, scaling, transitivity, and comparability. From these, they rigorously prove that there exists a unique (up to affine transformation) additive, extensive, and order-preserving entropy function S satisfying

$$a \prec b \iff S(a) \leq S(b). \tag{5.123}$$

Their formulation dispenses with differential geometry and provides a fully operational interpretation of entropy grounded in physical accessibility. In this sense, Lieb and Yngvason complete the axiomatic program that began with Carathéodory's postulate of inaccessibility and Giles's process algebra.

From a historical point of view, Carathéodory established the local differential condition for entropy while Giles globalized it into a compositional algebra of processes. Instead, Falk and Jung reinterpreted it in order-theoretic terms. More recently, Lieb and Yngvason restored physical interpretability and proved uniqueness formally. Together, these frameworks show that the second law is not merely empirical but a necessary consequence of the logical structure of thermodynamic transformations.

6

Kinetic Theory

6.1 Introduction to Kinetic Theory

The development of kinetic theory spans over two centuries, from Daniel Bernoulli's early ideas on molecular motion (1738) to the probabilistic framework of Maxwell and Boltzmann in the nineteenth century [Max60, Bol96].

The kinetic theory of gases arose as an attempt to provide a mechanical foundation for the macroscopic laws of thermodynamics. While thermodynamics describes quantities such as pressure, temperature, and entropy without referring to microscopic detail, kinetic theory seeks to interpret these variables in terms of the collective motion of atoms and molecules.

The idea that heat and pressure originate from the motion of particles first appeared in the work of Bernoulli, who explained pressure as the momentum transfer of molecules colliding with the container walls. In the nineteenth century, James Clerk Maxwell and Ludwig Boltzmann extended this view, introducing probabilistic reasoning and the notion of a *distribution function.* Their efforts established the link between mechanics and probability that ultimately gave rise to statistical mechanics.

Thermodynamics, despite its name, does not address the motion of particles or the dynamical mechanisms by which equilibrium is achieved. Its laws concern the relations between macroscopic observables—energy, pressure, temperature, and entropy—without describing the microscopic origin of these quantities. Kinetic theory bridges this conceptual gap by explaining how thermodynamic behavior emerges from microscopic dynamics, and by identifying the conditions under which a system *thermalizes.*

Thermalization is the process through which interacting particles redistribute their energy until a well-defined temperature characterizes the entire system. When this happens, the system

DOI: 10.1201/9781003603528-6

is said to be in equilibrium. If, over long times, the system's microscopic trajectories explore all accessible configurations in a manner such that time averages equal ensemble averages, the system is said to be *ergodic*. Ergodicity provides a dynamical justification for the statistical ensembles of equilibrium thermodynamics.

REMARK 6.1 The ergodic hypothesis is not required for the formulation of statistical mechanics, but it provides a deep conceptual bridge between deterministic microscopic motion and probabilistic macroscopic laws. Kinetic theory gives one of the first quantitative realizations of this bridge.

In thermodynamics, one begins with the macroscopic internal energy U, defined as a function of state variables such as entropy S, volume V, and particle number N. In statistical mechanics, by contrast, energy arises from the microscopic configuration of the system. For a monoatomic gas, the relevant microscopic variables are the particle positions $\vec{q}_i$ and momenta $\vec{p}_i$ $(i = 1, \ldots, N)$, and the total energy is specified by a Hamiltonian:

$$E = H(\{\vec{q}_i, \vec{p}_i\}). \tag{6.1}$$

Here, we distinguish between:

- U — the *macroscopic* internal energy (a thermodynamic function of state),
- E — the *microscopic* mechanical energy, defined by the Hamiltonian.

The two are related through statistical averaging: U corresponds to the ensemble mean of E over the accessible phase-space region.

The systematic distinction between microscopic and macroscopic energy appeared in the late nineteenth century in the works of Clausius, Maxwell, and Boltzmann, who sought to derive thermodynamics from mechanics.

The dynamics of an isolated system of N particles obey Hamilton's equations:

$$\begin{aligned}\frac{d\vec{p}_i}{dt} &= -\frac{\partial H}{\partial \vec{q}_i},\\ \frac{d\vec{q}_i}{dt} &= \frac{\partial H}{\partial \vec{p}_i},\end{aligned} \tag{6.2}$$

where H is the total energy. These equations encapsulate Newton's laws in symplectic form and guarantee energy conservation.

Since the system possesses $6N$ coupled degrees of freedom, direct integration of Eq. (6.2) quickly becomes intractable for large N. This difficulty motivates the transition from a deterministic to a statistical description of many-particle systems.

REMARK 6.2 Although the motion governed by Eq. (6.2) is deterministic and reversible, the complexity of many-body dynamics makes an explicit trajectory-based description useless for macroscopic systems. Statistical mechanics provides a coarse-grained language to deal with this complexity without abandoning determinism.

From Hamilton's equations, it follows that the flow in phase space is incompressible: the density of representative points $\rho(\Gamma, t)$ in the phase space Γ satisfies

$$\frac{d\rho}{dt} = 0. \tag{6.3}$$

This result, known as *Liouville's theorem*, ensures that phase-space volume is conserved under Hamiltonian evolution. It provides the mathematical foundation for the micro-canonical ensemble and underlies the derivation of kinetic equations.

We define the phase space Γ of an N-particle system as

$$\Gamma = \{(\vec{q}_1, \ldots, \vec{q}_N; \vec{p}_1, \ldots, \vec{p}_N)\}. \tag{6.4}$$

Each point in Γ corresponds to a complete specification of the system's microstate, and its time evolution corresponds to a trajectory determined by Eq. (6.2). Because the motion is continuous and deterministic, trajectories in Γ cannot intersect—otherwise, the future evolution would cease to be uniquely defined.

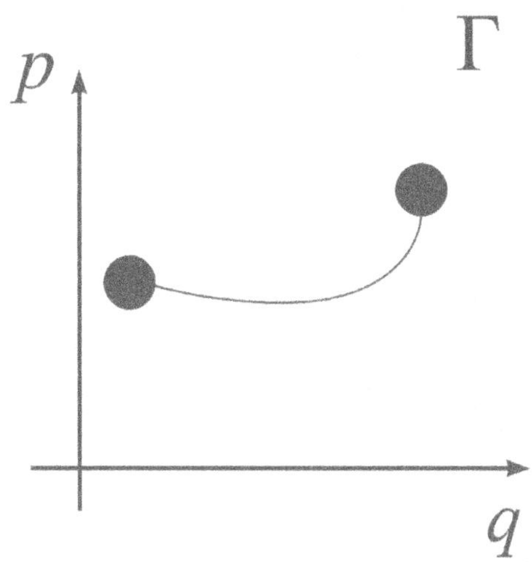

FIGURE 6.1 A trajectory in phase space for a single particle in one dimension. Each point represents a pair (q, p) of position and momentum. The curve shows the continuous evolution generated by Hamilton's equations.

Rather than following individual trajectories, kinetic theory introduces a statistical description through the *distribution function*:

$$f(\vec{q}, \vec{p}, t), \tag{6.5}$$

which represents the density of particles in phase space at time t. Macroscopic observables such as number density, pressure, and temperature can then be expressed as moments of f. The evolution of f encapsulates the collective behavior of all particles and forms the basis of transport theory.

REMARK 6.3 The introduction of $f(\vec{q}, \vec{p}, t)$ marks the transition from mechanics to statistical mechanics: it allows the description of macroscopic behavior as an emergent property of microscopic dynamics, while preserving the fundamental deterministic laws.

The passage from trajectories to distributions completes the conceptual bridge between microscopic dynamics and macroscopic laws. In the next section, we will derive the equation governing $f(\vec{q}, \vec{p}, t)$ from Liouville's theorem and the assumption of molecular chaos, obtaining the celebrated *Boltzmann equation*. This equation provides the cornerstone of kinetic theory and the foundation of modern non-equilibrium statistical mechanics.

6.2 From the Full Distribution to Reduced Descriptions

In the previous section, we introduced the concept of a phase-space trajectory for a system of N particles. Each microstate corresponds to a point in the $6N$-dimensional phase space Γ, characterized by the set of coordinates and momenta $\{\vec{q}_i, \vec{p}_i\}_{i=1}^N$. The time evolution of the system can therefore be represented by a trajectory $\Gamma(t)$ that obeys Hamilton's equations.

The Microscopic Distribution Function

For a deterministic Hamiltonian system, the most elementary phase-space distribution is obtained by following the trajectory of every particle. If $\vec{q}_i^{\,*}(t)$ and $\vec{p}_i^{\,*}(t)$ denote the exact positions and momenta of the i-th particle at time t, the corresponding N-particle distribution function is

$$f(\{\vec{p}, \vec{q}\}, t) = \prod_{i=1}^{N} \delta\big(\vec{p}_i - \vec{p}_i^{\,*}(t)\big)\, \delta\big(\vec{q}_i - \vec{q}_i^{\,*}(t)\big). \tag{6.6}$$

This distribution assigns unit probability to the precise microstate that the system occupies at time t and vanishes elsewhere.

In principle, one may consider a mixture of M deterministic trajectories, $f(\{\vec{p}, \vec{q}\}, t) = \sum_{k=1}^{M} c_k \prod_{i=1}^{N} \delta(\vec{p}_i - \vec{p}_i^{\,(k)*}(t))\delta(\vec{q}_i - \vec{q}_i^{\,(k)*}(t))$, with non-negative coefficients c_k satisfying $\sum_k c_k = 1$. This generalization corresponds to an ensemble of microstates, rather than a single trajectory.

Although formally exact, Eq. (6.6) is of little practical use: following the individual trajectories of N interacting particles is computationally impossible when N is macroscopic. The strategy of kinetic theory is therefore to consider *reduced distributions*, obtained by integrating out degrees of freedom that are not of direct interest.

Reduced Distribution Functions

We define the k-particle reduced distribution function f_k by integrating the full distribution f over the coordinates and momenta of all but k particles:

$$\begin{aligned}
f_1(\vec{q}_1, \vec{p}_1, t) &= \int d^3q_2 \cdots d^3q_N d^3p_2 \cdots d^3p_N\, f(\{\vec{p}, \vec{q}\}, t), \\
f_2(\vec{q}_1, \vec{p}_1, \vec{q}_2, \vec{p}_2, t) &= \int d^3q_3 \cdots d^3q_N d^3p_3 \cdots d^3p_N\, f(\{\vec{p}, \vec{q}\}, t), \\
&\ \ \vdots \\
f_k(\{\vec{q}, \vec{p}\}_k, t) &= \int d^3q_{k+1} \cdots d^3q_N d^3p_{k+1} \cdots d^3p_N\, f(\{\vec{p}, \vec{q}\}, t).
\end{aligned} \tag{6.7}$$

The function f_k represents the probability density of finding any k particles with given positions and momenta at time t, regardless of the state of the remaining $N - k$ particles.

REMARK 6.4 The reduced distributions encode correlation information. For instance, f_1 describes the single-particle statistics, while f_2 captures pair correlations arising from interactions. Knowledge of all f_k functions would be equivalent to knowing the full distribution f, but in practice only the first few are tractable.

The one-particle distribution $f_1(\vec{q}, \vec{p}, t)$ carries information about the *average momentum* and *average position* of particles in the gas, but not about individual trajectories. It will become the central object of kinetic theory and the quantity governed by the Boltzmann equation.

From Eq. (6.6), it follows that

$$\int f(\{\vec{p},\vec{q}\},t)\, d^{3N}q\, d^{3N}p = N \quad \forall t. \tag{6.8}$$

Alternatively, one may choose to normalize the distribution to the particle *number density*:

$$\int f(\{\vec{p},\vec{q}\},t)\, d^{3N}q\, d^{3N}p = \frac{N}{V} \quad \forall t, \tag{6.9}$$

a convention that will be used in what follows. For compactness, we introduce the notation

$$d\Gamma = d^{3N}p\, d^{3N}q \tag{6.10}$$

for integration over the full phase space, and $\{p,q\}_k$ to denote the subset of variables associated with the first k particles.

REMARK 6.5 Normalization ensures that total particle number (or density) is conserved under time evolution. In the Liouville picture, this conservation reflects the incompressibility of Hamiltonian flow.

Expectation Values and Observables

Once a normalized distribution f (or f_k) is known, any observable in phase space can be evaluated as a statistical average. Let $\mathcal{O}_k(\{p,q\}_k)$ be an observable depending on k particles. Its expectation value is

$$\langle \mathcal{O}_k(\{p,q\}_k,t)\rangle = \frac{\displaystyle\int d^{3(N-k)}p\, d^{3(N-k)}q\, \mathcal{O}_k(\{p,q\}_k)\, f_k(\{p,q\}_k,t)}{\displaystyle\int d^{3(N-k)}p\, d^{3(N-k)}q\, f_k(\{p,q\}_k,t)}. \tag{6.11}$$

This defines the correspondence between microscopic phase-space functions and measurable macroscopic quantities.

The average above generalizes the concept of ensemble averages introduced later in statistical mechanics. For an ergodic system, these ensemble averages coincide with time averages computed along trajectories. Ergodicity, in this formal sense, states that

$$\lim_{t\to\infty} \langle \mathcal{O}_k(\{p,q\}_k,t)\rangle_t = \langle \mathcal{O}_k(\{p,q\}_k)\rangle, \tag{6.12}$$

implying that the long-time behavior of dynamical observables equals their statistical expectation values. This equality provides the conceptual justification for equilibrium statistical mechanics: it allows time averages measured in experiments to be replaced by phase-space averages computed from stationary distributions.

Macroscopic Observables from the One-Particle Distribution

The introduction of the one-particle distribution function,

$$f_1(\vec{q},\vec{p},t), \tag{6.13}$$

marks the decisive step in connecting microscopic dynamics to macroscopic thermodynamics. Instead of following the individual motion of each of the N particles, kinetic theory describes the system statistically through f_1: the density of particles in phase space at position $\vec{q}$ with momentum $\vec{p}$ at time t.

REMARK 6.6 Conceptually, this represents a profound shift from the deterministic world-view of classical mechanics to a probabilistic one. As Boltzmann emphasized in his foundational work [Bol96], macroscopic order and irreversibility emerge not from the breakdown of mechanics, but from the statistical behavior of an enormous number of microscopic degrees of freedom.

Physical Meaning of $f_1(\vec{q}, \vec{p}, t)$

The quantity $f_1(\vec{q}, \vec{p}, t)\, d^3q\, d^3p$ represents the expected number of particles whose positions and momenta lie within the infinitesimal volume element $d^3q\, d^3p$ of phase space. Integrating f_1 over all momenta gives the local particle density in space, while integrating over all positions gives the momentum distribution. All macroscopic observables—pressure, temperature, energy, and fluxes—can thus be expressed as averages of microscopic quantities weighted by f_1.

This represents a major conceptual advance: macroscopic observables become *statistical moments* of the microscopic distribution, thereby providing a rigorous link between thermodynamics and mechanics.

In Maxwell's 1860 paper on the dynamical theory of gases [Max60], this idea first appeared in embryonic form. Boltzmann later systematized it, showing that macroscopic laws such as the ideal gas law or Fourier's law could be derived from such averages, rather than postulated.

Definition of Macroscopic Quantities

From the one-particle distribution $f_1(\vec{q}, \vec{p}, t)$, we can define the key macroscopic observables. Each is obtained by integrating over microscopic degrees of freedom and normalizing by the total number of particles (or total probability). These definitions provide the operational bridge between microscopic theory and measurable quantities.

Number Density

The local particle density $n(\vec{q}, t)$ gives the expected number of particles per unit volume at position $\vec{q}$:

$$n(\vec{q}, t) = \frac{\displaystyle\int d^3p\, f_1(\vec{q}, \vec{p}, t)}{\displaystyle\int d^3q\, d^3p\, f_1(\vec{q}, \vec{p}, t)}. \tag{6.14}$$

Physically, $n(\vec{q}, t)$ connects directly to measurable quantities such as mass density or local concentration in a gas.

Mean Velocity Field

The average velocity of particles at position $\vec{q}$ is obtained from the momentum distribution:

$$\vec{u}(\vec{q}, t) = \frac{\displaystyle\int d^3p\, \frac{\vec{p}}{m}\, f_1(\vec{q}, \vec{p}, t)}{\displaystyle\int d^3p\, f_1(\vec{q}, \vec{p}, t)}. \tag{6.15}$$

This field $\vec{u}(\vec{q}, t)$ represents the macroscopic flow velocity, connecting kinetic theory with fluid dynamics.

Pressure

The local pressure $P(\vec{q}, t)$ is derived from the average kinetic contribution of particles colliding with a surface element:

$$P(\vec{q}, t) = \frac{1}{3m} \frac{\int d^3p\, \vec{p}^2\, f_1(\vec{q}, \vec{p}, t)}{\int d^3q\, d^3p\, f_1(\vec{q}, \vec{p}, t)}. \tag{6.16}$$

This expression formalizes the molecular origin of pressure as the flux of momentum transfer—one of the central insights of kinetic theory.

Energy Density

Defining the microscopic kinetic energy as $E(\vec{p}) = \vec{p}^2/2m$, the local energy density reads

$$E(\vec{q}, t) = \frac{\int d^3p\, E(\vec{p})\, f_1(\vec{q}, \vec{p}, t)}{\int d^3q\, d^3p\, f_1(\vec{q}, \vec{p}, t)}. \tag{6.17}$$

This provides the link between microscopic motion and the internal energy of the gas.

Energy Flux

The rate of energy transport—per unit area and unit time—is given by the *energy flux* (or current density):

$$\vec{\mathcal{E}}(\vec{q}, t) = \frac{\int d^3p\, \frac{\vec{p}}{m}\, E(\vec{p})\, f_1(\vec{q}, \vec{p}, t)}{\int d^3q\, d^3p\, f_1(\vec{q}, \vec{p}, t)}. \tag{6.18}$$

This quantity describes how kinetic energy is transported across the system due to particle motion, and is central to transport theory and non-equilibrium thermodynamics.

The formalism above represents one of the great conceptual achievements of nineteenth-century physics. By introducing f_1, kinetic theory replaced the impossible task of tracking individual molecular trajectories with a statistical description governed by a distribution in phase space. This shift in viewpoint was revolutionary: it allowed mechanics to be reformulated in terms of probabilities, leading directly to statistical mechanics.

For a modern discussion of the historical and conceptual development of this transition, see Brush [Bru76] and Cercignani [Cer98]. Both provide detailed analyses of how Boltzmann's statistical perspective transformed the foundations of physics.

Boltzmann's introduction of the distribution function was not merely a technical trick, but it was a change in the scientific paradigm. Whereas classical mechanics focused on individual trajectories, kinetic theory replaced determinism with a statistical continuum, recognizing that macroscopic order and thermodynamic irreversibility are emergent properties of large ensembles of microscopic motions [Bru76, Cer98].

In this framework, the macroscopic state of matter is characterized not by exact coordinates and momenta, but by smooth fields—density, velocity, pressure, energy—obtained as statistical averages. This reduction from $6N$ microscopic variables to a handful of macroscopic observables encapsulates the entire spirit of statistical mechanics.

The next step in kinetic theory is to determine how $f_1(\vec{q}, \vec{p}, t)$ evolves in time. Starting from Liouville's equation for the full distribution f and applying a series of approximations, most notably the assumption of *molecular chaos*, one obtains the celebrated *Boltzmann equation*, which describes the irreversible relaxation toward equilibrium. We turn to this derivation next.

Liouville's Theorem and Hamiltonian Flow

Liouville's theorem is one of the central results of Hamiltonian mechanics and provides the bridge between microscopic determinism and statistical conservation laws. It states that the flow of representative points in phase space is *incompressible*: the number of phase-space trajectories within any volume element remains constant under time evolution. Consequently, the distribution function $f(\vec{q}, \vec{p}, t)$ behaves like the density of an incompressible fluid in the $6N$-dimensional phase space.

Liouville's theorem formalizes the idea that in Hamiltonian dynamics, no two trajectories can merge or diverge in phase space. This invariance under time evolution guarantees the conservation of probability and justifies the stationary nature of the micro-canonical ensemble.

This concept was first introduced by Joseph Liouville in 1838 in the context of measure-preserving transformations. Its application to mechanics was later emphasized by Gibbs and Boltzmann in their formulation of statistical ensembles.

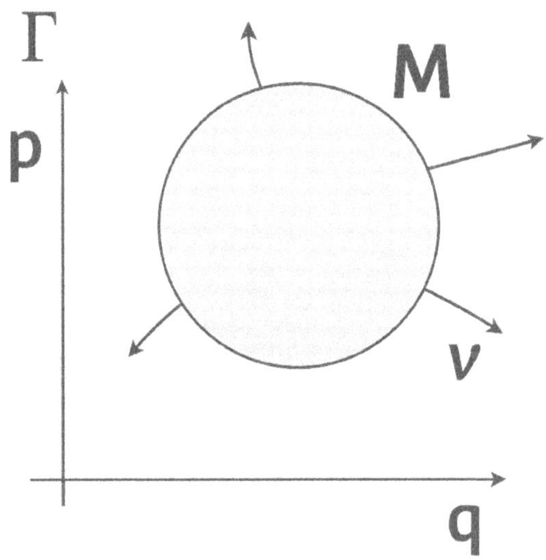

FIGURE 6.2 A region M in phase space and its boundary ∂M evolving under Hamiltonian evolution. The associated "velocity" field $(\dot{\vec{q}}, \dot{\vec{p}})$ represents the motion of representative points in phase space.

Let M be a finite region of phase space with boundary ∂M, and let $d\Gamma_M = d^{3N}p d^{3N}q$ denote the infinitesimal phase-space volume element within this region. The total probability of finding the system inside M at time t is given by

$$\int_M f(\{\vec{q}, \vec{p}\}, t)\, d\Gamma_M. \tag{6.19}$$

Liouville's theorem asserts that this probability remains constant as the region M evolves following the Hamiltonian flow:

$$\frac{d}{dt} \int_M f(\{\vec{q}, \vec{p}\}, t)\, d\Gamma_M = 0. \tag{6.20}$$

REMARK 6.7 The subscript M simply reminds us that the integration is performed over the specific region being followed in time. Since the region itself moves with the flow, $d\Gamma_M$ changes shape but not volume. This notation emphasizes that the integration is over a *material region* in phase space, analogous to the Lagrangian description of fluid motion.

Applying the Leibniz rule and the divergence theorem, Eq. (6.20) gives

$$\begin{aligned} 0 &= \frac{d}{dt}\int_M f\, d\Gamma_M = \int_M \left(\frac{\partial f}{\partial t} + \nabla \cdot (\dot{\Gamma} f)\right) d\Gamma_M, \\ &= \int_M \left(\frac{\partial f}{\partial t} + \sum_{i=1}^{N} \frac{\partial}{\partial \vec{q}_i} \cdot (\dot{\vec{q}}_i f) + \sum_{i=1}^{N} \frac{\partial}{\partial \vec{p}_i} \cdot (\dot{\vec{p}}_i f)\right) d\Gamma_M, \end{aligned} \tag{6.21}$$

where $\nabla = (\partial_{\vec{p}}, \partial_{\vec{q}})$ is the phase-space gradient operator and $\dot{\Gamma} = (\dot{\vec{q}}, \dot{\vec{p}})$ is the velocity field of points in phase space.

From Hamilton's equations,

$$\dot{\vec{q}}_i = \frac{\partial H}{\partial \vec{p}_i}, \qquad \dot{\vec{p}}_i = -\frac{\partial H}{\partial \vec{q}_i}, \tag{6.22}$$

we find

$$\frac{\partial \dot{\vec{q}}_i}{\partial \vec{q}_i} + \frac{\partial \dot{\vec{p}}_i}{\partial \vec{p}_i} = 0. \tag{6.23}$$

This expresses the *incompressibility condition* of the Hamiltonian flow:

$$\nabla \cdot \dot{\Gamma} = 0. \tag{6.24}$$

The divergence-free property implies that the phase-space "fluid" neither expands nor contracts—it merely rearranges itself as the system evolves. Equation (6.23) shows that the flow generated by H is measure-preserving. This is the fundamental reason why probability densities evolve without distortion in Hamiltonian systems.

Substituting this condition into Eq. (6.21), we obtain the local conservation law:

$$\frac{\partial f}{\partial t} + \sum_{i=1}^{N} \left(\dot{\vec{q}}_i \cdot \frac{\partial f}{\partial \vec{q}_i} + \dot{\vec{p}}_i \cdot \frac{\partial f}{\partial \vec{p}_i}\right) = 0. \tag{6.25}$$

This equation expresses the fact that f is constant along the trajectories of the Hamiltonian flow:

$$\frac{df}{dt} = 0. \tag{6.26}$$

Equation (6.25) is analogous to the continuity equation in fluid mechanics, where f plays the role of a density and $(\dot{\vec{q}}, \dot{\vec{p}})$ that of a velocity field in phase space.

Poisson Bracket Formulation

The Liouville equation can be written compactly using the *Poisson bracket*:

$$\{A, B\} = \frac{\partial A}{\partial \vec{q}} \cdot \frac{\partial B}{\partial \vec{p}} - \frac{\partial A}{\partial \vec{p}} \cdot \frac{\partial B}{\partial \vec{q}}. \tag{6.27}$$

With this notation, Eq. (6.25) becomes

$$\frac{\partial f}{\partial t} = -\{f, H\}. \tag{6.28}$$

This is the differential form of Liouville's theorem. It states that the total derivative of f along the phase-space trajectory vanishes:

$$\frac{df}{dt} = \frac{\partial f}{\partial t} + \{f, H\} = 0. \tag{6.29}$$

The Liouville equation, Eq. (6.28), is exact and fully equivalent to Hamilton's equations. It describes the evolution of the N-particle distribution function without approximation and forms the foundation for deriving the Boltzmann and BBGKY hierarchies.

Evolution of Phase-Space Averages

We can now examine how observables evolve under Hamiltonian flow. For an observable $\mathcal{O}(\{p,q\})$, the ensemble average is

$$\langle \mathcal{O} \rangle = \int d\Gamma \, \mathcal{O}(\{p,q\}) \, f(\{p,q\},t). \tag{6.30}$$

Differentiating with respect to time and using Eq. (6.28),

$$\begin{aligned} \frac{d\langle \mathcal{O} \rangle}{dt} &= \int d\Gamma \, \mathcal{O} \, \frac{\partial f}{\partial t} \\ &= \int d\Gamma \, \mathcal{O} \left(\frac{\partial f}{\partial \vec{p}} \cdot \frac{\partial H}{\partial \vec{q}} - \frac{\partial f}{\partial \vec{q}} \cdot \frac{\partial H}{\partial \vec{p}} \right). \end{aligned} \tag{6.31}$$

Integrating by parts and neglecting boundary terms (assuming $f \to 0$ at infinity), we obtain

$$\frac{d\langle \mathcal{O} \rangle}{dt} = - \int d\Gamma \, f(\{p,q\},t) \left(\frac{\partial \mathcal{O}}{\partial \vec{p}} \cdot \frac{\partial H}{\partial \vec{q}} - \frac{\partial \mathcal{O}}{\partial \vec{q}} \cdot \frac{\partial H}{\partial \vec{p}} \right), \tag{6.32}$$

or, equivalently,

$$\frac{d\langle \mathcal{O} \rangle}{dt} = \int d\Gamma \, f(\{p,q\},t) \, \{\mathcal{O}, H\} = \langle \{\mathcal{O}, H\} \rangle. \tag{6.33}$$

Equation (6.33) expresses the evolution of averages directly in terms of Poisson brackets. It shows that, in Hamiltonian dynamics, the time evolution of any observable is generated by its Poisson bracket with the Hamiltonian—an exact classical analogue of the Heisenberg equation of motion in quantum mechanics.

The framework above provides the basis for equilibrium statistical mechanics. At equilibrium, the distribution f must satisfy

$$\frac{\partial f}{\partial t} = 0, \tag{6.34}$$

which, from Eq. (6.28), implies $\{f, H\} = 0$. Thus, f must depend only on conserved quantities of the motion—energy, momentum, or other integrals of motion. In the next section, we will see how this principle leads naturally to the equilibrium ensembles of statistical mechanics.

A First View of the Boltzmann Transport Equation

We have seen that, for a Hamiltonian system, the total derivative of the N-particle distribution function vanishes,

$$\frac{df}{dt} = 0, \tag{6.35}$$

as a consequence of Liouville's theorem. However, this statement presumes that the particles evolve under a smooth Hamiltonian flow without abrupt interactions. In a real gas, microscopic collisions between particles lead to local redistributions of momenta, and the one-particle distribution $f_1(\vec{q}, \vec{p}, t)$ changes in time. The central idea of kinetic theory, due to James Clerk Maxwell [Max67] and Ludwig Boltzmann [Bol72, Bol96], is to describe this evolution statistically.

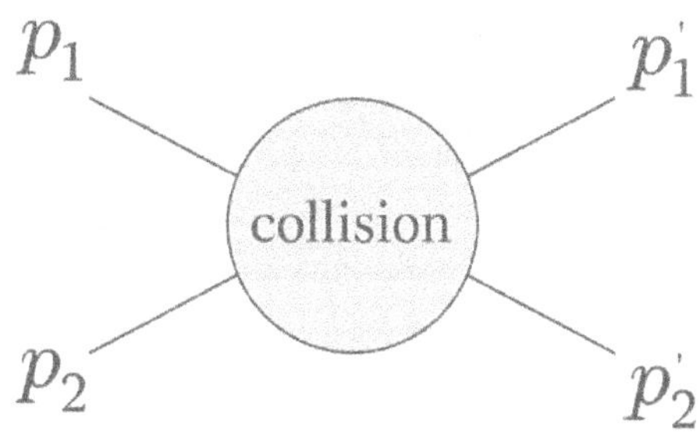

FIGURE 6.3 A binary collision between two gas particles. Incoming momenta $(\vec{p}_1, \vec{p}_2)$ are scattered into $(\vec{p}\,'_1, \vec{p}\,'_2)$, conserving total momentum and energy.

The Evolution of the One-Particle Distribution

Consider a dilute gas where only binary collisions between particles are relevant. Over a short time interval δt, a particle at position $\vec{q}$ and momentum $\vec{p}$ moves under the combined influence of an external force $\vec{F}$ and its velocity $\vec{v} = \vec{p}/m$. Neglecting collisions for the moment, the distribution function evolves smoothly according to

$$f_1(\vec{p} + \delta t\, \vec{F},\, \vec{q} + \delta t\, \vec{v},\, t + \delta t) = f_1(\vec{p}, \vec{q}, t) + \delta t \left(\frac{\partial f_1}{\partial t} + \vec{v} \cdot \nabla_q f_1 + \vec{F} \cdot \nabla_p f_1 \right). \tag{6.36}$$

If collisions are also present, then during the same time δt an additional change occurs because particles are scattered into or out of the momentum state $\vec{p}$. This effect is represented by a *collision term*, so that

$$\frac{df_1(\vec{q}, \vec{p}, t)}{dt} = \left(\frac{\partial f_1}{\partial t} \right)_{\text{coll}}. \tag{6.37}$$

Equation (6.37) expresses the core idea of the Boltzmann transport equation (BTE): the rate of change of f_1 is governed by collisions, which act as a source and sink in momentum space.

REMARK 6.8 The term $\left(\frac{\partial f_1}{\partial t}\right)_{\text{coll}}$ encapsulates all the microscopic interactions that redistribute particles among momentum states. Its precise form depends on the scattering cross section and the statistical assumptions about the gas (e.g., molecular chaos).

The Role of Collisions

In a dilute gas, collisions occur rarely and involve only pairs of particles. When two particles with incoming momenta $(\vec{p}_1, \vec{p}_2)$ interact via a short-range potential, they leave the collision with outgoing momenta $(\vec{p}\,'_1, \vec{p}\,'_2)$ satisfying the conservation laws:

$$\vec{p}_1 + \vec{p}_2 = \vec{p}\,'_1, \vec{p}\,'_2, \tag{6.38}$$

$$E_1 + E_2 = E_1' + E_2', \tag{6.39}$$

where $E_i = \vec{p}_i^{\,2}/2m$. The total momentum and kinetic energy of the pair are preserved, though their individual momenta are generally altered.

Not all collisions are equally probable: the likelihood of a specific scattering event depends on the interaction potential and on how frequently two particles with given momenta meet. This information is contained in the *two-particle distribution function* $f_2(\vec{q}_1, \vec{p}_1, \vec{q}_2, \vec{p}_2, t)$, which describes correlations between pairs of particles before collision.

REMARK 6.9 The transition from the exact Liouville description to an approximate kinetic equation involves two key assumptions:

1. The gas is sufficiently dilute that only binary collisions are important.
2. Pre-collision correlations are negligible—an assumption known as *molecular chaos* (*Stoßzahlansatz*) [Bol72].

Under these hypotheses, the collision integral can be expressed directly in terms of f_1 rather than f_2.

Physical Interpretation and Paradigm Shift

Boltzmann's formulation of Eq. (6.37) represented a radical conceptual shift. Before his work, the motion of molecules was seen as an intractable many-body mechanical problem. Boltzmann reinterpreted it statistically: instead of tracking individual trajectories, he studied the *evolution of probability distributions*. In doing so, he provided the first mechanical explanation for irreversible macroscopic behavior, unifying mechanics and thermodynamics.

REMARK 6.10 This perspective was the foundation for the later development of the *H-theorem*, which demonstrated how the Boltzmann equation leads to a monotonic increase of entropy toward equilibrium. It also laid the groundwork for the probabilistic interpretation of the second law of thermodynamics, a cornerstone of modern statistical mechanics [Bru76, Cer98].

We will now derive the explicit form of the collision term and obtain the full Boltzmann transport equation,

$$\frac{\partial f_1}{\partial t} + \vec{v} \cdot \nabla_{\vec{q}} f_1 + \vec{F} \cdot \nabla_{\vec{p}} f_1 = \left(\frac{\partial f_1}{\partial t} \right)_{\text{coll}} . \tag{6.40}$$

This equation governs the non-equilibrium dynamics of dilute gases and provides the foundation for the theory of transport phenomena such as viscosity, heat conduction, and diffusion. Above, we use the notation $\nabla_{\vec{f}} = (\partial_{f_1}, \cdots, \partial_{f_d})$ where $\vec{f}$ is a d dimensional vector.

The Collision Integral and the Molecular Chaos Hypothesis

Having introduced the Boltzmann transport equation qualitatively, we now examine how the effect of collisions enters the evolution of the one-particle distribution function $f_1(\vec{q}, \vec{p}, t)$. The goal is to express the rate of change of f_1 due to binary collisions between particles and to identify the conditions under which equilibrium is reached.

The Probability of Scattering Events

Let us denote by

$$M_{fi} \equiv M(\vec{p}_1, \vec{p}_2; \vec{p}_1\,', \vec{p}_2\,') \tag{6.41}$$

the transition probability density for a collision in which two particles with initial (incoming) momenta $\vec{p}_1, \vec{p}_2$ are scattered into final (outgoing) momenta $\vec{p}_1\,', \vec{p}_2\,'$. The subscript "$fi$" refers to "final–initial."

Because collisions are local in space, we consider interactions occurring within a small neighborhood of position $\vec{q}_1$. The rate of change of the one-particle distribution due to such events is then

$$\begin{aligned} \frac{d}{dt} f_1(\vec{q}_1, \vec{p}_1, t) &= \left(\frac{\partial f_1}{\partial t} \right)_{\text{coll}} \\ &= \int d^3p_2\, d^3p_1'\, d^3p_2'\, \delta^3(\vec{P}_i - \vec{P}_f)\, \delta(E_i - E_f)\, M_{fi}\, P, \end{aligned} \tag{6.42}$$

where $\vec{P}_i = \vec{p}_1 + \vec{p}_2$, $\vec{P}_f = \vec{p}_1{}' + \vec{p}_2{}'$, and $E_i = E_f$ are the total momentum and energy before and after the collision. The delta functions enforce the conservation of momentum and energy, and P is the joint probability of finding the two particles in the specified states.

Since the collision occurs in the vicinity of $\vec{q}_1$, spatial locality implies

$$P \propto \delta(\vec{q}_1 - \vec{q}_2)\, \delta(\vec{q}_1 - \vec{q}_1{}')\, \delta(\vec{q}_1 - \vec{q}_2{}'), \tag{6.43}$$

and the expression simplifies to

$$\frac{d}{dt} f_1(\vec{q}_1, \vec{p}_1, t) = \int d^3p_2\, d^3p_1'\, d^3p_2'\, \delta^3(\vec{P}_i - \vec{P}_f)\, \delta(E_i - E_f)\, M_{fi}\, P. \tag{6.44}$$

Above, we use the notation $\delta(\vec{f}) = \prod_i \delta(f_i)$.

Gain and Loss Terms

The probability P must include both contributions from collisions that *populate* the state $(\vec{q}_1, \vec{p}_1)$ and those that *deplete* it. The first are collisions producing particles with momentum $\vec{p}_1$ (gain term), and the second are collisions removing particles from this state (loss term). We may therefore write schematically

$$\frac{d}{dt} f_1(\vec{q}_1, \vec{p}_1, t) = \int d^3p_2\, d^3p_1'\, d^3p_2'\, \delta^3(\vec{P}_i - \vec{P}_f)\, \delta(E_i - E_f)\, M_{fi}\, \left[f_2(\vec{q}_1, \vec{p}_1{}', \vec{p}_2{}', t) - f_2(\vec{q}_1, \vec{p}_1, \vec{p}_2, t)\right]. \tag{6.45}$$

Equation (6.45) expresses that collisions continuously redistribute momentum among particles, changing f_1 according to the difference between the gain and loss processes.

The Molecular Chaos Approximation

Equation (6.45) is not closed, because it involves the two-particle distribution function f_2. At this point, Boltzmann introduced his crucial approximation: if the gas is sufficiently dilute, correlations between particles before a collision can be neglected. That is, the probability of finding two particles with given momenta at the same location factorizes as

$$f_2(\vec{q}_1, \vec{p}_1, \vec{p}_2, t) \approx f_1(\vec{q}_1, \vec{p}_1, t)\, f_1(\vec{q}_1, \vec{p}_2, t). \tag{6.46}$$

This assumption, known as the *molecular chaos hypothesis* or *Stoßzahlansatz* [Bol72], replaces correlated two-body probabilities with independent single-particle probabilities.

Inserting this factorization into Eq. (6.45) gives the celebrated Boltzmann collision integral:

$$\begin{aligned} \frac{d}{dt} f_1(\vec{q}_1, \vec{p}_1, t) &= \int d^3p_2\, d^3p_1'\, d^3p_2'\, \delta^3(\vec{P}_i - \vec{P}_f)\, \delta(E_i - E_f) M_{fi} \\ &\times \Big[f_1(\vec{q}_1, \vec{p}_1{}', t)\, f_1(\vec{q}_1, \vec{p}_2{}', t) - f_1(\vec{q}_1, \vec{p}_1, t)\, f_1(\vec{q}_1, \vec{p}_2, t)\Big]. \end{aligned} \tag{6.47}$$

REMARK 6.11 The first term in Eq. (6.47) is the *gain term*, representing collisions that create particles with momentum $\vec{p}_1$, while the second is the *loss term*, accounting for those that remove them. Their competition drives the system toward equilibrium.

At equilibrium, the distribution function is stationary, $df_1/dt = 0$, and the gain and loss terms exactly cancel. This implies the condition of *detailed balance*:

$$f_1(\vec{q}_1, \vec{p}_1{}', t)\, f_1(\vec{q}_1, \vec{p}_2{}', t) = f_1(\vec{q}_1, \vec{p}_1, t)\, f_1(\vec{q}_1, \vec{p}_2, t), \tag{6.48}$$

which states that every collision process is exactly balanced by its inverse.

Taking logarithms of Eq. (6.48) gives

$$\log f_1(\vec{q}_1, \vec{p}_1\,') + \log f_1(\vec{q}_1, \vec{p}_2\,') = \log f_1(\vec{q}_1, \vec{p}_1) + \log f_1(\vec{q}_1, \vec{p}_2), \tag{6.49}$$

showing that the function $\log f_1$ is conserved in a collision. Because energy and momentum are the only collisional invariants, $\log f_1$ must be a linear combination of these quantities:

$$\log f_1^{\mathrm{eq}} = A + B \cdot \vec{p} + C\, E(\vec{p}), \tag{6.50}$$

where A, B, C are constants determined by normalization and the macroscopic constraints. Rearranging yields the familiar *Maxwell–Boltzmann equilibrium distribution*:

$$f_1^{\mathrm{eq}}(\vec{p}) = C\, \exp\left[-A\,(\vec{p} - \vec{p}_0)^2\right], \tag{6.51}$$

where $\vec{p}_0 = m\vec{u}$ corresponds to the mean momentum associated with the macroscopic flow velocity $\vec{u}$.

REMARK 6.12 Equation (6.51) is the statistical expression of thermal equilibrium. It links the microscopic distribution of momenta to the macroscopic thermodynamic parameters—temperature and velocity—completing Boltzmann's program of deriving thermodynamics from mechanics.

The molecular chaos assumption and the ensuing $\mathcal{H}$-theorem appeared in Boltzmann's 1872 paper [Bol72]. For modern analyses and historical context, see Chapman and Cowling [CC70], Brush [Bru76], and Cercignani [Cer98].

Derivation of the Maxwell–Boltzmann Distribution

Let us determine the constants A and C in the equilibrium one–particle momentum distribution

$$f_1^{\mathrm{eq}}(\vec{p}) = C\, e^{-A(\vec{p}-\vec{p}_0)^2}, \tag{6.52}$$

for a dilute classical gas of identical particles of mass m in equilibrium. For simplicity, we assume that the average momentum vanishes, $\langle \vec{p} \rangle = \vec{p}_0 = 0$.

The total number of particles per unit volume is

$$\int d^3p\, f_1^{\mathrm{eq}}(\vec{p}) = \frac{N}{V}, \tag{6.53}$$

where N/V is the number density. The Gaussian integral

$$\int d^3p\, e^{-Ap^2} = \left(\frac{\pi}{A}\right)^{3/2} \tag{6.54}$$

implies

$$C = \frac{N}{V}\left(\frac{A}{\pi}\right)^{3/2}. \tag{6.55}$$

The mean kinetic energy per particle is

$$\langle E \rangle = \int d^3p\, \frac{p^2}{2m}\, f_1^{\mathrm{eq}}(\vec{p}) = \frac{C\,\pi^{3/2}}{2mA^{3/2}}. \tag{6.56}$$

Substituting Eq. (6.55) gives

$$\langle E\rangle = \frac{3}{4mA}. \tag{6.57}$$

From the equipartition theorem, $\langle E\rangle = 3k_BT/2$, which yields

$$A = \frac{1}{2mk_BT}. \tag{6.58}$$

Inserting this expression into Eq. (6.55), the normalization constant becomes

$$C = \frac{N}{V}\left(\frac{m}{2\pi k_BT}\right)^{3/2}, \tag{6.59}$$

and the equilibrium momentum distribution takes the familiar Maxwell–Boltzmann form

$$f_1^{\text{eq}}(\vec{p}) = \frac{N}{V}\left(\frac{m}{2\pi k_BT}\right)^{3/2} e^{-\frac{p^2}{2mk_BT}}. \tag{6.60}$$

Introducing the velocity variable $\vec{v} = \vec{p}/m$,

$$f_1^{\text{eq}}(\vec{v}) = \frac{N}{V}\left(\frac{m}{2\pi k_BT}\right)^{3/2} e^{-\frac{mv^2}{2k_BT}}. \tag{6.61}$$

Integrating over the solid angle gives the speed distribution,

$$f(v) = 4\pi v^2\left(\frac{m}{2\pi k_BT}\right)^{3/2} e^{-\frac{mv^2}{2k_BT}}, \tag{6.62}$$

from which one can compute the average and most probable speeds of a gas.

Finally, expressing the same result in terms of kinetic energy $E = mv^2/2$,

$$f(E) = 2\sqrt{\frac{E}{\pi(k_BT)^3}}\, e^{-E/k_BT}, \tag{6.63}$$

which shows that the kinetic energies follow an exponential (Boltzmann) distribution with mean value $\langle E\rangle = 3k_BT/2$.

REMARK 6.13 Equation (6.60) encapsulates a profound result: it connects temperature—a macroscopic thermodynamic quantity—with the statistical spread of microscopic kinetic energies. It also marks one of the first instances in physics where a universal exponential law emerged from purely mechanical reasoning.

The Maxwell–Boltzmann distribution is the cornerstone of classical statistical mechanics. It implies that:

- The kinetic energies of gas molecules follow an exponential law, with a mean value proportional to temperature.
- The most probable speed, mean speed, and root-mean-square speed all scale as $\sqrt{T/m}$.
- Macroscopic quantities such as pressure and internal energy can be derived directly from f_1^{eq}.

REMARK 6.14 Boltzmann's derivation of this result [Bol72, Bol96] established a new paradigm: macroscopic equilibrium arises from the statistical behavior of many independent microscopic components. This realization bridged mechanics and thermodynamics, and laid the foundation for the concept of entropy and the subsequent development of the H-theorem.

Boltzmann's Constant as a Bridge Between Scales

At the cost of being repetitive, let us highlight here again the role of the Boltzmann constant and the other quantities bridging the micro- to the macroscale. The Boltzmann constant k_B provides the essential link between the microscopic and macroscopic worlds. It converts temperature, expressed in Kelvins, into an energy scale measured in Joules, thereby connecting molecular motion to macroscopic thermodynamic quantities.
The constant

$$k_B = 1.380\,649 \times 10^{-23}\,\mathrm{J\,K^{-1}} \tag{6.64}$$

appears small and arbitrary at first glance. Its numerical value, however, is determined by the choice of human-scale units: it ensures that the thermal energy $k_B T$ has magnitudes consistent with everyday macroscopic energies.

Through k_B, temperature acquires a clear energetic interpretation. At room temperature, for example, $k_B T \approx 4 \times 10^{-21}\,\mathrm{J}$, corresponding to the average kinetic energy of a single particle. Thus, k_B serves as a conversion factor translating microscopic energy fluctuations into thermodynamic temperature.

Avogadro's number,

$$N_A \simeq 6.022 \times 10^{23}, \tag{6.65}$$

plays a complementary role. The product $R = N_A k_B$ yields the universal gas constant,

$$R = 8.314\,\mathrm{J\,mol^{-1}\,K^{-1}}, \tag{6.66}$$

demonstrating how the microscopic and macroscopic descriptions are connected. While k_B represents the energy per particle per Kelvin, R represents the energy per mole per Kelvin.

Boltzmann's constant is more than a numerical factor: it is a conceptual bridge between statistical mechanics and thermodynamics. It permits the statistical definition of entropy by Gibbs and Shannon,

$$S = -k_B \sum_i P_i \ln P_i, \tag{6.67}$$

to acquire physical units of Joules per Kelvin. It therefore defines the scale at which microscopic disorder becomes macroscopic entropy.

REMARK 6.15 For historical and modern discussions of the Maxwell–Boltzmann distribution and its implications, see Maxwell [Max67], Boltzmann [Bol72, Bol96], and the comprehensive accounts by Brush [Bru76] and Cercignani [Cer98].

Example: Maxwell–Boltzmann Average Speed and Collision Time

Average Speed from Maxwell–Boltzmann

For a monatomic ideal gas,

$$f(v) = 4\pi\left(\frac{m}{2\pi k_B T}\right)^{3/2} v^2 e^{-mv^2/(2k_B T)}, \qquad \langle v\rangle = \int_0^\infty v\, f(v)\, dv = \sqrt{\frac{8k_B T}{\pi m}}. \tag{6.68}$$

Writing the particle mass as $m = \mu\, m_u$ with μ in u and $m_u = 1.66054 \times 10^{-27}$ kg,

$$\langle v\rangle \approx 145.5\sqrt{\frac{T}{\mu}}\ \mathrm{m\,s^{-1}} \qquad (T \text{ in K}, \mu \text{ in u}). \tag{6.69}$$

Room-Temperature Speeds ($T = 300$ K)

Gas	μ (u)	$\langle v\rangle$ ($\mathrm{m\,s^{-1}}$)
He	4	≈ 1260
N_2	28	≈ 476
O_2	32	≈ 446
CO_2	44	≈ 380

(6.70)

If the number density is ρ (in $\mathrm{m^{-3}}$), a crude interparticle spacing is $\bar{L} = \rho^{-1/3}$. With kinetic energy $E = m\bar{v}^2/2$ and the hint $\tau = \bar{L}/\bar{v}$,

$$\tau = \frac{\rho^{-1/3}}{\bar{v}} = \sqrt{\frac{m}{2E\,\rho^{2/3}}} \qquad \left(\rho \text{ in m}^{-3}\right). \tag{6.71}$$

Remark. This spacing model captures scalings. In real gases the mean free path involves the molecular cross section σ: $\lambda \simeq (\sqrt{2}\,\rho\sigma)^{-1}$, and $\tau \simeq \lambda/\langle v\rangle$.

The Maxwell–Boltzmann distribution applies to dilute, classical gases where:

- particles are *distinguishable* and correlations are negligible,
- quantum effects are insignificant (the de Broglie wavelength is much smaller than the interparticle spacing),
- the gas is non-degenerate and obeys the ideal gas law.

At low temperatures or high densities, these assumptions break down, and quantum statistics must be used: *Fermi–Dirac* statistics for indistinguishable fermions, and *Bose–Einstein* statistics for bosons.

REMARK 6.16 The Maxwell–Boltzmann law represents the statistical foundation of classical thermodynamics. It connects the macroscopic concept of temperature with the microscopic dispersion of molecular velocities and provides the starting point for transport theory and the thermodynamic equilibrium of gases.

The derivation of Eq. (6.60) marked a turning point in physics. Boltzmann [Bol72, Bol96] demonstrated that the exponential form of f_1^{eq} is a consequence of mechanical laws combined with statistical reasoning. This result established the idea that macroscopic equilibrium emerges from the collective behavior of large numbers of microscopic constituents.

REMARK 6.17 Boltzmann's insight transformed mechanics from a theory of trajectories into a theory of distributions. It introduced probability as an essential tool of physics and prepared the conceptual ground for the statistical interpretation of entropy and the $\mathcal{H}$-theorem.

We now turn to one of Boltzmann's most profound results—the $\mathcal{H}$-theorem—which demonstrates how the distribution $f_1(\vec{q}, \vec{p}, t)$ evolves irreversibly toward the Maxwell–Boltzmann form of Eq. (6.60). This theorem provides the statistical foundation of the second law of thermodynamics and the microscopic meaning of entropy.

Historical Measurement of the Maxwell–Boltzmann Distribution

In 1919, Otto Stern performed one of the first direct experimental tests of the velocity distribution of gas atoms in a vacuum beam, the molecular beam experiment [Ste19]. He produced a beam of silver (Ag) atoms evaporated at high temperature and allowed them to travel through a series of narrow slits in a high-vacuum chamber. By measuring the spatial spread and arrival times of atoms from the beam, Stern deduced their velocity distribution and found it to match the theoretical form derived by James Clerk Maxwell and Ludwig Boltzmann:

$$f(v) \propto v^2 e^{-\frac{mv^2}{2k_B T}}. \tag{6.72}$$

This experiment provided key empirical validation of the Maxwell–Boltzmann distribution in real atomic beams, bridging kinetic theory and molecular-beam methods. Modern reconstructions of Stern's apparatus show the same essential result, underlining the robustness of the statistical-mechanics framework.

7

Boltzmann's Entropy Functional

7.1 The $\mathcal{H}$-Theorem

7.1.1 Boltzmann's Functional and Irreversibility

One of Boltzmann's most profound results is the $\mathcal{H}$*-theorem*, which establishes a quantitative measure of irreversibility within kinetic theory [Bol72, Bol96]. The theorem introduces a functional of the one-particle distribution function,

$$\mathcal{H}(t) = \int d^3p \, f_1(\vec{p}, t) \, \log f_1(\vec{p}, t), \tag{7.1}$$

and shows that, under the assumptions of molecular chaos and the Boltzmann equation, $\mathcal{H}(t)$ can only decrease in time:

$$\frac{d\mathcal{H}}{dt} \leq 0. \tag{7.2}$$

The stationary condition $d\mathcal{H}/dt = 0$ identifies equilibrium. Because $-\mathcal{H}(t)$ behaves like an increasing quantity, Boltzmann interpreted it as being proportional to entropy. Let us stress here that we call the functional $\mathcal{H}$ above for historical reasons, but it corresponds to $-S$, i.e., minus the entropy of the one-body distribution. Thus, it does not have to be confused with the *Hamiltonian* H.

REMARK 7.1 Historically, the $\mathcal{H}$-theorem represented the first explicit attempt to derive the second law of thermodynamics from mechanical principles. It shows that the irreversible approach to equilibrium emerges naturally from the statistical treatment of microscopic collisions.

Differentiating Eq. (7.1) with respect to time gives

$$\frac{d\mathcal{H}}{dt} = \int d^3p \, \frac{\partial f_1(\vec{p}, t)}{\partial t} \big(1 + \log f_1(\vec{p}, t)\big). \tag{7.3}$$

DOI: 10.1201/9781003603528-7

Substituting the collision term from the Boltzmann equation,

$$\frac{\partial f_1}{\partial t} = \int d^3p_2\, d^3p_1'\, d^3p_2'\, \delta^3(\vec{P}_i - \vec{P}_f)\, \delta(E_i - E_f)\, M_{fi}\, \big[f_1(\vec{p_1}', t)\, f_1(\vec{p_2}', t) - f_1(\vec{p}_1, t)\, f_1(\vec{p}_2, t)\big], \tag{7.4}$$

we obtain

$$\begin{aligned}\frac{d\mathcal{H}}{dt} &= \int d^3p_1\, d^3p_2\, d^3p_1'\, d^3p_2'\, (1 + \log f_1(\vec{p}_1, t)) \\ &\quad \times \big[f_1(\vec{p_1}', t) f_1(\vec{p_2}', t) - f_1(\vec{p}_1, t) f_1(\vec{p}_2, t)\big]\delta^3(\vec{P}_i - \vec{P}_f)\delta(E_i - E_f)\, M_{fi}.\end{aligned} \tag{7.5}$$

To simplify the expression, let us define

$$d\Gamma_P = \delta^3(\vec{P}_i - \vec{P}_f)\delta(E_i - E_f)\, M_{fi}\, d^3p_1\, d^3p_2\, d^3p_1'\, d^3p_2', \tag{7.6}$$

which represents the element of phase-space volume for all possible binary collisions.
Because the integrand is symmetric under the exchange $\vec{p}_1 \leftrightarrow \vec{p}_2$, we may write

$$\frac{d\mathcal{H}}{dt} = \frac{1}{2}\int d\Gamma_P \big[f_1(\vec{p_1}', t) f_1(\vec{p_2}', t) - f_1(\vec{p}_1, t) f_1(\vec{p}_2, t)\big]\big[2 + \log f_1(\vec{p}_1, t) f_1(\vec{p}_2, t)\big]. \tag{7.7}$$

By performing an additional exchange $\vec{p}_1 \leftrightarrow \vec{p_1}'$ and summing the two expressions, Boltzmann obtained

$$\begin{aligned}\frac{d\mathcal{H}}{dt} &= \frac{1}{4}\int d\Gamma_P \big[\log f_1(\vec{p}_1, t) f_1(\vec{p}_2, t) - \log f_1(\vec{p_1}', t) f_1(\vec{p_2}', t)\big] \\ &\quad \times \big[f_1(\vec{p_1}', t) f_1(\vec{p_2}', t) - f_1(\vec{p}_1, t) f_1(\vec{p}_2, t)\big].\end{aligned} \tag{7.8}$$

The integrand has the general form

$$(x - y)\,(\log x - \log y), \tag{7.9}$$

which is always non-positive, since $(x - y)(\log x - \log y) \geq 0$ for any positive x, y. Hence,

$$\frac{d\mathcal{H}}{dt} \leq 0, \tag{7.10}$$

with equality only at equilibrium, where detailed balance holds.

REMARK 7.2 The monotonic decrease of $\mathcal{H}$ implies that $\mathcal{H}$ acts as a Lyapunov functional for the Boltzmann equation. The system evolves irreversibly toward the stationary Maxwell–Boltzmann distribution.

From $\mathcal{H}$ to Entropy

Boltzmann noticed that the functional $\mathcal{H}(t)$ can be interpreted as an ensemble average:

$$\mathcal{H}(t) = \langle \log f_1(\vec{p}, t) \rangle = \int d^3p\, d^3q\, f_1(\vec{q}, \vec{p}, t)\, \log f_1(\vec{q}, \vec{p}, t). \tag{7.11}$$

The quantity

$$S_{\mathrm{B}}(t) = -k_B \mathcal{H}(t) \tag{7.12}$$

is therefore proportional to the entropy of the gas. The $\mathcal{H}$-theorem thus provides a dynamical justification for the increase of entropy in isolated systems.

Boltzmann's identification of $S_B = -k_B\mathcal{H}$ gave entropy a statistical meaning: it measures the logarithm of the number of microscopic configurations compatible with the macroscopic state [Bol72, Bru76, Cer98].

Discrete Phase Space and Equipartition

To make this idea more transparent, imagine discretizing phase space into finite cells of volume h^3. The continuous integral becomes

$$S_{\rm B}(t) \approx -k_B \sum_{\gamma \in \Gamma} h^3 f(\gamma, t) \log f(\gamma, t), \tag{7.13}$$

where each cell $\gamma = (\vec{q}_i, \vec{p}_j)$ represents a region in phase space. Maximizing $S_{\rm B}$ with respect to f under normalization leads to

$$\delta_f S_{\rm B} = 0 \quad \Rightarrow \quad f(\gamma) = \frac{1}{\Omega}, \tag{7.14}$$

where Ω is the total number of phase-space cells accessible to the system.

This result expresses the *equipartition of phase space*: in equilibrium, all microstates consistent with the macroscopic constraints are equally probable. This principle will later serve as one of the foundational postulates of statistical mechanics.

REMARK 7.3 The $\mathcal{H}$-theorem therefore not only explains the irreversible approach to equilibrium but also anticipates the statistical definition of entropy and the micro-canonical ensemble. It unifies mechanics, probability, and thermodynamics into a single conceptual framework.

A key conceptual step in Boltzmann's reasoning was the *discretization* of phase space. To connect the microscopic description to thermodynamic entropy, we must count the number of accessible microstates compatible with given macroscopic constraints.

If the phase space is divided into small cells of volume h^{3N}, the number of microstates compatible with a certain macroscopic configuration is approximately

$$\Omega \propto \frac{\int d\Gamma_V}{h^{3N}}, \tag{7.15}$$

where $d\Gamma_V$ represents the constrained phase-space volume (for instance, the region of phase space satisfying a fixed total energy). Different choices of constraints—fixed energy, fixed temperature, fixed chemical potential—lead to different statistical ensembles, which we will introduce formally in the next chapters.

If all microstates within this accessible region are equally probable (the *principle of equiprobability*), the entropy takes the familiar Boltzmann form:

$$S = -k_{\rm B} \sum_{i \in \Omega} \frac{1}{|\Omega|} \log\left(\frac{1}{|\Omega|}\right) = k_{\rm B} \log |\Omega|. \tag{7.16}$$

This expression encapsulates the statistical meaning of entropy: it is proportional to the logarithm of the number of microscopic configurations compatible with the macroscopic state.

REMARK 7.4 Equation (7.16) completes Boltzmann's bridge between microscopic dynamics and macroscopic thermodynamics. In the next chapters we will see that different ways of constraining the accessible region in phase space give rise to the *micro-canonical*, *canonical*, and *grand-canonical* ensembles.

Scattering and the Origin of Transport

Before discussing the BBGKY hierarchy, it is useful to recall the mechanical foundations of *scattering processes*, which underlie the microscopic description of collisions and transport phenomena. We consider two particles interacting via a *central potential* $V(|\vec{r}_1 - \vec{r}_2|)$. The total

FIGURE 7.1 The tombstone of Ludwig Boltzmann (1844–1906), whose work established the statistical foundations of thermodynamics. Located in Zentralfriedhof in Wien, Austria.

Hamiltonian is

$$H = \frac{p_1^2}{2m_1} + \frac{p_2^2}{2m_2} + V(|\vec{r}_1 - \vec{r}_2|). \tag{7.17}$$

To analyze the relative motion, we introduce the center-of-mass coordinates:

$$\vec{r} = \vec{r}_1 - \vec{r}_2, \qquad \vec{R} = \frac{m_1\vec{r}_1 + m_2\vec{r}_2}{m_1 + m_2}, \tag{7.18}$$

and the reduced mass

$$\mu = \frac{m_1 m_2}{m_1 + m_2}. \tag{7.19}$$

In the center-of-mass frame, the relative motion obeys

$$\mu\ddot{\vec{r}} = -\nabla V(r), \tag{7.20}$$

while the center of mass moves uniformly with velocity

$$\vec{V}_{\rm cm} = \frac{m_1\dot{\vec{r}}_1 + m_2\dot{\vec{r}}_2}{m_1 + m_2}. \tag{7.21}$$

We now focus on the relative motion described by Eq. (7.20).

If $V(r) = 0$, the two particles move along straight lines. In this case, the scattering is purely geometric and governed by angular momentum conservation. The motion is confined to a plane, which can be chosen as the y–z plane without loss of generality. Defining the *impact parameter* b as the perpendicular distance between the initial trajectory and the line through the target particle (see Fig. 7.2), the conserved angular momentum is

$$L = \mu\,|\vec{r} \times \dot{\vec{r}}| = \mu\, b\, v_\infty, \tag{7.22}$$

where $v_\infty = \lim_{t\to-\infty} |\dot{\vec{r}}(t)|$ is the relative velocity at large separations. The total energy is then

$$E = \frac{1}{2}\mu v_\infty^2. \tag{7.23}$$

From angular momentum conservation,

$$\dot{\theta} = -\frac{b v_\infty}{r^2}. \tag{7.24}$$

The quantity $\theta(b)$–the scattering angle as a function of the impact parameter—will play a central role in describing collision cross sections and transport coefficients.

For a general central potential $V(r) \neq 0$, the total energy of the relative motion is

$$E = \frac{\mu}{2}\dot{r}^2 + \frac{\mu r^2 \dot{\theta}^2}{2} + V(r). \tag{7.25}$$

Using $L = \mu r^2 \dot{\theta}$, this can be expressed as

$$H = \frac{\mu}{2}\dot{r}^2 + \frac{L^2}{2\mu r^2} + V(r) \equiv \frac{\mu}{2}\dot{r}^2 + V_{\text{eff}}(r), \tag{7.26}$$

where $V_{\text{eff}}(r)$ is the *effective potential*, incorporating both the central potential and the centrifugal barrier.

The equations of motion follow directly from Hamilton's equations:

$$\begin{aligned} \frac{dr}{dt} &= \pm\sqrt{\frac{2}{\mu}\left[E - V_{\text{eff}}(r)\right]}, \\ \frac{d\theta}{dt} &= \pm\frac{L}{\mu r^2}. \end{aligned} \tag{7.27}$$

These equations describe the trajectory of a particle under a central potential in the center-of-mass frame. They form the mechanical basis of the collision term introduced in the Boltzmann equation and, more generally, of the BBGKY hierarchy discussed next.

REMARK 7.5 The scattering problem provides the microscopic justification for the probabilistic description of collisions used in kinetic theory. Each collision event corresponds to a trajectory determined by $V_{\text{eff}}(r)$, and the scattering angle $\theta(b)$ connects microscopic dynamics to measurable cross sections. The systematic study of scattering processes dates back to the mid-nineteenth century. Rutherford's experiments (1911) provided the first direct evidence that atomic interactions could be understood as scattering by a central potential, later formalized within the framework of classical and quantum mechanics. The microscopic description of collisions was subsequently incorporated into kinetic theory by Maxwell and Boltzmann [Max67, Bol72, Bol96], who realized that macroscopic transport properties such as viscosity and thermal conductivity arise from molecular scattering events.

The rigorous connection between microscopic dynamics and macroscopic transport was later established through the *BBGKY hierarchy* (Bogoliubov–Born–Green–Kirkwood–Yvon) [Bog46, BG46, Kir46, Yvo35]. This framework formalized the reduction from the Liouville equation for N particles to coupled equations for reduced distribution functions, providing the modern foundation of non-equilibrium statistical mechanics.

For detailed historical accounts of the development of kinetic theory and scattering, see Brush [Bru76], Cercignani [Cer98], and Uhlenbeck and Ford [UF63].

The Deflection Function and Central-Force Scattering

Let us now consider in more detail the geometry of scattering in a *central potential*. For simplicity, we focus on repulsive interactions. After being deflected by the potential, the particle emerges along a straight-line trajectory forming a definite *scattering angle*, which we shall denote by the *deflection function* $\Theta(b)$, where b is the impact parameter.

The motion takes place in a plane, which we choose to be the y–z plane. Using Eqs. (7.27) and (7.26), the relation between r and θ can be expressed as

$$\frac{d\theta}{dr} = \frac{d\theta/dt}{dr/dt} = \pm\frac{L}{r^2\sqrt{2\mu\left[E - V_{\text{eff}}(r)\right]}}, \tag{7.28}$$

where the upper or lower sign corresponds to the two symmetric halves of the trajectory before and after the turning point.

The distance of closest approach, or *turning point*, $r_{\rm tp}$, is defined implicitly by

$$E = V_{\rm eff}(r_{\rm tp}), \tag{7.29}$$

so that the radial velocity vanishes at this point. During the approach ($t < 0$), $\dot{r} < 0$ and $\dot{\theta} < 0$; after scattering ($t > 0$), $\dot{r} > 0$ while $\dot{\theta}$ retains its sign. Hence, in Eq. (7.28), the negative sign corresponds to the incoming trajectory, and the positive to the outgoing one. Because the motion is symmetric about the line of closest approach, the total angular change between $t = -\infty$ and $t = +\infty$ is twice the angular displacement from infinity to $r_{\rm tp}$.

Integrating Eq. (7.28) gives

$$\Theta(b) = \pi - 2\int_{r_{\rm tp}}^{\infty} \frac{L\,dr}{r^2\sqrt{2\mu\left[E - V_{\rm eff}(r)\right]}}. \tag{7.30}$$

The angle $\Theta(b)$ measures the deviation of the outgoing trajectory from its initial direction and is determined entirely by the effective potential and the impact parameter b.

Using $L = \mu b v_\infty$ and $E = \frac{1}{2}\mu v_\infty^2$, Eq. (7.30) can be written in the equivalent form

$$\Theta(b) = \pi - 2b\int_{r_{\rm tp}}^{\infty} \frac{dr}{r^2\sqrt{1 - \frac{b^2}{r^2} - \frac{V(r)}{E}}}. \tag{7.31}$$

This expression provides the general deflection function for scattering in any central potential.

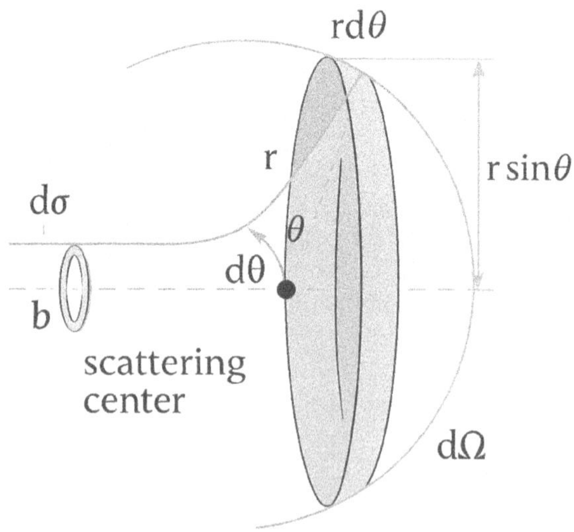

FIGURE 7.2 Geometry of a binary scattering event. The impact parameter b determines the deflection (or scattering) angle $\Theta(b)$.

Examples of Common Potentials

Several forms of $V(r)$ are of physical interest:

- *Coulomb potential:* $V(r) = \frac{C}{r}$,
- *Inverse-power potential:* $V(r) = \frac{C_\alpha}{r^\alpha}$,
- *Lennard–Jones potential:* $V(r) = \frac{C_{12}}{r^{12}} - \frac{C_6}{r^6}$.

For long-range repulsive potentials of the form $V(r) \simeq C_\alpha/r^\alpha$ as $r \gg 1$, it is convenient to introduce the dimensionless variable

$$\xi = \frac{r}{r_{\text{tp}}}, \tag{7.32}$$

and define

$$\epsilon = \frac{r_{\text{tp}}^2}{b^2} - 1. \tag{7.33}$$

Equation (7.31) can then be rewritten as

$$\Theta(b) = \pi - 2\int_1^\infty \frac{d\xi}{\sqrt{\xi^4 - \xi^2 + \epsilon\,(\xi^4 - \xi^{4-\alpha})}}. \tag{7.34}$$

This integral cannot generally be evaluated in closed form. However, for weak deflections ($b \gg r_{\text{tp}}$, or equivalently $\epsilon \ll 1$), an expansion yields the approximate result

$$\Theta(b) \simeq \frac{C_\alpha}{E\,b^\alpha}\,\frac{\sqrt{\pi}\,\Gamma\left(\frac{\alpha+1}{2}\right)}{\Gamma\left(\frac{\alpha}{2}\right)}, \tag{7.35}$$

where $\Gamma(x)$ denotes the Gamma function.

REMARK 7.6 Equation (7.35) shows that the deflection angle decreases as a power of the impact parameter, with the exponent determined by the range of the interaction. Such relations are fundamental in determining the differential scattering cross section, a quantity of central importance in both classical and quantum collision theory [Rut11, Bor26, UF63, Bru76].

Cross Section and Scattering Probability in a Gas

Let us now connect the geometric picture of scattering to the statistical description of collisions in a gas. In non-relativistic systems, the scattering process depends only on the *relative velocity* between two colliding particles,

$$\vec{v}_{\text{rel}} = \vec{v}_1 - \vec{v}_2. \tag{7.36}$$

The magnitude of this relative velocity determines the *flux of incident particles* per unit area per unit time. For a gas with number density $n = N/V$, the incident flux is therefore

$$\mathcal{I} = n\,|\vec{v}_1 - \vec{v}_2|. \tag{7.37}$$

We consider again Fig. 7.2. If the impact parameter b is known only within an infinitesimal range db, the corresponding uncertainty in the deflection (scattering) angle is $d\Theta(b)$. A small ring of impact parameters between b and $b + db$ corresponds to a solid angle element $d\Omega$ in the scattered directions.

The differential area perpendicular to the incident flux is

$$d\sigma = b\,db\,d\phi, \tag{7.38}$$

while the element of solid angle in the direction of scattering is

$$d\Omega = \sin\theta\,d\theta\,d\phi. \tag{7.39}$$

The number of particles scattered into $d\Omega$ per unit time is proportional to the incident flux times the corresponding effective area:

$$\mathcal{I}\,\frac{d\sigma}{d\Omega}\,d\Omega = \mathcal{I}\,b\,db\,d\phi. \tag{7.40}$$

It follows that the *differential scattering cross section* is given by

$$\frac{d\sigma}{d\Omega} = \frac{b}{\sin\theta}\left|\frac{db}{d\theta}\right| = \frac{1}{2}\left|\frac{db^2}{d(\cos\theta)}\right|. \tag{7.41}$$

The function $\Theta(b)$, determined earlier, provides the deflection angle as a function of the impact parameter. Since we often require $b(\theta) = \Theta^{-1}(\theta)$, we can write

$$\frac{db}{d\theta} = \frac{d\Theta^{-1}(\theta)}{d\theta} = \frac{1}{\Theta'(b)}. \tag{7.42}$$

This relation expresses the differential cross section entirely in terms of the deflection function.

We can now relate the differential cross section to the transition probability M_{fi} introduced previously. The probability that two particles with incoming momenta $(\vec{p}_1, \vec{p}_2)$ scatter into outgoing states $(\vec{p}_1{}', \vec{p}_2')$ within an element of solid angle $d\Omega$ is

$$M_{fi}\,\delta^3(\vec{P}_i - \vec{P}_f)\,\delta(E_i - E_f)\,d^3p_1'\,d^3p_2' = |\vec{v}_1 - \vec{v}_2|\left|\frac{d\sigma}{d\Omega}\right| d\Omega. \tag{7.43}$$

Equation (7.43) shows that the microscopic transition rate is proportional to the relative velocity and the differential cross section. This provides the missing link between the probabilistic collision term of the Boltzmann equation and the mechanical description of scattering.

At equilibrium, only the functional form of the Maxwell–Boltzmann distribution matters, and the detailed cross section plays no role. However, away from equilibrium—when studying *transport phenomena*, relaxation, or viscosity—the specific form of $d\sigma/d\Omega$ becomes crucial. It determines the rate at which the system approaches equilibrium and thus encodes all information about microscopic interactions [CC70, UF63].

7.2 Bogoliubov–Born–Green–Kirkwood–Yvon Hierarchy

The so-called *BBGKY hierarchy* (for Bogoliubov, Born, Green, Kirkwood, and Yvon) provides a rigorous bridge between microscopic Hamiltonian dynamics and the kinetic or hydrodynamic equations of macroscopic physics [Bog46, BG46, Kir46, Yvo35]. It generalizes Boltzmann's reasoning to an interacting many-body system and forms the mathematical foundation of modern transport theory and non-equilibrium statistical mechanics.

REMARK 7.7 While Boltzmann's transport equation rests on the assumption of molecular chaos and binary collisions, the BBGKY hierarchy retains all correlations between particles. In this sense, it is exact and constitutes a systematic framework from which approximate kinetic equations (Boltzmann, Landau, Vlasov, Fokker–Planck, etc.) can be derived.

Historically, the hierarchy was developed independently by Jacques Yvon (1935), Max Born and Herbert S. Green (1946), John G. Kirkwood (1946), and Nikolay N. Bogoliubov (1946), each building upon Liouville's theorem and the statistical interpretation of correlations. Its relevance extends far beyond statistical mechanics, influencing plasma physics, fluid dynamics, and quantum kinetic theory [UF63, Bru76, Cer98].

Motivation and Setting

We consider a system of N interacting particles described by the Hamiltonian

$$H = \sum_{i=1}^{N} \frac{\vec{p}_i^{\,2}}{2m} + \sum_{i=1}^{N} V(\vec{q}_i) + \frac{1}{2}\sum_{i\neq j} U(\vec{q}_i - \vec{q}_j) = \sum_i H_i + V_{\text{int}}. \tag{7.44}$$

Here, $V(\vec{q}_i)$ is an external potential, and $U(\vec{q}_i - \vec{q}_j)$ is the pairwise interaction potential between particles. If $U = 0$, the system is collisionless and particles evolve independently under the single-particle Hamiltonians H_i.

The dynamics of the system in phase space is governed by Hamilton's equations,

$$\begin{aligned} \frac{d\vec{q}_i}{dt} &= \frac{\partial H}{\partial \vec{p}_i} = \frac{\vec{p}_i}{m}, \\ \frac{d\vec{p}_i}{dt} &= -\frac{\partial H}{\partial \vec{q}_i} = -\nabla_{\vec{q}_i} V(\vec{q}_i) - \sum_{j\neq i} \nabla_{\vec{q}_i} U(\vec{q}_i - \vec{q}_j). \end{aligned} \tag{7.45}$$

The complete state of the system is described by the N-particle distribution function $f_N(\{\vec{q}_i, \vec{p}_i\}, t)$, normalized as

$$\int d\Gamma\, f_N(\{\vec{q}_i, \vec{p}_i\}, t) = 1, \qquad d\Gamma = \prod_{i=1}^{N} d^3q_i\, d^3p_i. \tag{7.46}$$

This distribution satisfies Liouville's equation,

$$\frac{\partial f_N}{\partial t} + \{f_N, H\} = 0, \tag{7.47}$$

where $\{\cdot,\cdot\}$ denotes the classical Poisson bracket.

Since it is impossible to follow all $6N$ variables in practice, we define the reduced k-particle distribution functions by integrating over the remaining coordinates:

$$f_k(\{\vec{q}_i, \vec{p}_i\}_{i=1}^{k}, t) = \int d\Gamma_{k+1\cdots N}\, f_N(\{\vec{q}_i, \vec{p}_i\}, t). \tag{7.48}$$

Thus, f_1 describes single-particle statistics, f_2 includes two-body correlations, and so on.

REMARK 7.8 The reduced distributions f_k contain increasingly detailed information about correlations among particles. In the limit $N \to \infty$, f_1 gives the mean-field (kinetic) behavior, while f_2 and higher terms encode fluctuations and collective effects.

Integrating Liouville's equation (7.47) over the coordinates of particles $2, 3, \ldots, N$ yields the evolution equation for f_1:

$$\begin{aligned} \frac{\partial f_1}{\partial t} &= -\int d\Gamma_{2\cdots N}\, \{f_N, H\} \\ &= \{f_1, H_1\} + \int d^3q_2\, d^3p_2\, \nabla_{\vec{q}_1} U(\vec{q}_1 - \vec{q}_2) \cdot \nabla_{\vec{p}_1} f_2(\vec{q}_1, \vec{p}_1, \vec{q}_2, \vec{p}_2, t). \end{aligned} \tag{7.49}$$

The first term describes free streaming under the single-particle Hamiltonian $H_1 = \frac{\vec{p}_1^{\,2}}{2m} + V(\vec{q}_1)$, while the second term accounts for interactions with other particles through the two-body distribution f_2. Equation (7.49) is the first equation in the BBGKY hierarchy.

Equation (7.50) forms the basis of kinetic and transport theories. For example, the Boltzmann equation corresponds to truncating the hierarchy at $k = 1$ with $f_2 = f_1 f_1$, while hydrodynamic and plasma models follow from more elaborate closures. In Volume II, we extend these ideas to quantum systems and collective phenomena such as Landau damping and transport in correlated media.

Similarly, integrating over all but k coordinates yields the general form:

$$\frac{\partial f_k}{\partial t} + \sum_{i=1}^{k} \left(\frac{\vec{p}_i}{m} \cdot \nabla_{\vec{q}_i} - \nabla_{\vec{q}_i} V(\vec{q}_i) \cdot \nabla_{\vec{p}_i} \right) f_k = \sum_{i=1}^{k} \int d^3 q_{k+1}\, d^3 p_{k+1}\, \nabla_{\vec{q}_i} U(\vec{q}_i - \vec{q}_{k+1}) \cdot \nabla_{\vec{p}_i} f_{k+1}. \tag{7.50}$$

Each equation in the hierarchy couples the k-particle distribution f_k to the next one f_{k+1}. This expresses the natural fact that the evolution of k-body correlations depends on interactions with additional particles. The hierarchy is infinite; truncations or approximations (e.g., the molecular chaos assumption $f_2 \approx f_1 f_1$) lead to the kinetic equations used in practice.

REMARK 7.9 It is often convenient to introduce the symmetrized quantities

$$\tilde{f}_k = \frac{N!}{(N-k)!} f_k, \tag{7.51}$$

which count the number of ways of selecting k particles from N. In terms of $\tilde{f}_k$, the hierarchy takes the compact form

$$\frac{\partial \tilde{f}_k}{\partial t} = \{\tilde{f}_k, H_k\} + \sum_{i=1}^{k} \int d^3 q_{k+1}\, d^3 p_{k+1}\, \nabla_{\vec{q}_i} U(\vec{q}_i - \vec{q}_{k+1}) \cdot \nabla_{\vec{p}_i} \tilde{f}_{k+1}. \tag{7.52}$$

This is the *Bogoliubov–Born–Green–Kirkwood–Yvon (BBGKY) hierarchy.* It is exact for any classical many-body system and serves as the starting point for all systematic derivations of kinetic and transport equations.

REMARK 7.10 The BBGKY hierarchy clarifies that irreversibility, transport, and relaxation emerge not from the fundamental dynamics (which are time-reversible), but from successive approximations and truncations of Eq. (7.52). It is therefore the modern framework for understanding how macroscopic laws arise from microscopic mechanics.

Truncation of the BBGKY Hierarchy and the Collision Term

We have seen that the BBGKY hierarchy provides an exact but infinite chain of coupled equations linking the k-particle distribution functions f_k and f_{k+1}. In practice, this hierarchy must be *truncated* at some finite order to obtain solvable kinetic equations. The simplest and most important truncation is at first order, which leads directly to the Boltzmann transport equation.

Physically, truncation corresponds to assuming that higher-order correlations are negligible or can be expressed in terms of lower-order ones. This is precisely the content of the molecular chaos assumption used by Boltzmann.

To visualize these dependencies, one may represent the k-particle distribution functions as *factor graphs*, where each node (or "blob") represents a particle and each edge represents a correlation or interaction, as illustrated in Fig. 7.3. Truncating the hierarchy corresponds to cutting these graphs at a given order—retaining only a finite number of "legs" in the diagrammatic expansion.

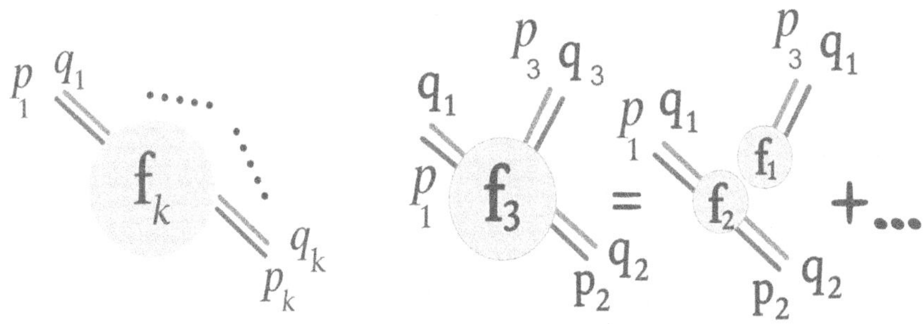

FIGURE 7.3 Schematic representation of k-particle distribution functions as factor graphs. Higher-order correlations correspond to graphs with more "legs," and truncations correspond to neglecting connections beyond a given order and decomposing interactions to lower-order corrections.

Let us examine the second level of the hierarchy. We consider the stationary condition for the two-particle distribution function,

$$\frac{df_2}{dt} = 0, \tag{7.53}$$

which implies, from Eq. (7.50),

$$\left(\frac{\partial}{\partial t} + \frac{1}{m}\left(\vec{p}_1 \cdot \nabla_{\vec{q}_1} + \vec{p}_2 \cdot \nabla_{\vec{q}_2}\right) - \nabla_{\vec{q}_1} U(\vec{q}_1 - \vec{q}_2) \cdot \left(\nabla_{\vec{p}_1} - \nabla_{\vec{p}_2}\right)\right) f_2 = 0. \tag{7.54}$$

Equation (7.54) describes the correlated motion of a pair of particles interacting via a potential $U(\vec{q}_1 - \vec{q}_2)$.

To simplify, we introduce the standard transformation to relative and center-of-mass variables:

$$\vec{p} = \frac{\vec{p}_1 - \vec{p}_2}{2}, \qquad \vec{P} = \frac{\vec{p}_1 + \vec{p}_2}{2}, \qquad \vec{r} = \frac{\vec{q}_2 - \vec{q}_1}{2}, \qquad \vec{R} = \frac{\vec{q}_1 + \vec{q}_2}{2}. \tag{7.55}$$

Assuming that variations in the center-of-mass position $\vec{R}$ and momentum $\vec{P}$ are slow compared with variations in $\vec{r}$ and $\vec{p}$, we can neglect their derivatives near equilibrium. Under this approximation, Eq. (7.54) reduces to the Liouville equation for two interacting particles:

$$\left(\frac{\vec{p}}{m} \cdot \frac{\partial}{\partial \vec{r}} - \frac{\partial U(\vec{r})}{\partial \vec{r}} \cdot \frac{\partial}{\partial \vec{p}}\right) f_2 = 0. \tag{7.56}$$

Using Eq. (7.56), the collisional contribution to the evolution of f_1 (from Eq. (7.49)) can be written as

$$\begin{aligned}\left(\frac{\partial f_1}{\partial t}\right)_{\text{coll}} &= \int d^3q_2\, d^3p_2\, \nabla_{\vec{q}_1} U(\vec{q}_1 - \vec{q}_2) \cdot \nabla_{\vec{p}_1} f_2(\vec{q}_1, \vec{p}_1, \vec{q}_2, \vec{p}_2, t) \\ &= \int d^3q_2\, d^3p_2\, \frac{\partial U(\vec{r})}{\partial \vec{r}} \cdot \left(\frac{\partial f_2}{\partial \vec{p}_1} - \frac{\partial f_2}{\partial \vec{p}_2}\right).\end{aligned} \tag{7.57}$$

For weak, short-range interactions, we approximate

$$\frac{\partial U(\vec{r})}{\partial \vec{r}} \cdot \frac{\partial f_2}{\partial \vec{p}} \simeq \frac{\vec{p}_1 - \vec{p}_2}{m} \cdot \frac{\partial f_2}{\partial \vec{r}}, \tag{7.58}$$

which leads to

$$\left(\frac{\partial f_1}{\partial t}\right)_{\text{coll}} \approx \int d^3p_2\, (\vec{v}_1 - \vec{v}_2) \cdot \int d^3r\, \frac{\partial f_2}{\partial \vec{r}}. \tag{7.59}$$

Transforming to cylindrical coordinates and using the definition of the differential cross section, this becomes

$$\left(\frac{\partial f_1}{\partial t}\right)_{\text{coll}} = \int d^3p_2\,|\vec{v}_1 - \vec{v}_2| \left|\frac{d\sigma}{d\Omega}\right| d\Omega \big[f(\vec{q}_1{}', \vec{p}_1{}', \vec{q}_2{}', \vec{p}_2{}') - f(\vec{q}_1, \vec{p}_1, \vec{q}_2, \vec{p}_2)\big]. \tag{7.60}$$

If the two particles are identical, the right-hand side of Eq. (7.60) must be multiplied by a factor 1/2, or equivalently by $1 - \frac{1}{2}\delta_{12}$, to avoid double counting of indistinguishable collisions.

Equation (7.60) is precisely the collision integral that appears in the Boltzmann transport equation, thus demonstrating how it emerges from the truncation of the BBGKY hierarchy.

Higher-Order Corrections

In principle, one could continue this procedure to higher orders in the hierarchy. At the level of f_3, three-body correlations begin to contribute, and the algebra rapidly becomes more complex. In diagrammatic terms, these higher-order terms correspond to graphs with additional "legs," representing more intricate collision and correlation processes. While the general formalism exists, practical solutions typically require closure approximations or perturbative methods, which are beyond the scope of the present Volume.

REMARK 7.11 The hierarchy truncation elucidates a deep conceptual point: all kinetic equations, from Boltzmann's to Vlasov's or Fokker–Planck's, are *approximations* to the BBGKY hierarchy. Their domain of validity depends on the nature of the interactions and the timescales of correlations. In Volume II, we will revisit these ideas in the context of quantum kinetic theory and transport phenomena in correlated systems.

7.3 Implications of the Maxwell–Boltzmann Distribution and Generalization

The formalism developed so far justifies the Maxwell–Boltzmann equilibrium distribution as the stationary solution of the Boltzmann equation. At equilibrium, the collision term (7.60) vanishes, corresponding to detailed balance. This provides a microscopic foundation for the equilibrium laws of thermodynamics derived earlier.

We can now make explicit the connection between Boltzmann's $\mathcal{H}$-function and the thermodynamic entropy of an ideal gas. Using the Maxwell–Boltzmann equilibrium distribution,

$$f_1^{\text{eq}}(\vec{p}) = \frac{N}{V}\left(\frac{m}{2\pi k_{\text{B}}T}\right)^{3/2} e^{-\frac{p^2}{2mk_{\text{B}}T}}, \tag{7.61}$$

we compute

$$\begin{aligned}\mathcal{H} = \langle \log f_1^{\text{eq}}\rangle &= \int d^3p\, f_1^{\text{eq}}(\vec{p})\,\log f_1^{\text{eq}}(\vec{p}) \\ &= \frac{N}{V}\log\left[\frac{N}{V(2\pi mk_{\text{B}}T)^{3/2}}\right] - \frac{3N}{2V}.\end{aligned} \tag{7.62}$$

Rearranging,

$$\mathcal{H} = -\frac{3N}{2V}\log\left(T\frac{V^{2/3}}{N^{2/3}}\right) - \frac{N}{V}\left[\frac{3}{2} + \log\left(k_{\text{B}}\sqrt{2\pi m^3}\right)\right]. \tag{7.63}$$

Using the ideal-gas equation of state $PV = Nk_\mathrm{B}T$, this becomes

$$\mathcal{H} = -\frac{3N}{2V}\log\left(\frac{PV^{5/3}}{N^{5/3}}\right) - \frac{N}{V}\left[\frac{3}{2} + \log\left(\sqrt{2\pi m^3}\right) - \frac{5}{2}\log N\right]. \tag{7.64}$$

Up to additive constants of order $\log N$, this expression coincides with the entropy density of an ideal gas.

For an ideal gas,

$$S = Nk_\mathrm{B}\log\left(T^{3/2}V\right) = \frac{3}{2}Nk_\mathrm{B}\log\left(PV^{5/3}\right) - \frac{3}{2}Nk_\mathrm{B}\log N, \tag{7.65}$$

and therefore, apart from additive constants,

$$S = -k_\mathrm{B}V\mathcal{H}. \tag{7.66}$$

Equation (7.66) identifies Boltzmann's $\mathcal{H}$-functional with the thermodynamic entropy density. The $\mathcal{H}$-theorem thus provides not only the microscopic origin of the second law ($dS/dt \geq 0$) but also the link between mechanics and thermodynamics.

Equation (7.66) is valid in the *thermodynamic limit*, $N \to \infty$ and $V \to \infty$ with N/V fixed. For finite or small systems, fluctuations of f_1 around its mean become important, and the correspondence between $\mathcal{H}$ and S breaks down. Such corrections are the domain of fluctuation theory and stochastic thermodynamics.

Since the internal energy is $U = N\langle E\rangle = 3Nk_\mathrm{B}T/2$, the differential form of the first law follows as usual,

$$dQ = dU + P\,dV, \tag{7.67}$$

while the $\mathcal{H}$-theorem ensures the second law, $dS \geq 0$, completing the derivation of thermodynamics from kinetic theory.

Inclusion of a Conservative Force Field

Up to now, we have considered a spatially homogeneous gas with no external forces ($\vec{F} = 0$). Let us now introduce a conservative external field characterized by a potential energy $\phi(\vec{q})$ such that

$$\vec{F} = -\nabla_{\vec{q}}\,\phi(\vec{q}). \tag{7.68}$$

At equilibrium, the stationary solution of the Boltzmann equation must satisfy

$$\left(\frac{\vec{p}}{m}\cdot\nabla_{\vec{q}} + \vec{F}\cdot\nabla_{\vec{p}}\right) f(\vec{q},\vec{p}) = 0. \tag{7.69}$$

We now look for a separable solution of the form

$$f(\vec{q},\vec{p}) = e^{-\frac{\phi(\vec{q})}{k_\mathrm{B}T}} f_1^{\mathrm{eq}}(\vec{p}), \tag{7.70}$$

where $f_1^{\mathrm{eq}}(\vec{p})$ is the homogeneous equilibrium distribution in the absence of forces.

Verification of the Ansatz

Let us verify that Eq. (7.70) satisfies Eq. (7.69). The two terms act as follows:

$$\begin{aligned}
\vec{F}\cdot\nabla_{\vec{p}}f &= f_1^{\mathrm{eq}}(\vec{p})e^{-\phi/k_\mathrm{B}T}\,\vec{F}\cdot\left(-\frac{\vec{p}}{mk_\mathrm{B}T}\right),\\
\frac{\vec{p}}{m}\cdot\nabla_{\vec{q}}f &= f_1^{\mathrm{eq}}(\vec{p})\frac{\vec{p}}{m}\cdot\nabla_{\vec{q}}\left(e^{-\phi/k_\mathrm{B}T}\right) = -f_1^{\mathrm{eq}}(\vec{p})e^{-\phi/k_\mathrm{B}T}\frac{\vec{p}}{mk_\mathrm{B}T}\cdot\vec{F}.
\end{aligned} \tag{7.71}$$

The two terms cancel exactly, and thus Eq. (7.70) is indeed a stationary solution of the Boltzmann equation.

Combining both factors, the equilibrium distribution in the presence of a conservative potential becomes

$$f(\vec{q},\vec{p}) = \frac{n_0}{(2\pi m k_{\mathrm{B}}T)^{3/2}} \exp\left[-\frac{1}{k_{\mathrm{B}}T}\left(\frac{p^2}{2m} + \phi(\vec{q})\right)\right], \tag{7.72}$$

with the normalization condition

$$\int d^3q\, d^3p\, f(\vec{q},\vec{p}) = \frac{N}{V}, \tag{7.73}$$

which fixes the constant n_0 given the external potential $\phi(\vec{q})$. Equation (7.72) is the full *Maxwell–Boltzmann distribution* in an external conservative field.

The term $\exp[-\phi(\vec{q})/(k_{\mathrm{B}}T)]$ is sometimes called *Boltzmann factor*, expressing the probability of finding a particle at position $\vec{q}$ in a potential $\phi(\vec{q})$. It shows that mechanical equilibrium corresponds to a spatial density $n(\vec{q}) \propto e^{-\phi(\vec{q})/(k_{\mathrm{B}}T)}$.

REMARK 7.12 The derivation above is valid only in the thermodynamic limit and for systems sufficiently close to equilibrium. If N is small or if the system exhibits large fluctuations (for example, near critical points, in rarefied gases, or in small-scale systems), the mean-field factorization implicit in Eq. (7.70) fails. In such cases, correlations and fluctuations modify the distribution, and one must use more refined approaches—such as the BBGKY hierarchy, Langevin dynamics, or stochastic thermodynamics—to describe the system accurately.

In Volume II we will revisit this result within the framework of quantum and non-equilibrium statistical mechanics, where fluctuations play a central role and the simple exponential form of Eq. (7.72) must be replaced by ensemble averages over fluctuating fields.

Maxwell–Boltzmann Distribution in an External Harmonic Field

Consider a dilute monoatomic gas of N particles at temperature T, confined near a point $\vec{x}_0$ by a weak external force

$$\vec{F} = -q(\vec{x} - \vec{x}_0), \tag{7.74}$$

corresponding to a harmonic potential $\phi(\vec{x}) = \frac{1}{2}q(\vec{x} - \vec{x}_0)^2$. Even though the gas is effectively non-interacting, the restoring force prevents it from dispersing infinitely.

The equilibrium phase-space distribution is the Maxwell–Boltzmann form in the presence of an external potential,

$$f(\vec{p},\vec{x}) = \frac{n(\vec{x})}{(2\pi m k_{\mathrm{B}}T)^{3/2}} \exp\left(-\frac{p^2}{2mk_{\mathrm{B}}T}\right), \tag{7.75}$$

with a spatial density obtained from the Boltzmann factor,

$$n(\vec{x}) = n_0\, e^{-\phi(\vec{x})/(k_{\mathrm{B}}T)} = n_0\, e^{-q(\vec{x}-\vec{x}_0)^2/(2k_{\mathrm{B}}T)}. \tag{7.76}$$

Normalization, $\int d^3x\, n(\vec{x}) = N$, yields

$$n_0 = \frac{N\, q^{3/2}}{(2\pi k_{\mathrm{B}}T)^{3/2}}, \tag{7.77}$$

so that

$$f(\vec{p},\vec{x}) = \frac{N q^{3/2}}{m^{3/2}(2\pi k_{\mathrm{B}}T)^2} \exp\left[-\frac{p^2}{2mk_{\mathrm{B}}T} - \frac{q(\vec{x}-\vec{x}_0)^2}{2k_{\mathrm{B}}T}\right]. \tag{7.78}$$

This is a separable Gaussian in both position and momentum, describing a thermal cloud confined by a harmonic trap—an archetype of equilibrium under an external conservative force.

A closely related physical situation occurs in modern cold-atom experiments. In magneto-optical traps (MOTs) and optical dipole traps, neutral atoms experience an approximately harmonic restoring force near the trap center, producing a spatial confinement similar to the potential considered here. However, unlike the purely mechanical force field of our model, these traps also rely on *laser cooling* to reduce the kinetic energy of the atoms to the micro-kelvin regime. Once cooled, the trapped atomic cloud settles into a Gaussian spatial distribution and a Maxwell–Boltzmann velocity profile, precisely as predicted by the theory above. The pioneering experiments by Steven Chu and collaborators, and later by William D. Phillips, demonstrated this behavior in detail, measuring spatial and velocity distributions that agree with the Maxwell–Boltzmann form [CHB+85, Phi98].

7.4 Onset of Relativistic Regimes

Our entire discussion so far has relied on the assumptions of classical, non-relativistic kinetic theory. Before concluding this section, it is useful to examine the limits of these approximations and the circumstances under which they fail, besides a small number of particles.

For a monoatomic gas described by the Maxwell–Boltzmann equilibrium distribution, the average kinetic energy per particle is

$$\langle E\rangle = \frac{3}{2}k_{\rm B}T, \tag{7.79}$$

and hence the mean-square velocity is

$$\langle v^2\rangle = \frac{3k_{\rm B}T}{m}, \qquad \Rightarrow \qquad \langle v\rangle = \sqrt{\langle v^2\rangle} = \sqrt{\frac{3k_{\rm B}T}{m}}. \tag{7.80}$$

This is the *root-mean-square speed* obtained by averaging over the Maxwell–Boltzmann distribution.

Another relevant quantity is the *most probable speed*, defined as the value of v (or $p = mv$) that maximizes the distribution $p^2 f_1(p)$. Setting $\partial_p[p^2 f_1(p)] = 0$, we find

$$2p\, f_1(p) + p^2\partial_p f_1(p) = f_1(p)\left(2p - \frac{p^2}{mk_{\rm B}T}\right) = 0, \tag{7.81}$$

which yields

$$\bar{p} = \sqrt{2mk_{\rm B}T}, \qquad \bar{v} = \frac{\bar{p}}{m} = \sqrt{\frac{2k_{\rm B}T}{m}}. \tag{7.82}$$

Thus, for a Maxwell–Boltzmann gas,

$$\bar{v} < \langle v\rangle < v_{\rm rms}, \tag{7.83}$$

the three characteristic velocities differing only by numerical factors of order unity.

The classical treatment assumes that particle speeds are small compared to the speed of light c. The onset of relativistic effects occurs when the most probable speed $\bar{v}$ becomes comparable to c, i.e.,

$$\bar{v} \sim c \quad \Longrightarrow \quad k_{\rm B}T_{\rm rel} \approx \frac{1}{2}mc^2. \tag{7.84}$$

For temperatures approaching $T_{\rm rel}$, relativistic corrections to the energy–momentum relation become significant, and the non-relativistic Maxwell–Boltzmann distribution is no longer valid. At these energies, one must replace

$$E = \frac{p^2}{2m} \quad \longrightarrow \quad E = \sqrt{p^2c^2 + m^2c^4}, \tag{7.85}$$

and reformulate the kinetic theory accordingly.

REMARK 7.13 Equation (7.84) provides only a rough scale for the onset of relativistic behavior. For electrons ($m_ec^2 \simeq 511\,\mathrm{keV}$), the transition occurs near $T_{\rm rel} \sim 3 \times 10^9\,\mathrm{K}$; for protons ($m_pc^2 \simeq 938\,\mathrm{MeV}$), the corresponding temperature is around $10^{13}\,\mathrm{K}$. In most laboratory or terrestrial conditions, $T \ll T_{\rm rel}$, but astrophysical environments (stellar interiors, supernovae, accretion disks) routinely reach and exceed these scales.

Even before reaching relativistic temperatures, classical kinetic theory can break down for other reasons:

- *Collective effects:* in plasmas or charged gases, long-range electromagnetic interactions invalidate the assumption of uncorrelated binary collisions. The dynamics is then governed by the *Vlasov* or *Landau* equations instead of Boltzmann's.
- *Quantum degeneracy:* at low temperatures or high densities, the de Broglie wavelength becomes comparable to the mean interparticle spacing, and quantum statistics (Fermi–Dirac or Bose–Einstein) replace Maxwell–Boltzmann.
- *Relativistic corrections:* when v/c is not negligible, both the expression for kinetic energy and the invariant phase-space measure must be modified. The correct equilibrium distribution is the *Jüttner distribution*:

$$f_{\rm J}(\vec{p}) = A \exp\left(-\frac{E(p)}{k_{\rm B}T}\right), \tag{7.86}$$

where $E(p) = \sqrt{p^2c^2 + m^2c^4}$ and A ensures normalization in relativistic phase space.

The Jüttner distribution (1911) [Jue11] is the relativistic generalization of Maxwell–Boltzmann statistics, also called Maxwell–Jüttner distribution. It reproduces the classical limit for $k_{\rm B}T \ll mc^2$ and converges to a purely exponential, ultrarelativistic form $f(p) \propto e^{-pc/k_{\rm B}T}$ when $k_{\rm B}T \gg mc^2$.

Beyond the Classical Kinetic Theory

The relativistic generalization of the Boltzmann equation is essential for describing high-energy astrophysical plasmas, cosmic rays, and early-universe cosmology. It retains the structure of the classical equation,

$$p^\mu \partial_\mu f = C[f], \tag{7.87}$$

but uses four-vectors $p^\mu = (E/c, \vec{p})$ and Lorentz-invariant phase-space measures. Similarly, in strongly ionized systems, one must include electromagnetic fields via the Lorentz force:

$$\frac{d\vec{p}}{dt} = q(\vec{E} + \vec{v} \times \vec{B}), \tag{7.88}$$

leading to the *Vlasov–Boltzmann* or *relativistic Boltzmann–Maxwell* equations.

For historical and modern accounts of relativistic kinetic theory, see Jüttner [Jue11], Synge [Syn57], de Groot *et al.* [dGvLvW80], and Cercignani and Kremer [CK02].

REMARK 7.14 The framework of kinetic theory thus extends far beyond the ideal gases discussed in this Volume. It provides the foundation for understanding transport, radiation, and collective phenomena in astrophysical, nuclear, and plasma systems. Although we do not discuss these regimes, it is worth knowing their existence.

Thermodynamics in Curved Spacetime

At a deeper level, thermodynamics itself must be reconsidered in curved spacetime [Tol34]. The advent of *black-hole thermodynamics*, beginning with Jacob Bekenstein's formulation of black-hole entropy [Bek73] and Stephen Hawking's discovery of black-hole radiation [Haw75], revealed a profound interplay between gravitation, quantum theory, and thermal physics. Here, entropy is proportional not to volume but to the horizon area, hinting at microscopic degrees of freedom associated with spacetime geometry.

Building on these insights, Thanu Padmanabhan's reviews [Pad10] develop the idea that gravity may be understood as an emergent thermodynamic phenomenon, while Ted Jacobson's work [Jac95] derives the Einstein field equations from the Clausius relation $\delta Q = T\,dS$ applied to local Rindler horizons. Together, these developments extend thermodynamics far beyond materials and gases, suggesting that spacetime itself possesses thermodynamic attributes and that gravitational dynamics may arise from underlying statistical principles.

8

Equilibration and Entropy

8.1 Can the $\mathcal{H}$-Theorem Be Violated?

Boltzmann's $\mathcal{H}$-theorem predicts that the quantity

$$\mathcal{H}(t) = \int f_1(\vec{p}, t) \log f_1(\vec{p}, t)\, d^3p \tag{8.1}$$

decreases monotonically in time, leading the system to thermal equilibrium. However, this conclusion relies on the assumptions of *molecular chaos* and the *ergodic hypothesis*. In this section, we examine situations where these assumptions may fail and explore what modern results tell us about equilibration.

From the Boltzmann Equation to Lanford's Theorem

The *Boltzmann transport equation* (BTE) can be rigorously derived for a gas of *hard spheres*. Under suitable conditions, Lanford's theorem (1975) [Lan75, Cer98] proves that, for short times, the solution to the full BBGKY hierarchy coincides with that obtained under the assumption of molecular chaos. Formally, this means that at equilibrium

$$f_k^{\mathrm{eq}}(\{\vec{p}, \vec{q}\}_k) = \prod_{i=1}^{k} f_1^{\mathrm{eq}}(\vec{p}_i, \vec{q}_i). \tag{8.2}$$

DOI: 10.1201/9781003603528-8

Lanford's result provides the first mathematical justification of the Boltzmann equation for a dilute gas, though only valid up to a fraction of the mean free time. Beyond this regime, correlations and recollisions can break the molecular chaos assumption.

REMARK 8.1 Lanford's proof connected kinetic theory to rigorous mathematical analysis and opened the modern field of non-equilibrium statistical mechanics. Lanford's theorem shows that Boltzmann's equation emerges as a short-time limit of the BBGKY hierarchy when correlations between particles are initially negligible. The long-time approach to equilibrium remains a deep mathematical problem. Later results by Cédric Villani and collaborators [Vil02, Vil10] established similar relaxation results for collisionless plasmas and mean-field systems, earning Villani the **2010 Fields Medal** for *his work involving the Boltzmann equation, which is central to mathematical physics.*

The Boltzmann equation also describes systems that are not in global equilibrium but can be considered in *local equilibrium*. In this regime, the distribution function is approximated by a locally Maxwellian form:

$$f(\vec{p},\vec{x},t) = \frac{n(\vec{x},t)}{\left[2\pi m k_{\mathrm{B}} T(\vec{x},t)\right]^{3/2}} \exp\left[-\frac{\left(\vec{p}-m\vec{v}(\vec{x},t)\right)^2}{2mk_{\mathrm{B}}T(\vec{x},t)}\right]. \tag{8.3}$$

Here $n(\vec{x},t)$, $T(\vec{x},t)$, and $\vec{v}(\vec{x},t)$ are slowly varying functions of space and time describing local density, temperature, and velocity.

Let us anticipate some results from Volume II to understand how the Boltzmann equation can be used to derive equations for gases and liquids out of equilibrium. Inserting Eq. (8.3) into the Boltzmann equation and performing an expansion in the small parameter corresponding to the mean free path yields, after lengthy algebra, the *Navier–Stokes equations* for a compressible fluid:

$$\left(\partial_t + \vec{v}\cdot\nabla_{\vec{x}}\right)\vec{v} = \frac{\vec{F}}{m} - \frac{1}{\rho(\vec{x})}\nabla_{\vec{x}}P + \frac{\eta}{\rho(\vec{x})}\nabla^2_{\vec{x}}\vec{v} + \frac{\eta}{3\rho(\vec{x})}\nabla_{\vec{x}}(\nabla_{\vec{x}}\cdot\vec{v}), \tag{8.4}$$

where $\rho = mn(\vec{x})$ is the local mass density and $\eta = nk_{\mathrm{B}}T\tau$ is the *shear viscosity*, with τ the characteristic relaxation time (mean free path divided by mean velocity).

REMARK 8.2 Viscosity, diffusion, and heat conduction are not imposed phenomenologically but emerge naturally from kinetic theory as consequences of the finite mean free path. Their values depend on the microscopic collision parameters, thus linking hydrodynamics to molecular dynamics.

Similarly, the kinetic theory predicts the *heat flux*:

$$\vec{q} = -K\,\nabla_{\vec{x}}T(\vec{x}), \tag{8.5}$$

where the thermal conductivity is

$$K = \frac{5}{2}\,\tau\, n(\vec{x})\, k_{\mathrm{B}}^2 T(\vec{x}). \tag{8.6}$$

The energy equation then takes the form of a heat diffusion law,

$$\rho\, k_{\mathrm{B}} \frac{\partial T}{\partial t} = \frac{2}{3}\,K\,\nabla^2 T, \tag{8.7}$$

recovering Fourier's law of thermal conduction, first formulated phenomenologically in the nineteenth century and later justified by kinetic theory.

REMARK 8.3 The Navier–Stokes and heat conduction equations represent the *hydrodynamic limit* of the Boltzmann equation—valid when the mean free path is small compared with the macroscopic gradients. They form the foundation of classical fluid dynamics and transport theory.

The derivation of the Navier–Stokes equations from kinetic theory was completed by Sydney Chapman and Thomas Cowling in the early twentieth century [CC70].

An Excursus on the Notion of Ergodicity

It is important to emphasize that not all Hamiltonian systems satisfy the assumptions underlying Boltzmann's statistical theory. A prominent class of counterexamples is provided by *Liouville-integrable systems*, in which the number of independent, Poisson-commuting conserved quantities equals the number of degrees of freedom. By the Liouville–Arnold theorem [Arn89], the regular level sets of such systems are smooth, compact *invariant tori* $\mathbb{T}^n$, and the Hamiltonian can be expressed locally in *action–angle coordinates* (I, θ), in which

$$H = H(I_1, \ldots, I_n), \qquad \dot{I}_k = 0, \qquad \dot{\theta}_k = \omega_k(I).$$

Thus, the dynamics reduces to uniform linear flow on $\mathbb{T}^n$ with frequencies $\omega_k(I)$, and the motion is quasi-periodic rather than ergodic. Because trajectories wind around these invariant tori without ever exploring the entire constant-energy hypersurface, the system fails to mix, produces no coarse-grained entropy, and therefore does not thermalize.

Famous manifestations of integrability include the Korteweg–de Vries and sine–Gordon equations, whose soliton solutions behave as localized, non-dispersive excitations that retain their identity after collisions. Despite interactions, the effective "particle" momenta remain invariant, a signature of the underlying integrable structure.

This phenomenon was famously anticipated in the pioneering numerical experiments of Enrico Fermi, John Pasta, Stanislaw Ulam, and Mary Tsingou (1955), who simulated a one-dimensional nonlinear oscillator chain, expecting equipartition, but instead observed a near-recurrence of the initial state [FPUT55, Dau08].

While Liouville–integrable systems provide counterexamples to Boltzmann's statistical assumptions, they also illuminate a subtle but important connection between dynamics and thermodynamics: under suitable conditions, entropy itself may behave as an *adiabatic invariant.*

Historically, the idea originates in the work of Paul Hertz (1910), who showed that for an *ergodic* Hamiltonian system, the phase-space volume

$$\Omega(E) = \int_{H<E} dq\, dp$$

is invariant under slow (adiabatic) deformations of the Hamiltonian. Since the micro-canonical entropy is defined as

$$S = k_{\mathrm{B}} \ln \Omega(E),$$

Hertz's analysis established that S is an *adiabatic invariant* for ergodic systems [Her10].

In integrable systems, however, the relevant invariants are not the full phase-space volume but the *actions* I_k. (see Arnold [Arn89]) show that I. Arnold, Mathematical Methods of Classical Mechanics, 2nd ed. (Springer, 1989) show that when a Hamiltonian depends slowly on external parameters, the action variables remain constant, and the dynamics stay confined to invariant tori. One may then define an entropy associated with the "volume" in action space,

$$S_{\mathrm{act}} = k_{\mathrm{B}} \ln \mathrm{Vol}(\mathcal{I}),$$

where $\mathcal{I}$ denotes the domain of allowed action variables. Because the actions are adiabatic invariants, this entropy is preserved under slow parameter changes.

Taken together, these works demonstrate that although integrable systems do not thermalize in the Boltzmann sense, they possess their own "adiabatic thermodynamics," in which entropy—defined geometrically via invariant tori— remains constant under slow deformations of the Hamiltonian. This illustrates the richness of Hamiltonian dynamics: both ergodic and integrable systems admit meaningful notions of entropy, but are grounded in distinct geometric structures.

Integrable systems, therefore, provide explicit counterexamples to the $\mathcal{H}$-theorem: irreversible behavior and entropy growth require mixing and ergodicity, properties absent in these finely constrained dynamical systems.

Integrable systems illustrate that the $\mathcal{H}$-theorem is not a universal statement: irreversibility and entropy production depend crucially on chaotic mixing and ergodicity, which fail in these special cases.

Action–Angle Coordinates and the Liouville–Arnold Theorem

Consider a Hamiltonian system with n degrees of freedom and canonical coordinates $(q, p) \in \mathbb{R}^{2n}$. Let $\{F_1, \ldots, F_n\}$ be n smooth, functionally independent integrals of motion in involution,

$$\{F_i, F_j\} = 0, \qquad i, j = 1, \ldots, n.$$

Such a system is called *Liouville-integrable.* The Liouville–Arnold theorem [Arn89] states that on every compact, connected regular level set

$$\mathcal{M}_{\mathbf{c}} = \{\, (q, p) \mid F_1 = c_1, \ldots, F_n = c_n \,\},$$

the phase-space manifold $\mathcal{M}_{\mathbf{c}}$ is diffeomorphic to an n-dimensional invariant torus $\mathbb{T}^n$, and there exists a canonical transformation to *action–angle coordinates*

$$(I_1, \ldots, I_n,\ \theta_1, \ldots, \theta_n) \in \mathbb{R}^n \times \mathbb{T}^n$$

such that the Hamiltonian depends only on the actions:

$$H = H(I_1, \ldots, I_n),$$

and the equations of motion reduce to

$$\dot{I}_k = 0, \qquad \dot{\theta}_k = \omega_k(I), \qquad k = 1, \ldots, n.$$

Thus, the dynamics consist of uniform linear flow on invariant tori with frequencies $\omega_k(I)$, and the system behaves locally as a set of free, non-interacting harmonic oscillators.

If, on the other hand, a system explores the entirety of phase space, then one can explain certain statistical properties of these systems. A central assumption underlying equilibrium statistical mechanics is that sufficiently complex Hamiltonian systems are *ergodic.* Ergodicity resolves the tension between the deterministic, reversible dynamics of classical mechanics and the probabilistic, irreversible framework of thermodynamics. The idea goes back to Boltzmann, who argued that the long-time evolution of a many-body system should "roam" throughout the entire constant-energy hypersurface, visiting any region compatible with the conserved quantities. In modern terms, a Hamiltonian flow Φ^t on phase space (Γ, μ) is ergodic if every invariant measurable set has either full measure or zero measure. Equivalently, by the Birkhoff ergodic theorem

[Bir31], for any integrable observable $A(\Gamma)$,

$$\lim_{T\to\infty}\frac{1}{T}\int_0^T A(\Phi^t\Gamma_0)\,dt = \int_{\Gamma_E} A(\Gamma)\,d\mu(\Gamma) \tag{8.8}$$

for almost every initial condition Γ_0 on the energy surface $\Gamma_E = \{\Gamma \in \Gamma \mid H(\Gamma) = E\}$. Thus, the *time average* of any observable equals its *ensemble (phase-space) average.*

In the language of statistical mechanics, this identity is precisely what is needed to justify replacing long-time measurements by equilibrium expectation values. If a system is ergodic, it eventually explores the entire region of phase space consistent with the conserved quantities, and no region is dynamically preferred. Hence, the long-time average of an observable $A(t)$,

$$\bar{A} = \lim_{T\to\infty}\frac{1}{T}\int_0^T A(t)\,dt, \tag{8.9}$$

coincides with its micro-canonical (or canonical, when appropriate) ensemble average $\langle A\rangle_{\mathrm{PS}}$,

$$\bar{A} = \langle A\rangle_{\mathrm{PS}}. \tag{8.10}$$

In physical systems, one typically observes that $A(t)$ approaches a stationary value A^* after a finite *relaxation time* T^*, which does not grow with the system size. Splitting the integral, one finds

$$\bar{A} = \lim_{T\to\infty}\frac{1}{T}\int_0^{T^*} A(t)\,dt + \lim_{T\to\infty}\frac{1}{T}\int_{T^*}^T A^*\,dt \tag{8.11}$$

$$= A^*. \tag{8.12}$$

Thus the experimentally measured stationary value A^* is equal to the ensemble average $\langle A\rangle_{\mathrm{PS}}$, providing the operational foundation of equilibrium thermodynamics.

Ergodicity therefore plays a crucial conceptual role: it justifies replacing an actual dynamical process—which occurs over time—by an ensemble description that assigns probabilities to microstates. Without ergodicity, as in integrable or nearly integrable systems, long-time dynamics need not converge to the predictions of statistical mechanics. Whether ergodicity holds for general interacting many-body systems is still an open problem in mathematical physics; nevertheless, for a wide class of chaotic Hamiltonian systems with many degrees of freedom, ergodicity (or the weaker property of mixing) provides a remarkably accurate description of physical reality.

For Mathematicians (or the Mathematically Inclined)

From the perspective of dynamical systems theory, ergodicity is a central concept that goes far beyond its physical interpretation. A measure-preserving dynamical system $(X, \mathcal{B}, \mu, T)$ is called *ergodic* if every T-invariant set has either full measure or zero measure. The importance of this property rests on Birkhoff's pointwise Ergodic Theorem [Bir31], which asserts that for any $A \in L^1(X, \mu)$, the time averages

$$\frac{1}{N}\sum_{n=0}^{N-1} A(T^n x)$$

converge for almost every x to the space average $\int_X A\,d\mu$. This identifies long-time behavior with statistical behavior and is the mathematical foundation underlying the physicist's replacement of temporal observables by ensemble averages.

Ergodicity is only the first step in a rich hierarchy of chaotic properties. The work of von Neumann [vN32] introduced the *mean ergodic theorem*, while Kolmogorov and Sinai developed the notions of *Kolmogorov–Sinai entropy* and *K-systems*, providing invariants that classify degrees of chaos. In hyperbolic dynamical systems, the Anosov and Axiom A paradigms demonstrate that ergodicity often arises from strong geometric structures, as in the classical works of Anosov and Sinai [Sin63, Sin70].

These developments show that ergodicity is not merely a physical hypothesis but a profound mathematical property connecting measure theory, probability, geometry, and chaos. While proving ergodicity for realistic many-body Hamiltonians remains an open challenge, the mathematical theory provides powerful insights into why statistical mechanics works so remarkably well for chaotic systems.

An even broader perspective reveals that the study of ergodicity has shaped and inspired the modern theory of chaos. The notion of chaotic dynamics emerged from the geometric program initiated by Stephen Smale, whose hyperbolic theory and construction of the Smale's horseshoe provided the first rigorous examples of structurally stable chaotic systems [Sma67]. This viewpoint was dramatically advanced by Sinai through his pioneering analysis of dispersing billiards and Anosov flows, establishing ergodicity and mixing in systems with strong hyperbolic structure [Sin70]. Building on these foundations, David Ruelle developed the notion of SRB (Sinai–Ruelle–Bowen) measures and introduced the framework of *thermodynamic formalism*, which provides a precise statistical description of invariant probability measures in chaotic systems and forms the basis for analyzing non-equilibrium steady states [Rue78, Rue89]; at about the same time, Rufus Bowen extended the construction of Markov partitions [Bow70].

Further advances in chaos theory provided powerful tools for understanding the statistical properties of nonlinear systems. The expositions of Edward Ott [Ott02] and the comprehensive text by Predrag Cvitanović and collaborators [CAM+16] developed practical and conceptual frameworks based on periodic-orbit theory, Lyapunov spectra, and computational methods for chaotic attractors. More recently, Stefano Luzzatto and coauthors [LMP05] have shown that even systems lacking uniform hyperbolicity—including those with intermittency or weak chaos—can still possess physical invariant measures and satisfy statistical limit laws. Together, these results illustrate how chaotic dynamics, far from being merely a source of irregular motion, supports robust statistical behavior, reinforcing ergodicity as a central concept linking geometry, probability, and the long-time behavior of deterministic systems.

8.2 Ensembles and the Emergence of Equiprobability

The long discussion on the notion of ergodicity that would be necessary to properly understand the emergence of a uniform measure will now be replaced with a much simpler explanation, or worse, a small toy model that is much simpler to understand. The postulate of *equal a priori probability*, e.g., the fact that all accessible microstates of an isolated system are equally likely at equilibrium, can be justified dynamically. Boltzmann's $\mathcal{H}$-theorem shows that collisions drive a gas toward equilibrium with increasing entropy, but how can we formalize this probabilistically without taking several courses in dynamical systems? A simple way is to view the microscopic dynamics as a random process in a discretized phase space. This section introduces a *toy model* based on Markov chains that illustrates how equiprobability naturally emerges from stochastic evolution.

We imagine dividing the phase space Γ into small cells or "pixels" labeled by $i = 1, 2, \ldots, N_\Gamma$, each representing a small region around $(\vec{p}_i, \vec{q}_i)$. A collision or small perturbation allows the system to *hop* from one cell to another with a certain probability. Let M_{ij} be the probability that, in one time step, the system moves from cell i to cell j:

$$\sum_j M_{ij} = 1, \tag{8.13}$$

meaning that each state i must evolve to some state j with total probability one.

Let $p(i, t)$ be the probability of finding the system in state i at time t. If $\vec{p}_t$ is the column vector of these probabilities, the time evolution is

$$\vec{p}_{t+1} = M\,\vec{p}_t, \qquad \text{or equivalently} \qquad \vec{p}_t = M^t\,\vec{p}_0, \tag{8.14}$$

where $\vec{p}_0$ encodes the initial condition (for instance, a system initially localized in state k with $p(i, 0) = \delta_{ik}$). This discrete stochastic evolution defines a *Markov chain* on the phase space.

A Markov chain is a stochastic process in which the probability of the next state depends only on the current state, not on the history of previous steps. In our context, the "state" of the system is its position in pixelated phase space, and the transition probabilities M_{ij} summarize all microscopic collision details into a single probabilistic rule.

REMARK 8.4 This construction replaces deterministic Hamiltonian evolution by a coarse-grained stochastic one. The idea is that, over macroscopic times, collisions randomize microstates so that only transition probabilities—not precise trajectories—matter. The Boltzmann equation can itself be viewed as a continuum limit of such a Markov process.

Because M is a *stochastic matrix* (all entries are non-negative and rows sum to one), its spectral properties determine the long-time evolution; see for instance [Beh00] for an introduction, but we will discus them in Volume II in more detail. Let λ_k be the eigenvalues of M and L^k, R^k the corresponding left and right eigenvectors:

$$M_{ij} = \sum_k \lambda_k\, L_i^{(k)}\, R_j^{(k)}. \tag{8.15}$$

The matrix raised to a power t evolves as

$$M_{ij}^t = \sum_k \lambda_k^t\, L_i^{(k)}\, R_j^{(k)}, \tag{8.16}$$

so that

$$\vec{p}_t = M^t \vec{p}_0 = \sum_k \lambda_k^t (\vec{R}^{(k)} \cdot \vec{p}_0)\, \vec{L}^{(k)}. \tag{8.17}$$

The notation M^t here denotes the matrix multiplied by itself t times, not the transpose.

From the normalization condition (8.13), the largest eigenvalue satisfies $\lambda_{\max} = 1$. All other eigenvalues satisfy $|\lambda_k| < 1$, implying that as $t \to \infty$,

$$M^t \longrightarrow \sum_{\{\lambda_k = 1\}} L^{(k)} R^{(k)}. \tag{8.18}$$

All modes corresponding to $\lambda_k < 1$ decay exponentially in time, so the system forgets its initial state and approaches a stationary distribution $\vec{p}_\infty$.

REMARK 8.5 The rate of convergence to equilibrium is determined by the *spectral gap* of M, defined as $1 - \max_{k>1} |\lambda_k|$. A large spectral gap corresponds to rapid mixing of states and hence faster entropy production.

The long-time limit depends on the structure of the Markov chain, that is, on which states communicate through transitions. If the underlying directed graph of the transitions is *strongly connected*—meaning any state can be reached from any other in a finite number of steps—then the *Perron–Frobenius theorem* guarantees [Beh00]:

(a) There is a unique largest eigenvalue $\lambda_1 = 1$, non-degenerate. (b) The corresponding left and right eigenvectors have strictly positive components. (c) The right eigenvector $\vec{L}^{(1)}$ represents the stationary distribution.

If the graph is also *aperiodic* (no exact cycles), the Markov chain is *ergodic* and

$$\vec{p}_\infty = \lim_{t \to \infty} M^t \vec{p}_0 = \vec{L}^{(1)}, \tag{8.19}$$

independent of the initial condition. Note that ergodic in Markov chains means that the system explores the entire phase space, e.g., the asymptotic distribution is the vector proportional to $\vec{1}$.

In the simplest case of a fully symmetric transition matrix—where each state is equally likely to hop to any other—every element of $\vec{L}^{(1)}$ is identical:

$$(p_\infty)_i = c = \text{constant}, \tag{8.20}$$

meaning that each microstate in phase space is equally probable. This reproduces the postulate of equal a priori probability derived earlier by maximizing entropy.

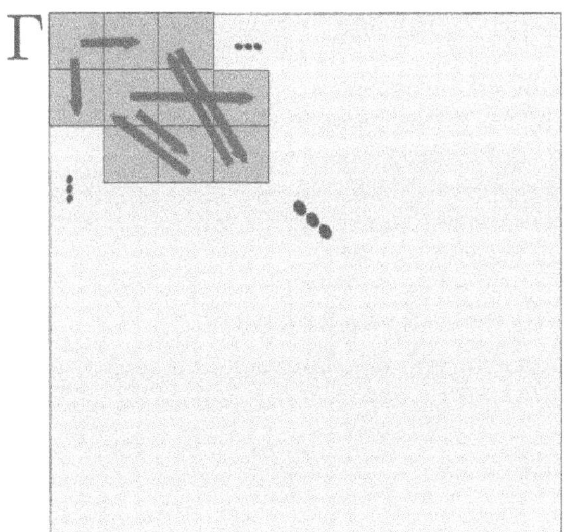

FIGURE 8.1 A schematic representation of "hopping" among discrete cells in phase space. Random transitions between cells lead to an equiprobable steady state, consistent with the postulate of equal a priori probability.

Strong connectivity in graph theory means that for any pair of nodes (i, j), there exists a finite sequence of transitions $i \to k_1 \to k_2 \to \cdots \to j$ with nonzero probability. Physically, this expresses the idea that the system can explore all regions of phase space through its dynamics. Such connectivity ensures ergodicity and the emergence of equilibrium.

Markov Chains and Entropy Production

The Markov process described above provides a concrete dynamical realization of entropy growth. At each time step, the probability distribution $\vec{p}_t$ becomes more uniform as correlations between successive states are washed out. A natural measure of disorder is the Shannon (or Gibbs) entropy,

$$S(t) = -k_{\rm B} \sum_i p_i(t) \log p_i(t). \tag{8.21}$$

Because M is a stochastic matrix with non-negative entries and rows summing to one, it can be shown that $S(t+1) \geq S(t)$ for all t if M is doubly stochastic (both rows and columns summing to one). In that case, the process drives the system toward a maximum-entropy stationary state.

REMARK 8.6 Markov-chain dynamics and its connection to entropy production underpin modern algorithms such as the *Metropolis–Hastings* method used in Monte Carlo simulations. There, one constructs transition matrices satisfying detailed balance to sample from desired equilibrium distributions. The Markov chain thus provides a microscopic model of equilibration: deterministic trajectories are replaced by stochastic hops that guarantee eventual mixing in phase space. The Perron–Frobenius stationary vector corresponds to the equilibrium ensemble, while the spectral gap determines the characteristic relaxation time.

In this simple model, the stationary distribution represents the micro-canonical ensemble, where all accessible states are equally likely.

Time Evolution and Invariant Measure in Hamiltonian Systems

In the previous section, we studied the probabilistic evolution of systems through Markov chains, where entropy maximization led naturally to a uniform stationary distribution. In Hamiltonian systems, the situation is different: the evolution is deterministic and reversible. Nevertheless, as a consequence of *Liouville's theorem*, there exists an invariant measure in phase space that plays the same role as the stationary distribution in stochastic dynamics. This property underlies the ergodic hypothesis and the foundations of ensemble theory.

For a classic discussion, see [Khi49], *Mathematical Foundations of Statistical Mechanics* (Dover, 1949), where the connection between Liouville's theorem and the ergodic hypothesis is established rigorously. See also [Cam24].

Hamiltonian Flow and Phase-Space Evolution

Consider a Hamiltonian system of N particles described by coordinates and momenta $(q_i, p_i) \in \Gamma$, where Γ is the $6N$-dimensional phase space. The equations of motion are

$$\dot{q}_i = \frac{\partial H}{\partial p_i}, \tag{8.22}$$

$$\dot{p}_i = -\frac{\partial H}{\partial q_i}. \tag{8.23}$$

These can be written generically as

$$\frac{dx_i}{dt} = f_i(x), \tag{8.24}$$

where x_i represents any phase-space coordinate, $x = (q_1, \ldots, q_N, p_1, \ldots, p_N)$, and $f_i(x)$ is the Hamiltonian vector field. The evolution of a point x_0 in phase space is determined by the time-evolution operator U_t:

$$U_t(x_0) = x(t), \qquad x(t) = (q_1(t), p_1(t), \ldots, q_N(t), p_N(t)). \tag{8.25}$$

Liouville's Theorem and Volume Preservation

Taking the divergence of the phase-space flow,

$$\nabla_x \cdot \dot{x} = \sum_i \frac{\partial f_i(x)}{\partial x_i}, \tag{8.26}$$

and substituting $f_i(x)$ from Hamilton's equations, we find

$$\sum_i \frac{\partial f_i(x)}{\partial x_i} = \sum_j \left(\frac{\partial^2 \mathcal{H}}{\partial q_j \partial p_j} - \frac{\partial^2 \mathcal{H}}{\partial p_j \partial q_j} \right) = 0. \tag{8.27}$$

Thus, the phase-space flow is *divergence-free*: infinitesimal volumes in phase space are conserved during the time evolution.

Let $A \subset \Gamma$ be a region of phase space, and define

$$M_t = \int_A d^{6N} x(t), \tag{8.28}$$

as the volume of this region at time t. We will show that M_t is constant in time, $\frac{dM_t}{dt} = 0$.

Consider the mapping between initial and final phase-space coordinates under the flow:

$$x(t) = U_t(x_0), \qquad x_0 = x(t'). \tag{8.29}$$

The infinitesimal volumes at times t and t' are related by the Jacobian determinant

$$J(t,t') = \left| \frac{\partial x_i(t)}{\partial x_j(t')} \right|. \tag{8.30}$$

Thus,

$$\int_A d^{6N} x(t) = \int_{A'} J(t,t')\, d^{6N} x(t'). \tag{8.31}$$

Differentiating with respect to time gives

$$\frac{dM_t}{dt} = \int_{A'} \frac{\partial J(t,t')}{\partial t}\, d^{6N} x(t'). \tag{8.32}$$

The time derivative of J satisfies

$$\frac{\partial J}{\partial t} = J \sum_i \frac{\partial f_i}{\partial x_i}, \tag{8.33}$$

as follows from the chain rule and the general formula for the derivative of a determinant. Because $\sum_i \partial f_i/\partial x_i = 0$ from Liouville's theorem, we obtain

$$\frac{\partial J}{\partial t} = 0 \quad \Rightarrow \quad J(t,t') = 1. \tag{8.34}$$

Hence the phase-space volume is preserved:

$$\frac{dM_t}{dt} = 0. \tag{8.35}$$

For any integrable function $f(x)$ defined on phase space, define the time-evolved function as $U_t f(x) = f(x_t)$, where $x_t = U_t(x)$. Liouville's theorem implies that integrals of such functions over any region of phase space are invariant under the flow:

$$\int_{A_t} f(x)\, d\Gamma_M = \int_A f(x_t)\, d\Gamma_M. \tag{8.36}$$

This means that averages of observables computed with the invariant measure $d\Gamma_M = d^{3N}p\, d^{3N}q$ do not change under time evolution.

In particular, if A is the entire accessible region of phase space (defined by fixed energy and particle number), then the uniform measure

$$\rho(x) = \text{const} \tag{8.37}$$

is invariant under the dynamics. This property forms the basis of the *micro-canonical ensemble*: all microstates compatible with the conserved quantities are equally probable.

REMARK 8.7 The equality

$$\int_{A_t} f(x)\, d\Gamma_M = \int_A f(x_t)\, d\Gamma_M \tag{8.38}$$

follows directly from the change of variables $x_t = U_t(x)$ with unit Jacobian. It expresses that phase-space averages are independent of the time parameter chosen to evaluate them—a statement of stationarity of the invariant measure.

The invariance of phase-space volume extends to any canonical transformation $(q_i, p_i) \to (Q_i, P_i)$. By definition, canonical transformations preserve Hamilton's equations and the symplectic two-form

$$\omega = \sum_i dq_i \wedge dp_i. \tag{8.39}$$

This implies that the Jacobian determinant of the transformation is unity:

$$\left| \frac{\partial(Q, P)}{\partial(q, p)} \right| = 1, \tag{8.40}$$

and therefore

$$\int d^{3N}q\, d^{3N}p = \int d^{3N}Q\, d^{3N}P. \tag{8.41}$$

For a Hamiltonian system of N particles in three dimensions, this preserves the $6N$-dimensional phase-space volume element. More generally, in any dimension D, the Jacobian determinant of a canonical transformation equals one, ensuring that the Liouville measure

$$d\Gamma = \prod_{i=1}^{N} dq_i\, dp_i \tag{8.42}$$

is invariant under both time evolution and canonical coordinate changes.

REMARK 8.8 The invariance of the Liouville measure establishes that statistical averages computed over canonical coordinates are physically meaningful and coordinate-independent. This geometric property—preservation of the symplectic structure—is the mathematical essence of Liouville's theorem.

In a Hamiltonian system, canonical transformations preserve the *symplectic structure* of phase space. This means that if J is the Jacobian matrix of a transformation from (q_i, p_i) to (Q_i, P_i), the transformation must satisfy

$$J^T \cdot \boldsymbol{\Omega} \cdot J = \boldsymbol{\Omega}, \tag{8.43}$$

where

$$\boldsymbol{\Omega} = \begin{pmatrix} 0 & I \\ -I & 0 \end{pmatrix} \tag{8.44}$$

is the standard *symplectic matrix*, with I the $3N \times 3N$ identity matrix. This condition ensures that Hamilton's equations retain their canonical form under the transformation. Since $\det(\boldsymbol{\Omega}) = 1$, it follows that

$$\det(J) = 1, \tag{8.45}$$

and therefore the phase-space volume element $d\Gamma = d^{3N}p\, d^{3N}q$ is invariant under canonical transformations. This invariance is a central result: all canonical coordinate systems describe the same phase-space measure and hence the same thermodynamic equilibrium.

REMARK 8.9 The invariance of phase-space integrals under canonical transformations expresses a fundamental symmetry of Hamiltonian mechanics: physical quantities and averages do not depend on the coordinate representation. From a thermodynamic perspective, it ensures that any canonical change of variables leaves the equilibrium measure unchanged.

The existence of an invariant measure allows us to relate the long-time behavior of Hamiltonian systems to statistical averages. This connection is formalized by *Birkhoff's ergodic theorem* which we described earlier [Bir31].

Birkhoff's Theorem. Let $f(x)$ be an integrable function on phase space, and let $x_t = U_t(x)$ denote its Hamiltonian evolution. Then:

1. The limit

$$\hat{f}(x) = \lim_{T\to\infty} \frac{1}{T} \int_0^T f(x_t)\, dt \tag{8.46}$$

exists for almost every initial condition $x \in \Gamma$.

2. If the system is *ergodic*—meaning that its trajectories explore the entire accessible region of phase space—then this time average equals the phase-space average:

$$\hat{f} = \frac{\displaystyle\int_V f(x)\, d\Gamma_M}{\displaystyle\int_V d\Gamma_M}. \tag{8.47}$$

Thus, in ergodic systems, the long-time average of any observable along a trajectory equals the ensemble average over the invariant measure. This result is the deterministic analog of the stationary limit of Markov chains discussed earlier.

REMARK 8.10 The term *indecomposable* in some older literature refers to what we now call *ergodic*: the phase space cannot be partitioned into smaller invariant subsets under the dynamics. Physically, ergodicity ensures that every region of the accessible phase space is visited with equal probability in the long run.

Energy Conservation and the Micro-Canonical Measure

In most physical systems, the total energy $E = H(p,q)$ is the only conserved quantity. We therefore consider averages over the subset of phase space defined by fixed energy. Let $\Omega(E)$ be the total phase-space volume enclosed by the surface of constant energy E, and let $\Sigma(E)$ denote the hypersurface where $H(p,q) = E$. We define

$$\mathcal{V}(E) = \{(p,q) \mid H(p,q) \le E\}, \qquad \Sigma(E) = \{(p,q) \mid H(p,q) = E\}. \tag{8.48}$$

We wish to show that

$$\frac{d}{dE}\,\mathcal{V}(E) = \int_{\Sigma(E)} \frac{d\Sigma_E}{|\nabla E|}, \tag{8.49}$$

where $|\nabla E|$ is the magnitude of the gradient of the Hamiltonian in phase space,

$$|\nabla E| = \sqrt{\sum_i \left[\left(\frac{\partial E}{\partial q_i}\right)^2 + \left(\frac{\partial E}{\partial p_i}\right)^2\right]}. \tag{8.50}$$

To derive Eq. (8.49), we use the fact that the function $E = H(p,q)$ foliates phase space into hypersurfaces of constant energy. We can express an integral over the entire accessible region as an integral over these hypersurfaces:

$$\int_{\mathcal{V}(E)} f(x)\, d\Gamma = \int_0^E d\epsilon \int_{\Sigma_\epsilon} f(x)\, \frac{d\Sigma_\epsilon}{|\nabla \epsilon|}. \tag{8.51}$$

Differentiating with respect to E gives

$$\frac{d}{dE}\int_{\mathcal{V}(E)} f(x)\, d\Gamma = \int_{\Sigma_E} f(x)\, \frac{d\Sigma_E}{|\nabla E|}, \tag{8.52}$$

which is the desired result.

Equation (8.51) geometrically expresses that the total phase-space volume can be viewed as the sum (integral) of infinitesimal "energy layers," each corresponding to a surface of constant Hamiltonian. This geometric picture underlies the definition of the *density of states*, given by

$$\Gamma(E) = \frac{d\mathcal{V}(E)}{dE}, \tag{8.53}$$

which plays a central role in statistical mechanics.

To make Eq. (8.51) explicit, consider locally parameterizing a point in phase space by coordinates along the constant-energy surface $\vec{\Sigma}_E$ and one coordinate normal to it, $\vec{n}_E$. We can write the position vector as

$$\vec{x} = \begin{pmatrix} \vec{\Sigma}_E \\ n_E \end{pmatrix}, \tag{8.54}$$

so that the infinitesimal volume element factorizes as $d\Gamma = d\Sigma\, dn$. The surface is defined by

$$E = f(\vec{x}) = H(p,q), \tag{8.55}$$

and its normal vector is $\vec{n} = \nabla f = \nabla E = \sum_i \hat{x}_i\,(\partial E/\partial x_i)$. Projecting the displacement $d\vec{x}$ along the normal direction gives

$$dE = \sum_i \frac{\partial E}{\partial x_i}\, dx_i = |\nabla E|\, dn, \tag{8.56}$$

so that

$$dn = \frac{dE}{|\nabla E|}. \tag{8.57}$$

Substituting this into the volume integral yields

$$\int_{\mathcal{V}(E)} f(x)\, d\Gamma = \int_0^E d\epsilon \int_{\Sigma_\epsilon} \frac{f(x)}{|\nabla \epsilon|}\, d\Sigma_\epsilon, \tag{8.58}$$

proving the relation.

The result above shows that the entire accessible region of phase space can be represented as a continuous family of constant-energy hypersurfaces. For isolated systems with fixed total energy E, the uniform distribution over $\Sigma(E)$,

$$\rho_E(p,q) = \frac{\delta(E - H(p,q))}{\Gamma(E)}, \tag{8.59}$$

defines the *micro-canonical ensemble*. It is the unique invariant measure consistent with Liouville's theorem and energy conservation.

REMARK 8.11 The foliation of phase space by constant-energy surfaces is the geometric backbone of statistical mechanics. The micro-canonical distribution is uniform over each such surface, and all other ensembles (canonical, grand-canonical) arise from it by considering subsystems in weak contact with large reservoirs.

We now make precise what we mean by the *measure* on phase space. By Liouville's theorem, the elementary volume element

$$d\Gamma = \prod_{i=1}^{N} d^3p_i \, d^3q_i \tag{8.60}$$

is invariant under time evolution and canonical transformations. However, when the energy of the system is fixed, we must restrict this measure to the infinitesimally thin region of phase space lying between two energy surfaces,

$$E - \frac{dE}{2} \leq H(p,q) \leq E + \frac{dE}{2}. \tag{8.61}$$

This defines a small "shell" in phase space—denoted D—whose thickness is dE. Integrating the Liouville measure over this shell gives the corresponding phase-space volume:

$$\int_D d\Gamma. \tag{8.62}$$

In the limit of an infinitesimally thin shell ($dE \to 0$), this volume element becomes the surface integral over the constant-energy hypersurface Σ_E:

$$\lim_{dE \to 0} \int_D d\Gamma = \int_{\Sigma_E} \frac{d\Sigma_E}{|\nabla E|}. \tag{8.63}$$

Here, $d\Sigma_E$ is the infinitesimal area element on the surface $\Sigma_E = \{(p,q) \mid H(p,q) = E\}$, and $|\nabla E|$ is the magnitude of the gradient of the Hamiltonian in phase space. This quantity defines the invariant measure on the constant-energy surface.

The same surface measure can be written equivalently using the Dirac delta function:

$$\delta(E - H(p,q)) = \int_{\Sigma_E} \frac{\delta(p - p^*)\,\delta(q - q^*)}{|\nabla E|} \, d\Sigma_E, \tag{8.64}$$

where the integral runs over all points (p^*, q^*) on the surface Σ_E. Hence, we can express the infinitesimal phase-space volume as

$$\lim_{dE \to 0} \int_D d\Gamma_M = \int d\Gamma_M \, \delta(E - H(p,q)). \tag{8.65}$$

The total accessible phase-space volume up to energy E is then

$$\Omega(E) = \int \theta(E - H(p,q)) \, d\Gamma_M, \tag{8.66}$$

where $\theta(x)$ is the Heaviside step function. The derivative of this volume with respect to energy defines the *density of states*,

$$\Gamma(E) = \frac{d\Omega(E)}{dE}. \tag{8.67}$$

This quantity measures the number of accessible microstates per unit energy interval near E.

The average of any observable $f(p,q)$ in the micro-canonical ensemble can now be expressed in terms of the density of states as

$$\bar{f} = \frac{1}{\Gamma(E)}\frac{d}{dE}\int f(p,q)\,\theta(E - H(p,q))\,d\Gamma. \tag{8.68}$$

Equation (8.68) is the standard expression used in statistical mechanics for ensemble averages at fixed energy. It is particularly convenient because it involves integrals over the entire phase space rather than explicit parameterization of the constant-energy surface.

REMARK 8.12 The micro-canonical average expresses the principle of equal a priori probability in quantitative form: all microstates compatible with the same macroscopic energy are equally likely. This principle is not a postulate of geometry but a consequence of Liouville's theorem and ergodicity: the dynamics uniformly explore the constant-energy surface.

A Recipe for Calculating Thermodynamic Quantities

The analysis above formalizes Boltzmann's *Postulate of Equal a Priori Probability*:

> When a macroscopic system is in thermodynamic equilibrium, it is equally likely to be found in any of the microstates compatible with the macroscopic constraints.

In practice, this means that the equilibrium probability density is constant within the thin shell of phase space corresponding to the fixed energy and zero elsewhere. This defines the *micro-canonical ensemble*:

$$\rho(p,q) = \begin{cases} \text{const}, & \text{if } E - \frac{dE}{2} \le H(p,q) \le E + \frac{dE}{2}, \\ 0, & \text{otherwise.} \end{cases} \tag{8.69}$$

This distribution corresponds to a uniform measure over the accessible region of phase space, with total probability normalized to one.

In this construction, $\Omega(E)$ represents the total number of microstates with energy less than or equal to E, while $\Gamma(E)$ represents the number of states per unit energy at E. The entropy is then defined as

$$S(E) = k_{\mathrm{B}} \log \Omega(E), \tag{8.70}$$

or equivalently (up to an additive constant),

$$S(E) = k_{\mathrm{B}} \log \Gamma(E), \tag{8.71}$$

depending on whether one counts states within an interval or exactly on the constant-energy surface. From these definitions, all thermodynamic quantities—temperature, pressure, and chemical potential—follow from derivatives of $S(E)$.

REMARK 8.13 This recipe establishes the logical chain from microscopic dynamics to macroscopic thermodynamics:

$$\begin{aligned} &\text{Liouville's theorem} \implies \text{invariant measure} \\ &\implies \text{equal a priori probability} \implies \text{ensembles and thermodynamics.} \end{aligned} \tag{8.72}$$

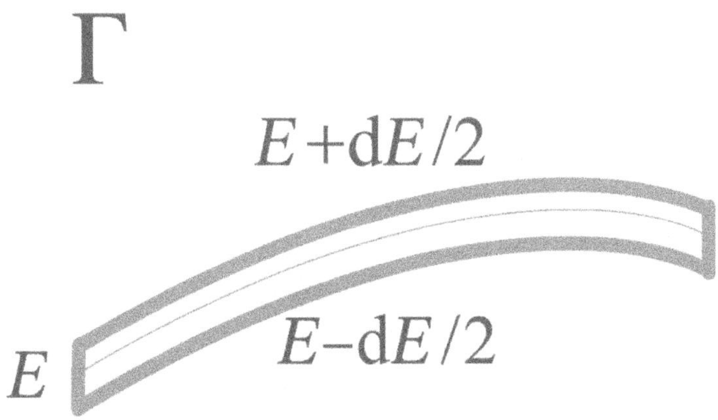

FIGURE 8.2 Illustration of the micro-canonical ensemble in phase space. The system is equally likely to occupy any microstate within the thin energy shell bounded by $E \pm dE/2$.

In the next chapters, we will extend this reasoning to systems exchanging energy or particles with reservoirs.

In the context of statistical mechanics, *thermalization* means that the system loses memory of its initial conditions while dynamically exploring the accessible portion of phase space. When the trajectories densely fill this region—i.e., when the system is *ergodic*—macroscopic observables become independent of microscopic details. The equilibrium distribution is then uniform over the constant-energy hypersurface, and all microstates compatible with the same macroscopic constraints are equally probable.

Once the density of states $\Gamma(E)$ is known, the (thermodynamic) *entropy* is defined as

$$S = k_{\mathrm{B}} \log \Gamma(E), \tag{8.73}$$

where k_{B} is Boltzmann's constant. This relation encapsulates the bridge between microscopic dynamics and macroscopic thermodynamics *at equilibrium.*

REMARK 8.14 Equation (8.73) expresses the fact that entropy measures the logarithm of the number of accessible microstates at fixed energy. Ergodicity ensures that each microstate contributes equally to the equilibrium ensemble.

Given $S(E, V)$, we can derive all thermodynamic quantities by following a systematic procedure. The steps are:

- *Step 1:* Compute $S(E, V) = k_{\mathrm{B}} \log \Gamma(E, V)$ from the micro-canonical ensemble.
- *Step 2:* Invert the relation to obtain $E \equiv U(S, V)$, the internal energy as a function of entropy and volume.
- *Step 3:* Define the thermodynamic quantities:

$$T = \left(\frac{\partial U}{\partial S}\right)_V, \qquad P = -\left(\frac{\partial U}{\partial V}\right)_S. \tag{8.74}$$

- *Step 4:* Introduce the thermodynamic potentials:

$$\begin{aligned} H &= U + PV && \text{(enthalpy)}, \\ F &= U - TS && \text{(Helmholtz free energy)}, \\ G &= H - TS && \text{(Gibbs free energy)}. \end{aligned} \tag{8.75}$$

This provides a complete "recipe" for deriving the full thermodynamic description from statistical mechanics once the entropy function is known. The heat capacity follows immediately as

$$C_V = \left(\frac{\partial U}{\partial T}\right)_V. \tag{8.76}$$

We have shown how these relations emerge from the micro-canonical ensemble, but the same logic applies to all ensembles.

Example: Entropy from Equilibrium Occupation Numbers

In this example, we show how the entropy of a system with discrete energy levels can be written in terms of the *equilibrium occupation numbers* $\tilde{N}_j$, and how the temperature dependence $s \sim k_B \log T$ emerges in the continuum limit.

We consider the total number of microstates

$$\Gamma = \sum_{N_1+\cdots+N_k=N} \frac{N!}{N_1!\cdots N_k!},$$

where N_j counts the number of particles in level j. In the thermodynamic limit, the sum is overwhelmingly dominated by the set of occupation numbers $\{\tilde{N}_j\}$ that *maximize* the term $N!/(N_1!\cdots N_k!)$:

$$\Gamma \approx \frac{N!}{\tilde{N}_1!\cdots \tilde{N}_k!}.$$

We now introduce an important trick that allows us to simplify calculations: *Stirling's formula*. It provides an asymptotic approximation for the factorial of a large integer n. It states that

$$n! \sim \sqrt{2\pi n}\, n^n e^{-n},$$

or, taking logarithms,

$$\log n! = n \log n - n + \tfrac{1}{2}\log(2\pi n) + O\left(\frac{1}{n}\right).$$

In statistical mechanics, one typically keeps only the leading terms,

$$\log n! \approx n \log n - n,$$

which becomes asymptotically exact as $n \to \infty$. Using Stirling's formula $\log n! \approx n \log n - n$, we obtain

$$\log \Gamma \approx N \log N - \sum_j \tilde{N}_j \log \tilde{N}_j.$$

Dividing by N and writing $P_j = \tilde{N}_j/N$, we obtain

$$\frac{1}{N}\log \Gamma = \log N - \sum_j P_j \log(N P_j) = -\sum_j P_j \log P_j,$$

since $1 = \sum_j P_j$. Multiplying by k_B then yields the entropy per particle

$$s = \frac{S}{N} = -k_B \sum_j P_j \log P_j.$$

To find P_j we maximize the entropy under the constraints

$$\sum_j P_j = 1, \qquad \sum_j P_j \epsilon_j = \bar{\epsilon},$$

introducing Lagrange multipliers. This yields the Boltzmann distribution

$$P_j = \frac{\exp(-\lambda\, \epsilon_j / k_B)}{\sum_k \exp(-\lambda\, \epsilon_k / k_B)},$$

where λ is proportional to the inverse temperature.

If the spectrum is dense, we approximate $\epsilon_j \to \epsilon(p) = p^2/(2m)$. The partition sum becomes an integral:

$$Z = \sum_j P_j \approx \int d^3p \, \exp\left(-\frac{\lambda}{k_B}\frac{p^2}{2m}\right) = \left(\sqrt{\frac{2\pi m k_B}{\lambda}}\right)^3.$$

From this Gaussian integral we identify $\lambda = 1/T$, recovering the usual Maxwell–Boltzmann distribution and confirming that the entropy contains a term of the form

$$s \sim k_B \log T.$$

This calculation shows how the familiar expression for entropy emerges from counting arguments and Stirling's approximation, and how the temperature dependence of the entropy follows from the continuum limit of the Boltzmann distribution. The structure of the ideal-gas partition function appears naturally once we identify the Lagrange multiplier λ with the inverse temperature $1/T$.

The Micro-Canonical Ensemble and the Density of States

As a warm-up for the canonical ensemble, let us return to the definition of the micro-canonical ensemble. We consider a system constrained to remain within a narrow *energy shell* in phase space:

$$E - \frac{dE}{2} \le H(p, q) \le E + \frac{dE}{2}. \tag{8.77}$$

Physically, this describes an isolated system whose total energy fluctuates within an infinitesimal range dE due to coarse-graining.

The probability density is uniform within this shell and zero elsewhere:

$$\rho(p, q) = \begin{cases} \text{const}, & \text{if } E - \frac{dE}{2} \le H(p, q) \le E + \frac{dE}{2}, \\ 0, & \text{otherwise.} \end{cases} \tag{8.78}$$

Despite its name, there is nothing particularly "micro" about the micro-canonical ensemble: it simply represents a uniform distribution over all accessible microstates at fixed macroscopic energy.

To normalize the distribution, we must compute the total phase-space volume of the shell. We define

$$\Gamma_{dE}(E) = \int_{E-dE/2\leq H\leq E+dE/2} d\Gamma$$
$$= \int \left[\theta\left(E + \tfrac{dE}{2} - H\right) - \theta\left(E - \tfrac{dE}{2} - H\right)\right] d\Gamma, \tag{8.79}$$

where $\theta(x)$ is the Heaviside function ($\theta(x) = 1$ for $x > 0$, 0 otherwise). The cumulative phase-space volume up to energy E is

$$\Omega(E) = \int \theta(E - H(p, q))\, d\Gamma = \int_{H\leq E} d\Gamma. \tag{8.80}$$

This function $\Omega(E)$ represents the total number of microstates with energy less than or equal to E.

Boltzmann's entropy is then defined as

$$S(E) = k_{\text{B}} \log \Gamma(E), \qquad \Gamma(E) = \frac{d\Omega(E)}{dE}, \tag{8.81}$$

where $\Gamma(E)$ is the *density of states*—the number of accessible microstates per unit energy interval.

The normalization condition for $\rho(p, q)$ reads

$$1 = \frac{1}{\mathcal{N}} \int_{\text{micro}} d\Gamma = \frac{\Omega(E + dE/2) - \Omega(E - dE/2)}{dE}, \tag{8.82}$$

from which

$$\mathcal{N} = \frac{dE}{\Omega(E + \frac{dE}{2}) - \Omega(E - \frac{dE}{2})}. \tag{8.83}$$

Taking the infinitesimal limit $dE \to 0$, we recover the differential relation

$$\lim_{dE\to 0} \frac{\Omega(E + \frac{dE}{2}) - \Omega(E - \frac{dE}{2})}{dE} = \frac{\partial\Omega(E)}{\partial E} = \Gamma(E), \tag{8.84}$$

which is equivalent to stating that

$$\Omega(E + dE/2) - \Omega(E - dE/2) = \Gamma(E)\, dE. \tag{8.85}$$

The micro-canonical ensemble describes a uniform probability distribution over all microstates with the same total energy. The function $\Omega(E)$ measures the accessible phase-space volume, while $\Gamma(E)$ gives the density of such states per energy interval. Entropy, temperature, and pressure follow as derived quantities:

$$\frac{1}{T} = \left(\frac{\partial S}{\partial E}\right)_V, \qquad \frac{P}{T} = \left(\frac{\partial S}{\partial V}\right)_E. \tag{8.86}$$

This structure is universal: any thermodynamic system—regardless of microscopic details—can be reconstructed once its density of states $\Gamma(E, V, N)$ is known.

The micro-canonical ensemble is one of the cornerstones of equilibrium statistical mechanics. From it, the canonical and grand-canonical ensembles follow by considering subsystems weakly coupled to large energy or particle reservoirs. These ideas will be developed in the following sections.

In the infinitesimal limit $dE \to 0$, the energy shell in phase space collapses to the constant-energy hypersurface. The corresponding *density of states* can be expressed directly in terms of a Dirac delta function:

$$\Gamma(E) = \int d\Gamma\, \delta(E - H(p,q)) \,. \tag{8.87}$$

This expression counts the number of accessible microstates with energy exactly equal to E. The delta function restricts the integration to points lying on the surface $\Sigma_E = \{(p,q) \mid H(p,q) = E\}$.

REMARK 8.15 Equation (8.87) is often referred to as the *surface integral representation* of the density of states. It is equivalent to integrating over an infinitesimal shell of thickness dE around E, normalized by dE.

Averages in the Micro-Canonical Ensemble

Once $\Gamma(E)$ is known, the ensemble average of any observable $A(p,q)$ can be computed as an integral over the constant-energy surface:

$$\langle A(p,q)\rangle_{\text{micro}} = \frac{1}{\Gamma(E)} \int d\Gamma\, A(p,q)\, \delta(E - H) \,. \tag{8.88}$$

This expression provides a practical way to calculate expectation values without explicitly defining a finite energy interval dE.

Because the delta function can be written as the derivative of the Heaviside step function,

$$\delta(E - H) = \frac{\partial}{\partial E}\theta(E - H), \tag{8.89}$$

Eq. (8.88) can be rewritten in an alternative but often more convenient form:

$$\langle A(p,q)\rangle_{\text{micro}} = \frac{1}{\Gamma(E)} \frac{\partial}{\partial E} \int d\Gamma\, A(p,q)\, \theta(E - H(p,q)). \tag{8.90}$$

REMARK 8.16 Equation (8.90) expresses micro-canonical averages as derivatives of cumulative phase-space integrals. This form is especially useful in analytic and numerical calculations, since it avoids the explicit handling of delta functions. Physically, it highlights that micro-canonical averages depend only on the geometry of the constant-energy surface and not on any arbitrary energy thickness dE.

Equations (8.88) and (8.90) are foundational in computational statistical mechanics. They appear, for example, in molecular dynamics simulations where averages are computed along trajectories constrained to a fixed total energy.

The Helmholtz Theorem, Equipartition, and the Mechanical Foundation of the Heat Theorem

The micro-canonical ensemble provides the fundamental link between Hamiltonian mechanics and thermodynamics. For a system of N interacting particles of masses m_i, the Hamiltonian is

$$H(q,p;\lambda) = K(p) + \phi(q;\lambda) = \sum_{i=1}^{s} \frac{p_i^2}{2m_i} + \phi(q;\lambda), \tag{8.91}$$

where $s = 3N$ is the number of degrees of freedom and λ denotes an external control parameter (such as volume, magnetic field, etc.).

The micro-canonical distribution is defined as

$$\rho_\mu(q,p;E,\lambda) = \frac{1}{\omega(E,\lambda)}\,\delta(E - H(q,p;\lambda))\,, \tag{8.92}$$

where the normalization factor

$$\omega(E,\lambda) = \int d^s q\, d^s p\, \delta(E - H(q,p;\lambda)) \tag{8.93}$$

is the *structure function*, also known as the *density of states* in the strict sense. The cumulative phase-space volume below energy E is

$$\Omega(E,\lambda) = \int_{H(q,p;\lambda)\le E} d^s q\, d^s p = \int d^s q\, d^s p\, \theta(E - H(q,p;\lambda))\,, \tag{8.94}$$

where $\theta(x)$ is the Heaviside function.

Ergodic Hypothesis

We now assume that the system is *ergodic*: for given E and λ, the Hamiltonian trajectory fills the entire constant-energy surface

$$\Sigma_{E,\lambda} = \{(q,p) \in \mathbb{R}^{2s} \mid H(q,p;\lambda) = E\}. \tag{8.95}$$

Then, for any observable $f(q,p)$, the long-time average equals its micro-canonical ensemble average:

$$\overline{f}_t = \lim_{\mathcal{T}\to\infty} \frac{1}{\mathcal{T}} \int_0^{\mathcal{T}} f(q(t),p(t))\, dt = \langle f\rangle_\mu = \int d^s q\, d^s p\, \rho_\mu(q,p;E,\lambda)\, f(q,p). \tag{8.96}$$

This assumption allows one to treat E and λ as a macroscopic state.

REMARK 8.17 The derivations of the micro-canonical equipartition theorem and the generalized Helmholtz theorem presented here follow Campisi [Cam24], Chapter 3, pp. 27–31. See also Helmholtz [vH84] for the original formulation, and Clausius [Cla65], Carathéodory [Car09], and Gibbs [Gib02] for historical treatments of the heat theorem.

We define the temperature and the generalized force conjugate to λ as time (or equivalently ensemble) averages [Cam24]:

$$T(E,\lambda) := \frac{2}{s}\,\langle K\rangle_t, \tag{8.97}$$

$$X(E,\lambda) := -\left\langle \frac{\partial H}{\partial \lambda} \right\rangle_t. \tag{8.98}$$

We now ask whether there exists a function $S(E,\lambda)$ such that

$$\frac{\partial S}{\partial E} = \frac{1}{T}, \qquad \frac{\partial S}{\partial \lambda} = \frac{X}{T}. \tag{8.99}$$

The existence of such a function is guaranteed by the *Generalized Helmholtz Theorem.*

The Generalized Helmholtz Theorem

We are in position now to connect to the heat theorem in thermodynamics. For an ergodic Hamiltonian system $H(q, p; \lambda)$, there exists a state function

$$S(E, \lambda) = k_{\mathrm{B}} \log \Omega(E, \lambda), \tag{8.100}$$

such that the partial derivatives of S satisfy Eq. (8.99).

The proof relies on the following result *Micro-Canonical Equipartition Theorem.* For any canonical coordinate or momentum $z_i \in \{q_1, \ldots, q_s, p_1, \ldots, p_s\}$,

$$\left\langle z_i \frac{\partial H}{\partial z_i} \right\rangle_\mu = \frac{\Omega(E, \lambda)}{\omega(E, \lambda)}. \tag{8.101}$$

To see this, consider

$$\begin{aligned}
\omega(E, \lambda) \left\langle z_i \frac{\partial H}{\partial z_i} \right\rangle_\mu &= \int dz\, z_i \frac{\partial H}{\partial z_i} \delta(E - H(z; \lambda)) \\
&= -\int dz\, z_i \frac{\partial}{\partial z_i} \theta(E - H(z; \lambda)) \\
&= -\int_{-\infty}^{\infty} dz_i \left[z_i \theta(E - H) \right]_{-\infty}^{+\infty} + \int dz\, \theta(E - H(z; \lambda)) \,.
\end{aligned} \tag{8.102}$$

The boundary term vanishes because H grows unboundedly as $z_i \to \infty$. Hence

$$\omega(E, \lambda) \left\langle z_i \frac{\partial H}{\partial z_i} \right\rangle_\mu = \Omega(E, \lambda), \tag{8.103}$$

which establishes Eq. (8.101).

Applying this to momenta gives $2\langle K \rangle_\mu = s\, \Omega / \omega$, leading to the *geometric definition of temperature*:

$$T(E, \lambda) = \frac{\Omega(E, \lambda)}{\omega(E, \lambda)}. \tag{8.104}$$

Equation (8.104) reveals that temperature is a purely geometric quantity: it measures how the phase-space volume $\Omega(E)$ changes with energy. In this sense, the concept of temperature arises naturally from the topology of the energy hypersurface in phase space.

Let us now prove Helmholtz's theorem [Cam24, vH84]. From Eq. (8.94) and $\omega(E, \lambda) = \partial_E \Omega(E, \lambda)$, we have

$$\frac{\partial S}{\partial E} = k_{\mathrm{B}} \frac{1}{\Omega} \frac{\partial \Omega}{\partial E} = k_{\mathrm{B}} \frac{\omega(E, \lambda)}{\Omega(E, \lambda)} = \frac{1}{T(E, \lambda)}, \tag{8.105}$$

establishing the first of Eq. (8.99). Similarly,

$$\begin{aligned}
\frac{\partial S}{\partial \lambda} &= k_{\mathrm{B}} \frac{1}{\Omega(E, \lambda)} \int d^s q\, d^s p\, \frac{\partial}{\partial \lambda} \theta(E - H(q, p; \lambda)) \\
&= -k_{\mathrm{B}} \frac{1}{\Omega(E, \lambda)} \int d^s q\, d^s p\, \frac{\partial H}{\partial \lambda} \delta(E - H(q, p; \lambda)) \\
&= \frac{X(E, \lambda)}{T(E, \lambda)},
\end{aligned} \tag{8.106}$$

where $X(E, \lambda) = -\langle \partial_\lambda H \rangle_\mu$ is the generalized force. Hence the differential form

$$dS = \frac{dE + X\, d\lambda}{T} \tag{8.107}$$

is exact, and $S(E, \lambda)$ is a bona fide state function. This completes the proof of the Helmholtz theorem.

The Helmholtz theorem provides a mechanical foundation for *Clausius's heat theorem.* It shows that for ergodic Hamiltonian systems, the combination

$$\frac{dE + X\,d\lambda}{T} \tag{8.108}$$

is an exact differential, implying the existence of entropy as a state function. Thus, the integrability of $\delta Q/T$ is not an empirical postulate but a mathematical consequence of Hamiltonian dynamics under the ergodic hypothesis [Cam24].

The theorem demonstrates that the conservative nature of thermodynamic fields is a direct consequence of microscopic energy conservation and ergodicity. Even for systems far from the thermodynamic limit, the heat theorem holds provided ergodicity applies to the relevant phase functions, typically the kinetic energy and the generalized forces.

Integrating Factors, Exact Differentials, and the Meaning of Temperature

From differential calculus, a first-order differential form

$$\omega = M(x, y)\,dx + N(x, y)\,dy \tag{8.109}$$

is said to be *exact* if there exists a potential function $G(x, y)$ such that

$$dG = M\,dx + N\,dy. \tag{8.110}$$

A necessary and sufficient condition for exactness is the equality of mixed partial derivatives,

$$\frac{\partial M}{\partial y} = \frac{\partial N}{\partial x}. \tag{8.111}$$

When this condition is not satisfied, it may still be possible to find a *multiplying function* $g(x, y)$, called an *integrating factor*, such that

$$g(M\,dx + N\,dy) = dG. \tag{8.112}$$

The function g renders the differential form exact, transforming a non-conservative field into a conservative one. In thermodynamics, this integrating factor corresponds to the *inverse temperature*, $g = 1/T$.

Non-Uniqueness of the Integrating Factor

If $g(M\,dx + N\,dy) = dG$, then for any strictly monotonic function $F(G)$ (with $F'(G) > 0$),

$$dF(G) = F'(G)\,dG = [F'(G)\,g]\,(M\,dx + N\,dy) \tag{8.113}$$

is also exact. This means that integrating factors are not unique: multiplying g by any positive function of G yields another valid integrating factor.

Thermodynamic Interpretation

In the context of the Helmholtz theorem [Cam24], the integrating factor $g = 1/T$ transforms the heat differential $\delta Q = dE + X\,d\lambda$ into an exact form:

$$dS = \frac{\delta Q_{\text{rev}}}{T}. \tag{8.114}$$

However, since any monotonic transformation of S is also valid, the absolute scale of T is not uniquely determined by mechanics alone. Only the *ordering* of temperatures—what it means for one body to be "hotter" than another—is fixed.

This geometric interpretation was made explicit by Carathéodory [Car09], who reformulated the second law in terms of the existence of integrating factors for inexact differentials. He showed that in the neighborhood of any equilibrium state, there exists a function $g(x, y)$ such that $g\,\delta Q$ is exact. In modern mechanical terms, this function corresponds precisely to the inverse temperature obtained in the Helmholtz theorem.

Temperature Scales and Thermometry

Because any monotonic function $F(T)$ defines an equally valid temperature scale, different thermometers correspond to different monotonic mappings between the same underlying thermal coordinate. For instance, ideal-gas thermometers obey $PV = Nk_{\text{B}}T$, establishing a linear relation between pressure, volume, and temperature. Other thermometers (e.g., resistance, voltage, or expansion-based) simply realize different monotonic mappings $F(T)$. All legitimate thermometers agree on equilibrium because they preserve the *order* of temperatures.

REMARK 8.18 The non-uniqueness of the integrating factor reflects a fundamental property of thermodynamics: *temperature is defined only up to a monotonic transformation.* The Kelvin scale is selected so that the Carnot efficiency

$$\eta = 1 - \frac{T_{\text{cold}}}{T_{\text{hot}}} \tag{8.115}$$

depends solely on the ratio of absolute temperatures. This establishes the modern absolute temperature scale.

The connection between integrating factors and the definition of temperature was first formalized by Carathéodory [Car09] in his geometric reformulation of the second law, and later reinterpreted mechanically in modern terms by Campisi [Cam24]. Both approaches identify $1/T$ as the integrating factor that renders δQ an exact differential, thereby defining entropy as a potential function.

Micro-Canonical Ensemble for an Ideal Gas

We now evaluate the density of states for an ideal (non-interacting) gas of N identical particles of mass m confined to a volume V. The Hamiltonian for this system is purely kinetic:

$$H = \sum_{i=1}^{N} \frac{\vec{p}_i^{\,2}}{2m}. \tag{8.116}$$

Since H depends only on the momenta, the spatial coordinates contribute a trivial factor of V^N to the phase-space volume.

To obtain a dimensionless measure, phase space is partitioned into elementary cells of volume h^{3N}, where h is Planck's constant. This factor ensures the correct counting of quantum states in the classical limit. Thus, the density of states is defined as

$$\Gamma(E) = \frac{1}{h^{3N}} \int d^{3N}q \int d^{3N}p\, \delta(E - H(p))$$
$$= \frac{V^N}{h^{3N}} \int d^{3N}p\, \delta\left(E - \frac{1}{2m}\sum_{i=1}^{3N} p_i^2\right). \tag{8.117}$$

The remaining momentum integral corresponds to the surface area of a $3N$-dimensional sphere in momentum space defined by

$$\sum_{i=1}^{3N} p_i^2 = 2mE, \tag{8.118}$$

that is, a hypersphere of radius $R = \sqrt{2mE}$.

To evaluate Eq. (8.117), we recall some standard results from geometry. The volume of an n-dimensional sphere of radius R is

$$V_n(R) = \frac{\pi^{n/2}}{\Gamma\left(\frac{n}{2}+1\right)} R^n, \tag{8.119}$$

and the corresponding surface area (the derivative of $V_n(R)$ with respect to R) is

$$S_n(R) = \frac{dV_n}{dR} = \frac{2\pi^{n/2}}{\Gamma\left(\frac{n}{2}\right)} R^{n-1}. \tag{8.120}$$

To confirm Eq. (8.119), note that

$$\left(\int_{-\infty}^{\infty} e^{-x^2}\, dx\right)^n = (\sqrt{\pi})^n = \int e^{-\sum_i x_i^2} d^n x = \int_0^{\infty} e^{-r^2} S_n(r)\, dr, \tag{8.121}$$

which, after performing the radial integration, gives $S_n = 2\pi^{n/2}/\Gamma(n/2)$ and $V_n = S_n R/n$. These expressions generalize the familiar $2\pi R$ (circumference of a circle) and $4\pi R^2/3$ (volume of a sphere) to arbitrary dimension.

Using Eqs. (8.119)–(8.120) with $n = 3N$, we can evaluate the integral over momenta:

$$\int d^{3N}p\, \delta\left(E - \frac{p^2}{2m}\right) = \int d^{3N}p\, \delta\left(\frac{p^2}{2m} - E\right)$$
$$= \frac{S_{3N}(R)}{\left|\frac{d(p^2/2m)}{dp}\right|_{p=R}} = S_{3N}(R)\,\frac{m}{R}. \tag{8.122}$$

Substituting $S_{3N}(R)$ from Eq. (8.120), we obtain

$$\int d^{3N}p\, \delta\left(E - \frac{p^2}{2m}\right) = \frac{2\pi^{3N/2}}{\Gamma(3N/2)} R^{3N-1}\,\frac{m}{R} = \frac{2\pi^{3N/2}}{\Gamma(3N/2)} (2mE)^{\frac{3N}{2}-1}\, m. \tag{8.123}$$

Simplifying,

$$\int d^{3N}p\, \delta\left(E - \frac{p^2}{2m}\right) = \frac{2\pi^{3N/2}}{\Gamma(3N/2)} (2mE)^{\frac{3N}{2}-1}. \tag{8.124}$$

Substituting Eq. (8.124) into Eq. (8.117) gives the density of states for an ideal gas:

$$\Gamma(E) = \frac{V^N}{h^{3N}}\,\frac{2\pi^{3N/2}}{\Gamma(3N/2)} (2mE)^{\frac{3N}{2}-1}. \tag{8.125}$$

REMARK 8.19 Equation (8.125) is the fundamental result for the micro-canonical ensemble of an ideal gas. It expresses the number of accessible microstates with energy E per unit energy interval.

Let us now provide some physical interpretation. 1. The factor $1/h^{3N}$ ensures that $\Gamma(E)$ is dimensionless and consistent with quantum counting of states in the classical limit. In quantum statistical mechanics, h^3 corresponds to the volume occupied by one state in phase space.

2. For identical particles, we must also divide by $N!$ to avoid overcounting permutations of indistinguishable particles:

$$\Gamma(E) = \frac{V^N}{N!\,h^{3N}} \frac{2\pi^{3N/2}}{\Gamma(3N/2)} (2mE)^{\frac{3N}{2}-1}. \tag{8.126}$$

This correction, known as the *Gibbs factor*, resolves the classical Gibbs paradox and ensures the extensivity of entropy.

3. The entropy of an ideal gas in the micro-canonical ensemble then follows as

$$S(E,V,N) = k_{\mathrm{B}} \log \Gamma(E) = k_{\mathrm{B}} \left[N \log V + \frac{3N}{2} \log E + \text{const} \right], \tag{8.127}$$

leading directly to the ideal-gas law $PV = Nk_{\mathrm{B}}T$ when one defines $1/T = (\partial S/\partial E)_V$.

REMARK 8.20 The expression for $\Gamma(E)$ provides the microscopic origin of the thermodynamic quantities derived earlier. It connects the geometry of high-dimensional phase space to macroscopic observables such as pressure, temperature, and entropy.

We can now compute the entropy of the ideal gas from the micro-canonical density of states. From Eq. (8.125), the entropy is

$$S = k_{\mathrm{B}} \log \Gamma(E) = Nk_{\mathrm{B}} \log v_{3N}^{1/N} + Nk_{\mathrm{B}} \log \left[\frac{V}{h^3} (2mE)^{3/2} \right], \tag{8.128}$$

where $v_n = \pi^{n/2}/\Gamma(n/2+1)$ is the volume of the unit n-sphere.

Using the identity

$$\log v_n = \frac{n}{2} \log \pi - \log \Gamma\left(\frac{n}{2} + 1\right), \tag{8.129}$$

and Stirling's approximation for large n,

$$\Gamma(n) \approx \sqrt{2\pi n}\, \frac{n^n}{e^n}, \tag{8.130}$$

we obtain

$$\log v_n \approx \frac{n}{2} \log \pi - \frac{n}{2} \log\left(\frac{n}{2}\right) + \frac{n}{2}, \tag{8.131}$$

valid for $n \gg 1$. Substituting $n = 3N$ and simplifying gives the entropy of the ideal gas in the micro-canonical ensemble:

$$S(E,V) = Nk_{\mathrm{B}} \log \left[V \left(\frac{4\pi mE}{3Nh^2} \right)^{3/2} \right] + \frac{3}{2} Nk_{\mathrm{B}}. \tag{8.132}$$

REMARK 8.21 Equation (8.132) is the celebrated *Sackur–Tetrode equation*, expressing the entropy of a classical ideal gas in the micro-canonical ensemble. It provides the microscopic foundation for the ideal-gas law.

Internal Energy, Temperature, and Heat Capacity

The energy as a function of S and V follows by inverting Eq. (8.132):

$$e^{\frac{S}{Nk_B}-\frac{3}{2}} = V\left(\frac{4\pi m}{3h^2}\frac{E}{N}\right)^{3/2}, \tag{8.133}$$

so that

$$U(V,S) = \frac{3h^2N}{4\pi m}\left(\frac{e^{\frac{S}{Nk_B}-\frac{3}{2}}}{V}\right)^{2/3}. \tag{8.134}$$

The temperature is obtained from

$$T = \left(\frac{\partial U}{\partial S}\right)_V = \frac{2U}{3Nk_B}, \tag{8.135}$$

and the constant-volume heat capacity is

$$C_V = \left(\frac{\partial U}{\partial T}\right)_V = \frac{3}{2}Nk_B. \tag{8.136}$$

Pressure and the Equation of State

At fixed entropy, the pressure follows from

$$P = -\left(\frac{\partial U}{\partial V}\right)_S = \frac{2U}{3V} = \frac{Nk_BT}{V}. \tag{8.137}$$

Thus, the micro-canonical formulation reproduces the ideal-gas equation of state $PV = Nk_BT$ directly from statistical mechanics.

REMARK 8.22 This derivation confirms that macroscopic thermodynamics emerges naturally from the microscopic description based on phase-space geometry and Liouville's theorem.

For completeness, differentiating the cumulative phase-space volume $\Omega(E)$ gives the density of states:

$$\begin{aligned}\Gamma(E) = \frac{d\Omega(E)}{dE} &= \frac{d}{dE}\left[\frac{\pi^{3N/2}}{\Gamma(3N/2+1)}(2mE)^{3N/2}\frac{V^N}{h^{3N}}\right] \\ &= \frac{2\pi^{3N/2}}{\Gamma(3N/2)}(2mE)^{\frac{3N}{2}-1}\frac{V^N}{h^{3N}},\end{aligned} \tag{8.138}$$

in agreement with Eq. (8.125). This expression fully determines the thermodynamic properties of the ideal gas in the micro-canonical ensemble.

8.3 The Gibbs Paradox and the Distinguishability of Particles

The *Gibbs paradox* arises when considering the entropy change that occurs upon mixing two gases. Suppose we have n_1 moles of gas type I in volume V_1 and n_2 moles of gas type II in volume V_2, as shown in Fig. 8.3. Both gases are ideal and at the same temperature T. If the ratio of their volumes equals the ratio of their mole numbers,

$$\frac{V_1}{V_2} = \frac{n_1}{n_2}, \tag{8.139}$$

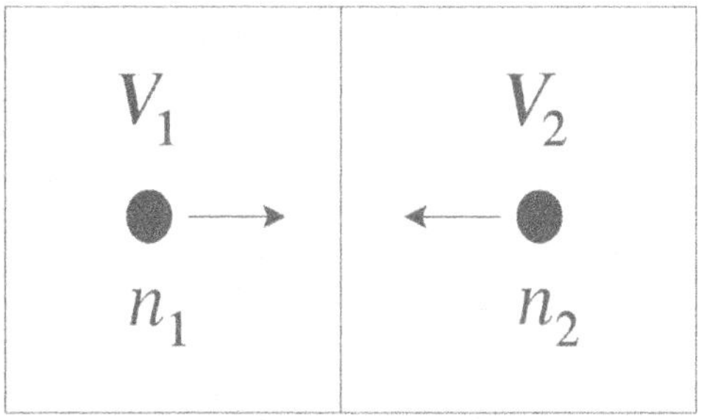

FIGURE 8.3 Setup of the Gibbs paradox. Two ideal gases, initially separated by a diaphragm, are allowed to mix once the barrier is removed. The question: What is the change in entropy?

then their pressures are initially equal,

$$P_1 = P_2 = \frac{n_1 RT}{V_1}. \tag{8.140}$$

When the partition separating them is removed, each gas expands to fill the total volume $V = V_1 + V_2$. The total change in entropy is then given by the entropy of expansion for each gas.

Entropy of Mixing for Different Gases

For two *different* ideal gases, the total entropy after mixing is

$$\begin{aligned}\Delta S &= S_{\text{final}} - S_{\text{initial}} \\ &= nR\log V - (n_1 R \log V_1 + n_2 R \log V_2) \\ &= -nR\left[f \log f + (1-f)\log(1-f)\right],\end{aligned} \tag{8.141}$$

where $n = n_1 + n_2$ and $f = n_1/n = V_1/V$. If the gases are present in equal amounts ($f = 1/2$), this simplifies to

$$\Delta S = nR\log 2. \tag{8.142}$$

This result means that mixing two *different* gases at the same temperature and pressure always increases the entropy, even though no heat is exchanged or work performed. The increase is purely combinatorial: there are more accessible microstates when the molecules can be of two distinct types.

The Paradox: Mixing the Same Gas

Now consider the same process but with *identical* gases on both sides of the partition. Naively applying the same formula (8.142) would still yield $\Delta S = nR\log 2$. However, this cannot be correct: when two samples of the same gas mix, there is no observable change in the system—its macroscopic state is unchanged. Therefore, the entropy should remain constant:

$$\Delta S = 0. \tag{8.143}$$

This apparent contradiction is the *Gibbs paradox*. It arises because the naive counting of microstates treats all particles as *distinguishable*. Under that assumption, exchanging particles between the two compartments artificially produces new microstates, even though physically nothing has changed.

REMARK 8.23 The Gibbs paradox highlights that entropy must be an *extensive* quantity: doubling both the number of particles and the volume should double the entropy. Without correcting for overcounting, the classical expression for S fails this test.

From the micro-canonical expression derived earlier,

$$S_{\text{micro}} = Nk_{\text{B}} \log\left(Vu^{3/2}\right) + Ns, \qquad u = \tfrac{3}{2}k_{\text{B}}T, \qquad s = \tfrac{3k_{\text{B}}}{2}\left[1 + \log\left(\frac{4\pi m}{3h^2}\right)\right], \tag{8.144}$$

it follows directly that if two subsystems contain *identical* particles, then the total entropy before and after removing the partition is identical:

$$S_{\text{final}} = S_{\text{initial}}. \tag{8.145}$$

This correctly predicts $\Delta S = 0$ for identical gases, resolving the paradox.

In the classical definition of the phase-space measure, we integrated over all N-particle coordinates and momenta as if each particle were uniquely labeled. This counts all $N!$ permutations of identical particles as distinct microstates, even though such permutations correspond to the same physical configuration. Consequently, the entropy is overestimated by a term $k_{\text{B}} \log N!$, making it non-extensive.

Quantum mechanics provides a natural resolution: indistinguishability is a fundamental property of identical particles. To correct the classical overcounting, one must divide the phase-space measure by $N!$, introducing the *Gibbs factor*:

$$\Gamma(E) = \frac{1}{N!\,h^{3N}} \int d^{3N}p\, d^{3N}q\, \delta(E - H(p,q))\,. \tag{8.146}$$

This modification ensures that entropy becomes an extensive quantity and that mixing identical gases produces no change in entropy.

REMARK 8.24 Although the introduction of the $1/N!$ factor can be justified formally only within quantum mechanics, its inclusion in the classical context restores the correct thermodynamic behavior. It ensures that the entropy depends only on physically distinguishable configurations.

A modern, information-theoretic interpretation views the paradox differently: the entropy increase upon mixing measures the *information gained* about which particle belongs to which gas. When the two gases are indistinguishable, no such information exists, and no entropy is produced. This perspective connects the Gibbs paradox to the concept of entropy as a measure of missing information, foreshadowing its central role in statistical inference and quantum information theory.

Before returning to the Gibbs paradox, let us briefly recall the distinction between *extensive* and *intensive* physical quantities.

A property is said to be *extensive* if it scales proportionally with the size or amount of the system. Volume, internal energy, total mass, and the number of particles are all extensive quantities. For example, in an ideal gas, the total internal energy is

$$U = N\frac{3}{2}k_{\text{B}}T, \tag{8.147}$$

so it doubles if the number of particles N is doubled at constant temperature.

By contrast, an *intensive* quantity does not depend on system size—examples include temperature, pressure, and density. These remain unchanged when two identical systems are combined.

In thermodynamics, entropy is expected to be an extensive property: if we combine two identical systems, the total entropy should be the sum of the individual entropies, $S_{\text{total}} = S_1 + S_2$. This requirement of additivity (or extensivity) will guide the resolution of the Gibbs paradox.

Resolution of the Gibbs Paradox Argued via Quantum Mechanics

As discussed earlier, the Gibbs paradox arises because the classical definition of entropy over-counts the number of microstates by treating identical particles as if they were distinguishable. Gibbs proposed that in counting microstates, one should divide by $N!$, the number of possible permutations of N identical particles, since exchanging two identical particles does not generate a new physical state.

Implementing this correction modifies the micro-canonical entropy to

$$S = k_{\mathrm{B}} \log \frac{\Omega(E)}{N!} \approx N k_{\mathrm{B}} \log\left(\frac{V\, u^{3/2}}{N}\right) + Ns + N k_{\mathrm{B}} \log N,$$
$$= N k_{\mathrm{B}} \log\left(\frac{V\, u^{3/2}}{N}\right) + \frac{5}{2} N k_{\mathrm{B}} \left[1 + \frac{3}{2} \log\left(\frac{4\pi m}{3h^2}\right)\right], \tag{8.148}$$

where $u = 3k_{\mathrm{B}}T/2$ and s collects additive constants. Here we have used Stirling's approximation, $\log N! \approx N \log N - N$, to simplify the expression.

Introducing the factor $1/N!$ does not alter measurable thermodynamic quantities such as pressure or temperature, because the correction depends only on N and thus drops out of derivatives with respect to E or V. Its effect is purely to make the entropy *extensive*, restoring additivity and resolving the Gibbs paradox.

Let us repeat the mixing thought experiment using the corrected entropy. The Sackur–Tetrode form of the entropy can be written as

$$S = N k_{\mathrm{B}} \log\left(\frac{V\, u^{3/2}}{N}\right) + \text{const.} \tag{8.149}$$

For two subsystems (1) and (2) with N_1, N_2 particles and V_1, V_2 volumes, the total change in entropy when the barrier is removed is

$$\frac{\Delta S}{k_{\mathrm{B}}} = (N_1 + N_2) \log\left(\frac{V_1 + V_2}{N_1 + N_2}\right) - N_1 \log\left(\frac{V_1}{N_1}\right) - N_2 \log\left(\frac{V_2}{N_2}\right). \tag{8.150}$$

Using the fact that the particle densities are equal before mixing,

$$n = \frac{N_1}{V_1} = \frac{N_2}{V_2} = \frac{N_1 + N_2}{V_1 + V_2}, \tag{8.151}$$

we find

$$\Delta S = 0. \tag{8.152}$$

Thus, when identical gases mix, there is no change in entropy—the paradox disappears. This correct counting of microstates is sometimes referred to as the *correct Boltzmann counting.*

Setting h equal to Planck's constant, the corrected formula yields the experimentally verified *Sackur–Tetrode equation* for the entropy of a monatomic ideal gas:

$$S(E, V, N) = N k_{\mathrm{B}} \left[\log\left(\frac{V}{N}\left(\frac{4\pi m E}{3 N h^2}\right)^{3/2}\right) + \frac{5}{2}\right]. \tag{8.153}$$

The value of the Planck's constant $h = 6.626 \times 10^{-34}$ J s ensures quantitative agreement between statistical mechanics and thermodynamic measurements of entropy.

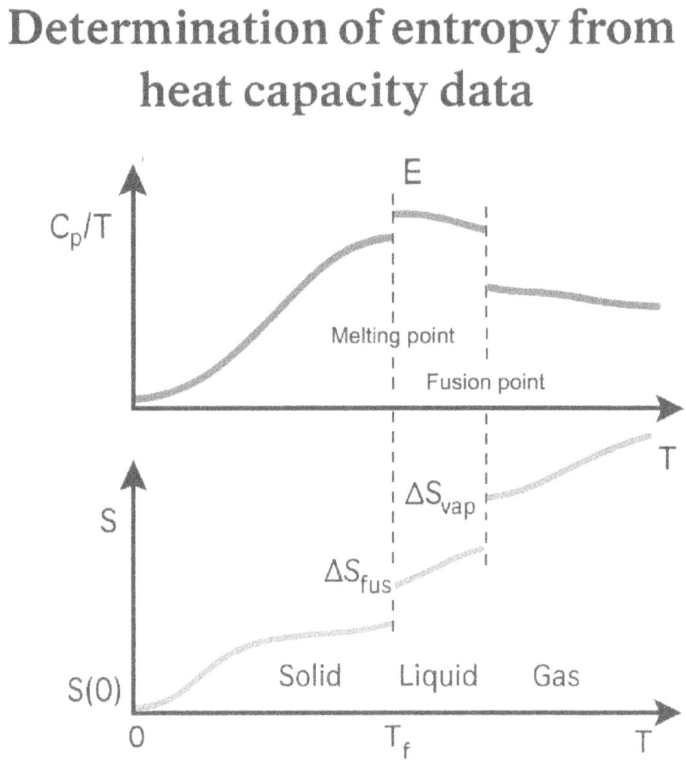

FIGURE 8.4 Experimental measurement of the Sackur–Tetrode entropy for water (after Rose Patruck, Brown University).

The $1/N!$ Factor

The historical success marked by the Sackur–Tetrode entropy lead the recognition that classical statistical mechanics is *incomplete* without quantum input, which is the standard in many statistical mechanics books.

The standard interpretation of the factor $1/N!$ is that it reflects the indistinguishability of particles; it is then a quite natural concept that this arises in quantum mechanics. Classically, labeling particles is an arbitrary human convention; quantum theory eliminates this ambiguity by assigning identical particles identical wavefunctions.

However, there are alternative interpretations. The concept of entropy has undergone a remarkable evolution from Gibbs's classical formulation to modern information-theoretic interpretations, which we will get into later in this and the next Volume. In his *Elementary Principles in Statistical Mechanics* (1902), J. Willard Gibbs [Gib02] established the ensemble approach and introduced entropy as the fundamental potential governing equilibrium, unifying the mechanical and thermodynamic descriptions. For Gibbs, entropy was already a property of macrostates (of sets of possible configurations consistent with macroscopic observables), not of any single microscopic state.

Eighty years later, E. T. Jaynes revisited these ideas in his essay *The Gibbs Paradox* (1992) [Jay92]. Jaynes argues that Gibbs himself had resolved the paradox of mixing long before quantum mechanics was developed, discussing why there is no mystery about "indistinguishability," and that it was needed to make entropy extensive. According to

Jaynes, Gibbs had recognized that thermodynamic quantities such as entropy are defined relative to the *macroscopic variables* chosen by the *observer*. Two experimenters can legitimately assign different entropies to the same physical microstate if they define their macrostates using different sets of observables. Entropy is thus a property of the observer's information about the system rather than a physical property of the microstate itself. This interpretation anticipates the modern view of entropy as a measure of information or uncertainty.

Jaynes further emphasized that this perspective restores the connection between thermodynamics and experiment: the apparent "increase of entropy" is not a metaphysical statement about disorder but an operational statement about what transformations are reproducible given our chosen macrovariables or order parameter. If new degrees of freedom become measurable or controllable, the associated entropy function changes accordingly. In his words, thermodynamics has a greater flexibility than usually recognized; it remains valid as long as we remain consistent with the macroscopic variables we use to describe a system [Jay92].

This broader interpretation finds resonance with modern statistical mechanics, in which Gibbs's ensembles are viewed as probability distributions representing incomplete knowledge. Jaynes's reformulation through the *maximum entropy principle* (1957) [Jay57] formalized this link: the equilibrium distribution is the least biased one consistent with the known constraints. In this way, the statistical entropy

$$S = -k_{\mathrm{B}} \sum_i p_i \log p_i \tag{8.154}$$

acquires a precise meaning as a measure of information, reducing to Boltzmann's $S = k_{\mathrm{B}} \log W$ when all accessible microstates are equally probable.

Finally, Carlo Cercignani [Cer98] has underlined that this synthesis between Gibbs's mechanical ensembles and Jaynes's informational viewpoint also restores Boltzmann's philosophical intent: entropy bridges our microscopic ignorance and macroscopic predictability. The second law then emerges not as a theorem of dynamics, but as a statement of overwhelming statistical likelihood grounded in Liouville's theorem. Entropy, in this light, becomes both a mechanical invariant and a measure of information—a dual role that remains central to modern physics.

8.4 Can Entropy Be Measured?

Up to this point, we have discussed entropy from a theoretical perspective—as a measure of the number of accessible microstates or, equivalently, as the logarithm of the density of states. But can the quantity $S = k_{\mathrm{B}} \log \Gamma(E)$ be *measured* experimentally?

Entropy cannot be observed directly; rather, it must be *inferred* from measurable quantities such as heat, temperature, or probabilities. The key thermodynamic relation is

$$dS = \frac{dQ_{\mathrm{rev}}}{T}, \tag{8.155}$$

which allows the determination of entropy differences through calorimetric measurements of the heat dQ_{rev} exchanged in a reversible process.

In practice, the entropy of a substance can be obtained by measuring the heat required to raise its temperature quasi-statically. If the process is reversible, the heat absorbed at each temperature increment dT is $dQ = C(T)\,dT$, where $C(T)$ is the specific heat. Integrating from $T = 0$ to a finite temperature T gives

$$S(T) = \int_0^T \frac{C(T')}{T'}\,dT' + S(0), \tag{8.156}$$

where $S(0)$ is the residual entropy at absolute zero.

According to the third law of thermodynamics, $S(0) = 0$ for a perfect crystal at $T = 0$. However, in many real systems, particularly in frustrated or disordered materials, the residual entropy can be nonzero. Such a finite value $S(0) \neq 0$ corresponds to a macroscopic degeneracy of the ground state.

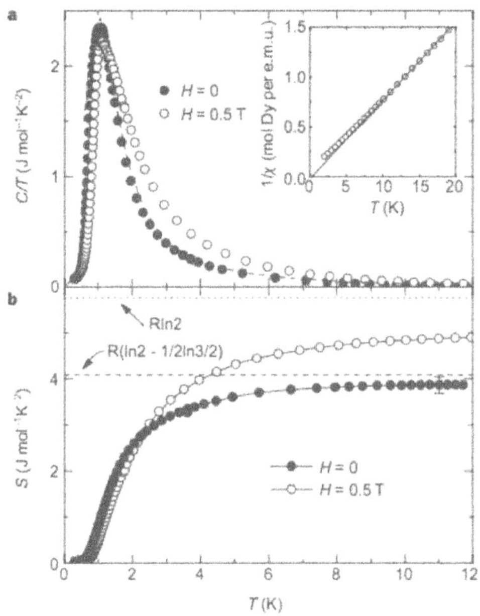

FIGURE 8.5 Experimental determination of entropy in spin ices. The integral of $C(T)/T$ reveals a residual "zero-point" entropy consistent with $S_0 = Nk_\mathrm{B}\ln 2$ [RHC+99].

Experimental Determination of Entropy

Although entropy is not an observable in the usual sense—it cannot be measured by a single instrument—the last century has shown that it is nevertheless an *experimentally accessible* and *physically meaningful* quantity.

Several complementary methods allow one to determine the entropy of a physical system:

- *Calorimetric Integration:* measuring the heat capacity as a function of temperature and integrating $C(T)/T\,dT$ provides the absolute entropy up to a constant. This approach is routinely used in low-temperature physics and was essential in identifying residual entropies in frustrated magnetic materials.
- *Direct Statistical Estimation:* in systems where the microscopic configurations can be counted or inferred—such as ice, spin ices, or certain lattice models—the entropy may be extracted from the combinatorial multiplicity of microstates.

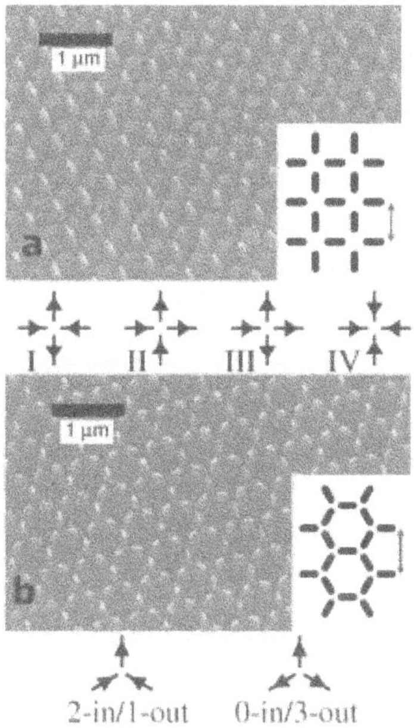

FIGURE 8.6 Visualization of microstates in an artificial spin-ice array (see [LKL+10], 2010). Each arrow represents a magnetic moment; direct imaging allows one to count microstates and evaluate the Shannon–Boltzmann entropy.

These techniques demonstrate that entropy, though not directly measurable, manifests itself through reproducible thermodynamic and statistical signatures. They also highlight the deep connection between thermodynamic and informational viewpoints: entropy quantifies the number of accessible microstates, whether counted microscopically or inferred thermodynamically.

Classic experimental realizations include the determination of the residual entropy of the triangular Ising antiferromagnet by Gregory Wannier [Wan50], the measurement of zero-point entropy in spin-ice compounds by Arthur Ramirez *et al.* [RHC+99], and later high-precision studies of disorder and frustration in magnetic materials by Schiffer [SR07].

A striking example of nonzero residual entropy is provided by *spin ices* and in *artificial spin ices*. In these magnetic materials, the microscopic spin orientations are constrained by "ice rules" analogous to the proton disorder in water ice. As a result, even at the lowest temperatures accessible experimentally, the system retains a large number of degenerate configurations. Measurements of specific heat, integrated according to Eq. (8.156), reveal a residual entropy

$$S_0 = N k_{\mathrm{B}} \ln 2, \tag{8.157}$$

indicating that half of the possible spin configurations remain disordered. This phenomenon, known as *zero-point entropy*, is one of the most striking experimental confirmations of the statistical definition of entropy.

In systems where individual microstates can be observed or sampled, entropy can also be estimated directly from the probabilities of microscopic configurations. Using the

Shannon–Boltzmann form,

$$S = -k_{\mathrm{B}} \sum_i p_i \log p_i, \tag{8.158}$$

one can determine the entropy by constructing the probability distribution p_i of observed states. This approach has been successfully applied to artificial magnetic lattices, granular media, and colloidal systems.

9

Entropy and Ensembles

In the previous chapter we saw that entropy, although not directly measurable, can be determined experimentally by integrating the reversible heat exchanged during quasi-static transformations. We now move from this empirical definition to its microscopic foundation in statistical mechanics. In particular, we show how different statistical definitions of entropy become equivalent in the thermodynamic limit and how the equality of temperature naturally arises when two systems are allowed to exchange energy.

9.1 Entropy Relations and Temperature

We have defined the entropy in the micro-canonical ensemble as

$$S = k_{\mathrm{B}} \log \Gamma(E), \tag{9.1}$$

DOI: 10.1201/9781003603528-9

where $\Gamma(E)$ is the number of accessible microstates per unit energy, or the *density of states.* However, in a previous chapter we also introduced two closely related quantities:

$$\Gamma(E) = \omega(E)\, dE = \frac{d\Omega(E)}{dE}\, dE, \tag{9.2}$$

where $\Omega(E)$ is the cumulative phase-space volume below energy E and $\omega(E)$ the "structure function" introduced in the Helmholtz theorem.

For small dE, one can write

$$\Gamma(E) = \Omega\left(E + \frac{dE}{2}\right) - \Omega\left(E - \frac{dE}{2}\right) = \frac{d\Omega(E)}{dE}\, dE. \tag{9.3}$$

We may ask: How does the infinitesimal interval dE scale with the system size? This question is analogous to asking how the surface and volume of an n-dimensional sphere scale with its radius R. The surface area scales as $S_n(R) \propto R^{n-1}$, while the enclosed volume scales as $V_n(R) \propto R^n$. In our analogy, $\Gamma(E)$ corresponds to the surface and $\Omega(E)$ to the volume.

This means that

$$\log \Omega(E) - \log \Gamma(E) \approx \log N, \tag{9.4}$$

where N is the number of degrees of freedom. Since entropy is extensive, differences of order $\log N$ are negligible for macroscopic systems. Therefore, in the thermodynamic limit $N \gg 1$, all three expressions for the entropy are equivalent:

$$S \approx k_{\mathrm{B}} \log \Gamma(E) \approx k_{\mathrm{B}} \log \Omega(E) \approx k_{\mathrm{B}} \log \omega(E). \tag{9.5}$$

REMARK 9.1 The three definitions differ only by subextensive terms of order $\log N$. For macroscopic systems, these are irrelevant, but for small ones, they can produce measurable deviations, as discussed in the statistical mechanics of small or non-extensive systems.

Let us now connect entropy to temperature by considering two weakly interacting systems, labeled (1) and (2), that can exchange energy but not particles or volume. The total energy $E = E_1 + E_2$ of the combined system is fixed, and there is no mechanical work exchange ($dW_1 = dW_2 = 0$).

Each subsystem is described by a density of states $\Gamma_1(E_1)$ and $\Gamma_2(E_2)$. The total number of accessible microstates for the composite system is

$$\Gamma(E) = \int_0^E \Gamma_1(\tilde{E})\, \Gamma_2(E - \tilde{E})\, d\tilde{E}. \tag{9.6}$$

The probability that subsystem 1 has energy E_1 is therefore

$$\rho(E_1) = \frac{\Gamma_1(E_1)\, \Gamma_2(E - E_1)}{\Gamma(E)}, \tag{9.7}$$

which satisfies the normalization condition $\int_0^E \rho(E_1)\, dE_1 = 1$.

The equilibrium distribution corresponds to the most probable value of E_1, obtained by maximizing $\rho(E_1)$:

$$\frac{d\rho(E_1)}{dE_1} = 0. \tag{9.8}$$

Setting $E_2 = E - E_1$, we find

$$\begin{aligned} 0 &= \frac{d\Gamma_1(E_1)}{dE_1}\Gamma_2(E - E_1) + \Gamma_1(E_1)\frac{d}{dE_1}\Gamma_2(E - E_1) \\ &= \Gamma_1(E_1)\, \Gamma_2(E - E_1)\left[\frac{1}{\Gamma_1(E_1)}\frac{d\Gamma_1(E_1)}{dE_1} - \frac{1}{\Gamma_2(E_2)}\frac{d\Gamma_2(E_2)}{dE_2}\right]. \end{aligned} \tag{9.9}$$

This condition is satisfied when

$$\frac{d\log\Gamma_1(E_1)}{dE_1} = \frac{d\log\Gamma_2(E_2)}{dE_2}. \tag{9.10}$$

Recalling that $S_i(E_i) = k_B \log\Gamma_i(E_i)$, this becomes

$$\frac{\partial S_1}{\partial E_1} = \frac{\partial S_2}{\partial E_2}. \tag{9.11}$$

By the micro-canonical definition of temperature,

$$\frac{1}{T_i} = \left(\frac{\partial S_i}{\partial E_i}\right)_{V_i}, \tag{9.12}$$

we obtain the condition for equilibrium:

$$\frac{1}{T_1} = \frac{1}{T_2}. \tag{9.13}$$

Entropy Maximization and Thermal Equilibrium

Two systems in weak thermal contact reach equilibrium when their temperatures are equal:

$$\frac{1}{T_1} = \frac{1}{T_2}. \tag{9.14}$$

At equilibrium, the total entropy

$$S_{\text{tot}} = S_1(E_1) + S_2(E - E_1) \tag{9.15}$$

is maximal, satisfying

$$\delta S_{\text{tot}} = 0, \qquad \delta^2 S_{\text{tot}} < 0. \tag{9.16}$$

This condition provides the statistical foundation of the zeroth law of thermodynamics: temperature equalizes between systems in thermal contact.

REMARK 9.2 The definition of temperature

$$\frac{1}{T} = \frac{\partial S}{\partial E} \tag{9.17}$$

follows directly from the condition of maximal entropy for two systems exchanging energy. This relation guarantees that heat flows spontaneously from the subsystem at higher temperature to that at lower temperature, increasing the total entropy until equilibrium is established.

Entropy changes can also be analyzed away from equilibrium. Consider two systems placed in thermal contact but not yet at equilibrium, such that the joint probability satisfies $\rho(E_1) \leq \rho(E_1^*)$, where E_1^* corresponds to the equilibrium configuration. From the definition of the phase-space volume,

$$\Gamma_1(E_1)\Gamma_2(E - E_1) \leq \Gamma_1(E_1^*)\Gamma_2(E - E_1^*), \tag{9.18}$$

which implies

$$1 \leq \frac{\Gamma_1(E_1^*)\,\Gamma_2(E - E_1^*)}{\Gamma_1(E_1)\,\Gamma_2(E - E_1)}. \tag{9.19}$$

Using $S = k_B \log \Gamma$, and noting that the logarithm preserves inequalities, one finds

$$0 \leq k_B \left[(S_1(E_1^*) - S_1(E_1)) + (S_2(E_2^*) - S_2(E_2)) \right], \tag{9.20}$$

which shows that for an isolated system the total entropy change satisfies $\Delta S \geq 0$.

When one of the systems (say, system 2) is much larger and acts as a thermal bath, the two systems remain isolated from the rest of the universe. For the bath, an infinitesimal energy change corresponds to

$$dS_2 = \frac{dE_2}{T_2} = \frac{dQ_2}{T_{\text{bath}}} = -\frac{dQ_1}{T_{\text{bath}}}, \tag{9.21}$$

which gives

$$dS_1 \geq \frac{dQ_1}{T_{\text{bath}}}. \tag{9.22}$$

This inequality expresses the second law of thermodynamics in differential form: for any process, the entropy change of the system is greater than or equal to the heat exchanged divided by the temperature of the bath.

9.2 Partition Functions in the Micro-Canonical and Canonical Ensembles

Direct calculations in the micro-canonical ensemble are often cumbersome. The canonical ensemble, introduced by Boltzmann and later formalized by Gibbs, simplifies these computations. Boltzmann's original terminology referred to the micro-canonical and canonical ensembles as the *ergode* and *holode*, while Gibbs later renamed them to their modern forms. In Gibbs's language, an ensemble was called a *monode* [Gib02].

The Canonical Ensemble

Consider again two systems in thermal contact, but assume one of them (system 2) is much larger than the other, acting as a heat reservoir. Let system 1 be small compared to the total, $E_1 \ll E$, and denote the bath temperature by T_{bath}.

The entropy of the reservoir can be expanded around $E_2 = E - E_1$:

$$\begin{aligned} k_B \log \Gamma_2(E_2) &\approx S_2(E_2) + E_2 \frac{\partial S_2}{\partial E_2} + \text{higher-order terms} \\ &= S_2(E_2) - E_1 \frac{\partial S_2}{\partial E_1} \qquad (\text{since } E = E_1 + E_2) \\ &= S_2(E) - \frac{E_1}{T_{\text{bath}}}, \qquad \text{because } \frac{\partial S_2}{\partial E_2} = \frac{1}{T_2} = \frac{1}{T_{\text{bath}}}. \end{aligned} \tag{9.23}$$

Hence,

$$\Gamma_2(E_2) \approx e^{S_2(E-E_1) - E_1/T_{\text{bath}}}. \tag{9.24}$$

The first term, $S_2(E)$, depends only on the total energy and is irrelevant for the subsystem's probability distribution. The probability that subsystem 1 has energy E_1 is therefore proportional to

$$\rho(E_1) \propto \Gamma_2(E - E_1) \approx e^{S_2(E) - E_1/T_{\text{bath}}}. \tag{9.25}$$

Consequently, the probability density in phase space for the small subsystem becomes

$$\rho_{\text{sub}}(p, q) = \frac{e^{-\beta H(p,q)}}{Z}, \tag{9.26}$$

where $\beta = 1/(k_B T_{\text{bath}})$ and

$$Z = \int d\Gamma \, e^{-\beta H(p,q)} \tag{9.27}$$

is the *partition function.* This distribution, known as the *Gibbs distribution*, defines the canonical ensemble. The energy distribution can be expressed as

$$p(E) = \frac{\Omega(E)}{Z} \, e^{-\beta E}, \tag{9.28}$$

where $\Omega(E)$ is the cumulative density of states and Z provides normalization.

Averages over the canonical ensemble are computed as

$$\langle A(p,q) \rangle_{\text{can}} = \frac{1}{h^{3N} N! \, Z} \int d^{3N}p \, d^{3N}q \, A(p,q) \, e^{-\beta H(p,q)}. \tag{9.29}$$

Since $E_1 \ll E$ and the reservoir is effectively infinite, the integration extends over the entire phase space without constraints.

Free Energy, Probabilities, and the Legendre Transform

In the canonical ensemble, a system of fixed temperature T and volume V is allowed to exchange energy with a thermal reservoir. The probability of finding the system in microstate i with energy E_i is assumed to depend only on E_i and T. The canonical distribution is then defined as

$$p_i = \frac{e^{-\beta E_i}}{Z}, \qquad \beta = \frac{1}{k_B T}, \qquad Z = \sum_i e^{-\beta E_i}, \tag{9.30}$$

where Z ensures normalization. The partition function Z is therefore the generating object for all thermodynamic quantities in the canonical ensemble.

Once a probability distribution $\{p_i\}$ over microstates is specified, the thermodynamic entropy must be a functional $S[\{p_i\}]$ satisfying two physical requirements:

- it is maximal for equiprobable states;
- it is additive for independent systems.

The unique (up to a multiplicative constant) function satisfying these conditions is the Gibbs(–Shannon) entropy,

$$S = -k_B \sum_i p_i \ln p_i, \tag{9.31}$$

which reduces to Boltzmann's $S = k_B \ln \Gamma$ when all accessible states are equally likely ($P_i = 1/\Gamma$).

Substituting the canonical probabilities from Eq. (9.30) into Eq. (9.31) yields

$$S = -k_B \sum_i \frac{e^{-\beta E_i}}{Z} \left[-\beta E_i - \ln Z\right] = k_B(\beta U + \ln Z), \qquad U = \langle E \rangle.$$

Using the thermodynamic relation $F = U - TS$, we obtain

$$F = U - T[k_B(\beta U + \ln Z)] = -k_B T \ln Z.$$

Thus the Helmholtz free energy arises directly from the canonical probability distribution:

$$F(T,V) = -\frac{1}{\beta} \ln Z. \tag{9.32}$$

Thermodynamically, the Helmholtz free energy is defined as the Legendre transform of the internal energy:

$$F(T, V) = U(S, V) - T\,S. \tag{9.33}$$

Differentiating with respect to T at fixed V gives the standard identities,

$$S = -\left(\frac{\partial F}{\partial T}\right)_V, \qquad U = F - T\left(\frac{\partial F}{\partial T}\right)_V. \tag{9.34}$$

To verify that the statistical-mechanical free energy

$$\tilde{F}(T, V) = -\frac{1}{\beta}\log Z$$

agrees with the thermodynamic Helmholtz free energy, we require it to satisfy the Legendre relation in Eq. (9.34). This leads to the consistency condition:

$$\tilde{F}(T, V) - T\left(\frac{\partial \tilde{F}}{\partial T}\right)_V = U(S, V), \tag{9.35}$$

which we now verify explicitly.

Differentiating $\tilde{F} = -k_{\mathrm{B}}T\ln Z$ gives

$$\left(\frac{\partial \tilde{F}}{\partial T}\right)_V = -k_{\mathrm{B}}\ln Z - \frac{k_{\mathrm{B}}T}{Z}\left(\frac{\partial Z}{\partial T}\right)_V.$$

Using

$$\frac{\partial Z}{\partial T} = -\frac{1}{k_{\mathrm{B}}T^2}Z\,\langle E\rangle,$$

we find

$$\tilde{F} - T\left(\frac{\partial \tilde{F}}{\partial T}\right)_V = \langle E\rangle = U.$$

Thus Eq. (9.35) holds and the statistical-mechanical expression

$$F = -k_{\mathrm{B}}T\ln Z$$

precisely reproduces the thermodynamic Helmholtz free energy.

Thus, once the canonical probabilities are introduced, the Gibbs entropy follows naturally from basic physical principles. Combined with the definition of the canonical distribution, these lead to the free energy identity $F = -k_{\mathrm{B}}T\ln Z$, fully consistent with the Legendre-transform structure of equilibrium thermodynamics.

Let us quickly rederive the same result without assuming the discreteness of energy. It is straightforward to verify that the internal energy can be expressed as the ensemble average

$$U(S, V) = \langle H\rangle = -\frac{\partial}{\partial\beta}\log Z. \tag{9.36}$$

To show that Eq. (9.35) holds, note that since $T\partial_T = -\beta\partial_\beta$, one must demonstrate that

$$-\beta\log Z - \partial_\beta\log Z + \beta\partial_\beta\log Z = 0, \tag{9.37}$$

which is equivalent to Eq. (9.35). Starting from $\tilde{F} = -\beta^{-1}\log Z$, the identity

$$\langle 1\rangle = 1 \tag{9.38}$$

can be written as

$$\frac{1}{h^{3N}N!}\int d\Gamma\, e^{-\beta H+\beta\tilde{F}} = 1. \tag{9.39}$$

Differentiating both sides with respect to β yields

$$\frac{1}{h^{3N}N!}\int d\Gamma\, e^{-\beta H+\beta\tilde{F}}(-H+\tilde{F}+\beta\partial_\beta\tilde{F}) = 0, \tag{9.40}$$

which implies

$$-\langle H\rangle + \tilde{F} + \beta\partial_\beta\tilde{F} = 0. \tag{9.41}$$

This result is precisely Eq. (9.35), confirming the consistency between the thermodynamic and statistical definitions of the Helmholtz free energy.

Consequently, all thermodynamic properties of a system at fixed volume can be derived from the partition function $Z(T,V)$. The temperature dependence of Z plays a central role and, contrary to some misconceptions, the partition function is not constant. Its dependence on temperature encodes essential physical information, particularly in the theory of phase transitions.

The Helmholtz free energy is additive for non-interacting systems. For two independent subsystems with Hamiltonians H_1 and H_2, the total partition function is

$$Z_N(T,V) = \frac{1}{h^{3N}N_1!N_2!}\int d\Gamma\, e^{-\beta(H_1+H_2)}. \tag{9.42}$$

Since the integrand factorizes,

$$Z_N(T,V) = Z_{N_1}(T,V_1)\, Z_{N_2}(T,V_2), \tag{9.43}$$

and therefore

$$F = -\beta^{-1}\log Z_N = F_1 + F_2, \tag{9.44}$$

confirming the extensivity of the free energy.

Grand-Canonical Ensemble

The grand-canonical ensemble describes systems in contact with a large reservoir that can exchange both energy and particles. The reservoir has temperature T_{bath}, chemical potential μ, and a much larger volume $V_2 \gg V_1$. The total Hamiltonian of the combined system is

$$H = H_1(p_1,q_1,N_1) + H_2(p_2,q_2,N_2), \tag{9.45}$$

with $N_1 + N_2 = N$ and $V_1 + V_2 = V$.

For the joint system, the total partition function is

$$Z_N = \sum_{N_1+N_2=N} \frac{N!}{h^{3N_1}N_1!N_2!h^{3N_2}} Z_{N_1} Z_{N_2}. \tag{9.46}$$

The probability of finding subsystem 1 in a state (p_1, q_1, N_1) is

$$\rho(p_1,q_1,N_1) = \frac{N!\, e^{-\beta H_1(p_1,q_1,N_1)}}{h^{3N_1}N_1!N_2!h^{3N_2}}\frac{Z_{N_2}}{Z_N}, \tag{9.47}$$

which satisfies the normalization condition

$$\sum_{N_1=0}^{N}\int d\Gamma_1\, \rho(p_1,q_1,N_1) = 1. \tag{9.48}$$

Writing $Z_{N_i} = e^{-\beta F(V_i,N_i,T)}$, and assuming $N_1 \ll N$ and $V_1 \ll V$, one finds

$$\frac{Z_{N_2}}{Z_N} = e^{-\beta[F(V_2,N_2,T)-F(V,N,T)]} = e^{-\beta[F(V-V_1,N-N_1,T)-F(V,N,T)]} = e^{-\beta\Delta F}. \tag{9.49}$$

Expanding ΔF to first order in N_1 and V_1 gives

$$\left(\frac{\partial F}{\partial V}\right)_{T,N} = -P,$$
$$\left(\frac{\partial F}{\partial N}\right)_{T,V} = \mu, \tag{9.50}$$

where μ is the *chemical potential*, measuring the energy cost of adding or removing a particle. Hence $\Delta F = -\mu N_1 + PV_1$.

The probability density of the grand-canonical ensemble becomes

$$\rho(p,q,N) = \frac{e^{\beta N\mu - \beta PV - \beta H(p,q,N)}}{N!\,h^{3N}}. \tag{9.51}$$

The *grand-canonical partition function* is defined as

$$\mathcal{G}(\zeta,P,T) = \sum_{N=0}^{\infty} \frac{\zeta^N}{h^{3N}N!} Z_N, \tag{9.52}$$

where $\zeta = e^{\beta\mu}$ is the *fugacity*. By construction,

$$\frac{PV}{k_{\mathrm{B}}T} = \log \mathcal{G}(\zeta,P,T). \tag{9.53}$$

When $\mu \neq 0$, the free energy must be modified to include the particle number:

$$F \to F + \mu N, \tag{9.54}$$

defining the *grand potential*. The average particle number is obtained from

$$\langle N \rangle = \frac{\sum_N N\,\zeta^N Z_N}{\sum_N \zeta^N Z_N} = \zeta \frac{\partial}{\partial \zeta} \log \mathcal{G}. \tag{9.55}$$

REMARK 9.3 The grand-canonical ensemble, introduced by Gibbs [Gib02], provides the natural framework for open systems exchanging both energy and particles with a reservoir. It plays a central role in chemical and condensed-matter systems, where the chemical potential and fugacity determine the equilibrium composition.

The average energy in the grand-canonical ensemble is obtained as

$$U = \langle H \rangle = \frac{\sum_N \zeta^N \int d\Gamma_N\, H\, e^{-\beta H}}{\sum_N \zeta^N Z_N} = -\frac{\partial}{\partial \beta} \log \mathcal{G}. \tag{9.56}$$

From this result, the generalized form of the first law follows directly. The generalized free energy is written as

$$F = U - PV + \mu N, \tag{9.57}$$

and its total differential becomes

$$dF = -P\,dV + T\,dS + \mu\,dN. \tag{9.58}$$

All thermodynamic quantities can then be obtained as derivatives of the appropriate potential.

Grand Potential and the Thermodynamic Meaning of $\log \mathcal{G}$

In the canonical ensemble, we showed that the Helmholtz free energy is obtained from the partition function through the relation $F = -k_{\rm B}T \log Z$. An analogous result holds in the grand-canonical ensemble, where the fundamental thermodynamic potential is the *grand potential* Ω.

By definition, the grand-canonical partition function is

$$\mathcal{G}(\zeta, V, T) = \sum_{N=0}^{\infty} \frac{\zeta^N}{h^{3N} N!} Z_N, \qquad \zeta = e^{\beta\mu}.$$

Once $\mathcal{G}$ is known, averages of all thermodynamic quantities follow by differentiation. In particular, the mean particle number and mean energy are

$$\langle N \rangle = \zeta \frac{\partial}{\partial \zeta} \log \mathcal{G}, \qquad U = -\frac{\partial}{\partial \beta} \log \mathcal{G}. \tag{9.59}$$

We now introduce the statistical-mechanical definition

$$\tilde{\Omega}(T, V, \mu) = -k_{\rm B}T \log \mathcal{G}. \tag{9.60}$$

To justify that $\tilde{\Omega}$ is indeed the thermodynamic grand potential, we must show that it satisfies the correct Legendre relations. Thermodynamics defines the grand potential by

$$\Omega = U - TS - \mu N,$$

and its natural variables are (T, V, μ). Hence it must satisfy

$$N = -\left(\frac{\partial \Omega}{\partial \mu}\right)_{T,V}, \qquad S = -\left(\frac{\partial \Omega}{\partial T}\right)_{V,\mu}, \qquad P = -\left(\frac{\partial \Omega}{\partial V}\right)_{T,\mu}. \tag{9.61}$$

To verify that the statistical potential $\tilde{\Omega}$ matches the thermodynamic one, we require the consistency condition

$$\tilde{\Omega} + TS + \mu\langle N \rangle = U, \tag{9.62}$$

which plays the same role as Eq. (9.35) in the canonical ensemble.

Differentiating Eq. (9.60) gives

$$\left(\frac{\partial \tilde{\Omega}}{\partial \mu}\right)_{T,V} = -k_{\rm B}T \frac{1}{\mathcal{G}} \frac{\partial \mathcal{G}}{\partial \mu} = -\langle N \rangle,$$

using Eq. (9.59). Similarly,

$$\left(\frac{\partial \tilde{\Omega}}{\partial T}\right)_{V,\mu} = -S, \qquad \left(\frac{\partial \tilde{\Omega}}{\partial V}\right)_{T,\mu} = -P,$$

so that all three Legendre identities in Eq. (9.61)) are reproduced. Finally, using Eqs. (9.59) and (9.31), one checks that Eq. (9.62) holds identically.

Thus,

$$\Omega(T, V, \mu) = -k_{\rm B}T \log \mathcal{G}(T, V, \mu) \tag{9.63}$$

is the correct thermodynamic grand potential, i.e., the Landau free energy introduced in Sec. 3.7, standing in precise analogy with $F = -k_{\rm B}T \log Z$ for the canonical ensemble. As before, all thermodynamic quantities follow from derivatives of the appropriate partition function.

Fluctuations in the Grand-Canonical Ensemble

Fluctuations of the number of particles can be derived in analogy with the canonical ensemble. The mean particle number is

$$\langle N \rangle = \frac{\sum_{N=0}^{\infty} N\,\zeta^N Z_N}{\sum_{N=0}^{\infty} \zeta^N Z_N} = \zeta\,\frac{\partial}{\partial \zeta}\log \mathcal{G}. \tag{9.64}$$

Differentiating again with respect to ζ gives

$$\zeta\,\partial_\zeta\,\zeta\,\partial_\zeta \log \mathcal{G} = \left(\frac{\zeta\,\partial_\zeta \mathcal{G}}{\mathcal{G}} - \left[\frac{\zeta\,\partial_\zeta \mathcal{G}}{\mathcal{G}}\right]^2 + \frac{\zeta^2\,\partial_\zeta^2 \mathcal{G}}{\mathcal{G}} \right) = \langle N^2 \rangle - \langle N \rangle^2. \tag{9.65}$$

Using Eq. (9.53), and recalling that $\mu = k_{\mathrm{B}}T \log\zeta$, we have $\zeta\,\partial_\zeta = k_{\mathrm{B}}T\,\partial_\mu$, which yields

$$\langle N^2 \rangle - \langle N \rangle^2 = k_{\mathrm{B}}T\,V\,\frac{\partial^2 P}{\partial \mu^2}. \tag{9.66}$$

The chemical potential and free energy are related through

$$\mu N = F - V\left(\frac{\partial F}{\partial V}\right)_T, \tag{9.67}$$

where the pressure is $P = -\left(\frac{\partial F}{\partial V}\right)_{T,N}$. From this, one finds

$$\frac{\partial \mu}{\partial V} = -V\frac{\partial^2 F}{\partial V^2}, \tag{9.68}$$

and consequently,

$$\frac{\partial P}{\partial \mu} = \frac{\frac{\partial P}{\partial V}}{\frac{\partial \mu}{\partial V}} = \frac{N}{V}. \tag{9.69}$$

Taking another derivative gives

$$\frac{\partial^2 P}{\partial \mu^2} = \frac{N^2}{V^2}\,K_T, \tag{9.70}$$

where K_T is the isothermal compressibility. The final relation between fluctuations and response functions is then

$$\langle N^2 \rangle - \langle N \rangle^2 = \bar{N}^2\,\frac{k_{\mathrm{B}}T\,K_T}{V}, \tag{9.71}$$

which is a fluctuation–dissipation relation linking particle-number fluctuations to the system's compressibility.

9.3 Thermodynamic Properties of the Ensembles

The micro-canonical, canonical, and grand-canonical ensembles provide equivalent descriptions of macroscopic systems in the thermodynamic limit but differ in the macroscopic constraints imposed. The choice of ensemble depends on which variables are externally controlled.

In the micro-canonical ensemble, the system is strictly isolated. The number of microstates,

$$\Gamma(E, V, N) = \sum_{\text{microstates at fixed } E,V,N} 1, \tag{9.72}$$

determines the entropy $S = k_{\mathrm{B}} \log \Gamma(E, V, N)$, with temperature given by $k_{\mathrm{B}}\beta = \partial_E S$.

In the canonical ensemble,

$$Z(\beta, V, N) = \sum_{\text{microstates } k, \text{ fixed } V,N} e^{-\beta E_k}, \tag{9.73}$$

while in the grand-canonical ensemble,

$$\mathcal{G}(\beta, V, \mu) = \sum_{\text{microstates } k, \text{ fixed } V} e^{-\beta E_k + \beta\mu N_k}. \tag{9.74}$$

Hence, the micro-canonical ensemble fixes (E, V, N), the canonical fixes (T, V, N), and the grand-canonical fixes (T, V, μ).

The grand-canonical ensemble is particularly important in chemistry, where the number of molecules is not conserved during reactions. For instance, in the oxidation of iron:

$$3\,\mathrm{Fe} + 4\,\mathrm{H_2O} \rightarrow \mathrm{Fe_3O_4} + 4\,\mathrm{H_2}, \tag{9.75}$$

the total number of atoms is conserved, but molecular species change. Describing such processes requires ensembles with variable particle number and well-defined chemical potentials.

The introduction of the *thermostat*—a device maintaining constant temperature—was historically crucial for experimental thermodynamics. Thermostats are ubiquitous in laboratory equipment such as incubators, ovens, and refrigerators, ensuring stable conditions for chemical and biological systems. Maintaining a constant temperature corresponds to fixing $\beta = 1/(k_B T)$. Keeping the volume constant corresponds to a rigid container, whereas keeping pressure constant corresponds to systems, like balloons, in contact with a large pressure reservoir (the atmosphere). Similarly, fixing the chemical potential requires exchange of specific particles with a reservoir. This is achieved experimentally using semipermeable membranes or selective barriers that allow only certain species to pass, maintaining the desired concentration and chemical potential on each side. Entropy plays a central role in all thermodynamic relations. Consider two systems brought into mutual contact. The total entropy is the sum of the entropies of the individual subsystems:

$$S_{\text{tot}} = S_1(E_1, N_1, V_1) + S_2(E_2, N_2, V_2). \tag{9.76}$$

At equilibrium, S_{tot} is maximal under the constraint of global conservation laws. If the total energy is conserved, $E = E_1 + E_2$, the total entropy becomes

$$S_{\text{tot}} = S_1(E_1, N_1, V_1) + S_2(E - E_1, N_2, V_2). \tag{9.77}$$

Maximization with respect to E_1 yields

$$0 = \frac{\partial S_{\text{tot}}}{\partial E_1} = \frac{1}{T_1} - \frac{1}{T_2}, \tag{9.78}$$

reproducing the thermal equilibrium condition $T_1 = T_2$.

Mechanical Equilibrium

Mechanical equilibrium is achieved when a movable wall or piston allows the systems to exchange volume while maintaining $E = E_1 + E_2$ and $V = V_1 + V_2$. The total entropy is then

$$S_{\text{tot}} = S_1(E_1, N_1, V_1) + S_2(E - E_1, N_2, V - V_1). \tag{9.79}$$

Maximization with respect to V_1 gives

$$0 = \frac{\partial S_{\text{tot}}}{\partial V_1} = \frac{\partial S_1(E_1, N_1, V_1)}{\partial V_1} - \frac{\partial S_2(E - E_1, N_2, V_2)}{\partial V_2}. \tag{9.80}$$

At fixed energy and particle number, the derivative of the entropy with respect to volume is

$$\left(\frac{\partial S}{\partial V}\right)_{E,N} = \frac{P}{T}, \tag{9.81}$$

so the equilibrium condition becomes

$$\frac{P_1}{T_1} - \frac{P_2}{T_2} = 0. \tag{9.82}$$

With $T_1 = T_2$, this reduces to $P_1 = P_2$, defining mechanical equilibrium.

At this point we can derive the first law. For a system with constant particle number N, the total differential of the entropy is

$$dS_{\text{tot}}(E, V, N) = \left(\frac{\partial S}{\partial E}\right)_{V,N} dE + \left(\frac{\partial S}{\partial V}\right)_{E,N} dV. \tag{9.83}$$

Using the relations

$$\left(\frac{\partial S}{\partial E}\right)_{V,N} = \frac{1}{T}, \qquad \left(\frac{\partial S}{\partial V}\right)_{E,N} = \frac{P}{T}, \tag{9.84}$$

we find

$$dS = \frac{1}{T}\, dE + \frac{P}{T}\, dV. \tag{9.85}$$

Multiplying by T gives

$$dE = T\, dS - P\, dV, \tag{9.86}$$

or, in finite differences,

$$\Delta E = Q - W, \tag{9.87}$$

which is the first law of thermodynamics for fixed N.

Chemical Equilibrium

When the number of particles is allowed to vary, particle exchange must also satisfy entropy maximization. At equilibrium, the total number of particles is conserved, $N = N_1 + N_2$, and hence $N_2 = N - N_1$. Maximizing the total entropy with respect to N_1 gives

$$0 = \frac{\partial S_{\text{tot}}}{\partial N_1} = \frac{\partial S_1}{\partial N_1} - \frac{\partial S_2}{\partial N_2}. \tag{9.88}$$

Using the thermodynamic identity

$$\left(\frac{\partial S}{\partial N}\right)_{E,V} = -\frac{\mu}{T}, \tag{9.89}$$

we obtain

$$0 = -\frac{\mu_1}{T_1} + \frac{\mu_2}{T_2}. \tag{9.90}$$

Since $T_1 = T_2$ at equilibrium, the result simplifies to

$$\mu_1 = \mu_2, \tag{9.91}$$

which defines chemical equilibrium.

The equality of chemical potentials across systems implies that particle exchange occurs until μ equalizes. To maintain a fixed chemical potential, a system must exchange particles with a reservoir while keeping temperature constant. This is realized experimentally by coupling the system to a particle bath or through the use of semipermeable membranes that selectively allow specific species to diffuse.

In the presence of chemical equilibrium, we can derive the generalized first law. Extending Eq. (9.83) to include the variable particle number yields

$$dS(E,V,N) = \left(\frac{\partial S}{\partial E}\right)_{V,N} dE + \left(\frac{\partial S}{\partial V}\right)_{E,N} dV + \left(\frac{\partial S}{\partial N}\right)_{E,V} dN. \tag{9.92}$$

Substituting the expressions for the derivatives gives

$$dS = \frac{1}{T}\,dE + \frac{P}{T}\,dV - \frac{\mu}{T}\,dN. \tag{9.93}$$

Multiplying both sides by T leads to

$$dE = T\,dS - P\,dV + \mu\,dN, \tag{9.94}$$

which is the general form of the first law of thermodynamics, valid for systems that can exchange energy, volume, and particles. The chemical potential μ quantifies the energy change associated with the addition or removal of a particle of a given species.

For a system with m distinct species, the total differential generalizes to

$$dE = T\,dS - P\,dV + \sum_{i=1}^{m} \mu_i\,dN_i, \tag{9.95}$$

with $S \equiv S(E, V, N_1, \ldots, N_m)$ evaluated at equilibrium.

9.4 The Gibbs–Duhem Relation and Maxwell Identities

For a single-component system, the Euler identity for homogeneous functions of first degree gives

$$E = TS - PV + \mu N. \tag{9.96}$$

Differentiating Eq. (9.96) yields

$$dE = T\,dS + S\,dT - P\,dV - V\,dP + \mu\,dN + N\,d\mu. \tag{9.97}$$

Subtracting the differential form of the first law, Eq. (9.94),

$$dE = T\,dS - P\,dV + \mu\,dN, \tag{9.98}$$

from the expression above eliminates dE, dS, dV, and dN, giving

$$S\,dT - V\,dP + N\,d\mu = 0. \tag{9.99}$$

This is the *Gibbs–Duhem relation* in differential form. It expresses that the intensive variables (T, P, μ) are not independent: for a single component, only two can be varied freely, while the third is determined by Eq. (9.99).

The Gibbs–Duhem relation also reveals that if a small portion of a system is in equilibrium with a larger reservoir, the intensive quantities must be identical in both systems. For example, in a grand-canonical ensemble where the temperature and chemical potential are fixed by the reservoir, the pressure adjusts accordingly to satisfy Eq. (3.119).

Thermodynamic Differentials and Maxwell Relations

All macroscopic quantities can be obtained from the differential relations between thermodynamic potentials. From Eq. (9.94),

$$\left(\frac{\partial S}{\partial E}\right)_{V,N} = \frac{1}{T}, \qquad \left(\frac{\partial S}{\partial N}\right)_{E,V} = -\frac{\mu}{T}, \tag{9.100}$$

$$\left(\frac{\partial E}{\partial S}\right)_{V,N} = T, \qquad \left(\frac{\partial E}{\partial V}\right)_{S,N} = -P. \tag{9.101}$$

Because mixed partial derivatives commute, the potentials satisfy additional cross-derivative identities known as the *Maxwell relations*. For instance,

$$\left(\frac{\partial T}{\partial V}\right)_{S,N} = -\left(\frac{\partial P}{\partial S}\right)_{V,N}, \tag{9.102}$$

and analogous relations follow from the other thermodynamic potentials.

From the Gibbs–Duhem relation, one further obtains

$$\left(\frac{dP}{dT}\right)_{\mu} = \frac{S}{V}, \qquad \left(\frac{dT}{d\mu}\right)_{P} = -\frac{N}{S}. \tag{9.103}$$

These results confirm that the macroscopic thermodynamic variables are deeply interdependent; specifying two intensive quantities fixes the third.

The differential relations of the principal thermodynamic potentials are summarized below. Each potential is a Legendre transform of the internal energy $E(S, V, N)$ and is naturally expressed in terms of a specific set of independent variables.

TABLE 9.1 Thermodynamic potentials, natural variables, and differentials.

Potential	Definition	Natural Variables	Differential Form
Internal energy	$E(S, V, N)$	(S, V, N)	$dE = T\,dS - P\,dV + \mu\,dN$
Helmholtz free energy	$F = E - TS$	(T, V, N)	$dF = -S\,dT - P\,dV + \mu\,dN$
Enthalpy	$H = E + PV$	(S, P, N)	$dH = T\,dS + V\,dP + \mu\,dN$
Gibbs free energy	$G = E - TS + PV$	(T, P, N)	$dG = -S\,dT + V\,dP + \mu\,dN$
Grand potential	$\Phi = E - TS - \mu N$	(T, V, μ)	$d\Phi = -S\,dT - P\,dV - N\,d\mu$

Each potential generates Maxwell relations by cross-differentiation of its differential form. For example, from $dF = -S\,dT - P\,dV + \mu\,dN$, one obtains

$$\left(\frac{\partial S}{\partial V}\right)_{T,N} = \left(\frac{\partial P}{\partial T}\right)_{V,N}, \quad \left(\frac{\partial \mu}{\partial T}\right)_{V,N} = -\left(\frac{\partial S}{\partial N}\right)_{T,V}, \quad \left(\frac{\partial \mu}{\partial V}\right)_{T,N} = \left(\frac{\partial P}{\partial N}\right)_{T,V}. \tag{9.104}$$

These relationships constitute the full set of Maxwell identities that connect measurable thermodynamic response functions. Together with the Gibbs–Duhem relation, they ensure the internal consistency of thermodynamics and provide a bridge between statistical mechanics and macroscopic observables. A summary is provided in Tab. 9.1.

The Clausius–Clapeyron Equation

As we have seen, the Gibbs–Duhem relation, Eq. (3.119), links variations of the intensive quantities (T, P, μ). An important application arises in the study of phase equilibria, where two phases of a substance coexist—such as a liquid and its vapor. In this case, both phases share the same temperature and pressure, and at equilibrium their chemical potentials must be equal:

$$\mu_l(T, P) = \mu_v(T, P), \tag{9.105}$$

where the subscripts l and v denote the liquid and vapor phases, respectively. This equality defines the coexistence curve $P(T)$ on the phase diagram.

Phase transitions of this type are central to chemistry and materials science. They determine vapor pressures, boiling and melting points, and the conditions for equilibrium between condensed and gaseous phases. The Clausius–Clapeyron relation provides a thermodynamic link between these measurable macroscopic quantities and microscopic properties such as the latent heat of vaporization.

The chemical potential for a single component in an ideal gas can be expressed as

$$\mu = \mu^\circ + RT \ln\left(\frac{P}{P^\circ}\right), \tag{9.106}$$

where μ° is the standard chemical potential, R is the gas constant, P the actual pressure, and P° a reference (standard) pressure.

During a phase transition, the difference in chemical potential between vapor and liquid is

$$\Delta\mu = \mu_v - \mu_l = \Delta\mu^\circ + RT \ln\left(\frac{P_v}{P^\circ}\right) - RT \ln\left(\frac{P_l}{P^\circ}\right), \tag{9.107}$$

where $\Delta\mu^\circ = \mu_v^\circ - \mu_l^\circ$ is the standard change in chemical potential. At equilibrium, $\mu_v = \mu_l$, and thus $\Delta\mu = 0$.

The Gibbs–Duhem relation for a single component reads

$$d\mu = -S_m\, dT + V_m\, dP, \tag{9.108}$$

where S_m and V_m are the molar entropy and molar volume, respectively. Taking the differential form for the two coexisting phases and using $\mu_l = \mu_v$, we obtain

$$d\mu_l = d\mu_v \quad \Rightarrow \quad -S_{m,l}\, dT + V_{m,l}\, dP = -S_{m,v}\, dT + V_{m,v}\, dP. \tag{9.109}$$

Rearranging gives

$$\frac{dP}{dT} = \frac{S_{m,v} - S_{m,l}}{V_{m,v} - V_{m,l}} = \frac{\Delta S_m}{\Delta V_m}. \tag{9.110}$$

Since the molar entropy change ΔS_m for a phase transition is related to the molar enthalpy change (latent heat) ΔH_m by

$$\Delta S_m = \frac{\Delta H_m}{T}, \tag{9.111}$$

substituting into Eq. (9.110) yields the *Clausius–Clapeyron equation*:

$$\frac{dP}{dT} = \frac{\Delta H_m}{T\, \Delta V_m}. \tag{9.112}$$

This relation connects the slope of the coexistence curve on a phase diagram to the latent heat and the volume change between the two phases.

Equation (9.112) shows that substances with a large latent heat or a large volume change between phases have a steep coexistence curve. For vaporization, $\Delta H_m > 0$ and $\Delta V_m > 0$, so $dP/dT > 0$: the vapor pressure increases with temperature.

In chemistry, the Clausius–Clapeyron equation provides the foundation for the determination of vapor pressures and boiling points. By integrating Eq. (9.112) and assuming that ΔH_m is approximately constant over a small temperature range, one obtains

$$\ln\left(\frac{P_2}{P_1}\right) = -\frac{\Delta H_m}{R}\left(\frac{1}{T_2} - \frac{1}{T_1}\right), \tag{9.113}$$

which describes how the vapor pressure P depends exponentially on temperature—an expression widely used in physical chemistry and chemical engineering.

9.5 Thermodynamic Stability Conditions

The relations derived in the previous sections—particularly the Gibbs–Duhem equation and the Maxwell identities—describe how intensive quantities vary under infinitesimal changes of state. These equations, however, presuppose that the system is at equilibrium. To determine whether a given equilibrium state is stable, one must appeal directly to Clausius's second law of thermodynamics:

"An isolated system evolves toward a state of maximum entropy."

At equilibrium, the entropy is stationary with respect to all infinitesimal, spontaneous redistributions of energy, volume, or matter among the system's subsystems. Thermodynamic *stability* requires more than stationarity: the entropy must reach a *maximum*. Any small fluctuation should decrease the total entropy, ensuring the system returns to equilibrium rather than moving farther away from it.

To understand whether a system is stable, one can follow a similar analysis to the one we used earlier, e.g., putting two systems in direct contact. Consider an isolated system composed of two subsystems (1) and (2), separated by a movable, diathermal, and permeable wall. The total internal energy $U = U_1 + U_2$, total volume $V = V_1 + V_2$, and total number of moles of each species $N_k = N_k^{(1)} + N_k^{(2)}$ are fixed. Nevertheless, infinitesimal transfers of energy, volume, and particles are allowed across the interface, subject to conservation:

$$\delta U_1 = -\delta U_2 = \delta U, \quad \delta V_1 = -\delta V_2 = \delta V, \quad \delta N_k^{(1)} = -\delta N_k^{(2)} = \delta N_k. \tag{9.114}$$

The total entropy is additive:

$$S = S_1(U_1, V_1, N_k^{(1)}) + S_2(U_2, V_2, N_k^{(2)}). \tag{9.115}$$

At equilibrium, S is stationary:

$$\delta S = 0, \qquad \delta^2 S < 0. \tag{9.116}$$

The first condition expresses entropy extremality; the second ensures the extremum corresponds to a maximum. To see this, one must analyze $\delta^2 S$ and check for its negativity.

Expanding each subsystem's entropy in a Taylor series around equilibrium and keeping terms up to second order gives

$$\begin{aligned}\delta^2 S = \sum_{a=1}^{2} \Bigg[& \frac{1}{2}\frac{\partial^2 S_a}{\partial U_a^2}(\delta U_a)^2 + \frac{1}{2}\frac{\partial^2 S_a}{\partial V_a^2}(\delta V_a)^2 + \sum_{i,j}\frac{1}{2}\frac{\partial^2 S_a}{\partial N_i^{(a)}\partial N_j^{(a)}}\delta N_i^{(a)}\delta N_j^{(a)} \\ & + \frac{\partial^2 S_a}{\partial U_a \partial V_a}\delta U_a \delta V_a + \sum_k \frac{\partial^2 S_a}{\partial U_a \partial N_k^{(a)}}\delta U_a \delta N_k^{(a)} + \sum_k \frac{\partial^2 S_a}{\partial V_a \partial N_k^{(a)}}\delta V_a \delta N_k^{(a)} \Bigg].\end{aligned} \tag{9.117}$$

At equilibrium, the first-order variations vanish, and because $\delta U_1 = -\delta U_2$ (and similarly for V and N_k), the cross-terms cancel for symmetric systems or when one subsystem acts as a reservoir. Evaluating derivatives at equilibrium yields

$$\delta^2 S = -\frac{1}{T^2}\, C_V^{-1}\, (\delta U)^2 - \frac{1}{T\kappa_T V}\, (\delta V)^2 - \sum_{i,j}\left(\frac{\partial}{\partial N_j}\frac{\mu_i}{T}\right)\delta N_i \delta N_j, \tag{9.118}$$

where $C_V = (\partial U/\partial T)_V$ is the heat capacity at constant volume, and κ_T is the isothermal compressibility:

$$\kappa_T = -\frac{1}{V}\left(\frac{\partial V}{\partial P}\right)_T. \tag{9.119}$$

The condition for stability is simply

$$\delta^2 S < 0 \quad \text{for all admissible fluctuations.} \tag{9.120}$$

This is a general form, but thermal, mechanical, and chemical stability are special cases of the expression above.

Thermal Stability

For purely thermal fluctuations ($\delta V = \delta N_k = 0$), one finds

$$\delta^2 S = -\frac{1}{T^2}\left(\frac{1}{C_{V,1}} + \frac{1}{C_{V,2}}\right)\frac{(\delta U)^2}{2} < 0, \tag{9.121}$$

implying $C_V > 0$ for stability. A negative heat capacity indicates that entropy would increase under a fluctuation—signaling thermodynamic instability, as observed in some non-equilibrium or self-gravitating systems.

Mechanical Stability

For purely mechanical perturbations ($\delta U = \delta N_k = 0$), and using $(\partial S/\partial V)_{U,N} = P/T$, the second variation becomes

$$\delta^2 S = -\left(\frac{1}{T\kappa_{T,1}V_1} + \frac{1}{T\kappa_{T,2}V_2}\right)\frac{(\delta V)^2}{2} < 0. \tag{9.122}$$

Assuming $T > 0$ and $V > 0$, stability requires a positive isothermal compressibility $\kappa_T > 0$. Negative compressibility corresponds to mechanical instability, often signaling a phase separation or a first-order transition.

Chemical Stability and the Duhem–Jouguet Theorem

When fluctuations involve particle exchange ($\delta U = \delta V = 0$), one obtains

$$\delta^2 S = -\sum_{i,j}\left(\frac{\partial}{\partial N_j}\frac{\mu_i}{T}\right)\delta N_i \delta N_j < 0. \tag{9.123}$$

The matrix $(\partial \mu_i/\partial N_j)_{T,V}$ must be positive definite, ensuring that entropy decreases for any fluctuation in composition. Diagonal elements such as $(\partial \mu_k/\partial N_k)_{T,V} > 0$ are required to suppress spontaneous instability of individual species.

This condition leads directly to the *Duhem–Jouguet theorem*, which states that a system stable with respect to diffusion is also stable with respect to chemical reactions. To illustrate, consider a chemical reaction of the general form

$$\nu_1\,\mathrm{A}_1 + \nu_2\,\mathrm{A}_2 \rightleftharpoons \nu_3\,\mathrm{A}_3 + \nu_4\,\mathrm{A}_4, \tag{9.124}$$

where ν_i are the stoichiometric coefficients (positive for products, negative for reactants). If ξ denotes the *extent of reaction*, then

$$dN_i = \nu_i\, d\xi. \tag{9.125}$$

Substituting into the entropy variation shows that stability with respect to particle diffusion implies stability with respect to chemical progress. For example, in

$$\mathrm{H_2} + \tfrac{1}{2}\mathrm{O_2} \rightleftharpoons \mathrm{H_2O}, \tag{9.126}$$

one mole of H_2 and half a mole of O_2 combine to form one mole of H_2O. At equilibrium, any infinitesimal change in ξ reduces the total entropy, confirming chemical stability.

Thermodynamic stability requires that the entropy of an isolated system be a *concave* function of its extensive variables (U, V, N_i). Equivalently, the internal energy $E(S, V, N_i)$ must be a *convex* function of the corresponding variables. The principal stability criteria are:

- $C_V > 0$ (thermal stability),
- $\kappa_T > 0$ (mechanical stability),
- $(\partial \mu_i / \partial N_j)_{T,V} > 0$ (chemical stability).

These conditions guarantee that small perturbations decay, restoring the system to equilibrium. Violations of these inequalities correspond to the onset of instabilities such as phase separation, runaway reactions, or self-organization phenomena observed in non-equilibrium thermodynamics.

Chemical Potentials, Rate Laws, and Kinetic Modeling

Thermodynamics identifies equilibrium via the condition $\mu_A(T,P) = \mu_B(T,P)$, but in practice chemical potentials are rarely measured directly. Instead, they are inferred from *reaction kinetics*, whose temperature dependence encodes free-energy differences between reactants and products. In classical chemical kinetics, the rate constants $k_{A\to B}(T)$ and $k_{B\to A}(T)$ typically obey an *Arrhenius form*,

$$k(T) = A \exp\left(-\frac{E_\mathrm{a}}{k_\mathrm{B}T}\right),$$

where A is the pre-exponential factor and E_a is the activation energy. This empirical law, first proposed by Svante Arrhenius, expresses the intuitive idea that a reaction proceeds only when thermal fluctuations provide sufficient energy to cross an energetic barrier. More refined theories, such as transition-state theory, relate E_a and A to enthalpies and entropies of activation and connect the rate constants to the underlying free-energy landscape [Hou07, Lai97]. We will develop these concepts rigorously in Volume II, where both the Arrhenius law and transition-state theory will be derived from microscopic statistical mechanics.

Thermodynamics and chemical kinetics become linked through the principle of detailed balance:

$$\frac{k_{A\to B}(T)}{k_{B\to A}(T)} = \exp\left[-\frac{\mu_B(T) - \mu_A(T)}{k_\mathrm{B}T}\right],$$

which allows chemical potential differences to be extracted from experimentally measured rates.

The differential equations derived in the main text for a simple reversible reaction,

$$\frac{dN_A}{dt} = -k_{A\to B}N_A + k_{B\to A}N_B, \qquad \frac{dN_B}{dt} = -k_{B\to A}N_B + k_{A\to B}N_A,$$

form the cornerstone of practical chemical kinetics. Once the rate law is experimentally determined—typically by varying initial concentrations and measuring initial rates, as taught in standard treatments of kinetics [Hou07, Lai97]—the resulting ordinary differential equations (ODEs) can be solved analytically or numerically to predict concentration profiles, half-lives, relaxation times, and approach to equilibrium.

In more complex systems, such as biochemical networks or catalytic reactions, these ODEs generalize to large coupled systems,

$$\frac{d\mathbf{X}}{dt} = \mathbf{R}(\mathbf{X}, T),$$

where $\mathbf{X}$ represents the vector of species concentrations and $\mathbf{R}$ encodes the stoichiometry and rate laws. Modern computational chemistry and systems biology rely heavily on numerical integration of such kinetic ODEs to identify rate-limiting steps, quantify bottlenecks, and predict dynamical responses to temperature or environmental changes. Single-molecule biophysics extends these ideas by measuring the stochastic analog of these ODEs. Rather than observing ensemble averages $N_A(t)$ and $N_B(t)$, one records the random dwell times that a *single* molecule spends in states A and B. From the exponential survival probabilities, one obtains the same rate constants $k_{A\to B}$ and $k_{B\to A}$ but without requiring synchronized bulk populations. This approach is especially powerful for studying protein folding and unfolding, where thermodynamic stability and kinetics are tightly linked [Fer99, XL$^+$18]. Temperature-dependent folding and unfolding rates yield estimates of $\Delta G(T)$, $\Delta H(T)$, and $\Delta S(T)$, offering direct insight into the underlying free-energy landscape.

Thus, the equations derived in this chapter—from differential conditions for chemical potentials to stability inequalities—have direct experimental consequences: they become the governing ODEs and kinetic laws through which thermodynamic information is extracted in the laboratory. The interplay between thermodynamics, kinetics, and ODE modeling forms one of the central pillars of modern chemical physics and biophysics.

9.6 Magnitude of Fluctuations

The stability conditions derived in the previous section ensure that a thermodynamic system returns to equilibrium after small perturbations. Nevertheless, in any finite system, small *spontaneous* deviations from equilibrium occur due to the microscopic dynamics of its constituents. Thermodynamics describes only the average, macroscopic behavior, while statistical mechanics reveals that fluctuations about equilibrium are inevitable. The magnitude of these fluctuations provides a quantitative measure of stability and forms a bridge between the deterministic macroscopic laws and the probabilistic nature of microscopic states.

Boltzmann first established the link between entropy and probability through his celebrated relation

$$S = k_{\mathrm{B}} \ln W, \tag{9.127}$$

where S is the entropy, k_{B} Boltzmann's constant, and W the number of microscopic configurations compatible with a given macroscopic state. In Boltzmann's view, probability was fundamental: the multiplicity W determines the likelihood of a macrostate, and entropy follows as its logarithmic measure. Equilibrium corresponds to the most probable macrostate, i.e., the one with maximal W.

In 1910, Einstein proposed a conceptual inversion of this reasoning in his study of critical opalescence [Ein10]. He regarded entropy, as defined in thermodynamics, as the primary quantity and derived from it the probability of observing a fluctuation. Instead of obtaining entropy from probability, he inferred probability from entropy itself.

Einstein argued that if an isolated system experiences a small fluctuation that changes its entropy by ΔS relative to its equilibrium value, then the probability of such a fluctuation is

$$dP = N^{-1} \exp\left(\frac{\Delta S}{k_{\mathrm{B}}}\right), \tag{9.128}$$

where N is a normalization constant ensuring that the total probability integrates to unity. This *Einstein formula* expresses that the probability of a fluctuation is exponentially weighted by the associated entropy change. Since equilibrium corresponds to maximum entropy, fluctuations that decrease S are exponentially unlikely.

Let the system be characterized by macroscopic variables $\{X_i\}$ (e.g., temperature, pressure, particle numbers). Expanding the entropy around its maximum at equilibrium gives

$$\Delta S = S - S_0 = \sum_i \left(\frac{\partial S}{\partial X_i}\right)_0 \delta X_i + \frac{1}{2}\sum_{i,j}\left(\frac{\partial^2 S}{\partial X_i \partial X_j}\right)_0 \delta X_i \delta X_j + \cdots . \tag{9.129}$$

Because the entropy is maximal at equilibrium, the first derivatives vanish, leaving the leading quadratic term

$$\Delta S = -\frac{1}{2}\sum_{i,j} g_{ij}\, \delta X_i \delta X_j, \tag{9.130}$$

with

$$g_{ij} = -\left(\frac{\partial^2 S}{\partial X_i \partial X_j}\right)_0, \tag{9.131}$$

a positive-definite matrix ensuring stability. Substituting into Einstein's formula yields the Gaussian probability density

$$dP = N^{-1} \exp\left[-\frac{1}{2k_{\mathrm{B}}}\sum_{i,j} g_{ij}\, \delta X_i \delta X_j\right]. \tag{9.132}$$

Fluctuations in thermodynamic variables are therefore distributed according to a multidimensional Gaussian, with variances determined by the curvature of the entropy surface at equilibrium.

Example: Fluctuations in an Ideal Gas Mixture

For an ideal mixture of gases, the entropy near equilibrium depends quadratically on small fluctuations of temperature, pressure, and particle numbers. A Taylor expansion of the entropy around its maximum yields

$$\Delta S = -\frac{C_V}{2T^2}(\delta T)^2 - \frac{1}{2K_T V}(\delta p)^2 + \sum_i \frac{R}{2N_i}(\delta N_i)^2,$$

where C_V is the heat capacity at constant volume, K_T the isothermal compressibility, V the volume, R the gas constant, and N_i the molar amount of species i.

Einstein's fluctuation formula states that the probability of a spontaneous fluctuation is weighted by the entropy change:

$$dP \propto \exp\left(\frac{\Delta S}{k_{\mathrm{B}}}\right).$$

Inserting the expression above gives a Gaussian probability distribution,

$$dP = N^{-1} \exp\left[-\frac{C_V(\delta T)^2}{2k_{\mathrm{B}}T^2} - \frac{(\delta p)^2}{2k_{\mathrm{B}}TK_TV} - \sum_i \frac{(\delta N_i)^2}{2N_i}\right],$$

showing that each fluctuating variable is independently Gaussian with a width determined by the corresponding thermodynamic susceptibility.

Since the distribution is Gaussian, the mean-square fluctuations follow immediately:

$$\langle(\delta T)^2\rangle = \frac{k_B T^2}{C_V}, \qquad \langle(\delta p)^2\rangle = \frac{k_B T}{K_T V}, \qquad \langle(\delta N_i)^2\rangle = N_i.$$

Thus, temperature fluctuations decrease with increasing heat capacity, pressure fluctuations scale inversely with system size and stiffness, and particle-number fluctuations obey Poisson statistics in an ideal gas.

In macroscopic systems these fluctuations are extremely small and generally unobservable. However, near critical points—where C_V or K_T diverge—or in mesoscopic systems with few particles, such fluctuations become significant and may dominate the observed behavior.

Einstein's analysis highlights a deep conceptual point: the probability of fluctuations is determined entirely by the macroscopic entropy function $S(X_i)$. Microscopic details enter only through the measurable response coefficients C_V, K_T, or N_i. This viewpoint is known as *Einstein's principle.*

Finally, the exponential weighting

$$P(X) \propto \exp\left(\frac{\Delta S}{k_B}\right)$$

anticipates the modern notion of a *large deviation principle*, where the probability of macroscopic fluctuations takes the asymptotic form

$$P(X) \propto \exp[-N\, I(X)],$$

with $I(X)$ the large-deviation functional associated with the entropy deficit. We will return to this interpretation in later chapters.

9.7 Equipartition Theorem

In the discussion of the Helmholtz theorem for the micro-canonical ensemble, we already encountered the fundamental result that for every quadratic degree of freedom of a Hamiltonian system, the average contribution to the internal energy is proportional to the temperature. This is the content of the *equipartition theorem*, one of the most important bridges between microscopic dynamics and macroscopic thermodynamics.

The theorem states that, in thermal equilibrium, each independent quadratic term in the Hamiltonian contributes an average energy of

$$\frac{1}{2}k_B T \tag{9.133}$$

per degree of freedom. For example, the mean kinetic energy per particle in a monoatomic ideal gas is

$$\frac{U(T)}{N} = \frac{3}{2}k_B T, \tag{9.134}$$

which we already derived from the canonical distribution. The equipartition theorem generalizes this observation to all ensembles and to any Hamiltonian that is quadratic in the canonical coordinates or momenta.

Formally, if the Hamiltonian of a system can be expressed as

$$H(p,q) = \sum_i a_i\, x_i^2 + \cdots\,, \tag{9.135}$$

then each quadratic variable x_i contributes an average energy of $k_B T/2$. Equivalently, for any canonical pair (p_i, q_i),

$$\langle p_i\,\frac{\partial H}{\partial p_i}\rangle = \langle q_i\,\frac{\partial H}{\partial q_i}\rangle = k_B T, \tag{9.136}$$

a result that follows directly from Hamilton's equations and holds in both the canonical and grand-canonical ensembles.

Connection with the Virial Theorem

To see the dynamical origin of this result, it is useful to recall the *virial theorem*, which relates the average kinetic energy to the average mechanical work performed by the forces in the system. Define the quantity

$$\mathcal{V} = p_i q_i, \tag{9.137}$$

and compute its time derivative using Hamilton's equations:

$$\dot{\mathcal{V}} = p_i \dot{q}_i + \dot{p}_i q_i. \tag{9.138}$$

Averaging over a long time T gives

$$\langle \dot{\mathcal{V}} \rangle_t = \frac{1}{T}\int_0^T \dot{\mathcal{V}}\,dt = \frac{\mathcal{V}(T) - \mathcal{V}(0)}{T}. \tag{9.139}$$

For a bounded or periodic system, $\mathcal{V}(T) \approx \mathcal{V}(0)$, so the time average of $\dot{\mathcal{V}}$ vanishes:

$$\langle p_i\, \dot{q}_i \rangle_t = -\langle q_i\, \dot{p}_i \rangle_t. \tag{9.140}$$

Using $\dot{q}_i = \partial H/\partial p_i$ and $\dot{p}_i = -\partial H/\partial q_i$, this leads to

$$\langle p_i\,\frac{\partial H}{\partial p_i}\rangle_t = \langle q_i\,\frac{\partial H}{\partial q_i}\rangle_t. \tag{9.141}$$

Summing over all degrees of freedom, one obtains the *virial theorem*:

$$2\langle T \rangle = -\left\langle \sum_i \vec{F}_i \cdot \vec{r}_i \right\rangle_t, \tag{9.142}$$

where $\vec{F}_i = -\vec{\nabla}_i U$ are the forces acting on the particles. This shows that the mean kinetic energy is directly linked to the average mechanical work performed by the internal forces.

In the following, we explore how this dynamical identity leads to the statistical equipartition theorem in different ensembles and provides a mechanical interpretation of temperature.

Experimental Significance of the Virial Theorem

The virial theorem provides a profound bridge between *microscopic degrees of freedom* and *macroscopic thermodynamic measurements*. For a classical system in equilibrium, it relates the average kinetic energy to the forces acting on the particles. When combined with the equipartition theorem, it implies that each quadratic degree of freedom contributes an average energy of

$$\frac{1}{2}k_B T.$$

Experimentally, this makes the virial theorem an indispensable tool for *counting active degrees of freedom* in real materials.

The molar heat capacity at constant volume,

$$C_V = \left(\frac{\partial U}{\partial T}\right)_V,$$

is one of the most direct experimental probes of the microscopic degrees of freedom accessible to a system at a given temperature. When divided by Nk_B, it yields

$$\frac{C_V}{Nk_\mathrm{B}} \approx \frac{f_\mathrm{active}}{2},$$

where f_active is the number of thermally excited quadratic degrees of freedom. Thus, by examining how C_V varies with temperature, one can determine whether translational, rotational, or vibrational modes are "frozen out" or active. This connection between heat capacity and accessible modes forms one of the most important experimental applications of the virial and equipartition theorems.

Early twentieth-century calorimetric experiments played a decisive role in revealing molecular structure. For example, at sufficiently high temperatures one expects a diatomic molecule to behave as if it has three translational and two rotational degrees of freedom, leading to $C_V \approx 5R/2$ (or $3R/2$ if rotation is frozen out). At even higher temperatures, vibrational modes contribute an additional R per mode, further increasing the heat capacity.

Measurements of the heat capacities of hydrogen and nitrogen by Arnold Eucken [Euc12], and later by Edwin Kemble and John van Vleck [KvV23], showed that at low temperatures only the three translational degrees of freedom contribute to the heat capacity. As temperature increases, C_V rises and approaches the value expected once the two rotational modes become thermally accessible (see Fig. 9.1). This behavior provided the first clear experimental evidence that rotational motion is "frozen out" at low temperature—a result that agreed with and supported early quantum-mechanical predictions.

In heavier diatomic molecules such as CO and HCl, a further rise in C_V at higher temperatures signaled the activation of vibrational modes. Detailed measurements by Peter Debye [Deb12] demonstrated the characteristic exponential onset expected from quantized harmonic oscillators. We will return to these ideas in later sections when discussing the quantum theory of molecular vibrations.

These classical experiments used nothing more than careful heat-capacity measurements and the virial theorem to deduce the existence of rotational and vibrational modes, the quantum nature of their activation and the internal structure of molecules, and the temperature scales associated with quantum-level spacing, which we will see in the next sections. Thus, long before modern spectroscopy, the virial theorem, combined with calorimetry, provided the earliest quantitative window into the microscopic degrees of freedom of matter.

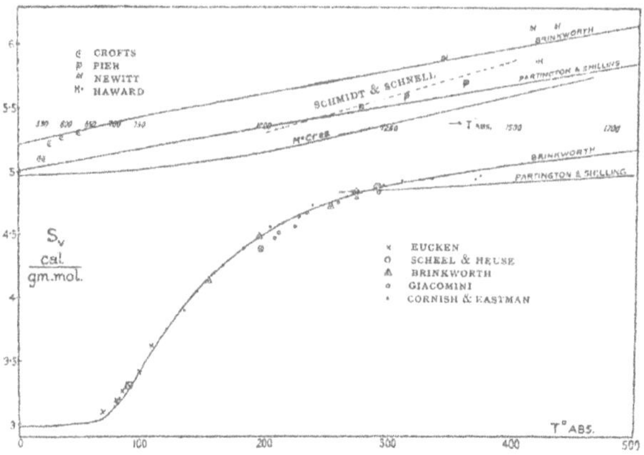

FIGURE 9.1 Heat capacity measurement for Hydrogen. Taken from Brinkworth, 1930 [Bri30].

Micro-Canonical Ensemble

The equipartition theorem can be derived directly within the micro-canonical ensemble, where the energy of the system is fixed and all accessible microstates are equally probable. For indistinguishable particles, the micro-canonical average of an observable $A(p, q)$ at total energy E is

$$\langle A \rangle_{\text{micr}} = \frac{1}{\omega(E)\, h^{3N} N!} \int d\Gamma\, A(p, q)\, \delta(E - H(p, q)), \tag{9.143}$$

where the cumulative and differential phase-space volumes are

$$\Omega(E) = \frac{1}{h^{3N} N!} \int d\Gamma\, \theta(E - H), \qquad \omega(E) = \frac{d\Omega(E)}{dE}. \tag{9.144}$$

We compute the quantity

$$\langle q_i \frac{\partial \mathcal{H}}{\partial q_j} \rangle_{\text{micr}}, \tag{9.145}$$

which represents the average product of a coordinate and its corresponding generalized force.

Derivation

Multiplying both sides by $h^{3N} N!$ and inserting the definition of the ensemble average:

$$\begin{aligned} h^{3N} N!\, \langle q_i \frac{\partial H}{\partial q_j} \rangle_{\text{micr}} &= \frac{1}{\omega(E)} \int d\Gamma\, q_i \frac{\partial H}{\partial q_j}\, \delta(E - H) \\ &= \frac{1}{\omega(E)} \frac{d}{dE} \int_{H \le E} d\Gamma\, q_i \frac{\partial H}{\partial q_j}. \end{aligned} \tag{9.146}$$

Next, note that

$$q_i \frac{\partial (E - H)}{\partial q_j} = \frac{\partial}{\partial q_j} [q_i (E - H)] - \delta_{ij} (E - H). \tag{9.147}$$

Hence,

$$\begin{aligned} \int_{H \le E} d\Gamma\, q_i \frac{\partial H}{\partial q_j} &= -\int_{H \le E} d\Gamma\, q_i \frac{\partial (E - H)}{\partial q_j} \\ &= -\int_{H \le E} d\Gamma \left[\frac{\partial}{\partial q_j} [q_i (E - H)] - \delta_{ij} (E - H) \right] \\ &= -\int_{H \le E} d\Gamma \frac{\partial}{\partial q_j} [q_i (E - H)] + \delta_{ij} \int_{H \le E} d\Gamma\, (E - H). \end{aligned} \tag{9.148}$$

The first term on the right-hand side is a total derivative over phase space and therefore reduces to a surface integral on the boundary $H = E$, where $E - H = 0$. Thus, it vanishes:

$$\int_{H\leq E} d\Gamma \, \frac{\partial}{\partial q_j}\big[q_i(E-H)\big] = 0. \tag{9.149}$$

Consequently,

$$\int_{H\leq E} d\Gamma \, q_i \, \frac{\partial H}{\partial q_j} = \delta_{ij} \int_{H\leq E} d\Gamma \, (E-H). \tag{9.150}$$

Differentiating the above expression with respect to E gives

$$\begin{aligned} h^{3N} N! \, \langle q_i \, \frac{\partial H}{\partial q_j} \rangle_{\text{micr}} &= \frac{\delta_{ij}}{\omega(E)} \frac{d}{dE} \int_{H\leq E} d\Gamma \, (E-H) \\ &= \frac{\delta_{ij}}{\omega(E)} \int_{H\leq E} d\Gamma = \delta_{ij} \, h^{3N} N! \, \frac{\Omega(E)}{\omega(E)}. \end{aligned} \tag{9.151}$$

In the micro-canonical ensemble, the temperature is defined through

$$\frac{1}{T} = \left(\frac{\partial S}{\partial E}\right)_{V,N} = k_{\mathrm{B}} \frac{d}{dE} \ln \Omega(E). \tag{9.152}$$

Therefore,

$$\frac{\Omega(E)}{\omega(E)} = \left(\frac{d}{dE} \ln \Omega(E)\right)^{-1} = k_{\mathrm{B}} T. \tag{9.153}$$

Substituting into the previous result yields the *micro-canonical form of the equipartition theorem:*

$$\langle q_i \, \frac{\partial H}{\partial q_j} \rangle_{\text{micr}} = \delta_{ij} \, k_{\mathrm{B}} T. \tag{9.154}$$

This is the same relation previously obtained from the Helmholtz theorem and expresses, in purely mechanical form, the equality between the average mechanical work associated with each quadratic coordinate and the thermal energy $k_{\mathrm{B}} T$.

Superadditivity of the Micro-Canonical Entropy

Consider two subsystems 1 and 2 with particle numbers N_1, N_2 and energies E_1, E_2, with

$$N_1 + N_2 = N, \qquad E_1 + E_2 = E.$$

Their cumulative state counts are $\Omega_a(E_a)$ and entropies

$$S_a(E_a) = k_{\mathrm{B}} \log \Omega_a(E_a) \qquad (a = 1, 2).$$

If the subsystems can exchange energy while the total energy E is fixed, the accessible phase-space volume of the composite system is the convolution

$$\Omega(E) = \int \Omega_1(E_1) \, \Omega_2(E - E_1) \, dE_1, \tag{9.155}$$

and the total entropy is $S(E) = k_{\mathrm{B}} \log \Omega(E)$.

For any particular split E_1° (hence $E_2^\circ = E - E_1^\circ$), the integrand in (9.155) is non-negative, so

$$\Omega(E) \geq \int_I \Omega_1(E_1) \, \Omega_2(E - E_1) \, dE_1 \geq |I| \, \Omega_1(E_1^\circ) \, \Omega_2(E_2^\circ),$$

where I is any interval containing E_1° on which the product is continuous and bounded below by its value at E_1°. Taking logs,

$$S(E) \;\geq\; S_1(E_1^\circ) + S_2(E_2^\circ) + k_{\rm B}\log|I|.$$

Since $|I|$ is an energy width independent of system size, this yields the *superadditivity* inequality

$$S(E) \;\geq\; S_1(E_1^\circ) + S_2(E_2^\circ) \qquad \text{for any split } E_1^\circ + E_2^\circ = E. \tag{9.156}$$

Write $S_a(E_a) = N_a s_a(\varepsilon_a)$ with $\varepsilon_a = E_a/N_a$. For large N, Laplace's method applied to (9.155) gives

$$\Omega(E) \;\asymp\; C\,\exp\Big(\tfrac{1}{k_{\rm B}}\,[S_1(E_1^*) + S_2(E_2^*)]\Big)\,N^{-1/2},$$

where E_1^* maximizes $S_1(E_1) + S_2(E - E_1)$ (equivalently, $T_1(E_1^*) = T_2(E_2^*)$). Hence

$$S(E) \;=\; \max_{E_1+E_2=E}\big[S_1(E_1) + S_2(E_2)\big] \;+\; O\big(k_{\rm B}\log N\big), \tag{9.157}$$

so additivity holds at leading order and deviations are at most $O(\log N)$, negligible per particle. We note that this is the first occurrence of the symbol $\asymp$ in this Volume, which implies that it is an asymptotic notion, e.g., for a large number of particles. This will be made precise in the section on large deviations later in the next Volume.

By definition,

$$\Omega(E) = \int_{\mathcal{H}(x)\leq E} d\Gamma, \qquad E' < E \;\Rightarrow\; \Omega(E') \leq \Omega(E),$$

so $\Omega(E)$ is non-decreasing and $S(E) = k_{\rm B}\log\Omega(E)$ is non-decreasing as well.

Let us now understand the role of entropy. Suppose initially the subsystems have energies $E_1^{(0)}, E_2^{(0)}$ (no exchange). Then

$$S^{(0)} = S_1(E_1^{(0)}) + S_2(E_2^{(0)}), \qquad \Omega^{(0)} = \Omega_1(E_1^{(0)})\,\Omega_2(E_2^{(0)}).$$

After contact and equilibration at the maximizing split E_1^*, E_2^*, Eq. (9.157) gives

$$S \;\geq\; S_1(E_1^*) + S_2(E_2^*) \;\geq\; S^{(0)},$$

so

$$\Delta S \equiv S - S^{(0)} \;\geq\; k_{\rm B}\log\frac{\Omega_1(E_1^*)\,\Omega_2(E_2^*)}{\Omega_1(E_1^{(0)})\,\Omega_2(E_2^{(0)})} \;\geq\; 0, \tag{9.158}$$

with equality if the initial split already satisfies the equilibrium condition $T_1(E_1^{(0)}) = T_2(E_2^{(0)})$.

If one discretizes energy into bins of width ΔE, the sum $\sum_{E_1}\Omega_1(E_1)\Omega_2(E - E_1)$ has $M \sim E/\Delta E = O(N)$ terms near the peak; then $\Omega(E) \leq M\,\max_{E_1}\Omega_1(E_1)\Omega_2(E - E_1)$, giving $S(E) \leq \max(S_1 + S_2) + k_{\rm B}\log M$. Together with (9.156), this pins $S(E)$ to the maximal sum $S_1 + S_2$ up to at most $O(k_{\rm B}\log N)$ corrections, confirming (9.157).

Canonical Ensemble

In the micro-canonical ensemble, we derived the equipartition theorem directly from the geometric properties of phase space. Here we show that the same result follows much more simply in the canonical ensemble, where the system is in thermal contact with a heat reservoir at fixed temperature T.

Let the total energy of the system be described by the Hamiltonian

$$E \equiv E(q_1, \ldots, q_N, p_1, \ldots, p_N), \tag{9.159}$$

with generalized coordinates $\{q_k\}$ and conjugate momenta $\{p_k\}$. Assume that the Hamiltonian separates as

$$E = \epsilon(p_1) + E(q_1, \dots, q_N, p_2, \dots, p_N), \tag{9.160}$$

so that the energy associated with one degree of freedom, $\epsilon(p_1)$, is quadratic in its momentum. These two assumptions—extensivity and quadratic momentum dependence—are the only ingredients required for the proof.

The canonical average of $\epsilon(p_1)$ is defined as

$$\langle \epsilon(p_1) \rangle = \frac{\displaystyle\int d\Gamma\, \epsilon(p_1)\, e^{-\beta E}}{\displaystyle\int d\Gamma\, e^{-\beta E}}, \tag{9.161}$$

where $\beta = 1/(k_B T)$ and $d\Gamma = d^{3N}p\, d^{3N}q$ is the phase-space measure.

Because E separates additively in p_1, the integration factorizes:

$$\begin{aligned} \langle \epsilon(p_1) \rangle &= \frac{\int dp_1\, \epsilon(p_1)\, e^{-\beta\epsilon(p_1)} \int d\Gamma_{2\cdots N}\, e^{-\beta E(q_1,\dots,q_N,p_2,\dots,p_N)}}{\int dp_1\, e^{-\beta\epsilon(p_1)} \int d\Gamma_{2\cdots N}\, e^{-\beta E(q_1,\dots,q_N,p_2,\dots,p_N)}} \\ &= \frac{\displaystyle\int dp_1\, \epsilon(p_1)\, e^{-\beta\epsilon(p_1)}}{\displaystyle\int dp_1\, e^{-\beta\epsilon(p_1)}}. \end{aligned} \tag{9.162}$$

For a quadratic energy term $\epsilon(p) = ap^2$ (with $a = 1/2m$ for a free particle), the average becomes

$$\langle \epsilon(p) \rangle = \frac{\displaystyle\int_{-\infty}^{\infty} dp\, (ap^2)\, e^{-\beta a p^2}}{\displaystyle\int_{-\infty}^{\infty} dp\, e^{-\beta a p^2}} = \frac{1}{2\beta} = \frac{1}{2} k_B T. \tag{9.163}$$

Thus, each quadratic term in the Hamiltonian contributes $k_B T/2$ to the mean energy. For a monoatomic ideal gas in three dimensions, with three translational momentum components per particle, this immediately gives

$$\langle H \rangle = \frac{3}{2} N k_B T, \tag{9.164}$$

in agreement with the result obtained previously from the Boltzmann distribution.

This derivation shows that in the canonical ensemble the equipartition theorem follows directly from the Gaussian character of the momentum distribution and is therefore simpler and more transparent than the corresponding micro-canonical proof.

Grand-Canonical Ensemble

In the grand-canonical ensemble, both the energy and the number of particles fluctuate, while the temperature T, chemical potential μ, and volume V remain fixed. Here, the chemical potential μ describes the energy associated with a single particle in the system. The probability density for a microstate with energy $H(p, q, N)$ and particle number N is

$$\rho(p, q, N) = \frac{1}{\mathcal{G}} \exp\big[-\beta(H(p, q, N) - \mu N)\big], \tag{9.165}$$

where $\mathcal{G}$ is the grand partition function:

$$\mathcal{G} = \sum_{N=0}^{\infty} e^{\beta\mu N} Z_N(T, V), \qquad Z_N(T, V) = \frac{1}{h^{3N} N!} \int d\Gamma_N\, e^{-\beta H(p,q,N)}. \tag{9.166}$$

In the literature, $\mathcal{G}$ is also called $\mathcal{Z}$ or sometimes Ξ. For a variable x_i that appears quadratically in the Hamiltonian—such as a component of momentum p_i or a harmonic coordinate q_i—the canonical ensemble gives

$$\langle x_i \frac{\partial H}{\partial x_i} \rangle_N = k_\mathrm{B} T. \tag{9.167}$$

The corresponding grand-canonical average is obtained by weighting this result over all particle numbers:

$$\langle x_i \frac{\partial H}{\partial x_i} \rangle_\mathrm{grand} = \sum_N P_N \, \langle x_i \frac{\partial H}{\partial x_i} \rangle_N, \tag{9.168}$$

$$P_N = \frac{e^{\beta \mu N} Z_N}{\mathcal{G}}. \tag{9.169}$$

Substituting Eq. (9.167) gives

$$\langle x_i \frac{\partial H}{\partial x_i} \rangle_\mathrm{grand} = \sum_N P_N \, k_\mathrm{B} T = k_\mathrm{B} T \sum_N P_N = k_\mathrm{B} T. \tag{9.170}$$

Thus, the equipartition theorem remains valid in the grand-canonical ensemble:

$$\langle x_i \frac{\partial H}{\partial x_i} \rangle_\mathrm{grand} = k_\mathrm{B} T, \tag{9.171}$$

identical in form to the canonical and micro-canonical results. The only difference is that the average now includes contributions from sectors with fluctuating particle number N.

In systems where H depends explicitly on N (for instance, through interaction terms), the same relation holds in the thermodynamic limit, as particle-number fluctuations are of order $\sqrt{N}$ and become negligible for large systems.

The persistence of the equipartition theorem across all ensembles reflects its purely local mechanical origin: it depends only on the quadratic form of the Hamiltonian and the existence of a well-defined temperature. Ensemble differences affect global fluctuations of extensive quantities, but not the average energy per quadratic degree of freedom.

9.8 Equivalence of Ensembles

In this section, we establish the equivalences between the ensembles.

Micro-Canonical ↔ Canonical

The equivalence between the canonical and micro-canonical ensembles rests on the fact that, in the thermodynamic limit, the canonical ensemble is sharply peaked around the same macroscopic state that dominates the micro-canonical ensemble. Although the canonical partition function involves an integral over the entire phase space, this integral is dominated by states whose energy corresponds to the micro-canonical expectation value at temperature T.

A particularly elegant argument, due to Melvin Lax [Lax55], shows how this equivalence follows directly from the properties of the entropy and the extensivity of thermodynamic variables.

We convert the phase-space integration into an integration over the energy shell. Introducing the density of states $\omega(E)$, defined through

$$\omega(E) \, dE = \frac{1}{h^{3N} N!} \int d^{3N}p \, d^{3N}q \, \delta(E - H(p, q)) \, dE, \tag{9.172}$$

we can rewrite the canonical partition function as

$$Z = \frac{1}{h^{3N} N!} \int d^{3N}p \, d^{3N}q \, e^{-\beta H(p,q)} = \int dE \, \omega(E) \, e^{-\beta E}. \tag{9.173}$$

Using the definition of the micro-canonical entropy,

$$S(E) = k_{\mathrm{B}} \ln \Omega(E), \qquad \omega(E) = \frac{d\Omega(E)}{dE}, \tag{9.174}$$

this expression becomes

$$Z = \int dE \, \exp\big[-\beta(E - TS(E))\big]. \tag{9.175}$$

Equation (9.175) already exhibits the Legendre-transform structure that connects the energy representation (micro-canonical) with the temperature representation (canonical).

Both the energy E and entropy $S(E)$ are extensive quantities. Writing them per particle as $E = Nu$ and $S(E) = Nk_{\mathrm{B}}s(u)$, we can express Eq. (9.175) as

$$Z = \int dE \, \exp[-\beta N(u - Ts(u))]. \tag{9.176}$$

In the thermodynamic limit $N \to \infty$, the integral can be evaluated by the saddle-point (Laplace) approximation. The integrand is dominated by the stationary point E^* satisfying

$$\frac{d}{dE}[E - TS(E)]_{E=E^*} = 0, \qquad \frac{d^2}{dE^2}[E - TS(E)]_{E=E^*} > 0. \tag{9.177}$$

The first condition gives

$$\left(\frac{\partial S}{\partial E}\right)_{E^*} = \frac{1}{T}, \tag{9.178}$$

i.e., the standard equilibrium condition connecting entropy and temperature in the micro-canonical ensemble. The second condition requires $\partial_E^2 S(E^*) < 0$, implying that $S(E)$ is concave, consistent with thermodynamic stability.

Expanding the exponent near E^*:

$$E - TS(E) \simeq E^* - TS(E^*) + \frac{1}{2}(E - E^*)^2 \left[-T\frac{d^2S}{dE^2}\right]_{E^*}, \tag{9.179}$$

and substituting into Eq. (9.175), we obtain

$$Z \approx e^{-\beta(E^* - TS(E^*))} \int dE \, \exp\left[-\frac{N\beta(E - E^*)^2}{2T^2 k_{\mathrm{B}} C_V}\right], \tag{9.180}$$

where we used

$$\frac{d^2S}{dE^2} = -\frac{1}{T^2 C_V}, \tag{9.181}$$

with $C_V = (\partial E/\partial T)_V$ the heat capacity at constant volume.

From the Gaussian integral, one finds that energy fluctuations in the canonical ensemble are given by

$$\langle (E - \langle E \rangle)^2 \rangle = k_{\mathrm{B}} T^2 C_V. \tag{9.182}$$

Thus, the relative fluctuations scale as

$$\frac{\sqrt{\langle (E - \langle E \rangle)^2 \rangle}}{\langle E \rangle} \sim \frac{1}{\sqrt{N}}, \tag{9.183}$$

and vanish in the thermodynamic limit. Consequently, the canonical ensemble becomes equivalent to the micro-canonical one when N is large: both ensembles describe the same macroscopic state.

This fluctuation–dissipation relation,

$$\langle (E - \langle E\rangle)^2\rangle = k_{\mathrm{B}}T^2 C_V, \tag{9.184}$$

provides an immediate connection between measurable energy fluctuations and the heat capacity at equilibrium. It is also the basis of many computational methods in statistical mechanics—for example, in Monte Carlo simulations, C_V can be extracted directly from fluctuations of the energy at constant temperature.

REMARK 9.4 The canonical and micro-canonical ensembles become equivalent when the entropy $S(E)$ is a concave, extensive function of E; energy fluctuations are small compared to the total energy, i.e., $\Delta E/E \sim 1/\sqrt{N}$; the system is away from critical points or long-range interactions, where concavity or extensivity may fail. When these conditions are met, all macroscopic observables take identical values in both ensembles.

Canonical ↔ Grand-Canonical

The equivalence between the canonical and grand-canonical ensembles also follows from the analysis of fluctuations. In the canonical ensemble, the particle number N is fixed, while in the grand-canonical ensemble it fluctuates about its mean value $\bar{N}$, determined by the chemical potential μ at equilibrium.

For a stable system, the relative particle-number fluctuations vanish in the thermodynamic limit. From the fluctuation–dissipation relation derived earlier,

$$\langle (\delta N)^2\rangle = k_{\mathrm{B}}T\, V \left(\frac{\partial n}{\partial \mu}\right)_{T,V} = \bar{N}^2\, \frac{k_{\mathrm{B}}T\, K_T}{V}, \tag{9.185}$$

where $n = \bar{N}/V$ is the particle density and K_T the isothermal compressibility defined by

$$K_T = -\frac{1}{V}\left(\frac{\partial V}{\partial P}\right)_{T,\mu}. \tag{9.186}$$

Mechanical stability requires $K_T > 0$, or equivalently $(\partial P/\partial V)_{T,N} < 0$. When this condition holds, the relative particle-number fluctuations scale as

$$\frac{\sqrt{\langle (\delta N)^2\rangle}}{\bar{N}} \sim \frac{1}{\sqrt{\bar{N}}}, \tag{9.187}$$

and vanish for large systems.

Thus, in the thermodynamic limit, both ensembles describe the same macroscopic state: the grand canonical distribution becomes sharply peaked around $\bar{N}$, reproducing the canonical results to all orders in $1/N$. Differences between the two ensembles appear only in finite systems or in the vicinity of phase transitions, where compressibility diverges and particle-number fluctuations become appreciable.

The $\mathcal{H}$-Theorem and Entropy

One of the most profound results of kinetic theory, due to Ludwig Boltzmann [Bol72, Bol96], is the $\mathcal{H}$-theorem. It provides a microscopic foundation for the second law by describing how the one-particle distribution function $f_t(p, q)$ evolves irreversibly toward equilibrium.

Boltzmann's theorem states that for a dilute gas obeying the Boltzmann transport equation,

$$\frac{\partial f}{\partial t} + \vec{v} \cdot \nabla_q f + \vec{F} \cdot \nabla_p f = C[f], \tag{9.188}$$

where $C[f]$ is the collision operator, the functional

$$\mathcal{H}(t) = \int f_t(p, q) \ln f_t(p, q)\, d^3p\, d^3q \tag{9.189}$$

satisfies

$$\frac{d\mathcal{H}}{dt} = \frac{d}{dt}\langle \ln f \rangle_f \leq 0. \tag{9.190}$$

The quantity $\mathcal{H}$ (introduced by Boltzmann for historical reasons) decreases monotonically with time and becomes stationary only at equilibrium, where the collision term vanishes. Up to a multiplicative constant, the negative of $\mathcal{H}$ coincides with the entropy.

At equilibrium, for an ideal gas, the stationary solution of the Boltzmann equation is the Maxwell–Boltzmann distribution,

$$f(p) = n\left(\frac{m}{2\pi k_{\mathrm{B}} T}\right)^{3/2} \exp\left(-\frac{mp^2}{2k_{\mathrm{B}} T}\right), \tag{9.191}$$

where n is the particle density. Substituting this expression into $\mathcal{H}$ gives

$$\ln f = \ln n + \frac{3}{2} \ln\left(\frac{m}{2\pi k_{\mathrm{B}} T}\right) - \frac{mp^2}{2k_{\mathrm{B}} T}, \tag{9.192}$$

$$H = n \ln n + \frac{3n}{2} \ln\left(\frac{m}{2\pi k_{\mathrm{B}} T}\right) - \frac{1}{k_{\mathrm{B}} T}\left\langle \tfrac{1}{2} mp^2 \right\rangle_f. \tag{9.193}$$

Since $\langle mp^2 \rangle_f / 2 = 3k_{\mathrm{B}} T/2$, one finds

$$\mathcal{H} = n \ln n, \tag{9.194}$$

so that the entropy per particle $s = -k_{\mathrm{B}}\mathcal{H}/n$ is constant at equilibrium. The $\mathcal{H}$-theorem thus provides a dynamical proof of the tendency of an isolated system to evolve toward the maximum-entropy Maxwellian distribution, uniting microscopic reversibility with macroscopic irreversibility.

Origin of Ensemble Inequivalence: Loss of Concavity

The equivalence between ensembles relies on the concavity of the entropy function $S(E, V, N)$. In the canonical and grand-canonical ensembles, the stability conditions derived earlier require, for instance, that the isothermal compressibility be positive and that the entropy be a concave function of energy. When this property fails—such as across a first-order phase transition—the canonical and micro-canonical ensembles become inequivalent.

In the grand-canonical to canonical correspondence, mechanical stability demands

$$\left(\frac{\partial P}{\partial V}\right)_{T,N} < 0, \tag{9.195}$$

which may fail or become discontinuous at a phase transition, as in the van der Waals gas which we discuss in the next chapter. Similarly, for the micro-canonical to canonical equivalence, the saddle-point argument requires that the entropy be concave. If $S(E)$ develops non-concave regions, the micro-canonical ensemble can exhibit negative heat capacities or other anomalies absent in the canonical ensemble.

Concavity of entropy is a general property of well-behaved thermodynamic systems and follows from the convexity of the exponential in the definition of the partition function. Mathematically, for any two normalized probability distributions $\vec{p}_1$ and $\vec{p}_2$, and for $0 \leq a \leq 1$,

$$S(a\,\vec{p}_1 + (1-a)\,\vec{p}_2) \geq a\,S(\vec{p}_1) + (1-a)\,S(\vec{p}_2), \tag{9.196}$$

which is the *Jensen inequality*. When the inequality is strict, S is strictly concave, ensuring ensemble equivalence. Loss of concavity signals the emergence of phase coexistence and inequivalence between ensembles.

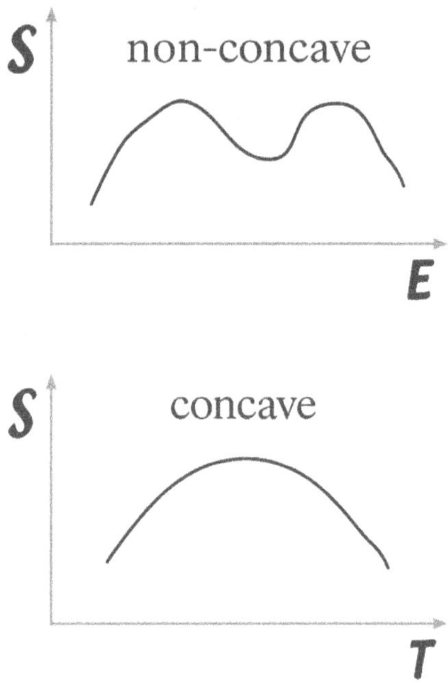

FIGURE 9.2 Examples of concave (stable) and non-concave (unstable) entropy functions as a function of energy.

In the thermodynamic limit,

$$N \to \infty, \qquad V \to \infty, \qquad \frac{N}{V} = \text{const}, \tag{9.197}$$

the average values of observables become ensemble-independent, but the magnitude of fluctuations may remain ensemble-specific. Breakdown of concavity, and thus of ensemble equivalence, occurs most notably in systems with long-range interactions or near phase transitions, topics that will be examined later in this Volume.

9.9 Legendre and Laplace Transforms: Connection Between Micro-Canonical and Canonical Ensembles

The connection between the micro-canonical and canonical ensembles can be understood mathematically through the relation between the Laplace and Legendre transforms. In the same way that the various thermodynamic potentials are related by Legendre transforms, the canonical partition function arises as a Laplace transform of the micro-canonical density of states.

The Laplace transform of a function $f(x)$ is defined as

$$\tilde{f}(\lambda) = \int_0^\infty e^{-\lambda x} f(x)\, dx. \tag{9.198}$$

In the context of statistical mechanics, this transform relates the micro-canonical and canonical ensembles.

As we have seen, the micro-canonical ensemble is characterized by the probability density

$$\mu(x) = \frac{\delta(H(x) - E)}{\Omega(E)}, \qquad d\mu(x) = \frac{\delta(H(x) - E)}{\Omega(E)}\, dx, \tag{9.199}$$

where $x = (p, q)$ represents a point in phase space, and

$$\Omega(E) = \frac{1}{h^{3N} N!} \int dx\, \theta(E - H(x)) \tag{9.200}$$

is the cumulative number of microstates with energy below E. The density of states is then $\omega(E) = d\Omega(E)/dE$.

The canonical partition function can now be expressed as the Laplace transform of $\omega(E)$:

$$Z(\beta) = \frac{1}{h^{3N} N!} \int dx\, e^{-\beta H(x)} = \int_0^\infty \omega(E)\, e^{-\beta E}\, dE. \tag{9.201}$$

The Laplace transform thus acts as the mathematical device that couples an isolated system (described by $\omega(E)$) to a thermal reservoir at temperature $T = 1/(k_B \beta)$.

Conversely, since the Laplace transform is invertible,

$$\omega(E) = \frac{1}{2\pi i} \int_{\mathcal{C}} Z(\beta)\, e^{\beta E}\, d\beta, \tag{9.202}$$

where $\mathcal{C}$ is a contour parallel to the imaginary axis in the right half-plane.

Define the dimensionless entropy and free energy as

$$\mathcal{S}(E) = \frac{S(E)}{k_B} = \ln \Omega(E), \qquad \mathcal{F}(\beta) = \beta F(\beta) = -\ln Z(\beta), \tag{9.203}$$

so that the Laplace relation becomes

$$e^{\mathcal{S}(E)} = \int_{\mathcal{C}} e^{-\mathcal{F}(\beta) + \beta E}\, d\beta. \tag{9.204}$$

In the thermodynamic limit, this integral can be evaluated by the saddle-point method. Expanding the exponent around its stationary point β^* satisfying

$$\frac{d}{d\beta}[\beta E - \mathcal{F}(\beta)]_{\beta = \beta^*} = 0, \tag{9.205}$$

we obtain, to leading order,

$$e^{\mathcal{S}(E)} \simeq e^{-\mathcal{F}(\beta^*) + \beta^* E}. \tag{9.206}$$

Equation (9.204) immediately implies the Legendre-transform relationship between entropy and free energy:

$$\mathcal{S}(E) + \mathcal{F}(\beta) = \beta E, \tag{9.207}$$

with differential relations

$$\frac{d\mathcal{S}}{dE} = \beta, \qquad \frac{d\mathcal{F}}{d\beta} = E. \tag{9.208}$$

These are precisely the equilibrium conditions obtained from thermodynamics, showing that the canonical and micro-canonical descriptions encode the same information in different conjugate variables.

Physically, the Legendre transform expresses the duality between an isolated system and a system in contact with a reservoir. If the energy E is fixed, the relevant potential is the entropy $S(E)$, and variations in E yield the response $\partial_E S = 1/T$. If the temperature T is fixed by a reservoir, the appropriate potential is the Helmholtz free energy $F(T)$, and variations in T yield the response $\partial_T F = -S$.

The same reasoning extends to other pairs of conjugate variables. Fixing the particle number N corresponds to using the chemical potential μ as the response function, while fixing μ corresponds to using the grand potential as the relevant thermodynamic potential. Thus, the Laplace and Legendre transforms provide the mathematical backbone of ensemble equivalence and of the relationships among thermodynamic potentials.

9.10 Generating Functions and the Canonical ↔ Grand-Canonical Transform

The relationship between the canonical and grand-canonical ensembles has a mathematical structure analogous to that between the micro-canonical and canonical ensembles. While the latter involves a *continuous* Laplace transform over the energy variable, the former involves a *discrete* Laplace (or generating-function) transform over the particle number N.

In the canonical ensemble, the partition function at fixed temperature T, volume V, and particle number N is

$$Z_N(T,V) = \frac{1}{h^{3N}N!}\int d\Gamma_N\, e^{-\beta H(p,q,N)}. \tag{9.209}$$

The grand-canonical ensemble generalizes this by allowing the system to exchange particles with a reservoir at fixed chemical potential μ. The grand-canonical partition function is defined as

$$\mathcal{G}(T,V,\mu) = \sum_{N=0}^{\infty} e^{\beta\mu N} Z_N(T,V), \tag{9.210}$$

which is a discrete Laplace transform of Z_N with respect to N, where the conjugate variable is $-\beta\mu$. This transform couples the system to a particle reservoir in exactly the same way that the canonical Laplace transform couples it to a heat reservoir.

Conversely, the canonical partition function can be recovered by inverting this discrete transform:

$$Z_N(T,V) = \frac{1}{2\pi i}\int_{\mathcal{C}} \mathcal{G}(T,V,\mu)\, e^{-\beta\mu N}\, d(\beta\mu), \tag{9.211}$$

where $\mathcal{C}$ is a contour in the complex $\beta\mu$ plane parallel to the imaginary axis.

Defining the dimensionless potentials

$$\mathcal{F}_N = \beta F_N = -\ln Z_N, \qquad \Phi = \beta\Omega = -\ln\mathcal{G}, \tag{9.212}$$

where F_N is the Helmholtz free energy and $\Omega = F_N - \mu N$ is the grand potential, we can rewrite the transform as

$$e^{-\Phi(T,V,\mu)} = \sum_N e^{-\mathcal{F}_N(T,V)+\beta\mu N}. \tag{9.213}$$

In the thermodynamic limit, the sum is dominated by its largest term and can be evaluated by a saddle-point approximation. The stationary point N^* satisfies

$$\frac{\partial}{\partial N}[\mathcal{F}_N - \beta\mu N]_{N=N^*} = 0, \tag{9.214}$$

which gives

$$\left(\frac{\partial F_N}{\partial N}\right)_{T,V} = \mu. \tag{9.215}$$

This is precisely the Legendre-transform relation between the Helmholtz free energy $F(T, V, N)$ and the grand potential $\Omega(T, V, \mu)$.

Substituting the stationary value back into the exponent yields

$$e^{-\Phi(T,V,\mu)} \simeq e^{-\mathcal{F}_{N^*}(T,V)+\beta\mu N^*}, \tag{9.216}$$

or equivalently,

$$\Phi(T, V, \mu) + \mathcal{F}_{N^*}(T, V) = \beta\mu N^*. \tag{9.217}$$

Differentiating with respect to the conjugate variables gives the standard thermodynamic relations:

$$\frac{\partial \mathcal{F}_N}{\partial N} = \beta\mu, \qquad \frac{\partial \Phi}{\partial(\beta\mu)} = N. \tag{9.218}$$

Thus, the canonical and grand-canonical potentials form a Legendre pair, just as $S(E)$ and $F(T)$ do in the micro-canonical–canonical correspondence.

From the thermodynamic point of view, the discrete Laplace transform connecting Z_N and $\mathcal{G}$ expresses the exchange of particles with a reservoir. Fixing the particle number corresponds to isolating the system with respect to matter exchange; fixing the chemical potential corresponds to placing it in contact with a particle bath.

As before, the ensemble equivalence follows from the saddle-point argument: the distribution of particle numbers in the grand canonical ensemble is sharply peaked around its mean $\bar{N}$, with relative fluctuations of order $1/\sqrt{N}$. Therefore, the canonical and grand-canonical ensembles describe the same macroscopic state in the thermodynamic limit, while the Legendre structure between F and Ω ensures that they encode the same physical information in conjugate variables.

9.11 The Pressure–Temperature or Gibbs Ensemble

The *pressure–temperature ensemble*, also called the *isothermal–isobaric* or *Gibbs ensemble*, describes a system in contact with both a heat bath at temperature T and a pressure reservoir at pressure P. In this configuration, both energy and volume fluctuate, while temperature, pressure, and particle number remain fixed. This ensemble is appropriate for systems capable of exchanging heat and mechanical work with their surroundings, such as a fluid confined by a movable piston under constant external pressure.

Microscopic Definition and Distribution Function

Consider a system characterized by a Hamiltonian $H(x, V)$, where $x = (p, q)$ denotes the microstate variables. When coupled to a large reservoir imposing fixed T and P, the probability for the system to occupy microstate x is

$$P(x) = \frac{1}{\Delta(T, P, N)} \exp[-\beta(H(x, V) + PV)], \tag{9.219}$$

where the normalization factor

$$\Delta(T, P, N) = \sum_x e^{-\beta(H(x,V)+PV)} = \frac{1}{h^{3N} N!} \int_0^\infty dV \int d\Gamma_N \, e^{-\beta(H(p,q,V)+PV)} \tag{9.220}$$

is the *isothermal–isobaric partition function* [LL80, PB11]. The exponential weight $\exp[-\beta PV]$ expresses the mechanical coupling to the pressure reservoir and replaces the fixed-volume constraint of the canonical ensemble. The integration over V embodies the fact that the system's volume can fluctuate in this ensemble.

Defining the Gibbs free energy as

$$G(T, P, N) = -k_B T \ln \Delta(T, P, N), \tag{9.221}$$

one obtains the relations

$$\left(\frac{\partial G}{\partial P}\right)_{T,N} = V, \qquad \left(\frac{\partial G}{\partial T}\right)_{P,N} = -S, \tag{9.222}$$

which confirm that $G = H - TS$, where $H = E + PV$ is the enthalpy. Hence, $G(T, P, N)$ is the appropriate thermodynamic potential for systems at fixed temperature and pressure, and the Gibbs ensemble is its natural statistical representation.

The average values of energy and volume follow as

$$\langle H\rangle = -\frac{\partial \ln \Delta}{\partial \beta} - P\langle V\rangle, \qquad \langle V\rangle = -\frac{1}{\beta}\left(\frac{\partial \ln \Delta}{\partial P}\right)_{T,N}, \tag{9.223}$$

in full agreement with the thermodynamic differential

$$dG = -S\,dT + V\,dP + \mu\,dN. \tag{9.224}$$

Because V is not fixed, its fluctuations are related to the isothermal compressibility K_T:

$$\langle(\delta V)^2\rangle = k_B T\, V\, K_T, \qquad K_T = -\frac{1}{V}\left(\frac{\partial V}{\partial P}\right)_{T,N}. \tag{9.225}$$

As $N \to \infty$, the relative fluctuations scale as $1/\sqrt{N}$, ensuring ensemble equivalence in the thermodynamic limit. Near critical points, where K_T diverges, these fluctuations become macroscopic, signaling the breakdown of ensemble equivalence.

Example: Ideal Gas in the Isothermal–Isobaric (Gibbs) Ensemble

For an ideal gas, the canonical partition function is

$$Z(T, V, N) = \frac{V^N}{N!\,\lambda^{3N}}, \qquad \lambda = \frac{h}{\sqrt{2\pi m k_B T}},$$

where λ is the thermal de Broglie wavelength. The Gibbs (isothermal–isobaric) partition function is obtained by a Laplace transform over the volume:

$$\Delta(T, P, N) = \int_0^\infty dV\, e^{-\beta PV} Z(T, V, N).$$

Inserting the ideal-gas expression,

$$\Delta = \frac{1}{N!\,\lambda^{3N}} \int_0^\infty dV\, V^N e^{-\beta PV} = \frac{\Gamma(N+1)}{N!\,\lambda^{3N}(\beta P)^{N+1}} = \frac{1}{\lambda^{3N}(\beta P)^{N+1}}.$$

Thus the Gibbs free energy,

$$G(T, P, N) = -k_B T \ln \Delta = N k_B T \ln\left(\frac{P\lambda^3}{k_B T}\right) + \mathcal{O}(1),$$

is extensive up to a negligible N–independent term.

Using the equation of state $PV = Nk_BT$, the Helmholtz and Gibbs free energies become

$$F = Nk_BT \ln\left(\frac{n\lambda^3}{e}\right), \qquad G = F + PV = Nk_BT \ln\left(\frac{P\lambda^3}{k_BT}\right),$$

recovering the standard ideal-gas chemical potential

$$\mu = \frac{\partial G}{\partial N} = k_BT \ln\left(\frac{P\lambda^3}{k_BT}\right).$$

The Gibbs ensemble is the natural theoretical description of systems held at fixed temperature and pressure—the most common experimental conditions in chemistry, biology, and materials science. Geometrically, the relation

$$\Delta(T, P, N) = \mathcal{L}_V[Z(T, V, N)]$$

is the analog of the Legendre transform connecting the Helmholtz and Gibbs free energies:

$$G = F + PV.$$

Together with $S = -\partial F/\partial T$ and $\beta = \partial S/\partial E$, this yields the hierarchy

$$S(E, V, N) \longleftrightarrow_{\beta} F(T, V, N) \longleftrightarrow_{\beta P} G(T, P, N),$$

showing explicitly how the micro-canonical, canonical, and isothermal–isobaric ensembles arise as successive Laplace–Legendre transforms over their conjugate variables.

REMARK 9.5 The Gibbs or *isothermal–isobaric* ensemble completes the chain of statistical descriptions connecting isolated, closed, and open systems. It arises from a Laplace transform of the canonical partition function with respect to the volume variable, mirroring the thermodynamic Legendre transform between the Helmholtz and Gibbs free energies. Physically, this transformation replaces the fixed-volume constraint by a fixed-pressure condition, reflecting the exchange of mechanical work with an external bath.

When both temperature and pressure are controlled, the natural potential is the Gibbs free energy $G(T, P, N)$, whose minimum determines equilibrium at fixed T and P. In this sense, the Gibbs ensemble provides the most experimentally relevant statistical framework: chemical reactions, phase coexistence, and most biological and geophysical processes occur under these constraints.

10

Properties of Classical Gases

10.1 Helmholtz Free Energy and the Gibbs–Jaynes Formulation of the Partition Function

Statistical mechanics was historically built around the question: *How does one determine the probability distribution over microscopic states that best represents a macroscopic system in thermal equilibrium?* Boltzmann's original answer was rooted in counting microstates of isolated systems, but the extension to systems in contact with a heat bath required a more general and principled framework.

A decisive step came from the work of Gibbs, who introduced ensembles as probability distributions over microstates who recognized that equilibrium distributions can be obtained from a *variational principle*: the physical distribution is the one that maximizes the Gibbs–Shannon entropy subject to the known macroscopic constraints. This unifies thermodynamics and probability theory: the canonical distribution emerges not from mechanical assumptions but from an inference principle.

To formalize this idea, consider a system that may occupy microstates i with energies E_i. The probabilities $\{P_i\}$ must satisfy the normalization and fixed-average energy constraints

$$\sum_i P_i = 1, \qquad \sum_i P_i E_i = U.$$

Among all probability distributions that satisfy these conditions, the equilibrium distribution is the one that maximizes the Gibbs–Shannon entropy,

$$S = -k_B \sum_i P_i \ln P_i.$$

DOI: 10.1201/9781003603528-10

Introducing Lagrange multipliers α and β for the two constraints, we extremize

$$-k_B \sum_j P_j \ln P_j - \alpha \left(\sum_j P_j - 1 \right) - \beta \left(\sum_j P_j E_j - U \right),$$

leading to

$$\frac{\partial}{\partial P_i} \left[-k_B \sum_j P_j \ln P_j - \alpha \left(\sum_j P_j - 1 \right) - \beta \left(\sum_j P_j E_j - U \right) \right] = 0.$$

Solving for P_i yields the canonical equilibrium distribution

$$P_i = \frac{e^{-\beta E_i}}{Z}, \qquad Z = \sum_i e^{-\beta E_i},$$

where $\beta = 1/(k_B T)$ and Z is the *partition function.* This is the Gibbs–Jaynes formulation of the canonical ensemble, and it provides a clean variational foundation for the Helmholtz free energy.

REMARK 10.1 In the Gibbs formulation, the canonical ensemble arises naturally as the stationary distribution that extremizes entropy under the energy constraint. In Jaynes's reinterpretation, this procedure embodies a principle of inference: it selects the least biased probability distribution compatible with the information available about the system. Jaynes provided a more modern information-theoretical reinterpretation. We will discuss it more carefully in the next Volume.

Historical Note: From Gibbs to Jaynes

The statistical definition of entropy,

$$S = -k_{\mathrm{B}} \sum_i P_i \ln P_i, \tag{10.1}$$

was introduced by J. Willard Gibbs (1902) as the fundamental measure of the system's statistical uncertainty over its microstates.

Half a century later, Claude Shannon (1948) introduced an identical functional form in information theory, interpreting [Sha48]

$$H = -\sum_i p_i \log p_i \tag{10.2}$$

as the measure of information or "uncertainty" in a communication source. Building on this insight, Edwin T. Jaynes (1957) reinterpreted Gibbs's ensembles as the result of an *inference principle*: among all possible probability distributions consistent with known macroscopic constraints, the physical equilibrium distribution is the one that *maximizes entropy* [Jay57]. This is now known as the *maximum entropy principle.*

Entropy, Free Energy, and the Meaning of the Partition Function

The purpose of introducing the partition function is not merely algebraic: it provides the bridge between the microscopic probabilities defined by Gibbs and the macroscopic thermodynamic potentials. From Gibbs's viewpoint, the equilibrium ensemble maximizes entropy under fixed constraints; from the thermodynamic viewpoint, equilibrium minimizes an appropriate potential. The connection between these two principles becomes explicit once we relate the entropy S and the free energy F through the partition function Z.

Starting from the equilibrium Gibbs distribution,

$$P_i = \frac{e^{-\beta E_i}}{Z}, \qquad Z = \sum_i e^{-\beta E_i}, \tag{10.3}$$

we substitute this expression into the statistical definition of entropy,

$$S = -k_{\mathrm{B}} \sum_i P_i \ln P_i, \tag{10.4}$$

to obtain

$$\begin{aligned} S &= -k_{\mathrm{B}} \sum_i \frac{e^{-\beta E_i}}{Z} \ln\left(\frac{e^{-\beta E_i}}{Z}\right) \\ &= k_{\mathrm{B}}\beta \sum_i E_i P_i + k_{\mathrm{B}} \ln Z. \end{aligned} \tag{10.5}$$

The first term represents the energetic contribution to entropy, while the second encodes the multiplicity of accessible microstates through Z. Recognizing the mean internal energy as

$$U = \sum_i E_i P_i, \tag{10.6}$$

we can write

$$S = \frac{U}{T} + k_{\mathrm{B}} \ln Z. \tag{10.7}$$

The Helmholtz free energy, defined thermodynamically as

$$F = U - TS, \tag{10.8}$$

then provides the natural potential for systems at fixed T, V, N. Substituting the expression for S leads to

$$\begin{aligned} F &= U - T\left(\frac{U}{T} + k_{\mathrm{B}} \ln Z\right) \\ &= -k_{\mathrm{B}} T \ln Z. \end{aligned} \tag{10.9}$$

Hence,

$$F(T, V, N) = -k_{\mathrm{B}} T \ln Z(T, V, N). \tag{10.10}$$

REMARK 10.2 This result gives physical meaning to the partition function: Z is not merely a normalization constant, but the mathematical object that links the microscopic statistical weights of all configurations to the macroscopic free energy that governs equilibrium. Minimizing F in thermodynamics is equivalent to maximizing S in Gibbs's statistical formulation—the two are dual expressions of the same equilibrium condition.

The Helmholtz free energy provides the link between statistical and thermodynamic descriptions. Once Z is known, all macroscopic observables follow:

$$U = -\frac{\partial \ln Z}{\partial \beta}, \qquad S = -\left(\frac{\partial F}{\partial T}\right)_{V,N}, \qquad P = -\left(\frac{\partial F}{\partial V}\right)_{T,N}. \tag{10.11}$$

From the Gibbs–Jaynes point of view, the free energy represents the minimum of the functional

$$F[P] = \sum_i P_i E_i + k_B T \sum_i P_i \ln P_i, \tag{10.12}$$

whose extremization yields both the canonical ensemble and the equilibrium thermodynamic potential.

Key Relation: Gibbs–Jaynes Partition Function and Helmholtz Free Energy

$$F(T, V, N) = -k_B T \ln Z(T, V, N), \tag{10.13}$$

$$Z(T, V, N) = \sum_i e^{-\beta E_i}. \tag{10.14}$$

The partition function Z encapsulates all equilibrium properties. In the Gibbs picture, it normalizes the canonical ensemble; in the Jaynes interpretation, it plays the role of the generating function that encodes all macroscopic predictions given limited microscopic information.

The Free Energy as a Variational Bound on the Mean Energy

One of the most important properties of the Helmholtz free energy is that it provides a *variational bound* on the mean internal energy. This property expresses, in statistical form, the principle that equilibrium minimizes the free energy among all admissible probability distributions. It forms the conceptual bridge between equilibrium thermodynamics and the variational methods used to approximate complex systems, such as in quantum and mean-field theories.

From the definition of the canonical partition function,

$$Z = \sum_i e^{-\beta E_i}, \qquad F = -k_B T \ln Z, \tag{10.15}$$

we observe that the logarithm is a concave function. Applying *Jensen's inequality* to the random variable E distributed according to the normalized probabilities $P_i = e^{-\beta E_i}/Z$, we have

$$\ln\langle e^{-\beta E}\rangle \geq \langle \ln e^{-\beta E}\rangle = -\beta\langle E\rangle, \tag{10.16}$$

where $\langle\cdot\rangle$ denotes an ensemble average. Recalling that $Z = \langle e^{-\beta E}\rangle_{\text{microstates}}$, we obtain

$$\ln Z \geq -\beta U, \tag{10.17}$$

or equivalently,

$$F = -k_B T \ln Z \leq U. \tag{10.18}$$

REMARK 10.3 The inequality $F \leq U$ expresses the fact that the free energy represents the smallest possible mean energy compatible with the thermal fluctuations imposed by the environment. At equilibrium, the system chooses the probability distribution that minimizes $F[P]$,

$$F[P] = \sum_i P_i E_i + k_B T \sum_i P_i \ln P_i, \tag{10.19}$$

subject to normalization. This functional attains its minimum for the Gibbs distribution $P_i = e^{-\beta E_i}/Z$. In this sense, the Helmholtz free energy is the *variational functional* of statistical mechanics, and the canonical distribution is the stationary point of this variational problem.

The inequality $F \leq U$ has deep implications beyond equilibrium thermodynamics. In many-body and quantum statistical physics, it underlies the *Gibbs–Bogoliubov* or *Feynman variational principle*, which states that for any trial Hamiltonian H_0,

$$F \leq F_0 + \langle H - H_0 \rangle_0, \tag{10.20}$$

where F_0 and $\langle \cdot \rangle_0$ denote averages with respect to the trial ensemble defined by H_0. Minimizing the right-hand side provides the best approximation to the true free energy within the chosen class of trial distributions. This variational framework generalizes the zero-temperature variational principle for ground states, making the Helmholtz free energy the natural extension of the energy functional to finite-temperature systems.

REMARK 10.4 In the limit $T \to 0$, thermal fluctuations vanish, and $F \to U_{\min}$, the ground-state energy. Thus, the statistical variational principle embodied by $F \leq U$ smoothly connects equilibrium statistical mechanics to the variational principles of quantum mechanics and condensed matter theory.

10.2 The Ideal Gas and the Free Energy

As we have seen in the previous chapters, and from a variety of perspectives, the ideal gas represents the simplest system in statistical mechanics—one for which the integrals defining the partition function can be performed exactly. The canonical partition function for N identical particles is given by

$$Z(N, V, T) = \frac{1}{N!} \left(\prod_{i=1}^{N} \int d^3 x_i \right) \left(\prod_{i=1}^{N} \int d^3 p_i \, e^{-\beta H_{\text{kin}}} \right), \tag{10.21}$$

where the first factor accounts for the positions of the particles and the second for their momenta.

Factorization of Kinetic and Configurational Parts

The partition function naturally separates into a kinetic and a configurational contribution,

$$Z(N, V, T) = Z_{\text{k}}(N, T)\, Z_{\text{c}}(N, V, T), \tag{10.22}$$

where Z_{k} contains the momentum integrals and Z_{c} encodes the configurational dependence on interparticle interactions and on the accessible volume. For a wide class of systems—including the ideal gas—the Hamiltonian separates as $H = H_{\text{kin}} + H_{\text{pot}}$, leading to the same factorization in Z. In the absence of interactions, $H_{\text{pot}} = 0$, and the configurational part reduces to a purely geometric volume factor.

In general, one can write

$$Z_{\text{c}}(N, V, T) = V^N \, M(N, V), \tag{10.23}$$

where $M(N, V)$ accounts for possible corrections due to correlations or boundary effects. Using the definition of the Helmholtz free energy,

$$F = -k_{\text{B}} T \ln Z, \tag{10.24}$$

we obtain

$$F = -k_{\text{B}} T N \ln V + F_0(M, T), \tag{10.25}$$

with

$$F_0(M, T) = -k_{\text{B}} T \big[\ln M(N, T) + \ln Z_{\text{k}}(N, T) \big], \tag{10.26}$$

which is constant with respect to the system's volume.

The pressure now follows from the thermodynamic identity

$$P = -\left(\frac{\partial F}{\partial V}\right)_{T,N} = k_{\mathrm{B}}T\,\frac{1}{Z_{\mathrm{c}}}\,\frac{\partial Z_{\mathrm{c}}}{\partial V}. \tag{10.27}$$

For an ideal gas, $M = 1$ and $Z_{\mathrm{c}} = V^N$, yielding immediately

$$P = \frac{Nk_{\mathrm{B}}T}{V}, \tag{10.28}$$

which is the familiar ideal-gas law.

REMARK 10.5 Although this derivation of pressure appears purely algebraic, it conceals the underlying physical mechanism: pressure originates from the transfer of momentum between particles and the system's boundaries. The statistical expression $P = k_{\mathrm{B}}T\,\partial_V \ln Z_{\mathrm{c}}$ encapsulates this microscopic picture by relating macroscopic force to configurational changes in phase space volume.

Excess Free Energy

In many applications, one is not interested in the absolute value of the free energy but in the deviation from an ideal reference system. If the partition function factorizes, it is convenient to define the *excess free energy* as

$$\tilde{F} = F - F_0, \tag{10.29}$$

which isolates the configurational contribution. For the ideal gas, where $M = 1$, $\tilde{F} = 0$; but for real systems, $\tilde{F}$ quantifies the effect of interactions or structural correlations. In practice, this quantity is often what is measured or computed, for instance, when determining pressures, phase equilibria, or surface effects.

REMARK 10.6 The decomposition $F = F_0 + \tilde{F}$ parallels the separation between kinetic and potential energies in the microscopic Hamiltonian. In both cases, the ideal part provides a universal reference, while the excess part contains the specific physical information about interactions and structure.

10.3 Pressure as an Entropic Force

For an ideal gas, the internal energy depends only on temperature and not on the system's volume. This implies that the pressure exerted by the gas cannot originate from its internal energy alone. Instead, it arises from entropy: pressure in an ideal gas has an *entropic origin*. Thermal fluctuations drive the system toward the macroscopic state corresponding to the largest number of compatible microstates—that is, the state of maximum entropy. From this viewpoint, pressure expresses the system's intrinsic statistical tendency to occupy larger volumes, not because of energetic preference, but because of the combinatorial abundance of accessible configurations.

REMARK 10.7 In the ideal gas, the mechanical picture of pressure as molecules colliding with the walls is equivalent to the entropic picture in which pressure arises from the gradient of the system's entropy with respect to volume. Both descriptions are consistent: the microscopic dynamics enforce the same equilibrium that maximizes the number of accessible states.

Potential of Mean Force

The concept of the *potential of mean force* provides a precise way to express these entropic effects within statistical mechanics. For a system described by coordinate x, the probability density of finding the system at position x is

$$P(x) = \frac{e^{-\beta U(x)}}{Z_c}, \tag{10.30}$$

where $U(x)$ is the potential energy, $\beta = 1/(k_B T)$, and Z_c is the configurational integral. The *potential of mean force*, denoted $W(x)$, is defined as

$$W(x) = -k_B T \ln P(x), \tag{10.31}$$

and represents the effective free-energy landscape along a reaction coordinate x, incorporating both energetic and entropic contributions.

Substituting the expression for $P(x)$ gives

$$W(x) = U(x) + k_B T \ln Z_c, \tag{10.32}$$

up to an irrelevant additive constant. The corresponding *mean force* acting on the coordinate x is

$$F_m(x) = -\frac{dW(x)}{dx}. \tag{10.33}$$

This average force reflects the effective, thermally averaged interactions that govern the system's macroscopic behavior.

For a system of N interacting particles in one dimension, with pairwise potentials $U_{ij}(x_i - x_j)$, the total potential energy is

$$U(\mathbf{x}) = \sum_{i=1}^{N} \sum_{j>i}^{N} U_{ij}(x_i - x_j), \tag{10.34}$$

where $\mathbf{x} = (x_1, \ldots, x_N)$. The potential of mean force for particle i, keeping its position x_i fixed, is

$$W_i(x_i) = -k_B T \ln \int e^{-\beta U(\mathbf{x})}\, dx_1 \cdots dx_{i-1}\, dx_{i+1} \cdots dx_N, \tag{10.35}$$

and the corresponding mean force is

$$F_i(x_i) = -\frac{dW_i(x_i)}{dx_i}. \tag{10.36}$$

In practice, these quantities are difficult to compute analytically for interacting systems, but they can be evaluated numerically or through coarse-grained models. This construction is fundamental in molecular and soft-matter physics, where potentials of mean force describe effective interactions in colloids, biomolecules, and polymers [Sok10].

Entropy and the Mean Force

To relate this formulation directly to entropy, consider a macroscopic coordinate X defining a subset of microstates. In the canonical ensemble,

$$P(X) = \langle \delta(X - x) \rangle = \frac{1}{Z} \int d\Gamma\, \delta(X - x)\, e^{-\beta H}, \tag{10.37}$$

so that the potential of mean force can be expressed as

$$W(X) = -k_B T \ln P(X) = F - k_B T \ln P(X), \tag{10.38}$$

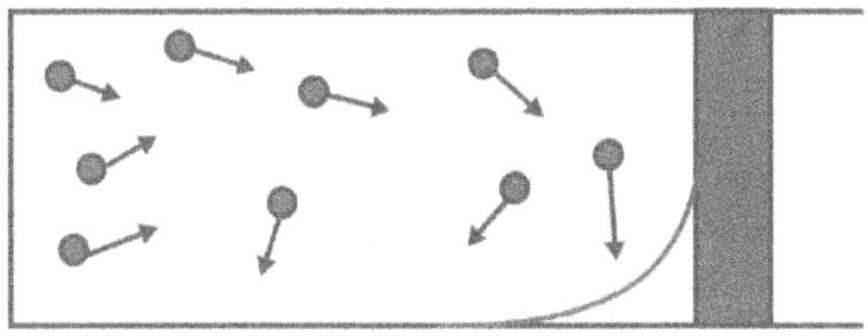

FIGURE 10.1 The model of gas confinement is considered here, with the potential highlighted in blue.

where F is the Helmholtz free energy of the system. The quantity $\langle\delta(X-x)\rangle$ represents the fraction of configurations compatible with the macrostate X. Hence,

$$P(X) \propto e^{S(X)/k_{\mathrm{B}}}, \tag{10.39}$$

where $S(X)$ is the entropy associated with the constraint X. Taking the derivative of this expression yields

$$F_X = T\,\frac{\partial S(X)}{\partial X}. \tag{10.40}$$

This shows that the mean force is the gradient of entropy multiplied by temperature: an *entropic force.*

REMARK 10.8 The concept of entropic force captures the essence of thermodynamic equilibrium: systems evolve toward configurations of higher entropy, and the macroscopic forces we observe can arise purely from this statistical tendency. In the ideal gas, the pressure is one such entropic force—it reflects the drive of the system to explore larger volumes, increasing the number of accessible microstates rather than lowering its energy [Hil60, Sok10].

We consider a simple model of a wall.

A Simple Model of a Wall

To make the concept of entropic pressure more concrete, consider a one-dimensional model of a gas particle confined by a soft wall. The wall exerts a restoring potential $U(x)$ acting along the x-direction, where $0 < x < L$ is the region accessible to the particle, and $U(x)$ increases beyond $x = L$. Let the potential near the wall be harmonic,

$$U(x) = \frac{1}{2}kx^2, \tag{10.41}$$

with k a stiffness constant.

The probability density for the particle's position is then given by the Boltzmann distribution,

$$p(x) = C\,e^{-\frac{kx^2}{2k_{\mathrm{B}}T}}, \tag{10.42}$$

where C is a normalization constant. Integrating over the accessible region $0 < x < L$ yields

$$C^{-1} = A\left(L + \sqrt{\frac{\pi k_{\mathrm{B}}T}{2k}}\right), \tag{10.43}$$

where A is the cross-sectional area of the container. Hence,

$$p(x) = \frac{e^{-\frac{kx^2}{2k_{\mathrm{B}}T}}}{A\left(L + \sqrt{\frac{\pi k_{\mathrm{B}}T}{2k}}\right)}. \tag{10.44}$$

The mean force exerted by the wall on the particle is

$$f_x = \langle -\partial_x U(x) \rangle = \langle -kx \rangle. \tag{10.45}$$

In the low-temperature limit $T \ll kL^2/k_\mathrm{B}$, the exponential factor strongly localizes the particle near the wall, and the average force becomes

$$f_x = -\frac{k_\mathrm{B}T}{L}. \tag{10.46}$$

This force acts in the negative x-direction, opposite to the wall.

The macroscopic pressure is then obtained by dividing by the wall area A and multiplying by the total number of particles N,

$$P = N\,\frac{-f_x}{A} = \frac{Nk_\mathrm{B}T}{AL} = \frac{Nk_\mathrm{B}T}{V}, \tag{10.47}$$

recovering the ideal gas law. The result is independent of the stiffness parameter k: the microscopic details of the wall potential are irrelevant to the macroscopic pressure. Thus, the pressure here emerges as an entropic mean force, consistent with the thermodynamic definition

$$f_x = -k_\mathrm{B}T\,\frac{\partial \ln p(x)}{\partial x}. \tag{10.48}$$

REMARK 10.9 This simple model illustrates that the macroscopic pressure does not originate from mechanical energy stored in the wall, but from entropy. Thermal motion biases the distribution of particles in such a way that the system exerts an average outward force—an entropic pressure—on its boundaries.

We now refine the previous argument by introducing a movable wall or "screen" located at position X, modeled by a local one-body potential. Let

$$U(x_1, \ldots, x_N) = \sum_{i=1}^{N} \frac{k}{2}\,(x_i - X)^2\,\theta(x_i - X), \tag{10.49}$$

where $\theta(x)$ is the Heaviside step function. The force on the wall is

$$f(\mathbf{x}) = -\sum_{i=1}^{N} \frac{\partial U}{\partial x_i} = -\sum_{i=1}^{N} k(x_i - X)\,\theta(x_i - X). \tag{10.50}$$

The mean force acting on the wall is then obtained from the canonical average,

$$F_X = \frac{1}{Z}\int d\Gamma\, f(\mathbf{x})\, e^{-\beta H}, \qquad H = \sum_i \frac{p_i^2}{2m} + U(\mathbf{x}), \tag{10.51}$$

with Z the partition function of the full system.

Because the total derivative of the potential with respect to the wall position satisfies

$$\frac{dU}{dX} = -\sum_i \frac{\partial U}{\partial x_i} = k_\mathrm{B}T\, e^{-\beta U}, \tag{10.52}$$

we find that the mean force can be expressed as

$$F_X = k_\mathrm{B}T\,\frac{d}{dX}\ln Z. \tag{10.53}$$

In the hard-wall limit $k \to \infty$, the exponential factor

$$e^{-\frac{k}{2k_BT}(x_i-X)^2\theta(x_i-X)} \to \theta(X - x_i), \tag{10.54}$$

restricts integration to the accessible region of configuration space. Dividing by the wall area A, and recognizing $dM = A\,dX$ as the change in system volume, yields

$$P = k_BT\,\frac{d}{dM}\ln Z. \tag{10.55}$$

This expression is identical to the thermodynamic definition of pressure derived from the free energy:

$$P = -\left(\frac{\partial F}{\partial V}\right)_{T,N}. \tag{10.56}$$

REMARK 10.10 Both in the soft-wall and in the hard-wall limits, pressure emerges as an entropic mean force arising from the statistical tendency of the system to explore configurations that maximize the available phase-space volume. The microscopic model of a flexible boundary provides a mechanical realization of this entropic principle, showing that macroscopic pressure is fundamentally a consequence of entropy.

10.4 Beyond Ideal Gases

Up to this point, we have considered an ideal gas in which particles do not interact. Real gases, however, exhibit both short-range repulsion and intermediate-range attraction. For neutral species, a widely used model of intermolecular interaction is the Lennard–Jones potential,

$$U_{\rm LJ}(r) = 4\varepsilon\left[\left(\frac{\sigma}{r}\right)^{12} - \left(\frac{\sigma}{r}\right)^{6}\right], \tag{10.57}$$

where $r = |\vec{r}_1 - \vec{r}_2|$, ε is the depth of the attractive well, and σ is the distance at which the potential vanishes. For charged species, the Coulomb interaction dominates at long distances while the Lennard–Jones form controls short-range repulsion.

The systematic treatment of interactions in a classical gas is provided by the *Mayer cluster expansion.* For a system of N classical particles with Hamiltonian

$$H = \sum_{i=1}^{N}\frac{\vec{p}_i^{\,2}}{2m} + \sum_{i<j}U(|\vec{r}_i - \vec{r}_j|) \equiv H^{(p)} + H^{(q)},$$

the canonical partition function is

$$Z_N = \frac{1}{N!h^{3N}}\int d^{3N}p\,e^{-\beta H^{(p)}}\int d^{3N}r\,e^{-\beta H^{(q)}} = Z_N^{(p)}\,Z_N^{(q)}. \tag{10.58}$$

Using the thermal wavelength λ,

$$Z_N^{(p)} = \frac{1}{\lambda^{3N}}, \qquad \lambda = \sqrt{\frac{2\pi\hbar^2}{mk_BT}},$$

the configurational integral becomes

$$Z_N^{(q)} = \int d^{3N}r\,e^{-\beta\sum_{i<j}U_{ij}}.$$

To proceed, write the exponential of interactions as a product over all pairs:

$$e^{-\beta U} = \prod_{i<j} e^{-\beta U_{ij}} = \prod_{i<j} (1 + f_{ij}), \qquad f_{ij} = e^{-\beta U_{ij}} - 1. \tag{10.59}$$

The quantity f_{ij} is the *Mayer f-function.* At high temperature or weak coupling,

$$f_{ij} \approx -\beta U_{ij} \ll 1.$$

Expanding the product in Eq. (10.59) gives

$$\prod_{i<j}(1 + f_{ij}) = 1 + \sum_{i<j} f_{ij} + \sum_{i<j,\, k<l} f_{ij} f_{kl} + \cdots .$$

Each term corresponds to a *cluster diagram.* For $N = 3$, the diagrams are shown in Fig. 10.2, where each link corresponds to a factor f_{ij}.

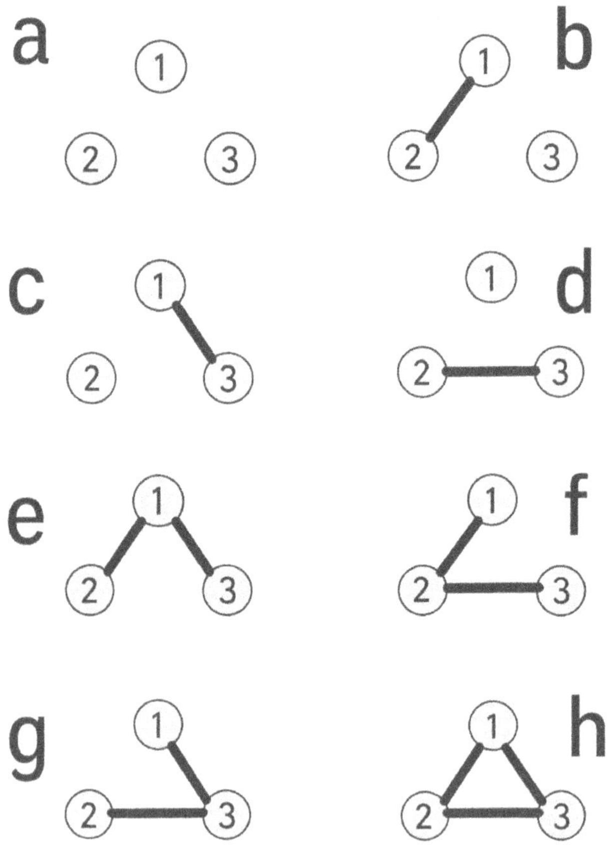

FIGURE 10.2 All cluster diagrams for three particles. Each link corresponds to a Mayer function f_{ij}.

For three particles, the corresponding integrals are

$$
\begin{aligned}
&(a) \int d^3r_1\, d^3r_2\, d^3r_3 = V^3, \\
&(b) \int d^3r_1\, d^3r_2\, d^3r_3\, f_{12} = V^2C_1, \\
&(c) \int d^3r_1\, d^3r_2\, d^3r_3\, f_{13} = V^2C_1, \\
&(d) \int d^3r_1\, d^3r_2\, d^3r_3\, f_{23} = V^2C_1, \\
&(e) \int d^3r_1\, d^3r_2\, d^3r_3\, f_{12}f_{13} = VC_2, \\
&(f) \int d^3r_1\, d^3r_2\, d^3r_3\, f_{12}f_{23} = VC_2, \\
&(g) \int d^3r_1\, d^3r_2\, d^3r_3\, f_{13}f_{23} = VC_2, \\
&(h) \int d^3r_1\, d^3r_2\, d^3r_3\, f_{12}f_{23}f_{31} = C_3.
\end{aligned}
$$

Counting multiplicities,

$$
Z_3^{(q)} = V^3 + 3C_1V^2 + 3C_2V + C_3 = V^3\left(1 + \frac{3C_1}{V} + \frac{3C_2}{V^2} + \frac{C_3}{V^3}\right).
$$

More generally, for arbitrary N,

$$
Z_N^{(q)} = \left\langle \prod_{i<j}(1+f_{ij}) \right\rangle_0 = \int d^{3N}r \left(1 + \sum_{i<j} f_{ij} + \sum_{i<j,\, k<l} f_{ij}f_{kl} + \cdots\right),
$$

where the average $\langle \cdots \rangle_0$ is over the ideal-gas measure.

Expanding the first nontrivial terms gives

$$
\begin{aligned}
Z_N &= \frac{1}{N!\lambda^{3N}} \int d^{3N}r \left(1 + \sum_{j<k} f_{jk} + \cdots\right) \\
&= \frac{V^N}{N!\lambda^{3N}} \left(1 + \frac{N(N-1)}{2V} \int d^3r\, f(r) + \cdots\right). \qquad (10.60)
\end{aligned}
$$

This expansion resums as

$$
Z_N = \frac{V^N}{N!\lambda^{3N}} \left(1 + \frac{1}{2V} \int d^3r\, f(r) + \cdots\right)^N,
$$

anticipating the virial expansion.

A crucial observation is that many diagrams correspond to identical integrals. For example, in Fig. 10.2, diagrams (b), (c), and (d) all yield the same contribution C_1. For $N > 3$, diagrams may also contain multiple disconnected components, leading to terms of the form C_ℓ^r, corresponding to clusters of ℓ links repeated r times.

Thus the general expansion of $Z_N^{(q)}$ may be written as a sum over all collections of clusters with a fixed total number of links L:

$$
Z_L^{(q)} = \sum_{\{n_\ell\}} \left[\prod_\ell C_\ell^{n_\ell}\right] W(\{n_\ell\}), \qquad (10.61)
$$

where n_ℓ is the number of clusters with ℓ links, constrained by $\sum_\ell \ell n_\ell = L$. The combinatorial factor

$$W(\{n_\ell\}) = \frac{L!}{\prod_\ell n_\ell!(\ell!)^{n_\ell}}$$

counts the number of distinct assignments of particle labels to the specified collection of clusters: the factor $L!$ accounts for permutations of links, $n_\ell!$ corrects for identical clusters, and each $\ell!$ corresponds to internal permutations of a cluster with ℓ links.

We will now show that the full series Eq. (10.61) exponentiates, leading to the cluster expansion for the free energy and the virial coefficients.

Having organized the configurational contribution $Z_N^{(q)}$ in terms of cluster integrals, we now turn to the thermodynamic quantity of real interest,

$$\log Z = \log\langle e^{-\beta V}\rangle.$$

As in all equilibrium statistical-mechanics calculations, the free energy

$$F = -k_{\mathrm{B}}T\log Z$$

is determined by the logarithm of the partition function. The natural question is: *How do we extract* $\log Z$ *perturbatively from the full cluster expansion?*

The key fact is that the logarithm of the partition function receives contributions only from *connected* diagrams. Disconnected diagrams factorize into products of lower-order terms, and these products cancel against the denominators that arise when expanding $\log\langle e^{-\beta V}\rangle$. This is the combinatorial essence of the *cumulant expansion*: only those clusters that remain connected under removal of any single link (so-called "1-connected" diagrams) contribute to $\log Z$.

Cumulants involve fluctuations of the form

$$\langle (A - \langle A\rangle)^n\rangle.$$

Any disconnected contribution factorizes and therefore cancels in the connected part. The Mayer expansion exhibits this same structure: disconnected diagrams are products of smaller cluster integrals and do not contribute to the connected cumulant, i.e., to $\log Z$.

To formalize this idea, it is convenient to use a standard device—the *replica trick* (which will appear repeatedly in Volume II):

$$\log x = \lim_{n\to 0}\frac{x^n - 1}{n}.$$

Expanding $e^{n\log x}$ for small n confirms:

$$\frac{x^n-1}{n} = \frac{1 + n\log x + \frac{1}{2}n^2(\log x)^2 + \cdots - 1}{n} \xrightarrow[n\to 0]{} \log x.$$

We apply this identity to the partition function by considering n independent replicas of the system:

$$\langle Z^n\rangle = \left\langle \exp\left[-\beta\sum_{\alpha=1}^{n}\sum_{i<j} V\left(|\vec{r}_i^{\,\alpha} - \vec{r}_j^{\,\alpha}|\right)\right]\right\rangle.$$

Each particle now carries a label i and a replica index α. In the graphical expansion, links f_{ij} can only connect particles within the *same* replica, since interactions do not couple distinct copies.

Therefore, a diagram with M disconnected components requires at least M independent replica indices. The expansion takes the schematic form

$$1 + \sum_{\alpha_1} M_1 + \sum_{\alpha_1,\alpha_2} M_2 + \sum_{\alpha_1,\alpha_2,\alpha_3} M_3 + \cdots,$$

where M_k denotes a diagram consisting of k disconnected clusters.

Because each connected component is independent of the replica label, the sums over replica indices give

$$1 + nM_1 + n^2 M_2 + n^3 M_3 + \cdots .$$

Using the replica identity,

$$\log Z = \lim_{n \to 0} \frac{\langle Z^n \rangle - 1}{n},$$

only the term *linear in n* survives. All higher powers vanish in the limit. Thus only diagrams with exactly one connected component contribute:

$$\log Z \equiv \sum_{\text{1-connected diagrams}} .$$

This is the classical *linked cluster expansion.* It represents one of the earliest appearances of graph-theoretical methods in statistical mechanics and forms the foundation for diagrammatic approaches in quantum field theory, many-body physics, and modern statistical mechanical methods.

Second Virial Coefficient

From the Mayer expansion derived above, we found that for weak interactions and low densities,

$$Z(N, V, T) = Z_{\text{ideal}}^N \left(1 + \frac{N}{2V} \int d^3 r \, f(r) + \cdots \right)^N .$$

Taking the logarithm gives the free energy

$$F = -k_B T \log Z = F_{\text{ideal}} - N k_B T \log \left(1 + \frac{N}{2V} \int d^3 r \, f(r) \right) .$$

The pressure follows as

$$P = -\frac{\partial F}{\partial V} = P_{\text{ideal}} + N k_B T \frac{\partial}{\partial V} \log \left(1 + \frac{N}{2V} \int d^3 r \, f(r) \right) ,$$

with $P_{\text{ideal}} = N k_B T / V$.

In the dilute limit $N/V \ll 1$, using $\log(1 + x) \simeq x$ yields

$$P = \frac{N k_B T}{V} - N k_B T \frac{N}{2V^2} \int d^3 r \, f(r),$$

or equivalently

$$\frac{PV}{N k_B T} = 1 - \frac{N}{2V} \int d^3 r \, f(r).$$

This is the virial expansion to second order. The first term is the ideal-gas contribution (first virial coefficient), and the correction proportional to $\int f(r) \, d^3 r$ is the *second virial coefficient.* Its sign reflects whether interactions are predominantly attractive or repulsive:

- $U(r) > 0$ (repulsive) $\implies f(r) < 0$, giving $PV/(N k_B T) < 1$;
- $U(r) < 0$ (attractive) $\implies f(r) > 0$, giving $PV/(N k_B T) > 1$.

In realistic systems, the intermolecular potential has both repulsive and attractive parts. The Lennard–Jones potential (Fig. 10.3)

$$U(r) = \left(\frac{r_0}{r} \right)^{12} - \left(\frac{r_0}{r} \right)^6 ,$$

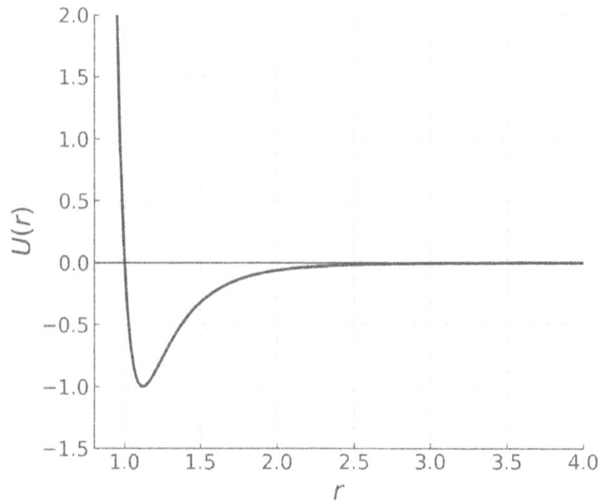

FIGURE 10.3 The Lennard–Jones potential for $r_0 = 5$.

is strongly repulsive at short range and attractive at intermediate range; thus the overall virial correction reflects a balance between these contributions.

Let us now consider a short–range interaction of the form

$$U(r) = \begin{cases} \infty, & r \le r_0, \\ U_0(r), & r > r_0, \end{cases}$$

shown in Fig. 10.4. The Mayer function is then

$$f(r) = e^{-\beta U(r)} - 1 = \begin{cases} -1, & r \le r_0, \\ e^{-\beta U_0(r)} - 1, & r > r_0. \end{cases}$$

Thus

$$\int d^3r\, f(r) = 4\pi \int_0^{r_0} r^2(-1)\, dr + 4\pi \int_{r_0}^{\infty} r^2 \left(e^{-\beta U_0(r)} - 1\right) dr.$$

At high temperature ($\beta \ll 1$),

$$e^{-\beta U_0(r)} \approx 1 - \beta U_0(r),$$

giving

$$\int d^3r\, f(r) \approx -\frac{4\pi r_0^3}{3} - 4\pi\beta \int_{r_0}^{\infty} r^2 U_0(r)\, dr.$$

As a concrete example, take

$$U_0(r) = -U_0 \frac{r_0^6}{r^6}.$$

Then

$$\int_{r_0}^{\infty} r^2 U_0(r)\, dr = -U_0 r_0^6 \int_{r_0}^{\infty} r^{-4}\, dr = -\frac{U_0 r_0^3}{3}.$$

Therefore,

$$\int d^3r\, f(r) \approx \frac{4\pi r_0^3}{3} \left(\frac{U_0}{k_B T} - 1\right).$$

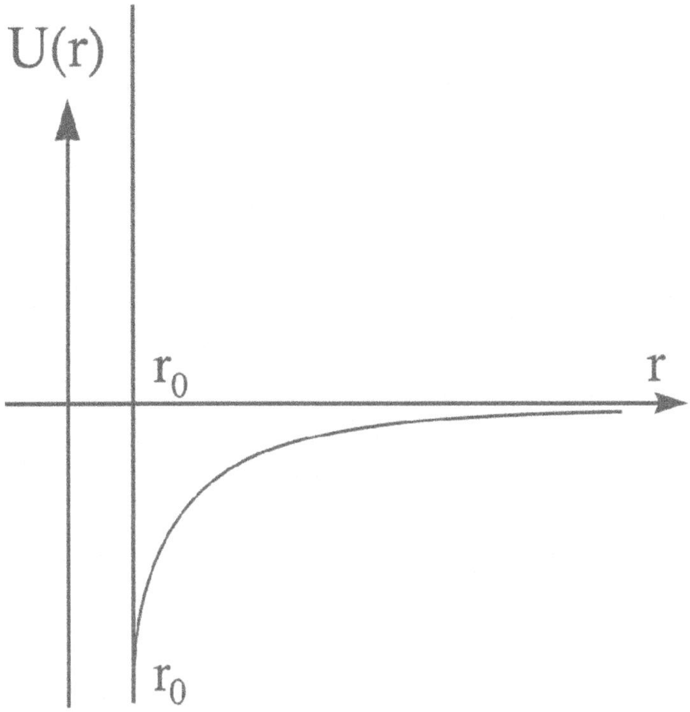

FIGURE 10.4 A short-range repulsive core of radius r_0 with an outer potential $U_0(r)$.

Define

$$b = \frac{2\pi r_0^3}{3}, \qquad a = \frac{2\pi U_0 r_0^3}{3},$$

so that

$$\frac{PV}{Nk_BT} = 1 - \frac{N}{V}\left(\frac{a}{k_BT} - b\right).$$

Multiplying by k_BT,

$$P\frac{V}{N} = k_BT - \frac{N}{V}(a - bk_BT).$$

Using $\rho = N/V$ and $(1+x)^{-1} \simeq 1 - x$, this can be arranged into the van der Waals–like form

$$k_BT \simeq (P + \rho^2 a)\left(\frac{1}{\rho} - b\right),$$

or equivalently,

$$k_BT = \frac{\tilde{P}}{\tilde{\rho}}, \qquad \tilde{P} = P + \rho^2 a, \qquad \tilde{\rho} = \frac{1}{1/\rho - b}.$$

We see that $a > 0$ represents attractive interactions (increasing the effective pressure $P+\rho^2 a$); $b > 0$ represents excluded volume, and the second virial coefficient reconstructs the structure of the van der Waals equation of state.

As long as $\rho r_0^3 \ll 1$, the expansion remains valid. Thus, the second virial coefficient captures the leading correction to the ideal-gas behavior in terms of microscopic interaction parameters.

Grand-Canonical Mayer Expansion

The Mayer expansion becomes even more transparent in the grand-canonical ensemble, where the particle number fluctuates and the fugacity

$$\zeta = e^{\beta\mu}$$

weights each configuration. The grand-canonical partition function is

$$\mathcal{G}(\mu, V, T) = \sum_{N=0}^{\infty} \zeta^N Z_N(V, T).$$

Inserting the configurational expansion obtained previously,

$$Z_N(V,T) = \frac{1}{N!\lambda^{3N}} \int \prod_{i=1}^{N} d^3 r_i \left(1 + \sum_{j<k} f_{jk} + \sum_{j<k,\, l<m} f_{jk} f_{lm} + \cdots \right),$$

shows that $\mathcal{G}$ is a sum over all possible graphs with N nodes, weighted by ζ^N.

From the canonical analysis, any term of $Z_N^{(q)}$ can be written as a product of cluster integrals C_ℓ with multiplicities n_ℓ, obeying

$$\sum_{\ell=1}^{L} \ell\, n_\ell = N.$$

Thus,

$$\mathcal{G} = \sum_{N=0}^{\infty} \frac{\zeta^N}{\lambda^{3N}} \sum_{\{n_\ell\}} \prod_{\ell} \frac{C_\ell^{n_\ell}}{n_\ell!\,(\ell!)^{n_\ell}}.$$

We now trade the sum over N for a direct sum over cluster multiplicities $\{n_\ell\}$. Since $N = \sum_\ell \ell n_\ell$, the fugacity factor becomes

$$\zeta^N = \prod_\ell (\zeta^\ell)^{n_\ell}.$$

Therefore,

$$\mathcal{G} = \sum_{\{n_\ell\}} \prod_{\ell=1}^{\infty} \left[\frac{1}{n_\ell!} \left(\frac{\zeta^\ell}{\lambda^{3\ell}} \frac{C_\ell}{\ell!} \right)^{n_\ell} \right].$$

Each factor is an exponential series, so the full product exponentiates:

$$\mathcal{G}(\mu, V, T) = \prod_{\ell=1}^{\infty} \exp\left[\frac{\zeta^\ell}{\lambda^{3\ell}} \frac{C_\ell}{\ell!} \right] = \exp\left[\sum_{\ell=1}^{\infty} \frac{\zeta^\ell}{\lambda^{3\ell}} \frac{C_\ell}{\ell!} \right].$$

It is convenient to define the (dimensionless) cluster coefficients

$$b_\ell = \frac{C_\ell}{\ell!\,\lambda^{3(\ell-1)}} \frac{1}{V},$$

so that the grand-canonical partition function becomes

$$\mathcal{G}(\mu, V, T) = \exp\left[\frac{V}{\lambda^3} \sum_{\ell=1}^{\infty} b_\ell\, \zeta^\ell \right], \qquad \zeta = e^{\beta\mu}.$$

Thus, exactly as in the canonical ensemble, *only connected diagrams contribute to* $\log \mathcal{G}$:

$$\frac{PV}{k_B T} = \log \mathcal{G} = \frac{V}{\lambda^3} \sum_{\ell=1}^{\infty} b_\ell \zeta^\ell.$$

The average particle number follows by differentiating with respect to ζ:

$$N = \zeta\, \partial_\zeta \log \mathcal{G} = \frac{V}{\lambda^3} \sum_{\ell=1}^{\infty} \ell\, b_\ell \zeta^\ell.$$

Hence,

$$\frac{PV}{N k_B T} = \frac{\sum_{\ell=1}^{\infty} b_\ell \zeta^\ell}{\sum_{\ell=1}^{\infty} \ell\, b_\ell \zeta^\ell}.$$

This gives a direct route to the virial coefficients in terms of the connected cluster integrals b_ℓ, completing the connection between the Mayer expansion and the virial equation of state.

At low densities, it is convenient to rewrite the equation of state as a virial series in the number density $\rho = N/V$:

$$\frac{P}{k_B T \rho} = 1 + B_2(T)\, \rho + B_3(T)\, \rho^2 + \cdots, \tag{10.62}$$

where the coefficients $B_n(T)$ are the *virial coefficients*. In the Mayer expansion, these are expressed in terms of the cluster integrals $b_\ell(T)$; for example,

$$B_2(T) = -b_2(T), \qquad B_3(T) = 4b_2(T)^2 - 2b_3(T),$$

and similarly for higher orders. Thus, the virial coefficients encode the collective effect of two-body, three-body, and higher-order correlations generated by the intermolecular potential.

The van der Waals equation of state provides a simple phenomenological model that effectively truncates this hierarchy. Starting from

$$\left(P + a\rho^2\right)\left(\frac{1}{\rho} - b\right) = k_B T,$$

and expanding at low density $\rho \to 0$, one finds

$$\begin{aligned}
\frac{P}{k_B T \rho} &= \frac{1}{1 - b\rho} - \frac{a}{k_B T}\rho \\
&= \left(1 + b\rho + b^2\rho^2 + \cdots\right) - \frac{a}{k_B T}\rho \\
&= 1 + \left(b - \frac{a}{k_B T}\right)\rho + b^2\rho^2 + \cdots.
\end{aligned} \tag{10.63}$$

Comparing with Eq. (10.62), we obtain the van der Waals predictions for the first virial coefficients,

$$B_2^{\mathrm{vdW}}(T) = b - \frac{a}{k_B T}, \qquad B_3^{\mathrm{vdW}}(T) = b^2, \tag{10.64}$$

with all higher B_n determined by the same phenomenological parameters a and b. From the microscopic point of view, a captures the net attractive effect of intermolecular forces, while b represents the excluded-volume contribution.

van der Waals Gases

In the macroscopic (molar) formulation, the van der Waals equation of state reads

$$\left(P + a\frac{n^2}{V^2}\right)\left(\frac{V}{n} - b\right) = RT,$$

where n is the number of moles and V/n is the molar volume. In this form, the parameters a and b are defined *per mole* of substance: a (units $\mathrm{Pa\,m^6/mol^2}$) quantifies the net attractive part of the interaction, while b (units $\mathrm{m^3/mol}$) represents the excluded volume per mole due to short-range repulsion.

Representative experimental values are listed in Table 10.1.

The parameter b has a direct microscopic interpretation as an excluded volume. If each molecule occupies an effective volume V_{mol}, then

$$V_{\mathrm{mol}} = \frac{b}{N_{\mathrm{A}}},$$

where N_{A} is Avogadro's number. Approximating a molecule as a sphere of radius r_0,

$$V_{\mathrm{mol}} = \frac{4\pi r_0^3}{3},$$

allows one to extract a molecular length scale from thermodynamic measurements: a macroscopic constraint on b translates directly into a microscopic estimate for r_0.

TABLE 10.1 van der Waals parameters for selected gases (adapted from Fishbane, Gasiorowicz, and Thornton, *Physics for Scientists and Engineers*, 3rd ed. [FGT05]).

Gas	a ($\mathrm{Pa\ m^6/mol^2}$)	b ($\mathrm{m^3/mol}$)
Helium	3.46×10^{-3}	23.71×10^{-6}
Neon	2.12×10^{-2}	17.10×10^{-6}
Hydrogen	2.45×10^{-2}	26.61×10^{-6}
Carbon dioxide	3.96×10^{-1}	42.69×10^{-6}
Water vapor	5.47×10^{-1}	30.52×10^{-6}

The van der Waals Equation

The van der Waals equation of state represents one of the earliest and most successful attempts to incorporate microscopic physics into macroscopic thermodynamics. In his 1873 doctoral thesis [vdW73b], Johannes Diderik van der Waals introduced the parameters a and b to account for intermolecular attractions and finite molecular size, generalizing the ideal gas law to real substances. This framework captured, in a remarkably simple form, the qualitative features of liquid–gas coexistence and the existence of a critical point. The importance of this achievement was recognized with the 1910 Nobel Prize in Physics for *his work on the equation of state for gases and liquids* [vdW10].

Beyond its theoretical importance, the van der Waals equation had a profound impact on *thermometry*. As mentioned at the beginning of this Volume, practical thermometers rely on a gas whose pressure–temperature relationship is linear and independent of density. This requires the working fluid to behave as closely as possible to an *ideal gas*. The corrections identified by van der Waals explain precisely why real gases deviate from ideality: at too high a pressure, excluded-volume effects (the parameter b) distort the temperature dependence of pressure, while at lower temperatures or higher densities, attractive forces (the parameter a), further perturb the law $PV = Nk_BT$. As a result, accurate gas thermometry is achieved only in the dilute limit where both $a\rho^2$ and $b\rho$ become negligibly small.

Thus, the van der Waals theory not only advanced the conceptual understanding of real fluids, but also clarified the operational requirements for building a thermometer: the working gas must be sufficiently dilute that the ideal gas law holds to an excellent approximation. Modern temperature scales and primary thermometric standards continue to exploit this principle, using gases whose behavior is deliberately pushed toward the ideal regime.

10.5 van der Waals Gases and Phase Transitions

Before moving to the properties of quantum gases, we discuss *classical interacting gases*, which can undergo macroscopic phase transitions such as the gas–liquid transition. Although a complete microscopic theory of liquids remains elusive, the phenomenological model proposed by van der Waals (1873), which we discussed earlier, captures many essential features of such transitions and provides the foundation for modern theories of critical phenomena. As we saw, the coefficients a and b of the van der Waals gas can be derived from the virial expansion.

It is worth noting that the *classical liquid state* is still not fully understood from first principles. Unlike gases or crystals, liquids lack a small expansion parameter: their particles are densely packed and strongly correlated. Nevertheless, mean-field approaches such as the van der Waals model yield a remarkably accurate qualitative description of coexistence and criticality.

The van der Waals equation introduces two empirical corrections to the ideal-gas law to account for (i) the finite size of molecules and (ii) the attractive forces between them. The resulting equation of state for one mole of gas reads

$$\left(P + \frac{a}{V^2}\right)(V - b) = RT, \tag{10.65}$$

where $a > 0$ quantifies the mean attractive interaction and $b > 0$ represents the excluded molecular volume. These parameters are characteristic of each substance and can be determined experimentally.

Expanding Eq. (10.65), we obtain

$$P(V,T) = \frac{RT}{V - b} - \frac{a}{V^2}. \tag{10.66}$$

At high temperatures or low densities ($V \gg b$), the equation reduces to the ideal-gas law. At lower temperatures, however, the pressure–volume relation develops an inflection, signaling the emergence of nontrivial thermodynamic behavior.

Critical Point

The critical point corresponds to the temperature and pressure at which the isotherm in the P–V plane exhibits an inflection:

$$\left(\frac{\partial P}{\partial V}\right)_{T_c} = 0, \qquad \left(\frac{\partial^2 P}{\partial V^2}\right)_{T_c} = 0.$$

Applying these conditions to Eq. (10.66), we find

$$-\frac{RT_c}{(V_c-b)^2}+\frac{2a}{V_c^3}=0, \tag{10.67}$$

$$\frac{2RT_c}{(V_c-b)^3}-\frac{6a}{V_c^4}=0. \tag{10.68}$$

Solving simultaneously gives the critical parameters:

$$V_c=3b, \qquad P_c=\frac{a}{27b^2}, \qquad T_c=\frac{8a}{27bR}. \tag{10.69}$$

The constants a and b can thus be extracted from experimental measurements of P_c, T_c, and V_c.

It is convenient to express Eq. (10.66) in terms of the dimensionless *reduced variables*

$$P_r=\frac{P}{P_c}, \qquad V_r=\frac{V}{V_c}, \qquad T_r=\frac{T}{T_c}, \tag{10.70}$$

which measure pressure, volume, and temperature in units of their critical values.

Substituting the relations Eq. (10.69) into the equation of state Eq. (10.66), we obtain the universal form

$$\left(P_r+\frac{3}{V_r^2}\right)(3V_r-1)=8T_r. \tag{10.71}$$

This is the *law of corresponding states*: all substances obeying the van der Waals equation have the same reduced equation of state, independent of the specific values of a and b. When plotted in terms of P_r, V_r, and T_r, their phase diagrams are therefore identical.

Near the critical point, the reduced isotherm ($T_r = 1$) may be expanded as

$$P_r=1+4(T_r-1)-6(T_r-1)(V_r-1)-3(V_r-1)^3+\cdots, \tag{10.72}$$

showing that the pressure deviation is cubic in $(V_r - 1)$. This mean-field result predicts classical critical exponents for the compressibility and order parameter (density difference between coexisting phases), though experiments reveal small deviations due to fluctuations—an important motivation for the development of modern renormalization-group theory.

Phase Coexistence and the Maxwell Construction

For $T < T_c$, the van der Waals isotherms display an unphysical loop in which the pressure decreases as the volume decreases (see Fig. 10.5). This region corresponds to mechanical instability ($\partial P/\partial V > 0$). In reality, the system undergoes phase separation: gas and liquid phases coexist at constant pressure between volumes V_1 and V_2.

Thermodynamically, this is corrected by the *Maxwell equal-area construction*, which replaces the oscillating part of the isotherm by a horizontal line such that

$$\int_{V_1}^{V_2}[P(V,T)-P_{\text{coex}}]\,dV=0.$$

This ensures equality of chemical potentials (or Gibbs free energies) between the two phases, corresponding to equilibrium coexistence. In the grand-canonical ensemble, the same condition manifests as a vanishing isothermal compressibility, identifying the phase transition point.

Below the critical temperature, the system minimizes its free energy by separating into two phases of different densities but equal pressure and chemical potential. As the gas is compressed, droplets of the liquid phase nucleate and grow, keeping the pressure constant until the transformation is complete. This process constitutes a *first-order phase transition*, characterized by a discontinuity in density and by latent heat.

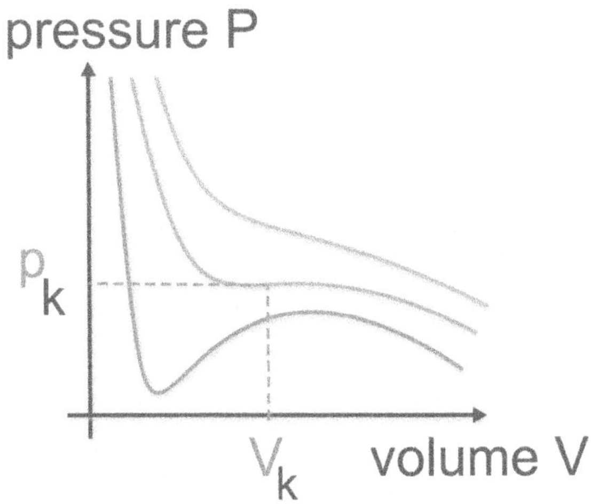

FIGURE 10.5 Isotherms of the van der Waals gas in the pressure–volume plane. At $T = T_c$, the inflection point marks the critical point (P_k, V_k). For $T < T_c$, the dashed loop indicates an unstable region, replaced in equilibrium by the horizontal Maxwell line representing phase coexistence between gas and liquid phases.

The van der Waals model therefore provides a simple but insightful picture of how molecular interactions can give rise to phase transitions and critical phenomena. Although it is a mean-field theory and neglects fluctuations, its qualitative predictions—phase coexistence, metastability, and critical behavior—remain remarkably robust and form the basis for more refined statistical-mechanical approaches.

REMARK 10.11 A detailed mathematical and physical discussion of the van der Waals equation, its critical behavior, and the Maxwell construction can be found in standard references such as Kerson Huang's *Statistical Mechanics* [Hua87], R. K. Pathria and P. D. Beale's *Statistical Mechanics* [PB11], and H. B. Callen's *Thermodynamics and an Introduction to Thermostatistics* [Cal85]. The model's simplicity and generality make it an enduring cornerstone for understanding first-order phase transitions and the law of corresponding states.

Universality of Phase Transitions

The liquid–gas transition is one instance of a wider class of critical phenomena. Ferromagnets near the Curie point exhibit the same universal behavior under the mapping $M \leftrightarrow (V - V_c)$ and $H \leftrightarrow (P - P_c)$. Similarly, percolation, random graphs, and network-formation models display continuous transitions governed by analogous scaling laws. The ubiquity of these patterns motivates the development of a unified framework for phase transitions and critical exponents, which we turn to in the next chapter.

The phenomena discussed so far, i.e., the gas transition of the van der Waals gas, are both examples of *phase transitions*, in which a system exhibits abrupt changes in macroscopic properties as external parameters (temperature, pressure, or density) are varied.

From Gases to Fluids

Classically, the distinction between a gas and a liquid is largely one of density and compressibility. Gases are easily compressed, while liquids are nearly incompressible. From a statistical-mechanical standpoint, however, a liquid can be regarded as a *dense gas*, in which intermolecular interactions become dominant. The mathematical framework for describing this regime is the *cluster expansion*, where correlations between particles are systematically incorporated.

The first thermodynamic quantity that reflects the difference between a gas and a liquid is the *isothermal compressibility*,

$$\kappa_T = -\frac{1}{V}\left(\frac{\partial V}{\partial P}\right)_T, \tag{10.73}$$

which decreases sharply as the system becomes denser and more "fluidlike."

To study such dense states, one introduces the local density field

$$\rho(\mathbf{r}) = \frac{1}{N}\sum_i \langle \delta(\mathbf{r} - \mathbf{r}_i)\rangle, \tag{10.74}$$

and the radial distribution function

$$\rho\, g(r) = \frac{1}{N}\sum_{i<j} \langle \delta(\mathbf{r} + \mathbf{r}_j - \mathbf{r}_i)\rangle. \tag{10.75}$$

This function describes the probability of finding two particles separated by a distance r relative to an ideal gas.

Using the classical cluster expansion (see, e.g., Goodstein [Goo75]), one finds to leading order

$$g(r) = e^{-\beta u(r)}, \tag{10.76}$$

where $u(r)$ is the pair potential between particles. For an ideal gas, $u(r) = 0$ and hence $g(r) = 1$. The deviation of $g(r)$ from unity quantifies the spatial correlations induced by interactions.

Pressure and Correlations

The pressure can be expressed in terms of $g(r)$ and the interparticle potential $u(r)$ as

$$\frac{PV}{Nk_{\mathrm{B}}T} = 1 - \frac{\rho}{6k_{\mathrm{B}}T}\int d^3r\, r\frac{du(r)}{dr}\, g(r), \tag{10.77}$$

showing explicitly how the deviation from ideal behavior arises from interactions. Discontinuities in the compressibility or pressure as functions of T or V thus signal changes in the "fluidity" of the system and mark the occurrence of a phase transition.

At higher orders in the cluster expansion, corrections to $g(r)$ can be expressed through the potential of mean force,

$$\ln g(r) + \beta u(r) = W(r), \tag{10.78}$$

which encodes the effective many-body correlations between particles.

10.5.1 Phase Transitions and the Ehrenfest Classification

From a thermodynamic perspective, a phase transition occurs when some derivative of the free energy becomes discontinuous as an external parameter changes. In statistical mechanics, this corresponds to a non-analyticity of the partition function [Jae98].

For the grand-canonical ensemble, recall that

$$\frac{P}{k_{\mathrm{B}}T} = \frac{1}{V}\ln Z, \tag{10.79}$$

$$\frac{N}{V} = \frac{1}{V}\zeta\frac{\partial}{\partial\zeta}\ln Z. \tag{10.80}$$

A discontinuity in the pressure implies a discontinuity in $\ln Z$ itself, whereas a discontinuity in its derivatives (e.g., in compressibility or specific heat) corresponds to higher-order singularities of $\ln Z$. Hence, phase transitions are identified with non-analytic points of the thermodynamic potential.

Ehrenfest's Classification

In the original classification by Paul Ehrenfest (1933), phase transitions were distinguished by the lowest derivative of the free energy F that exhibits a discontinuity:

- *First-order transitions:* discontinuity in the first derivatives of F, such as entropy $S = -(\partial F/\partial T)_V$ or volume $V = (\partial F/\partial P)_T$. Examples: melting, boiling, and the liquid–gas transition.
- *Second-order transitions:* continuity of first derivatives but discontinuity in second derivatives such as heat capacity, compressibility, or magnetic susceptibility. Examples: superconducting transitions and ferromagnetic transitions at the Curie point.

In modern statistical mechanics, the Ehrenfest classification has been superseded by a more general understanding of phase transitions as singularities of the free energy, often described by scaling and universality near critical points. Nevertheless, Ehrenfest's criterion remains a valuable historical and conceptual tool for distinguishing the thermodynamic signatures of different kinds of transitions.

REMARK 10.12 For a modern discussion of the thermodynamics and microscopic origin of phase transitions, see Goodstein, *States of Matter* [Goo75], Huang, *Statistical Mechanics* [Hua87], and Pathria and Beale, *Statistical Mechanics* [PB11]. These works develop the formal connection between discontinuities in thermodynamic potentials and the analytical structure of the partition function.

Order of Phase Transitions and Critical Phenomena

A *phase transition* is said to be of the *nth order* if the nth derivative of the free energy (or equivalently, of the logarithm of the partition function) with respect to a thermodynamic variable is discontinuous. For example, the liquid–gas transition described by the van der Waals equation is a *first-order transition*: below the critical temperature T_c, the pressure exhibits a finite jump in its first derivative with respect to volume. Above T_c, this discontinuity disappears and the system crosses smoothly between gas and liquid phases.

Because the coexistence line between liquid and gas terminates at the critical point, it is always possible to move continuously from one phase to the other without encountering a discontinuity by following a path in the (P, T) plane that passes above T_c. This means that, from a fundamental point of view, there is no sharp distinction between gases and liquids—both belong to the same phase, connected by a *continuous phase transition.* Such continuous transitions are often referred to as *critical phenomena.* A schematic phase diagram illustrating this behavior is shown in Fig. 10.6.

In contrast, transitions between liquid or gas and solid phases involve a change of symmetry: the crystalline solid breaks translational invariance, while fluids remain isotropic. Such transitions are therefore qualitatively different and will be analyzed later in the context of symmetry breaking and order parameters.

Gases are paradigmatic examples of weakly interacting, thermodynamically large systems. When interactions are included—whether attractive, repulsive, or long-range—new forms of collective behavior emerge. The study of these collective phenomena constitutes the domain of *many-body physics*, which will be the focus of the following chapters.

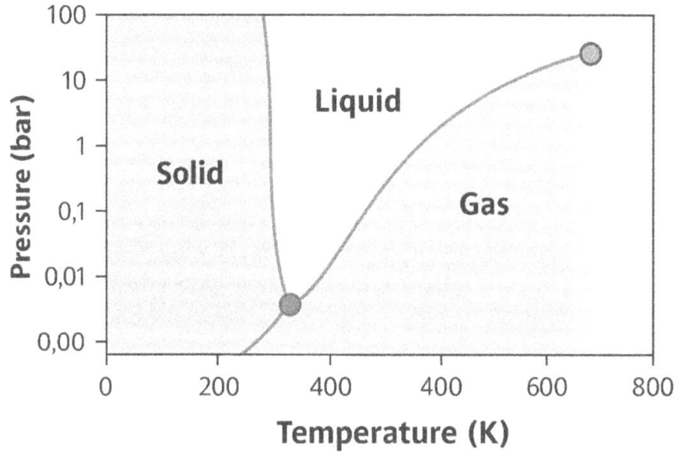

FIGURE 10.6 Schematic phase diagram of water in the pressure–temperature plane. The liquid–gas coexistence line terminates at the critical point (T_c, P_c). Above this temperature, liquid and gas phases become indistinguishable, allowing a continuous transformation between them.

11

Simulating a Particle Gas: A Primer

11.1 A Brief Historical Prelude

The numerical exploration of gases and liquids has roots in the kinetic theories of the nineteenth century, developed by Ludwig Boltzmann, James Clerk Maxwell, and Josef Loschmidt [Bol72, Max60, Los76]. Their studies of the hard-sphere gas—a system of particles interacting only through elastic collisions—established the microscopic basis of thermodynamics and introduced the ideas of ergodicity, reversibility, and the statistical nature of equilibrium.

The advent of electronic computation in the mid-twentieth century transformed these theoretical constructs into numerical experiments. In the early 1950s, Berni J. Alder and Thomas E. Wainwright, working at the newly founded Lawrence Livermore National Laboratory, carried out the first molecular-dynamics (MD) simulations of hard spheres, aided by the programming expertise of Mary Ann Mansigh Karlsen. Their pioneering work [AW57] demonstrated that Newtonian dynamics, executed on a digital computer, could reproduce macroscopic thermodynamic behavior and reveal emergent collective effects such as diffusion and hydrodynamic correlations. The simulations established that even purely repulsive hard-sphere systems undergo a first-order transition from fluid to solid, a striking confirmation that entropy alone can drive crystallization.

At roughly the same time, researchers at Los Alamos National Laboratory introduced a complementary computational approach. Nicholas Metropolis, and Augusta and Edward Teller formulated the *Monte Carlo method* [MRR+53], which replaced deterministic trajectories with stochastic sampling of phase space according to the Boltzmann distribution. Together, the MD and MC paradigms became the dual pillars of computational statistical mechanics.

Another milestone was reached with the celebrated *Fermi–Pasta–Ulam–Tsingou* experiment (1955) [FPU55],[Tsi]55, performed at Los Alamos by Enrico Fermi, John Pasta, Stanislaw Ulam, and programmer Mary Tsingou. Using one of the first electronic computers, they investigated the dynamics of a nonlinear atomic chain and discovered the unexpected recurrence of energy—a phenomenon that defied classical expectations of thermalization and laid the foundation for

DOI: 10.1201/9781003603528-11

chaos and soliton theory. Like Karlsen at Livermore, Tsingou's role was essential, as she translated theoretical ideas into executable code, transforming abstract mathematics into a numerical experiment.

Historical Note: The Birth of Molecular Simulation

The 1950s marked the convergence of theory, experiment, and computation into a new discipline—*computational statistical mechanics.* At Livermore, Berni Alder, Thomas Wainwright, and Mary Ann Mansigh Karlsen implemented the first molecular-dynamics simulations of hard spheres and disks, capturing phase transitions and transport phenomena directly from Newton's laws. At Los Alamos, Nicholas Metropolis, and Augusta and Edward Teller introduced the stochastic Monte Carlo method, while Enrico Fermi, John Pasta, Stanislaw Ulam, and Mary Tsingou performed the first numerical study of nonlinear dynamics.

These projects were among the first large-scale collaborations between theorists and programmers, many of them women, whose expertise in the nascent art of coding made simulation possible. Their collective efforts demonstrated that computers could serve not merely as calculators, but as instruments of discovery. The enduring impact of these early pioneers is visible in every field where numerical experiments complement theory and experiment to unravel the collective behavior of matter.

11.2 The Gas in One Dimension: The Tonks Gas

Before considering gases in higher dimensions, it is instructive to examine the simplest nontrivial system: a gas of particles constrained to move along a single spatial axis. This model, introduced by Lewi Tonks in 1936 [Ton36], is known as the *Tonks gas.* It consists of N identical, impenetrable particles of length σ, confined to a one-dimensional segment of total length L. Because the particles cannot overlap, their centers must satisfy the hard–core exclusion condition

$$|x_i - x_j| \geq \sigma, \qquad \text{for all } i \neq j. \tag{11.1}$$

Despite its apparent simplicity, the Tonks gas exhibits rich statistical behavior and provides a direct link between kinetic theory, classical fluid models, and modern simulation methods. It was one of the first exactly solvable interacting systems in statistical mechanics, and it explicitly demonstrated that excluded–volume effects alone can generate nontrivial thermodynamic properties, even in the absence of attractive forces. In this sense, the Tonks gas is the one–dimensional ancestor of the hard–sphere and hard–disk models that later became the workhorses of molecular dynamics simulations.

Tonks's work preceded the first computational studies of particle gases by several decades, but the same principles of impenetrability and geometric exclusion reappeared in the numerical hard–sphere simulations of Alder and Wainwright (1953) and in later analyses of transport and correlation phenomena in fluids.

Exact Partition Function

The configurational part of the partition function can be evaluated exactly because the exclusion condition simply reduces the accessible length available to each particle. For N identical particles of length σ, the total free length is $L - N\sigma$. Accounting for indistinguishability introduces the combinatorial factor $1/N!$. Thus, the canonical partition function for the Tonks gas is

$$Z_N(T, L) = \frac{1}{N!\lambda^N} (L - N\sigma)^N, \tag{11.2}$$

where

$$\lambda = \frac{h}{\sqrt{2\pi m k_B T}} \tag{11.3}$$

is the thermal de Broglie wavelength.

Equation of State

From the Helmholtz free energy

$$F = -k_B T \ln Z_N, \tag{11.4}$$

the pressure follows from the standard thermodynamic relation

$$P = -\left(\frac{\partial F}{\partial L}\right)_{T,N}. \tag{11.5}$$

Evaluating this derivative gives

$$P = \frac{N k_B T}{L - N\sigma}. \tag{11.6}$$

This simple equation of state resembles the van der Waals equation without the attractive term. It clearly shows how the finite particle size σ reduces the available volume and thereby increases the pressure at fixed temperature and density.

REMARK 11.1 The Tonks gas provides one of the rare examples of a classical interacting system that is exactly solvable. It demonstrates how geometric constraints alone can endow a system with nontrivial thermodynamics, foreshadowing the importance of excluded–volume effects in dense fluids, colloids, and granular materials.

Thermodynamic Properties

From the canonical partition function of the Tonks gas,

$$Z_N(T, L) = \frac{1}{N!\lambda^N} (L - N\sigma)^N, \tag{11.7}$$

the entropy per particle follows directly as

$$\frac{S}{N k_B} = \ln\left[\frac{L/N - \sigma}{\lambda}\right] + 1. \tag{11.8}$$

The internal energy, being purely kinetic in one dimension, obeys the equipartition theorem,

$$U = \frac{1}{2} N k_B T, \tag{11.9}$$

and is independent of the excluded volume parameter σ.

Despite its simplicity, the Tonks gas encapsulates the essential interplay between entropy and excluded volume. As the density increases and the available free length $L - N\sigma$ shrinks, the entropy decreases logarithmically, reflecting the reduction in configurational freedom caused by crowding. The pressure, given by

$$P = \frac{N k_B T}{L - N\sigma}, \tag{11.10}$$

can thus be interpreted as an *entropic pressure*: it increases as the system seeks to restore the lost configurational entropy by expanding the available volume.

REMARK 11.2 The Tonks gas remains a fundamental analytic benchmark in one-dimensional statistical mechanics. It has been extended and generalized to study transport and hydrodynamic behavior in One-dimensional systems [Per76, Jep65, Fra13]. More recent works connect it to integrable models [LL63, Len64], to non-equilibrium steady states [BO05], and to experiments on colloidal particles confined in narrow channels [WBL00, LR05]. These developments underscore the enduring relevance of Tonks's original insight: that geometry and excluded volume alone can generate complex collective behavior.

Numerical Experiment: Simulating the Tonks Gas in MATLAB

Although the Tonks gas is exactly solvable, it is enlightening to simulate it numerically. A simple implementation illustrates how excluded volume and elastic collisions shape the microscopic dynamics.

In one dimension, particles can be represented by their positions x_i and velocities v_i within an interval $[0, L]$, with hard-core diameter σ. The dynamics are purely ballistic between collisions, and collisions exchange velocities between neighboring particles. This simple rule is sufficient to preserve total energy and momentum, reproducing the behavior of an ideal hard-rod gas.

Below is a minimal MATLAB script implementing such a simulation:

Listing 11.1: **Minimal MATLAB simulation of the Tonks gas**

```
%% Tonks gas simulation (1D hard rods)
clear; clf; rng(1);
% reproducible random seed

% --- Parameters ------------------------------------------
%number of parameters
N      = 10;
% system length
L      = 10;
% rod length
sigma  = 0.5;
dt     = 0.001;
tmax   = 10;
steps  = round(tmax/dt);

% --- Initialization --------------------------------------
x = linspace(sigma/2, L - sigma/2, N);
% initial non-overlapping positions
v = randn(1,N);
% Gaussian velocities (kBT/m = 1)

Ekin = zeros(1,steps);
% kinetic-energy record
x_record = zeros(N,steps);
% store positions for trajectories
```

```
% —— Main simulation loop ————————————————
for t = 1:steps
    % Update positions
    x = x + v*dt;

    % Reflect at boundaries
    for i = 1:N
        if x(i) < sigma/2
            x(i) = sigma - x(i);
            v(i) = -v(i);
        elseif x(i) > L - sigma/2
            x(i) = 2*L - sigma - x(i);
            v(i) = -v(i);
        end
    end

    % Check for collisions between nearest neighbors
    for i = 1:N-1
        if (x(i+1) - x(i)) < sigma
            % Elastic collision: exchange velocities
            [v(i), v(i+1)] = deal(v(i+1), v(i));
            % Separate overlapping rods
            overlap = sigma - (x(i+1) - x(i));
            x(i)   = x(i)   - overlap/2;
            x(i+1) = x(i+1) + overlap/2;
        end
    end

    % Record data
    Ekin(t)       = 0.5 * sum(v.^2);
    x_record(:,t) = x(:);
end

% —— Analysis ————————————————————————
time = (0:steps-1)*dt;
x_mean = mean(x_record,2);
% mean positions (stationary)
v_all  = diff(x_record,1,2)/dt;
% approximate velocities for histogram

% —— Visualization: four-panel summary ————————
figure('Color','w','Position',[100 100 950 700]);

% (1) Particle trajectories
subplot(2,2,1)
plot(time, x_record','LineWidth',0.8)
xlabel('time'); ylabel('position  x_i(t)');
title('Particle trajectories')
```

```
xlim([0 tmax]); ylim([0 L]); box on

% (2) Total kinetic energy vs time
subplot(2,2,2)
plot(time, Ekin,'k','LineWidth',1.2)
xlabel('time'); ylabel('E_{kin}(t)')
title('Instantaneous kinetic energy')
box on

% (3) Average spatial distribution
subplot(2,2,3)
histogram(x_record(:),40,'Normalization',...
'pdf','FaceColor',[0.2 0.4 0.8])

xlabel('x'); ylabel('P(x)')
title('Spatial distribution of particles')
xlim([0 L]); box on

% (4) Kinetic-energy distribution
Ek = 0.5*v_all(:).^2;
subplot(2,2,4)
histogram(Ek,30,'Normalization','pdf','FaceColor',[0.8 0.3 0.3])
xlabel('E_k'); ylabel('P(E_k)')
title('Distribution of kinetic energies')
box on

sgtitle('Tonks Gas Simulation: 1D Hard Rods','FontWeight','bold');
set(findall(gcf,'-property','FontName'),'FontName','Times');
set(findall(gcf,'-property','FontSize'),'FontSize',12);

% Optionally save figure
% print('-dpng','-r300','tonks_gas_simulation.png');
```

The code tracks the deterministic dynamics of N hard rods under periodic reflection at the boundaries. When the system evolves for a sufficiently long time, the histogram of velocities approaches a Maxwell–Boltzmann distribution, and the pressure can be estimated from the average momentum transfer on the walls:

$$P = \frac{1}{L} \langle 2mv_{\text{wall}} \rangle . \tag{11.11}$$

This simple algorithm captures the essential physics of the one-dimensional hard-rod gas: energy conservation, microscopic reversibility, and the emergence of macroscopic equilibrium from deterministic motion. It also provides a foundation for the numerical exploration of two-dimensional hard disks, discussed in the next section.

11.3 The Gas in Two Dimensions

The extension of the Tonks gas to two dimensions introduces a rich geometrical structure. Particles are now represented by hard disks of diameter σ confined within a square box of side L.

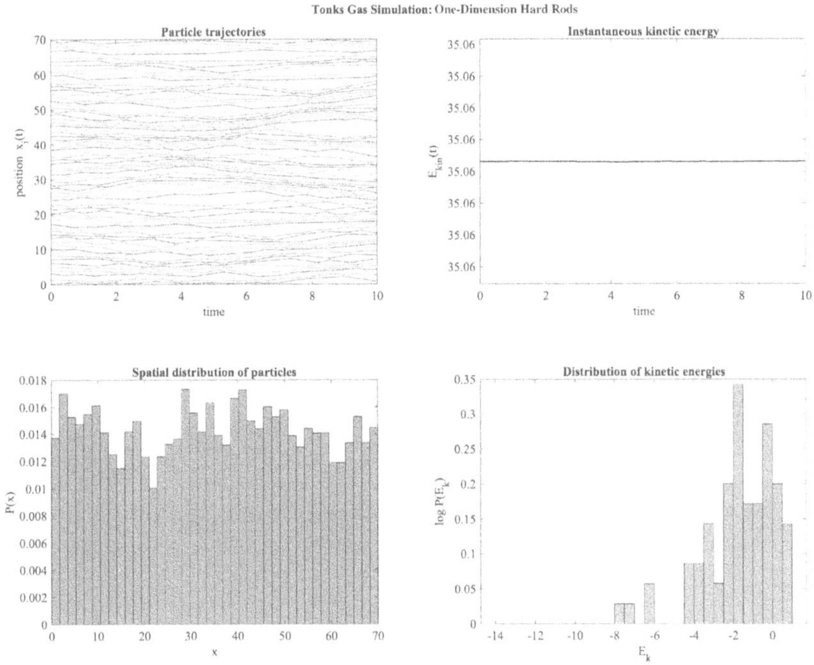

FIGURE 11.1 Tonks gas simulation with $N = 70$ hard rods in one dimension. The four panels illustrate the microscopic and macroscopic behavior of an idealized one-dimensional hard-rod gas confined in a box of length $L = 70$. (**a**) Particle trajectories $x_i(t)$ obtained from event-driven dynamics show elastic collisions and reflections at the walls. (**b**) Instantaneous total kinetic energy $E_{\text{kin}}(t)$, which fluctuates around a constant mean, confirming energy conservation and thermal equilibrium. (**c**) Spatial probability density $P(x)$ displaying a uniform distribution, consistent with ergodic sampling of all accessible configurations. (**d**) Histogram of single-particle kinetic energies $\log P(E_k)$, approaching the Maxwell–Boltzmann distribution expected for the canonical ensemble. The simulation demonstrates how purely repulsive hard-core interactions lead to macroscopic thermodynamic behavior through microscopic deterministic dynamics.

Each disk moves in straight lines between collisions, undergoing instantaneous, elastic reflections when it collides with another disk or with the boundary of the box. This simple model, known as the *hard disk gas*, is the two-dimensional analog of the hard-sphere model and constitutes one of the earliest and most studied systems in statistical mechanics.

From Tonks to Hard Disks

While the Tonks gas captures the effect of excluded volume in one dimension, in two dimensions collisions change not only the magnitudes but also the *directions* of particle velocities. This additional degree of freedom gives rise to momentum transport, viscosity, and diffusion—phenomena absent in one dimension. It is this geometry of collisions that makes the hard disk gas a fundamental testing ground for the microscopic foundations of thermodynamics.

The first numerical experiments on the hard disk gas were performed by Berni Alder and Thomas Wainwright in 1957, only a few years after their hard-sphere simulations. These studies revealed collective phenomena such as the emergence of hydrodynamic modes and the onset of the liquid-solid transition purely from elastic collisions. Their work also demonstrated that irreversible macroscopic behavior can arise from time-reversible microscopic dynamics, a striking realization at the time.

Since then, the hard disk gas has become a canonical model in non-equilibrium statistical mechanics, underpinning modern developments such as the Green–Kubo relations, Lyapunov analysis of chaos, and kinetic theory of transport.

In analogy with the Tonks gas, we can approximate the equation of state of N hard disks in a box of area $A = L^2$ by accounting for the excluded area per particle. At low densities, the system behaves as an ideal gas with small corrections:

$$PA = Nk_{\mathrm{B}}T\left(1 + B_2\frac{N}{A} + \cdots\right), \tag{11.12}$$

where B_2 is the second virial coefficient, arising from the excluded area $A_{\mathrm{ex}} = \pi\sigma^2$. For hard disks,

$$B_2 = \frac{1}{2}A_{\mathrm{ex}} = \frac{\pi\sigma^2}{2}, \tag{11.13}$$

so that, to first order,

$$P = \frac{Nk_{\mathrm{B}}T}{A - N\frac{\pi\sigma^2}{2}}. \tag{11.14}$$

This result mirrors the Tonks equation of state in two dimensions, showing again how excluded volume increases the pressure relative to the ideal gas.

Numerical Simulation: Molecular Dynamics of Hard Disks

The microscopic dynamics of the hard disk gas can be simulated using a simple event-driven algorithm. Between collisions, each particle moves freely:

$$\vec{r}_i(t + \Delta t) = \vec{r}_i(t) + \vec{v}_i\,\Delta t. \tag{11.15}$$

When two disks i and j collide, their relative velocity is reversed along the line of centers:

$$\vec{v}_i' = \vec{v}_i - (\vec{v}_{ij}\cdot\hat{n})\,\hat{n}, \qquad \vec{v}_j' = \vec{v}_j + (\vec{v}_{ij}\cdot\hat{n})\,\hat{n}, \tag{11.16}$$

where $\vec{v}_{ij} = \vec{v}_i - \vec{v}_j$ and $\hat{n}$ is the unit vector joining the centers of the two disks at contact.

For simplicity, periodic boundary conditions can be implemented to mimic an infinite gas, allowing each particle leaving one side of the box to reenter from the opposite side. The kinetic temperature is defined from the velocity fluctuations as

$$k_{\mathrm{B}}T = m\langle v^2\rangle/2, \tag{11.17}$$

and the pressure can be computed from the average momentum transfer per unit time and length on the walls or, in periodic systems, from the virial relation.

Minimal MATLAB Implementation

A pedagogical two-dimensional simulation can be constructed using MATLAB or Python, where particles are advanced in discrete time steps and collisions are resolved pairwise.

Listing 11.2: **Skeleton of a 2D hard disk gas simulation**

```
N = 20; L = 10; sigma = 0.3;
x = L*rand(N,1); y = L*rand(N,1);
vx = randn(N,1); vy = randn(N,1);
dt = 0.001; tmax = 50;

for t = 0:dt:tmax
    x = x + vx*dt;   y = y + vy*dt;

    % Reflective boundaries
    vx(x<sigma/2 | x>L-sigma/2) = -vx(x<sigma/2 | x>L-sigma/2);
    vy(y<sigma/2 | y>L-sigma/2) = -vy(y<sigma/2 | y>L-sigma/2);

    % Check for collisions between pairs
    for i = 1:N-1
      for j = i+1:N
        dx = x(j)-x(i); dy = y(j)-y(i);
        r2 = dx^2 + dy^2;
        if r2 < sigma^2
            % Unit normal
            n = [dx dy]/sqrt(r2);
            vrel = [vx(i)-vx(j), vy(i)-vy(j)];
            vn = dot(vrel,n);
            % Update velocities (elastic collision)
            vx(i) = vx(i) - vn*n(1); vy(i) = vy(i) - vn*n(2);
            vx(j) = vx(j) + vn*n(1); vy(j) = vy(j) + vn*n(2);
        end
      end
    end

    % Plot particles
    if mod(round(t/dt),200)==0
        plot(x,y,'ko','MarkerFaceColor','k');
        axis([0 L 0 L]); axis square;
        title(['t = ',num2str(t)]);
        drawnow;
    end
end
```

This algorithm reproduces the essential phenomenology of the hard disk gas: the Maxwell–Boltzmann velocity distribution, pressure fluctuations, and the approach to equilibrium from random initial conditions. More sophisticated event-driven schemes can compute the exact collision times analytically, ensuring perfect energy conservation and allowing the study of transport coefficients.

The two-dimensional hard disk gas provides a bridge between exactly solvable systems and realistic molecular simulations. It exhibits the key ingredients of thermodynamics—energy conservation, momentum transport, and entropy increase—while remaining conceptually transparent and computationally manageable.

11.4 Event-Driven Molecular Dynamics

In the simulations described so far, particles are advanced in small discrete time steps, and potential overlaps are resolved approximately at each iteration. This procedure, often called *time-driven molecular dynamics*, is simple to implement but introduces numerical errors: finite steps may cause small violations of energy conservation, and collisions can be missed or resolved inconsistently when the time step is too large.

A different approach, both conceptually elegant and computationally exact for systems with instantaneous interactions, is the *event-driven* scheme. Instead of integrating the equations of motion at fixed time intervals, we advance the system from one collision event to the next, solving analytically for the times at which events occur.

The basic idea of event-driven molecular dynamics is quite simple. Between collisions, the motion of each particle is free:

$$\vec{r}_i(t) = \vec{r}_i^{\,0} + \vec{v}_i\, t, \tag{11.18}$$

and all forces vanish except at the instant of contact. For every pair of particles (i, j), one can compute the time τ_{ij} at which their trajectories will bring them into contact, by solving for the root of

$$|\mathbf{r}_i^0 + \mathbf{v}_i\,\tau - (\mathbf{r}_j^0 + \mathbf{v}_j\,\tau)| = \sigma. \tag{11.19}$$

Squaring both sides gives a quadratic equation in τ:

$$(\mathbf{v}_{ij}\cdot\mathbf{v}_{ij})\,\tau^2 + 2\,(\mathbf{r}_{ij}\cdot\mathbf{v}_{ij})\,\tau + (\mathbf{r}_{ij}\cdot\mathbf{r}_{ij} - \sigma^2) = 0, \tag{11.20}$$

where $\mathbf{r}_{ij} = \mathbf{r}_i^0 - \mathbf{r}_j^0$ and $\mathbf{v}_{ij} = \mathbf{v}_i - \mathbf{v}_j$. The smallest positive root of this equation (if it exists) gives the time until the next collision between disks i and j. The global next event is the smallest of all these times:

$$\tau_{\text{next}} = \min_{i<j} \tau_{ij}. \tag{11.21}$$

The simulation clock then advances by τ_{next}, all positions are updated using the ballistic motion, and the corresponding colliding pair is updated using the exact collision rules:

$$\mathbf{v}_i' = \mathbf{v}_i - (\mathbf{v}_{ij}\cdot\hat{\mathbf{n}})\,\hat{\mathbf{n}}, \qquad \mathbf{v}_j' = \mathbf{v}_j + (\mathbf{v}_{ij}\cdot\hat{\mathbf{n}})\,\hat{\mathbf{n}}, \tag{11.22}$$

where $\hat{\mathbf{n}}$ is the unit vector along the line of centers at impact.

After the event, new collision times involving particles i and j are recalculated, and the process repeats. In practice, event-driven algorithms maintain a *priority queue* (often implemented as a binary heap) that stores the scheduled times of all future events, allowing the next event to be retrieved efficiently.

In addition to particle-particle collisions, particles may collide with the walls of the simulation box or cross a periodic boundary. The time to the next wall collision can be computed analytically as

$$\tau_{i,\text{wall}} = \min_{\alpha=x,y} \begin{cases} (\frac{\sigma}{2} - r_{i,\alpha})/v_{i,\alpha}, & v_{i,\alpha} < 0, \\ (L - \frac{\sigma}{2} - r_{i,\alpha})/v_{i,\alpha}, & v_{i,\alpha} > 0, \end{cases} \tag{11.23}$$

and treated in exactly the same way as a pairwise collision. In periodic systems, no reflection occurs; positions are simply wrapped modulo L, ensuring translational invariance.

Advantages of Event-Driven Simulations

The advantages of the event-driven approach are both conceptual and practical. First, because collisions are resolved analytically, kinetic energy and momentum are conserved to machine precision, with no need for infinitesimal time steps. The simulation automatically adapts its temporal resolution to the actual dynamics: long intervals of free flight require no computation, while dense configurations naturally generate more frequent events.

In dilute gases, collisions are rare, so event-driven algorithms are far more efficient than time-driven schemes, which waste many updates on free motion.

Because each collision is handled exactly, statistical quantities such as diffusion coefficients, velocity autocorrelation functions, and transport coefficients can be computed without numerical artifacts.

Thus, event-driven methods are ideally suited for systems with instantaneous, discontinuous interactions (hard spheres, hard rods, or billiards). For continuous potentials—such as Lennard–Jones or Coulomb interactions—forces act at all times, and standard time-driven integrators are required. Hybrid algorithms have been developed to combine both approaches, and sophisticated event-driven schemes can also incorporate thermostats or stochastic boundaries.

The first event-driven molecular dynamics algorithms were formulated in the late 1950s by Berni Alder and Thomas Wainwright for systems of hard spheres, which they used to demonstrate the existence of phase transitions in purely repulsive fluids. Since then, molecular dynamics (MD) has become one of the most powerful methods for studying condensed-matter systems, with refinements that make it one of the most elegant demonstrations of how deterministic mechanics can reproduce the statistical laws of thermodynamics.

The algorithms discussed so far operate in the *micro-canonical ensemble* (constant N, V, E), where the total energy is conserved and the system evolves according to Newton's equations of motion without external exchange of heat or matter. While such simulations reveal fundamental properties of isolated systems, most real physical situations correspond to systems in contact with an environment at a fixed temperature or pressure. To reproduce these conditions, molecular dynamics simulations employ additional dynamical mechanisms known as *thermostats* (for constant temperature) and *barostats* (for constant pressure).

Thermostatting introduces an effective coupling between the simulated system and an idealized heat bath, allowing the total kinetic energy—and hence the temperature—to fluctuate around a prescribed value. It is an advanced method we do not discuss here, but various algorithms achieve this in different ways: stochastic schemes such as the *Andersen thermostat* randomize particle velocities through occasional collisions with a virtual heat bath [And80], while deterministic ones such as the *Nosé–Hoover* [Nos84, Hoo85] or *Berendsen* thermostat [BPvG$^+$84] modify the equations of motion to enforce the canonical ensemble (N, V, T). More sophisticated methods, such as the *Nosé–Hoover chain* thermostat [MKT92], extend this approach to reproduce the correct thermodynamic fluctuations in large systems and ensure ergodic sampling of phase space.

These extensions make it possible to simulate not only isolated systems but also open systems at fixed temperature or pressure, thereby bridging the gap between deterministic molecular dynamics and the statistical ensembles of equilibrium thermodynamics. For detailed discussions of these techniques, the reader is referred to standard texts on molecular dynamics and simulation methods [Hai97, AT17, FS02].

For further details and practical insight into molecular simulations and related numerical techniques, the reader is referred to standard texts such as Haile's *Molecular Dynamics Simulation: Elementary Methods* [Hai97].

Excluded Volume and Collective Phenomena in Two Dimensions

In two dimensions, excluded volume acquires a profound geometrical significance. Unlike the one-dimensional Tonks gas, where particles cannot bypass one another, disks in two dimensions can rearrange collectively, giving rise to rich structural and dynamical phenomena. At moderate densities, the *hard-disk fluid* behaves as a simple liquid whose structure is dominated by short-range order. As the packing fraction increases, however, local constraints due to excluded volume lead to the emergence of cooperative motion, transient cages, and eventually to the onset of *jamming* and *glassy dynamics* [BB11, Par20].

At even higher densities, the hard-disk system exhibits a continuous transition to a quasi-ordered *hexatic phase*, followed by crystallization into a triangular lattice. This two-step melting scenario, first proposed by Kosterlitz, Thouless, Halperin, Nelson, and Young (the KTHNY theory) [KT73, HN78, You79], and later confirmed in large-scale simulations [BK11, Kra12], reveals how purely entropic interactions can produce long-range correlations and topological defect unbinding.

At even larger packing fractions ($\phi \gtrsim 0.84$), the system becomes mechanically rigid: motion ceases without energy barriers—a phenomenon known as *jamming*. This behavior, central to the physics of granular matter and colloidal suspensions, is characterized by a dramatic slowdown of dynamics and the formation of force-bearing contact networks [LN98, vH10].

Modern computational and experimental studies have connected these phenomena to the broader framework of non-equilibrium statistical mechanics. Hard-disk and hard-sphere systems remain paradigmatic models for exploring how collective order, frustration, and entropy compete to produce emergent mechanical rigidity, glass formation, and complex transport behavior.

12

Quantum Statistical Mechanics

12.1 Basic Postulates of Quantum Mechanics

The development of quantum mechanics in the early twentieth century marked one of the most profound conceptual revolutions in physics. Beginning with Planck's quantization of radiation and Einstein's interpretation of the photoelectric effect, followed by Bohr's atomic model, and later formalized by Heisenberg and Schrödinger, the theory provided a new description of nature—one in which probability amplitudes replaced deterministic trajectories. Within this framework, microscopic systems such as atoms, electrons, and photons are characterized by wavefunctions and operators rather than by points in phase space.

Since the formal structure of quantum mechanics is assumed to be familiar to the reader, we summarize here only its essential postulates and notation, which will serve as the foundation for the construction of *quantum statistical mechanics*. The goal of this brief review is not to rederive the theory, but to establish the language and conventions that will be used in the subsequent sections.

DOI: 10.1201/9781003603528-12

The quantum description of a physical system rests on a small number of fundamental postulates that generalize classical mechanics from phase space to Hilbert space.

- *State space.* The phase space of classical mechanics is replaced by a complex Hilbert space $\mathcal{H}$. The state of a system is represented by a normalized vector (or ray) $|\psi\rangle \in \mathcal{H}$, which for a composite system is the tensor product of the spaces associated with its subsystems.
- *Observables.* Physical observables correspond to self-adjoint linear operators $\hat{A}$ acting on $\mathcal{H}$. They are obtained from the classical quantities through a quantization rule,

$$q \longrightarrow \hat{q}, \qquad p \longrightarrow \hat{p} = -i\hbar\,\partial_q, \tag{12.1}$$

so that the fundamental commutation relation

$$[\hat{q}, \hat{p}] = i\hbar\,\mathbb{I} \tag{12.2}$$

replaces the classical Poisson bracket $\{q, p\} = 1$.
- *Measurement.* The possible outcomes of a measurement of an observable $\hat{A}$ are its eigenvalues. Immediately after a measurement yielding the value a_n, the system is projected onto the corresponding eigenstate $|a_n\rangle$, a process known as the *collapse of the wavefunction.*
- *Dynamics.* The time evolution of a closed system is generated by the Hamiltonian operator $\hat{H}$. In the Schrödinger picture,

$$i\hbar\,\frac{d}{dt}|\psi(t)\rangle = \hat{H}\,|\psi(t)\rangle, \tag{12.3}$$

whereas in the Heisenberg picture, the states are fixed and observables evolve according to

$$i\hbar\,\frac{d\hat{B}(t)}{dt} = [\hat{B}(t), \hat{H}]. \tag{12.4}$$

The two formulations are equivalent and related by a unitary transformation generated by $U(t) = e^{-i\hat{H}t/\hbar}$. The correspondence principle between classical and quantum mechanics is summarized by the rule

$$\{A, B\}_{\text{classical}} \;\longrightarrow\; \frac{1}{i\hbar}[\hat{A}, \hat{B}]. \tag{12.5}$$

12.2 Operators and Uncertainty Relations

The transition from classical to quantum mechanics is implemented through a correspondence between observables and operators, and between Poisson brackets and commutators. In this operator formalism, the state of the system is represented by a vector $|\psi\rangle$ in Hilbert space, and measurable quantities correspond to self-adjoint operators acting on that space.

From Classical to Quantum Hamiltonians

For a classical system described by canonical coordinates (q_i, p_i) and Hamiltonian

$$H(q,p) = \sum_i \frac{p_i^2}{2m} + V(q), \tag{12.6}$$

quantization promotes q_i and p_i to operators $\hat{q}_i$ and $\hat{p}_i$ satisfying the canonical commutation relations

$$[\hat{q}_i, \hat{p}_j] = i\hbar\,\delta_{ij}\,\mathbb{I}. \tag{12.7}$$

The corresponding Hamiltonian operator is then

$$\hat{H} = \sum_i \left(-\frac{\hbar^2}{2m} \nabla_i^2 \right) + V(\hat{q}), \tag{12.8}$$

where the kinetic energy operator is obtained by substituting $\hat{p}_i = -i\hbar \nabla_i$.

Expectation Values and Variances

For any observable represented by an operator $\hat{A}$, its expectation value in a normalized state $|\psi\rangle$ is

$$\langle \hat{A} \rangle_\psi = \langle \psi | \hat{A} | \psi \rangle, \tag{12.9}$$

and the corresponding variance (or squared uncertainty) is

$$(\Delta_\psi A)^2 = \langle \psi | (\hat{A} - \langle \hat{A} \rangle_\psi)^2 | \psi \rangle. \tag{12.10}$$

If two operators $\hat{A}$ and $\hat{B}$ fail to commute, i.e.,

$$[\hat{A}, \hat{B}] = i\hat{C}, \tag{12.11}$$

then their uncertainties obey the general inequality

$$\Delta_\psi A \, \Delta_\psi B \geq \frac{1}{2} |\langle \hat{C} \rangle_\psi|. \tag{12.12}$$

This is a direct consequence of the Cauchy–Schwarz inequality applied to the inner product structure of Hilbert space.

Heisenberg Uncertainty Principle

For position and momentum, where $[\hat{q}, \hat{p}] = i\hbar \mathbb{I}$, the inequality above gives

$$\Delta_\psi \hat{q} \, \Delta_\psi \hat{p} \geq \frac{\hbar}{2}. \tag{12.13}$$

This is the *Heisenberg uncertainty principle.* Analogous relations exist for other pairs of non-commuting observables, such as different components of angular momentum:

$$\Delta_\psi \hat{L}_x \, \Delta_\psi \hat{L}_y \geq \frac{\hbar}{2} |\langle \hat{L}_z \rangle_\psi|. \tag{12.14}$$

In each case, the right-hand side reflects the magnitude of the commutator, which quantifies the degree of incompatibility between the observables.

REMARK 12.1 The uncertainty principle does not arise from measurement imperfections, but from the non-commutative structure of quantum observables themselves. It expresses the fact that certain physical quantities cannot be simultaneously specified beyond the limits set by $\hbar$, which plays the role of a universal quantum of action. In the classical limit $\hbar \to 0$, commutators vanish, the inequalities collapse, and deterministic trajectories in phase space are recovered.

Mathematical Framework of Quantum Mechanics

A quantum mechanical system is defined by two essential mathematical objects: a Hilbert space and a Hamiltonian operator.

A Hilbert space $\mathcal{H}$ is a complex vector space endowed with a scalar product $\langle a|b\rangle$ that satisfies $\langle a|b\rangle = \langle b|a\rangle^*$ and $\langle a|a\rangle \geq 0$, with equality only if $|a\rangle = 0$. Kets $|a\rangle$ represent vectors in $\mathcal{H}$, while bras $\langle a|$ denote their duals. The linearity of quantum mechanics allows any state to be written as a linear superposition,

$$|b\rangle = c_1|a\rangle + c_2|t\rangle, \qquad c_1, c_2 \in \mathbb{C}. \tag{12.15}$$

Given an orthonormal basis $\{|i\rangle\}$, any state can be expanded as

$$|v\rangle = \sum_j a_j|j\rangle, \qquad \langle i|j\rangle = \delta_{ij}. \tag{12.16}$$

The dimension of $\mathcal{H}$ is the number of independent basis states required to describe the system.

The Hamiltonian $\hat{H} : \mathcal{H} \to \mathcal{H}$ is a linear, self-adjoint operator that determines both the energy spectrum and the dynamics of the system. Self-adjointness ensures that its eigenvalues are real and that the time evolution it generates is unitary. In a given basis $\{|i\rangle\}$, its matrix elements are

$$H_{ij} = \langle i|\, \hat{H}\, |j\rangle. \tag{12.17}$$

The state $|v(t)\rangle$ at time t evolves from an initial state $|v(t_0)\rangle$ under the unitary operator $\hat{U}(t, t_0)$,

$$|v(t)\rangle = \hat{U}(t, t_0)\, |v(t_0)\rangle. \tag{12.18}$$

For a time-independent Hamiltonian, the propagator is

$$\hat{U}(t, t_0) = \exp\left[-\frac{i}{\hbar}(t - t_0)\hat{H}\right], \tag{12.19}$$

which satisfies the differential equation

$$\frac{d\hat{U}(t, t_0)}{dt} = -\frac{i}{\hbar}\, \hat{H}\, \hat{U}(t, t_0). \tag{12.20}$$

Consequently, the state obeys the time-dependent Schrödinger equation,

$$i\hbar \frac{d}{dt}|v(t)\rangle = \hat{H}\, |v(t)\rangle. \tag{12.21}$$

If $|v(0)\rangle$ is an energy eigenstate $|E_i\rangle$ of $\hat{H}$, then the time evolution introduces only a phase factor,

$$|E_i(t)\rangle = e^{-iE_it/\hbar}\, |E_i(0)\rangle. \tag{12.22}$$

This property will play a central role when constructing quantum statistical ensembles, where stationary states form the basis of the equilibrium density operator.

12.3 Properties of Quantum Particles, Ladder Operators, and Spin

The algebraic structure underlying many quantum systems is revealed most clearly through the use of *ladder operators* (or *creation* and *annihilation* operators). We encountered this formalism earlier in the quantization of the harmonic oscillator, where the energy spectrum emerged from the repeated action of a raising operator on the ground state. In this section we review this operator method and extend it to angular momentum and spin systems—both of which will play a central role in describing quantum gases. A detailed and elegant treatment of the ladder-operator formalism and angular momentum algebra can be found in Sakurai's classic text on quantum mechanics [SN94].

Ladder-Operator Formalism and the Harmonic Oscillator

For a particle of mass m attached to a linear restoring force of constant k, the Hamiltonian is

$$H = \frac{p^2}{2m} + \frac{1}{2}kx^2 = \frac{p^2}{2m} + \frac{1}{2}m\omega^2 x^2, \tag{12.23}$$

where $\omega = \sqrt{k/m}$ is the oscillator frequency. Hamilton's equations give

$$\begin{aligned} \dot{x} &= \frac{\partial H}{\partial p} = \frac{p}{m}, \\ \dot{p} &= -\frac{\partial H}{\partial x} = -m\omega^2 x, \end{aligned} \tag{12.24}$$

which yield the familiar sinusoidal solutions

$$x(t) = A\sin(\omega t + \delta), \qquad p(t) = Am\omega\cos(\omega t + \delta).$$

The total energy $E = (1/2)m\omega^2 A^2$ remains constant during the motion.

Canonical Quantization

Quantization proceeds by promoting x and p to operators obeying the canonical commutation relation

$$[\hat{x}, \hat{p}] = i\hbar. \tag{12.25}$$

In the position representation, $\hat{x}\psi(x) = x\psi(x)$ and $\hat{p}\psi(x) = -i\hbar\,\partial_x\psi(x)$, so that the Hamiltonian operator becomes

$$\hat{\mathcal{H}} = \frac{\hat{p}^2}{2m} + \frac{1}{2}m\omega^2\hat{x}^2 = -\frac{\hbar^2}{2m}\frac{d^2}{dx^2} + \frac{1}{2}m\omega^2 x^2. \tag{12.26}$$

The stationary Schrödinger equation,

$$\hat{\mathcal{H}}\phi_n(x) = E_n\phi_n(x), \tag{12.27}$$

has well-known Gaussian–Hermite polynomial solutions. The spectrum can also be derived more elegantly through the algebraic (Dirac) method.

We recall the solution of the harmonic oscillator. We introduce dimensionless position and momentum operators,

$$\hat{Q} = \sqrt{\frac{m\omega}{\hbar}}\,\hat{x}, \qquad \hat{P} = \frac{\hat{p}}{\sqrt{m\hbar\omega}}, \tag{12.28}$$

which satisfy $[\hat{Q}, \hat{P}] = i$. In these variables, the Hamiltonian reads

$$\hat{H} = \frac{\hbar\omega}{2}(\hat{P}^2 + \hat{Q}^2). \tag{12.29}$$

For a particle of mass m in a harmonic potential $V(x) = \frac{1}{2}kx^2 = \frac{1}{2}m\omega^2x^2$, the Hamiltonian operator is

$$\hat{H} = \frac{\hat{p}^2}{2m} + \frac{1}{2}m\omega^2\hat{x}^2, \qquad [\hat{x}, \hat{p}] = i\hbar. \tag{12.30}$$

To solve the eigenvalue problem $\hat{H}\psi_n = E_n\psi_n$, we introduce the dimensionless operators

$$\hat{a} = \frac{1}{\sqrt{2\hbar m\omega}}(m\omega\hat{x} + i\hat{p}), \qquad \hat{a}^\dagger = \frac{1}{\sqrt{2\hbar m\omega}}(m\omega\hat{x} - i\hat{p}), \tag{12.31}$$

which satisfy the canonical commutation relation $[\hat{a}, \hat{a}^\dagger] = 1$. These are bosonic ladder operators: $\hat{a}^\dagger$ raises and $\hat{a}$ lowers the energy of a quantum state by one quantum $\hbar\omega$.

In terms of these operators, the Hamiltonian takes the compact form

$$\hat{H} = \hbar\omega\left(\hat{a}^\dagger\hat{a} + \frac{1}{2}\right) = \hbar\omega\left(\hat{N} + \frac{1}{2}\right), \qquad \hat{N} = \hat{a}^\dagger\hat{a}. \tag{12.32}$$

The operator $\hat{N}$ is called the *number operator*; its eigenvalues $n = 0, 1, 2, \ldots$ correspond to the excitation number of the oscillator. Acting on the normalized eigenstates $|n\rangle$,

$$\begin{aligned} \hat{a}^\dagger|n\rangle &= \sqrt{n+1}\,|n+1\rangle, \\ \hat{a}|n\rangle &= \sqrt{n}\,|n-1\rangle, \\ \hat{N}|n\rangle &= n|n\rangle, \end{aligned} \tag{12.33}$$

so that $\hat{H}|n\rangle = \hbar\omega(n+\frac{1}{2})|n\rangle$. The ground state $|0\rangle$ satisfies $\hat{a}|0\rangle = 0$ and has energy $E_0 = \frac{1}{2}\hbar\omega$. All higher eigenstates are generated algebraically:

$$|n\rangle = \frac{(\hat{a}^\dagger)^n}{\sqrt{n!}}|0\rangle. \tag{12.34}$$

This ladder-operator formalism is not only elegant but also universal: the same algebraic structures reappear in the quantization of fields, many-body systems, and angular momentum.

Angular Momentum and Ladder Operators

Classically, the angular momentum vector is $\vec{L} = \vec{r} \times \vec{p}$. In quantum mechanics we promote these components to operators,

$$\hat{L}_x = \hat{y}\hat{p}_z - \hat{z}\hat{p}_y, \qquad \hat{L}_y = \hat{z}\hat{p}_x - \hat{x}\hat{p}_z, \qquad \hat{L}_z = \hat{x}\hat{p}_y - \hat{y}\hat{p}_x,$$

which satisfy the commutation relations

$$[\hat{L}_x, \hat{L}_y] = i\hbar\hat{L}_z, \quad [\hat{L}_y, \hat{L}_z] = i\hbar\hat{L}_x, \quad [\hat{L}_z, \hat{L}_x] = i\hbar\hat{L}_y. \tag{12.35}$$

These relations define the $\mathfrak{su}(2)$ Lie algebra. Since the components of $\hat{\vec{L}}$ do not commute, they cannot be simultaneously diagonalized. However, the total angular momentum

$$\hat{L}^2 = \hat{L}_x^2 + \hat{L}_y^2 + \hat{L}_z^2, \tag{12.36}$$

commutes with each component, $[\hat{L}^2, \hat{L}_i] = 0$, and thus the simultaneous eigenstates of $\hat{L}^2$ and $\hat{L}_z$,

$$\hat{L}^2|l,m\rangle = \hbar^2 l(l+1)|l,m\rangle, \qquad \hat{L}_z|l,m\rangle = \hbar m|l,m\rangle,$$

form a complete basis with integers or half-integers $l \geq 0$ and $m = -l, \ldots, l$.

To move between states with different m, we define the ladder operators

$$\hat{L}_\pm = \hat{L}_x \pm i\hat{L}_y, \tag{12.37}$$

which satisfy

$$\hat{L}_\pm|l,m\rangle = \hbar\sqrt{l(l+1) - m(m\pm 1)}\,|l,m\pm 1\rangle. \tag{12.38}$$

These operators raise or lower the magnetic quantum number m while keeping l fixed, closely paralleling the algebra of the harmonic oscillator.

Spin and Intrinsic Angular Momentum

We also wish to recall the formalism of the spin operators. Experiments by Otto Stern and Walther Gerlach (1922) revealed that electrons possess an intrinsic form of angular momentum, *spin*, which is quantized even in the absence of orbital motion [SG22]. Unlike orbital angular momentum, spin does not originate from spatial coordinates and momenta; it is an intrinsic property of particles [Edm57].

The spin operators $\hat{S}_x, \hat{S}_y, \hat{S}_z$ satisfy the same algebra as $\hat{L}_i$:

$$[\hat{S}_x, \hat{S}_y] = i\hbar\hat{S}_z, \qquad [\hat{S}_y, \hat{S}_z] = i\hbar\hat{S}_x, \qquad [\hat{S}_z, \hat{S}_x] = i\hbar\hat{S}_y. \tag{12.39}$$

The total spin operator $\hat{S}^2 = \hat{S}_x^2 + \hat{S}_y^2 + \hat{S}_z^2$ has eigenvalues $\hbar^2 s(s+1)$, where s is the spin quantum number. For the electron, $s = \frac{1}{2}$, giving two possible spin projections $s_z = \pm\frac{1}{2}$.

We define spin states $|s, m_s\rangle$, often written simply as $|+\rangle$ and $|-\rangle$:

$$\hat{S}_z|+\rangle = \tfrac{\hbar}{2}|+\rangle, \qquad \hat{S}_z|-\rangle = -\tfrac{\hbar}{2}|-\rangle, \tag{12.40}$$

$$\hat{S}^+|-\rangle = \hbar|+\rangle, \qquad \hat{S}^-|+\rangle = \hbar|-\rangle, \tag{12.41}$$

with the ladder operators $\hat{S}^\pm = \hat{S}_x \pm i\hat{S}_y$. The two spin states form a complete orthonormal basis:

$$\langle +|+\rangle = \langle -|-\rangle = 1, \quad \langle +|-\rangle = 0, \quad \mathbb{I} = |+\rangle\langle +| + |-\rangle\langle -|. \tag{12.42}$$

In this representation,

$$\hat{S}^+ = \hbar|+\rangle\langle -|, \quad \hat{S}^- = \hbar|-\rangle\langle +|, \quad \hat{S}_z = \tfrac{\hbar}{2}(|+\rangle\langle +| - |-\rangle\langle -|). \tag{12.43}$$

The spin operators generate the group SU(2), the Lie group of 2×2 unitary matrices with unit determinant, which describes rotations in spin space. Because SU(2) is non-Abelian, spin components do not commute, and combining two spins requires the *Clebsch–Gordan coefficients* for decomposing tensor products of spin representations [Cle66, Gor68].

Spin projection operators such as $\hat{S}_z$ correspond to measurable observables. In contrast, total spin quantum numbers such as $s = \frac{1}{2}, 1, \frac{3}{2}, \ldots$ classify the representation of SU(2) to which the particle belongs. The distinction between spin and its projection is central to understanding magnetic properties and quantum statistics.

REMARK 12.2 The ladder-operator formalism is a unifying language for describing quantized systems:

- For the harmonic oscillator, $\hat{a}^\dagger$ and $\hat{a}$ create and destroy energy quanta.
- For angular momentum and spin, $\hat{L}_\pm$ and $\hat{S}_\pm$ raise and lower the magnetic quantum number.

These algebraic structures form the foundation for quantum field theory, where creation and annihilation operators generate and annihilate particles, and for quantum statistical mechanics, where they underlie Bose and Fermi statistics in many-body systems.

12.4 Limitations of Classical Statistical Mechanics

Before turning to quantum statistical mechanics, it is useful to mention the limitations of its classical predecessor, which we just described. Although classical statistical mechanics provided deep insight into the relationship between microscopic motion and macroscopic thermodynamics, it fails in describing phenomena where energy quantization is essential.

The Heat Capacity Dilemma

The decisive impetus for the quantization of statistical mechanics came from the failure of classical theory to explain the temperature dependence of the specific heat of solids. According to classical equipartition, each quadratic degree of freedom contributes $\frac{1}{2}k_{\mathrm{B}}T$ to the mean energy, so a three-dimensional harmonic oscillator has an average energy $3k_{\mathrm{B}}T$. For a crystalline solid containing N atoms, this leads to

$$U = 3Nk_{\mathrm{B}}T, \qquad C_V = \frac{\partial U}{\partial T} = 3Nk_{\mathrm{B}} = 3R, \tag{12.44}$$

in apparent agreement with the empirical relation discovered in 1819 by Pierre-Louis Dulong and Alexis-Thérèse Petit [DP19]. This so-called *Dulong–Petit law* accurately describes the high-temperature behavior of many solids and was long regarded as a triumph of classical statistical reasoning.

However, precise calorimetric measurements soon revealed a dramatic breakdown of the law at low temperatures: the specific heat of solids decreases rapidly and tends to zero as $T \to 0$. Classical mechanics, which assumes continuous vibrational motion, offers no explanation for this behavior. If every vibrational mode could store energy continuously, the heat capacity would remain constant at all temperatures—a prediction in striking contradiction with experiment.

The resolution required a fundamental change in the microscopic description of matter. Max Planck's quantization of radiation (1900) had already introduced the concept that energy is exchanged in discrete packets of size $\hbar\omega$. Building on this idea, Albert Einstein (1907) proposed that each atom in a crystal behaves as an independent quantum harmonic oscillator with quantized energy levels

$$E_n = \left(n + \tfrac{1}{2}\right)\hbar\omega. \tag{12.45}$$

This assumption replaces the classical equipartition result by a *Bose–Einstein occupation factor* that suppresses excitation of high-frequency modes when $k_{\mathrm{B}}T \ll \hbar\omega$. As a consequence, the heat capacity decreases exponentially at low temperatures, in qualitative agreement with experimental observations.

A few years later, Peter Debye (1912) refined the model by recognizing that the atoms in a crystal do not all vibrate with the same frequency. Treating the solid as a continuum elastic medium, he introduced a *density of vibrational modes* up to a maximum cutoff frequency ω_D, the *Debye frequency*. The resulting expression for the heat capacity,

$$C_V = 9R\left(\frac{T}{\Theta_D}\right)^3 \int_0^{\Theta_D/T} \frac{x^4 e^x}{(e^x - 1)^2}\, dx, \tag{12.46}$$

where $\Theta_D = \hbar\omega_D/k_{\mathrm{B}}$ is the *Debye temperature*, yields the correct limiting behaviors,

$$C_V \to 3R \quad (T \gg \Theta_D), \qquad C_V \propto T^3 \quad (T \ll \Theta_D), \tag{12.47}$$

and reproduces experimental results with remarkable accuracy across all temperatures.

REMARK 12.3 The heat capacity dilemma marks the conceptual birth of *quantum statistical mechanics.* Classical equipartition, based on continuous phase-space motion, fails once the thermal energy $k_{\mathrm{B}}T$ becomes comparable to the discrete spacing of energy levels $\hbar\omega$. Quantization of lattice vibrations—*phonons*—resolved this inconsistency and unified the thermodynamics of solids with the emerging quantum theory of energy. The Einstein and Debye models together established that discreteness of microscopic states is not confined to radiation but is a universal property of matter.

From Temperature to Entropy: The Need for a Quantum Reformulation

The failure of classical statistical mechanics to describe the heat capacities of solids, the spectrum of black-body radiation, and similar phenomena points to a deeper issue. Temperature, as Boltzmann understood, is not a fundamental microscopic variable but an *emergent* quantity arising from the collective behavior of many degrees of freedom. In thermodynamics it is defined operationally, while in statistical mechanics it emerges as a parameter connecting microscopic disorder to macroscopic equilibrium.

In classical statistical mechanics, the canonical distribution

$$p(E) \propto e^{-\beta H}, \qquad \beta = \frac{1}{k_{\mathrm{B}}T}, \tag{12.48}$$

assigns probabilities to energy levels, but it hides the microscopic origin of temperature. The variable T enters as an external parameter controlling the exponential weighting of phase-space configurations, rather than as a dynamical property derived from the system's microstates. Entropy, in contrast, is the truly fundamental quantity. Boltzmann's celebrated formula,

$$S = k_{\mathrm{B}} \ln N, \tag{12.49}$$

relates entropy directly to the number N of accessible microstates and remains valid, in essence, even after the advent of quantum mechanics. What changes between the classical and quantum worlds is not the definition of entropy itself, but how we *count* microstates.

In the classical picture, the counting is continuous: phase space is divided into cells of volume h^f for a system with f degrees of freedom. The constant h appears as a normalization with no intrinsic meaning—it is introduced only to render the argument of the logarithm dimensionless. Yet its numerical value must be fixed empirically to match observations. Planck recognized that this constant carries profound physical significance: by introducing the same h into the calculation of the black-body spectrum, he found exact agreement with experiment. The "quantum of action" was thus promoted from a mere dimensional parameter to a fundamental constant of nature, defining the elementary volume of phase space that corresponds to one quantum state.

This recognition led to the modern formulation of quantum statistical mechanics, where the discreteness of phase-space cells—and thus of energy levels—arises naturally from the structure of quantum theory itself. The statistical formalism that follows is therefore not a modification of thermodynamics, but its necessary completion at the microscopic level. The present chapter develops the standard operator and density-matrix framework; in the second Volume, we will return to the dynamical foundations that *attempts* to justify it.

12.5 Quantum Statistical Mechanics via Density Matrices

In quantum statistical mechanics, the notion of a *state* must be extended beyond a single wavefunction. Ordinary quantum mechanics describes isolated systems through a state vector that evolves unitarily under the Schrödinger equation. However, in statistical mechanics, we deal with ensembles of systems—or subsystems interacting weakly with large environments—where a probabilistic description is unavoidable. In these situations, the appropriate formal object is the *density operator* (or *density matrix*), which captures both quantum superposition and statistical uncertainty.

For an isolated system with Hamiltonian $\hat{H}$, any state can be expanded in an orthonormal energy basis:

$$|\psi(t)\rangle = \sum_j c_j(t)\,|\phi_j\rangle, \qquad \hat{H}|\phi_j\rangle = E_j|\phi_j\rangle, \tag{12.50}$$

with the time evolution of the coefficients given by the Schrödinger equation $i\hbar\,\dot{c}_j = E_j c_j$. If the system interacts with an external environment, the Hilbert space becomes a tensor product:

$$\mathcal{S} = \mathcal{R} \otimes \mathcal{B}, \tag{12.51}$$

where $\mathcal{R}$ represents the subsystem of interest and $\mathcal{B}$ the environment (or bath). A general pure state of the full system is then

$$|\Psi\rangle = \sum_{r,b} c_{rb}\,|\psi_r\rangle_{\mathcal{R}} \otimes |\psi_b\rangle_{\mathcal{B}}, \qquad \sum_{r,b} |c_{rb}|^2 = 1. \tag{12.52}$$

When we are interested only in observables acting on the subsystem, we trace out the bath degrees of freedom. For an operator $\hat{Q} = \hat{Q}_{\mathcal{R}} \otimes \mathbb{I}_{\mathcal{B}}$,

$$\begin{aligned} \langle \hat{Q} \rangle &= \langle \Psi | (\hat{Q}_{\mathcal{R}} \otimes \mathbb{I}_{\mathcal{B}}) | \Psi \rangle \\ &= \sum_{r,r',b} c_{rb}^* c_{r'b}\, \langle \psi_r | \hat{Q}_{\mathcal{R}} | \psi_{r'} \rangle. \end{aligned} \tag{12.53}$$

This motivates the definition of the *reduced density operator*:

$$\hat{\rho}_{\mathcal{R}} = \sum_{r,r',b} c_{rb} c_{r'b}^*\, |\psi_r\rangle\langle\psi_{r'}| = \mathrm{Tr}_{\mathcal{B}}(|\Psi\rangle\langle\Psi|)\,, \tag{12.54}$$

so that expectation values take the compact form

$$\langle \hat{Q} \rangle = \mathrm{Tr}_{\mathcal{R}}\left(\hat{\rho}_{\mathcal{R}} \hat{Q}_{\mathcal{R}}\right). \tag{12.55}$$

The reduced density operator encodes all measurable information about the subsystem, even though the total system may remain in a pure state.

The density operator evolves unitarily under the Hamiltonian dynamics of the full system. For an isolated system governed by $\hat{H}$,

$$i\hbar \frac{d\hat{\rho}}{dt} = [\hat{H}, \hat{\rho}], \tag{12.56}$$

with $\hat{\rho}(0)$ as the initial density matrix. This is the quantum analog of Liouville's theorem in classical mechanics, expressing the conservation of total probability in Hilbert space.

The formal solution is

$$\hat{\rho}(t) = \hat{U}(t)\,\hat{\rho}(0)\,\hat{U}^\dagger(t), \qquad \hat{U}(t) = e^{-i\hat{H}t/\hbar}. \tag{12.57}$$

Because the trace is invariant under cyclic permutations,

$$\mathrm{Tr}\big(\hat{U}^\dagger \hat{A} \hat{U}\big) = \mathrm{Tr}(\hat{A}), \tag{12.58}$$

the normalization of $\hat{\rho}$ is preserved:

$$\mathrm{Tr}[\hat{\rho}(t)] = \mathrm{Tr}[\hat{\rho}(0)] = 1. \tag{12.59}$$

Expectation values can be equivalently written in either the Schrödinger or the Heisenberg picture:

$$\langle \hat{Q}(t) \rangle = \mathrm{Tr}\big(\hat{\rho}(t)\hat{Q}(0)\big) = \mathrm{Tr}\big(\hat{\rho}(0)\hat{Q}(t)\big)\,, \tag{12.60}$$

where the observable in the Heisenberg picture evolves as $\hat{Q}(t) = e^{i\hat{H}t/\hbar}\hat{Q}(0)e^{-i\hat{H}t/\hbar}$. This duality underscores that only the relative time dependence between $\hat{\rho}$ and observables carries physical significance.

A density operator can represent both pure and mixed ensembles. In a general statistical mixture of states $|\phi_j\rangle$ with probabilities λ_j,

$$\hat{\rho} = \sum_j \lambda_j \, |\phi_j\rangle\langle\phi_j|, \qquad \sum_j \lambda_j = 1. \tag{12.61}$$

It satisfies the properties

$$\hat{\rho} = \hat{\rho}^\dagger, \qquad \mathrm{Tr}(\hat{\rho}) = 1, \qquad \langle\psi|\hat{\rho}|\psi\rangle \geq 0, \tag{12.62}$$

where † represents the hermitian conjugate. If only one λ_j is nonzero, the ensemble is *pure* and $\hat{\rho}^2 = \hat{\rho}$; otherwise, it is *mixed* and $\mathrm{Tr}(\hat{\rho}^2) < 1$.

Under a unitary change of basis $\hat{U}$, $\hat{\rho}' = \hat{U}\hat{\rho}\hat{U}^\dagger$, but since the trace is basis independent, the expectation value $\langle\hat{A}\rangle = \mathrm{Tr}(\hat{\rho}\hat{A})$ is invariant. In the basis that diagonalizes $\hat{\rho}$, the eigenvalues represent occupation probabilities of the corresponding quantum states.

The Random-Phase Postulate and the Loss of Coherence

In quantum statistical mechanics we face an apparent paradox. Microscopic quantum evolution is strictly unitary and reversible, yet macroscopic systems exhibit irreversible behavior and approach thermal equilibrium. Where does this loss of memory come from? Why can we replace coherent superpositions by classical probability mixtures? The *random-phase postulate* provides the minimal physical assumption that resolves this tension. It formalizes the idea that, for systems interacting weakly with large and complex environments, phase information is effectively lost on experimentally relevant timescales.

From Coherence to Dephasing

Consider a subsystem $\mathcal{R}$ interacting with its environment $\mathcal{B}$, described by the total pure state

$$|\Psi\rangle = \sum_{r,b} c_{rb} \, |r\rangle_{\mathcal{R}} \otimes |b\rangle_{\mathcal{B}}, \qquad \sum_{r,b} |c_{rb}|^2 = 1. \tag{12.63}$$

The coefficients c_{rb} encode both the amplitudes and the *relative phases* between the components of the wavefunction. For a time-independent Hamiltonian

$$\hat{H} = \hat{H}_{\mathcal{R}} + \hat{H}_{\mathcal{B}} + \hat{V}, \tag{12.64}$$

these coefficients evolve as

$$c_{rb}(t) = c_{rb}(0) e^{-iE_{rb}t/\hbar}, \tag{12.65}$$

where E_{rb} are energy eigenvalues of the full system. Expectation values of observables contain oscillatory cross-terms of the form $e^{-i(E_{rb}-E_{r'b'})t/\hbar}$, which represent quantum interference between distinct components of the state. If nothing suppresses these oscillations, the subsystem never settles to stationary behavior, and time averages do not converge—contradicting the empirical stability of thermodynamic equilibrium.

Two generic mechanisms eliminate the effect of these cross-terms:

(i) Dephasing by frequency dispersion. When the system's spectrum is dense and nondegenerate, the energy gaps $E_{rb} - E_{r'b'}$ are incommensurate. For observation times $t \gg \tau_\varphi \sim \hbar/\Delta E$, where ΔE is a typical spacing, the oscillatory phases become effectively uniformly distributed in $[0, 2\pi)$, and their time average vanishes (Riemann–Lebesgue lemma intuition). Thus,

$$\overline{c_{rb}(t) c^*_{r'b'}(t)} \to 0 \quad \text{for } (r,b) \neq (r',b'). \tag{12.66}$$

(ii) Environmental decoherence. When the subsystem is entangled with many degrees of freedom in its environment through the coupling $\hat{V}$, the off-diagonal elements of its reduced

density matrix, $\rho_{\mathcal{R}}(r,r') = \sum_b c_{rb}c^*_{r'b}$, decay on a characteristic *decoherence time* τ_{D}. This timescale is often many orders of magnitude shorter than relaxation times [Zur03, JZK$^+$03]. In the Markovian limit, the same process is captured by the Lindblad master equation, which describes exponential damping of coherences [Lin76, GKS76].

Pauli's Random-Phase Postulate

On these physical grounds, Pauli proposed in 1928 the *random-phase postulate* [Pau28]: after coarse-graining in time or over environmental degrees of freedom,

$$\overline{c_{rb}c^*_{r'b'}} = 0 \quad \text{for } (r,b) \neq (r',b'), \tag{12.67}$$

so that only diagonal terms contribute to statistical averages. Intuitively, microscopic phase relations fluctuate too rapidly to be observable in macroscopic measurements, and are therefore replaced by effective randomness. As a consequence, any observable acting solely on the subsystem, $\hat{Q} = \hat{Q}_{\mathcal{R}} \otimes \mathbb{I}_{\mathcal{B}}$, yields

$$\begin{aligned} \langle \hat{Q} \rangle = \langle \Psi | (\hat{Q}_{\mathcal{R}} \otimes \mathbb{I}) | \Psi \rangle &= \sum_{r,r',b} c^*_{rb}c_{r'b} \langle r|\hat{Q}_{\mathcal{R}}|r'\rangle \\ \xrightarrow{\text{random phases}} &\sum_r \Big(\sum_b |c_{rb}|^2 \Big) \langle r|\hat{Q}_{\mathcal{R}}|r\rangle = \sum_r \lambda_r \langle r|\hat{Q}_{\mathcal{R}}|r\rangle, \end{aligned} \tag{12.68}$$

where the classical probabilities $\lambda_r = \sum_b |c_{rb}|^2$ satisfy $\sum_r \lambda_r = 1$. The reduced density operator becomes effectively diagonal:

$$\hat{\rho}_{\mathcal{R}} \approx \sum_r \lambda_r |r\rangle\langle r|, \qquad \langle \hat{Q} \rangle = \mathrm{Tr}_{\mathcal{R}}(\hat{\rho}_{\mathcal{R}}\hat{Q}_{\mathcal{R}}). \tag{12.69}$$

This marks the transition from a quantum coherent description to a statistical mixture.

Choice of Basis and the Onset of Classicality

If the coupling $\hat{V}$ is weak and the bath is broadband on the scale of level spacings, the energy eigenbasis of $\hat{H}_{\mathcal{R}}$ is typically the natural diagonal basis, giving rise to the so-called *diagonal ensemble* [vN29, Deu91, Sre94]. For stronger or structured coupling, decoherence selects a *pointer basis*—the eigenstates of the subsystem operators that couple most directly to the environment (for instance, position under position-coupled noise) [Zur03, JZK$^+$03]. In realistic systems one often finds block-diagonal structures corresponding to degenerate subspaces or conserved quantities, within which the same reasoning applies.

The random-phase approximation is justified when (i) the dephasing or decoherence time is much shorter than both the relaxation and observation times, and (ii) recurrences are negligible within the relevant timeframe (i.e., for systems with very large Hilbert spaces). Under these conditions, interference terms average to zero and only the populations of stationary states contribute to observables, producing effectively classical probabilities.

When Is the Random-Phase Approximation Justified?

- *Large environment and weak coupling:* rapid entanglement and dephasing; $\tau_{\mathrm{D}} \ll \tau_{\mathrm{R}}$ (relaxation).
- *Nondegenerate spectra:* incommensurate energy gaps cause dephasing; degeneracies require block-diagonal treatment.
- *Coarse graining:* averaging over time windows $\Delta t \gg \tau_\varphi$ suppresses oscillatory terms.

- *Markovian noise:* Lindblad dynamics exponentially damps coherences in the decoherence basis.

Caveats: small or integrable systems can exhibit revivals; near-integrable or many-body localized systems may retain long-lived coherences; strong memory effects (non-Markovian baths) can reintroduce phase correlations.

Once the reduced density matrix has become diagonal—or block-diagonal—in the relevant basis, statistical mechanics determines the weights $\{\lambda_r\}$ through macroscopic constraints. Maximizing entropy subject to fixed normalization and energy yields the canonical form

$$\hat{\rho}_\beta = \frac{e^{-\beta \hat{H}_{\mathcal{R}}}}{Z}, \qquad Z = \operatorname{Tr} e^{-\beta \hat{H}_{\mathcal{R}}}. \tag{12.70}$$

In isolated but chaotic quantum systems, the same conclusion follows from the *eigenstate thermalization hypothesis* (ETH): matrix elements of few-body observables become effectively diagonal in the energy basis in the thermodynamic limit [Deu91, Sre94]. Thus, Pauli's statistical postulate, born as a pragmatic assumption, finds modern justification in the physics of dephasing, decoherence, and quantum chaos.

12.6 Quantum Ensembles

Having introduced the density operator as the central object of quantum statistical mechanics, we now formulate the quantum analogues of the classical ensembles. All statistical averages of physical quantities are expressed as traces of observables weighted by the density matrix. The coefficients λ_j introduced earlier represent the probabilities with which the system occupies the eigenstates of its Hamiltonian—or more generally, the eigenstates of the stationary density operator $\hat{\rho}$.

From Classical to Quantum Averaging

In the quantum formalism, the expectation value of an observable $\hat{Q}$ is written as

$$\langle \hat{Q} \rangle = \operatorname{Tr}(\hat{\rho}\, \hat{Q}), \tag{12.71}$$

which generalizes the classical ensemble average $\langle A \rangle = \frac{1}{N! h^{3N}} \int A(\Gamma)\, p(\Gamma)\, d\Gamma$. The replacement

$$\frac{1}{N!\, h^{3N}} \int d\Gamma_N(\,\cdot\,) \quad \longrightarrow \quad \operatorname{Tr}(\hat{\rho}\,\cdot\,) \tag{12.72}$$

summarizes the transition from classical to quantum statistical mechanics: the continuous integration over phase space becomes a discrete trace over Hilbert-space states, and indistinguishability is accounted for by the factor $1/N!$.

The trace operation is invariant under unitary transformations and under cyclic permutations of operators. For any invertible operator C,

$$\operatorname{Tr}(C\, \hat{\rho}\, C^{-1}) = \operatorname{Tr}(\hat{\rho}), \qquad \operatorname{Tr}(ABC) = \operatorname{Tr}(BCA) = \operatorname{Tr}(CAB). \tag{12.73}$$

Hence, ensemble averages are representation independent:

$$\operatorname{Tr}(\hat{\rho}\, \hat{B}_1 \hat{B}_2 \cdots \hat{B}_n) = \operatorname{Tr}(\hat{\rho}_C\, \hat{B}_{1,C} \hat{B}_{2,C} \cdots \hat{B}_{n,C}), \tag{12.74}$$

where the subscript C denotes quantities in any convenient basis—whether the eigenbasis of $\hat{\mathcal{H}}$, of $\hat{\rho}$, or of a set of commuting observables. This invariance parallels the coordinate independence of phase-space integrals in classical mechanics.

Quantum ensembles fall naturally into two categories—*pure* and *mixed*—depending on whether they can be represented by a single state vector or by a statistical mixture of several. A pure ensemble consists of identically prepared systems, each in the same quantum state $|\psi\rangle$, so that

$$\hat{\rho}_{\text{pure}} = |\psi\rangle\langle\psi|. \tag{12.75}$$

In a mixed ensemble, different states $|\psi_i\rangle$ occur with probabilities P_i, leading to

$$\hat{\rho}_{\text{mixed}} = \sum_i P_i \, |\psi_i\rangle\langle\psi_i|, \qquad \sum_i P_i = 1, \tag{12.76}$$

as derived in the formal definition of the density operator.

The distinction between the two is quantitative and basis independent. Consider the trace of $\hat{\rho}^2$:

$$\mathrm{Tr}(\hat{\rho}^2) = 1 \quad \text{for a pure ensemble}, \qquad \mathrm{Tr}(\hat{\rho}^2) < 1 \quad \text{for a mixed ensemble}. \tag{12.77}$$

The quantity

$$\mathcal{P} = \mathrm{Tr}(\hat{\rho}^2) \tag{12.78}$$

is called the *purity* of the ensemble and ranges between 1 (complete coherence) and $1/d$ (maximal mixing for a system of Hilbert-space dimension d). It provides a direct measure of how much information about the quantum state is lost due to mixing. Purity remains invariant under unitary transformations, making it an intrinsic property of the ensemble itself.

REMARK 12.4 In any representation, the diagonal elements of $\hat{\rho}$ correspond to the occupation probabilities of the basis states, while the off-diagonal elements represent quantum coherences. A pure ensemble may appear as a "superposition" in one basis and diagonal in another, but its purity $\mathrm{Tr}(\hat{\rho}^2) = 1$ is invariant. By contrast, a mixed ensemble cannot be reduced to a single state vector in any representation—it truly represents statistical ignorance rather than quantum interference. Physically, $\mathcal{P} < 1$ signals partial decoherence or entanglement with unobserved degrees of freedom.

Stationary Ensembles and Their Justification

The classical ensembles acquire their quantum counterparts through the condition of *stationarity*. In thermodynamic equilibrium, macroscopic observables are constant in time. For an observable represented by a Hermitian operator $\hat{A}$ without explicit time dependence, this requirement translates to

$$\frac{d}{dt}\langle \hat{A}\rangle = \frac{d}{dt}\mathrm{Tr}(\hat{\rho}\,\hat{A}) = 0.$$

Using the Liouville–von Neumann equation

$$i\hbar\,\frac{d\hat{\rho}}{dt} = [\hat{H}, \hat{\rho}],$$

we see that this condition is satisfied whenever

$$[\hat{H}, \hat{\rho}] = 0. \tag{12.79}$$

Hence, at equilibrium the density operator commutes with the Hamiltonian and can therefore be expressed as a function of it,

$$\hat{\rho} = f(\hat{H}), \tag{12.80}$$

where f is a real, positive, and monotonically decreasing function ensuring normalization and positivity. Different choices of f correspond to different macroscopic constraints and thus define distinct statistical ensembles.

The Micro-Canonical Ensemble

For an isolated system with fixed total energy E, all accessible eigenstates within a narrow energy band are equally probable. The corresponding density operator is

$$\hat{\rho}_{\text{micro}} = \frac{\delta(E - \hat{H})}{\Omega(E)}, \qquad \Omega(E) = \operatorname{Tr} \delta(E - \hat{H}), \tag{12.81}$$

which satisfies Eq. (12.79) trivially because it depends solely on $\hat{H}$.

The Canonical Ensemble

When the system can exchange energy (but not particles) with a thermal reservoir at temperature T, the average energy is fixed while the microscopic energy fluctuates. The maximum-entropy principle subject to the constraints $\operatorname{Tr} \hat{\rho} = 1$ and $\operatorname{Tr}(\hat{\rho}\hat{H}) = U$ yields

$$\hat{\rho}_{\text{can}} = \frac{e^{-\beta \hat{H}}}{Z}, \qquad Z = \operatorname{Tr} e^{-\beta \hat{H}}. \tag{12.82}$$

Since $e^{-\beta \hat{H}}$ is an explicit function of $\hat{H}$, it also satisfies $[\hat{H}, \hat{\rho}] = 0$ and is therefore stationary.

The Grand-Canonical Ensemble

If the system exchanges both energy and particles with a reservoir, the additional constraint on the mean particle number $\langle \hat{N} \rangle$ leads to

$$\hat{\rho}_{\text{grand}} = \frac{e^{-\beta(\hat{H} - \mu \hat{N})}}{\mathcal{G}}, \qquad \mathcal{G} = \operatorname{Tr}(\, e^{-\beta(\hat{H} - \mu \hat{N})}), \tag{12.83}$$

which again commutes with $\hat{H}$ and is thus stationary.

Canonical Ensemble for a Spin-$\frac{1}{2}$ Particle

Consider a spin-$\frac{1}{2}$ magnetic moment $\boldsymbol{\mu}$ in a uniform magnetic field $B\hat{z}$, described by the Hamiltonian

$$\hat{H} = -\mu B \, \hat{\sigma}_z,$$

where $\hat{\sigma}_z$ is the Pauli operator. The canonical density operator is

$$\hat{\rho} = \frac{e^{-\beta \hat{H}}}{Z} = \frac{1}{2}\Big(\mathbb{I} + P_z \, \hat{\sigma}_z\Big), \qquad P_z = \tanh(\beta \mu B),$$

with partition function $Z = 2\cosh(\beta \mu B)$. The vector $\vec{P} = (0, 0, P_z)$ represents the average polarization of the ensemble: $\vec{P} = \operatorname{Tr}(\hat{\rho}\,\boldsymbol{\sigma})$.

Purity and Stationarity

For this system,

$$\operatorname{Tr}(\hat{\rho}^2) = \tfrac{1}{2}\big(1 + P_z^2\big),$$

so the ensemble is pure ($\operatorname{Tr} \hat{\rho}^2 = 1$) at zero temperature, where all spins align along the field, and maximally mixed ($\operatorname{Tr} \hat{\rho}^2 = \frac{1}{2}$) at infinite temperature, where spin orientations are equally likely. Because $\hat{\rho}$ depends only on $\hat{\sigma}_z$ and hence on $\hat{H}$, it satisfies $[\hat{H}, \hat{\rho}] = 0$: the spin ensemble is stationary.

REMARK 12.5 Stationarity is the unifying feature of all equilibrium ensembles: macroscopic observables remain constant because the density operator commutes with the Hamiltonian. The specific functional form of $\hat{\rho} = f(\hat{H})$ is determined by the physical constraints—energy conservation for the micro-canonical ensemble, mean energy for the canonical, and both mean energy and particle number for the grand canonical. Each ensemble thus corresponds to a different way of embedding a subsystem within a larger environment while maintaining the condition of equilibrium.

Quantum Micro-Canonical Ensemble

The quantum micro-canonical ensemble represents the direct translation of Boltzmann's *equal a priori probability* postulate into the quantum domain. It describes an isolated system with fixed energy, particle number, and volume, whose microscopic state can be any stationary eigenstate of the Hamiltonian compatible with those macroscopic constraints.

At equilibrium, the density operator must satisfy the stationarity condition

$$[\hat{H}, \hat{\rho}] = 0, \tag{12.84}$$

so that $\hat{\rho}$ can be expressed as a function of the Hamiltonian operator, $\hat{\rho} = f(\hat{H})$. Let $\{|\phi_n\rangle\}$ denote the complete set of energy eigenstates,

$$\hat{H}|\phi_n\rangle = E_n|\phi_n\rangle. \tag{12.85}$$

The most general stationary density operator diagonal in this basis reads

$$\hat{\rho} = \sum_n \lambda_n \, |\phi_n\rangle\langle\phi_n|. \tag{12.86}$$

The micro-canonical hypothesis now asserts that all energy eigenstates within a small window $E - \frac{\delta E}{2} \leq E_n \leq E + \frac{\delta E}{2}$ are equally probable, while all others have zero weight:

$$\lambda_n = \begin{cases} \dfrac{1}{\Gamma(E)} & \text{if } |E_n - E| \leq \frac{\delta E}{2}, \\ 0 & \text{otherwise,} \end{cases} \tag{12.87}$$

where $\Gamma(E)$ denotes the number of eigenstates in that narrow energy band. Normalization of the density matrix, $\mathrm{Tr}\,\hat{\rho} = 1$, fixes the proportionality constant as $1/\Gamma(E)$.

Thus the quantum micro-canonical density operator is

$$\hat{\rho}_{\text{micro}} = \frac{1}{\Gamma(E)} \sum_{|E_n - E| < \delta E/2} |\phi_n\rangle\langle\phi_n|. \tag{12.88}$$

All states within the shell contribute equally, embodying the quantum version of Boltzmann's equal probability principle.

In analogy with the classical formulation, the micro-canonical entropy is defined through the number of accessible states:

$$S(E) = k_{\mathrm{B}} \ln \Gamma(E), \tag{12.89}$$

where $\Gamma(E)$ counts the eigenstates in the chosen energy window. The density of states is then

$$\omega(E) = \frac{\partial \Gamma(E)}{\partial E}, \qquad \Gamma(E) = \int_{E-\delta E/2}^{E+\delta E/2} \omega(E')\, dE'. \tag{12.90}$$

This definition ensures that all classical relations between thermodynamic quantities remain valid: the temperature follows from the derivative

$$\frac{1}{T} = \left(\frac{\partial S}{\partial E}\right)_{V,N}, \tag{12.91}$$

and other potentials, such as the Helmholtz free energy $F = U - TS$, can be obtained through Legendre transforms as usual.

Thermodynamic consistency requires that the entropy be extensive. For two uncorrelated systems with Hamiltonian $\hat{\mathcal{H}} = \hat{\mathcal{H}}_1 \otimes \mathbb{I}_2 + \mathbb{I}_1 \otimes \hat{\mathcal{H}}_2$, the corresponding density matrix factorizes as

$$\hat{\rho} = \hat{\rho}_1 \otimes \hat{\rho}_2. \tag{12.92}$$

The entropy of the composite system should then satisfy

$$S(\hat{\rho}_1 \otimes \hat{\rho}_2) = S(\hat{\rho}_1) + S(\hat{\rho}_2). \tag{12.93}$$

The logarithmic form of the entropy guarantees this property, as can be verified directly.

Von Neumann Entropy: Origin, Definition, and Meaning

The proper generalization of thermodynamic entropy to quantum mechanics was first formulated by John von Neumann in 1927–1929 [vN27, vN29], building upon Boltzmann's combinatorial definition $S = k_B \ln W$ and Gibbs's ensemble formulation. In classical statistical mechanics, the Gibbs entropy

$$S_G = -k_B \int p(\Gamma) \, \ln p(\Gamma) \, d\Gamma$$

quantifies the missing information about the precise microstate Γ of the system. Von Neumann recognized that this same concept extends naturally to quantum systems if probabilities are replaced by the eigenvalues of the density operator $\hat{\rho}$, leading to

$$S(\hat{\rho}) = -k_B \operatorname{Tr}(\hat{\rho} \ln \hat{\rho}). \tag{12.94}$$

This expression defines the *von Neumann entropy*, which measures the degree of mixedness or uncertainty associated with a quantum ensemble. It reduces to the classical Gibbs form when $\hat{\rho}$ and $\hat{\mathcal{H}}$ commute and can be simultaneously diagonalized, in which case the density matrix becomes a probability distribution over discrete states. In this sense, the von Neumann entropy is both a thermodynamic quantity and a measure of quantum information—later reinterpreted by Claude Shannon (1948) as the information entropy of a probability distribution [Sha48].

Properties of the von Neumann Entropy

The von Neumann entropy enjoys several key properties that make it the unique consistent generalization of classical entropy:

- *Unitary invariance:* $S(\hat{\rho}) = S(U\hat{\rho}U^\dagger)$ for any unitary U; entropy depends only on the eigenvalues of $\hat{\rho}$.
- *Additivity:* For independent systems $\hat{\rho} = \hat{\rho}_1 \otimes \hat{\rho}_2$,

$$S(\hat{\rho}_1 \otimes \hat{\rho}_2) = S(\hat{\rho}_1) + S(\hat{\rho}_2).$$

- *Purity limits:* $S = 0$ for a pure state ($\hat{\rho}^2 = \hat{\rho}$), while S reaches its maximum value $S_{\text{max}} = k_B \ln d$ for a completely mixed state in a Hilbert space of dimension d.
- *Concavity:* Entropy is a concave functional of $\hat{\rho}$, ensuring that statistical mixing always increases entropy.

Because S depends only on the spectrum of $\hat{\rho}$, one may diagonalize $\hat{\rho}$ as $\hat{\rho} = \sum_i p_i |\psi_i\rangle\langle\psi_i|$, where p_i are its non-negative eigenvalues satisfying $\sum_i p_i = 1$. Equation (12.94) then becomes

$$S(\hat{\rho}) = -k_{\mathrm{B}} \sum_i p_i \ln p_i, \tag{12.95}$$

which coincides formally with the Shannon entropy of the probability distribution $\{p_i\}$.

Relation to the Micro-Canonical Ensemble

For the micro-canonical ensemble, all nonzero eigenvalues of $\hat{\rho}$ are equal, $p_i = 1/\Gamma(E)$, where $\Gamma(E)$ counts the number of states within the allowed energy band. Substituting this into Eq. (12.95) yields

$$S(\hat{\rho}_{\text{micro}}) = -k_{\mathrm{B}}\Gamma(E)\left(\frac{1}{\Gamma(E)} \ln \frac{1}{\Gamma(E)}\right) = k_{\mathrm{B}} \ln \Gamma(E),$$

which reproduces exactly the Boltzmann formula and confirms that the von Neumann entropy generalizes classical thermodynamic entropy to the quantum regime.

The operator logarithm in Eq. (12.94) is defined via its spectral decomposition or equivalently by a convergent series expansion:

$$\ln(\mathbb{I} + A) = \sum_{n=1}^{\infty} (-1)^{n+1} \frac{A^n}{n}, \qquad \|A\| < 1. \tag{12.96}$$

In the eigenbasis of $\hat{\rho}$, the logarithm acts elementwise on its eigenvalues. Thus, even though $\hat{\rho}$ is an operator, its entropy is a scalar quantity depending solely on its spectrum.

The von Neumann entropy provides a unified measure of both statistical and quantum uncertainty. In thermodynamics it quantifies the irretrievable loss of information about microstates; in quantum mechanics it measures the loss of coherence or the degree of entanglement between a subsystem and its environment. For an isolated pure state, $S = 0$; for a subsystem entangled with its surroundings, tracing out the environment yields a mixed state with $S > 0$. Hence, the von Neumann entropy not only reproduces classical thermodynamics but also extends it to situations where quantum correlations and entanglement dominate.

REMARK 12.6 Equation (12.93) follows directly from the multiplicative property of the tensor product under the trace:

$$\ln(\hat{\rho}_1 \otimes \hat{\rho}_2) = \ln \hat{\rho}_1 \otimes \mathbb{I} + \mathbb{I} \otimes \ln \hat{\rho}_2,$$

which implies

$$S(\hat{\rho}_1 \otimes \hat{\rho}_2) = -k_{\mathrm{B}} \operatorname{Tr}[(\hat{\rho}_1 \otimes \hat{\rho}_2) \ln(\hat{\rho}_1 \otimes \hat{\rho}_2)] = S(\hat{\rho}_1) + S(\hat{\rho}_2).$$

Thus, the von Neumann entropy preserves extensivity and reduces to Boltzmann's entropy in the micro-canonical limit. It remains the natural quantum generalization of classical statistical entropy and the foundation for all subsequent ensembles.

Entropy Production and the Second Law in Open Quantum Systems

The von Neumann entropy defined by

$$S(\hat{\rho}) = -k_{\mathrm{B}} \operatorname{Tr}(\hat{\rho} \ln \hat{\rho}) \tag{12.97}$$

provides a unified measure of disorder and information loss in quantum systems. For an isolated system evolving unitarily under the Liouville–von Neumann equation,

$$i\hbar \frac{d\hat{\rho}}{dt} = [\hat{H}, \hat{\rho}],$$

the eigenvalues of $\hat{\rho}$ are constant in time; hence $S(\hat{\rho})$ is conserved. Unitary dynamics is entropy preserving—a reversible flow on the space of density operators.

However, for a subsystem interacting with a large environment, the reduced density operator obeys an effective, non-unitary dynamics that leads to decoherence and dissipation. In this case the von Neumann entropy increases, expressing the quantum form of the second law.

When a quantum system interacts weakly with a large environment, its evolution is no longer unitary: information leaks irreversibly into the bath. If environmental correlations decay much faster than the characteristic timescales of the system—a condition known as the *Markov approximation*—the dynamics of the system's reduced density operator $\hat{\rho}(t)$ can be described by a local-in-time master equation of the *Lindblad form* [Lin76, GKS76, BP02]:

$$\frac{d\hat{\rho}}{dt} = -\frac{i}{\hbar}[\hat{\mathcal{H}}, \hat{\rho}] + \sum_{\alpha} \left(\hat{L}_\alpha \hat{\rho} \hat{L}_\alpha^\dagger - \frac{1}{2}\{\hat{L}_\alpha^\dagger \hat{L}_\alpha, \hat{\rho}\} \right). \tag{12.98}$$

Here, the first term describes the unitary part of the evolution generated by the system Hamiltonian $\hat{\mathcal{H}}$, while the second term represents the dissipative, non-unitary contribution arising from the interaction with the environment. The operators $\hat{L}_\alpha$, called *Lindblad* or *jump operators*, encode the specific channels through which the system exchanges energy, momentum, or information with its surroundings. Each term $\hat{L}_\alpha \hat{\rho} \hat{L}_\alpha^\dagger$ describes a stochastic quantum jump associated with the process labeled by α, while the anticommutator ensures overall trace preservation:

$$\frac{d}{dt}\mathrm{Tr}\,\hat{\rho}(t) = 0, \qquad \mathrm{Tr}\,\hat{\rho}(t) = 1 \ \forall t. \tag{12.99}$$

The Lindblad form is the most general generator of a *completely positive, trace-preserving (CPTP) semigroup* of dynamical maps,

$$\Phi_t : \hat{\rho}(0) \mapsto \hat{\rho}(t) = \Phi_t[\hat{\rho}(0)], \qquad \Phi_{t+s} = \Phi_t \circ \Phi_s,$$

which guarantees the physical consistency of the reduced dynamics for all initial states, even when the system is entangled with an external ancilla.

The jump operators satisfy no universal algebraic relations, as they depend on the microscopic details of the coupling to the environment. However, they are typically normalized such that $\sum_\alpha \hat{L}_\alpha^\dagger \hat{L}_\alpha$ has the dimension of a rate operator, ensuring that the second term in Eq. (12.98) preserves the positivity and normalization of $\hat{\rho}$. In specific physical situations, each $\hat{L}_\alpha$ corresponds to a measurable process: for instance, spontaneous emission, dephasing, thermal excitation, or particle loss. The Lindblad master equation therefore provides the universal description of all Markovian open quantum systems, combining unitary evolution with irreversible dissipative dynamics in a single trace-preserving framework.

REMARK 12.7 The derivation of Eq. (12.98) rests on two key approximations: (i) the *Born approximation*, assuming weak system–bath coupling so that the environment remains nearly stationary; and (ii) the *Markov approximation*, neglecting memory effects in the bath correlations. Under these conditions, the generator in Eq. (12.98) is the unique form that ensures complete positivity and physical consistency of the reduced dynamics [BP02].

To examine the entropy production, we differentiate Eq. (12.97) using the Lindblad equation. Since $\mathrm{Tr}\,\hat{\rho} = 1$, we obtain

$$\frac{dS}{dt} = -k_{\mathrm{B}}\mathrm{Tr}\left(\frac{d\hat{\rho}}{dt} \ln \hat{\rho} \right). \tag{12.100}$$

The Hamiltonian term in Eq. (12.98) gives no contribution, because $\mathrm{Tr}([\hat{H},\hat{\rho}]\ln\hat{\rho}) = 0$ by cyclicity of the trace. Only the dissipative part contributes to entropy production:

$$\frac{dS}{dt} = -k_{\mathrm{B}} \sum_{\alpha} \mathrm{Tr}\left[\left(\hat{L}_{\alpha}\hat{\rho}\hat{L}_{\alpha}^{\dagger} - \frac{1}{2}\{\hat{L}_{\alpha}^{\dagger}\hat{L}_{\alpha},\hat{\rho}\}\right)\ln\hat{\rho}\right]. \tag{12.101}$$

The proof of the non-negativity of entropy production in Eq. (12.101) rests on three foundational concepts of quantum operator theory: (i) the notion of operator inequalities, (ii) convex functions of operators, and (iii) completely positive trace-preserving (CPTP) maps, also known as quantum channels. We briefly review these here.

Let $\mathcal{B}(\mathcal{H})$ denote the set of bounded linear operators acting on a Hilbert space $\mathcal{H}$. For any Hermitian operators $A, B \in \mathcal{B}(\mathcal{H})$, we write

$$A \geq B \qquad \Longleftrightarrow \qquad \langle\psi|(A-B)|\psi\rangle \geq 0 \quad \forall\,|\psi\rangle \in \mathcal{H}. \tag{12.102}$$

The notation $A \geq 0$ means that A is a *positive semidefinite operator*. This generalizes the familiar scalar inequality $a \geq b$ to operators while preserving order relations under expectation values. An operator function $f(A)$ is said to be *monotone increasing* if $A \geq B$ implies $f(A) \geq f(B)$.

A real-valued function $f(x)$ defined on an interval $I \subset \mathbb{R}$ is *operator convex* if for all Hermitian A, B with spectra in I and for all $\lambda \in [0,1]$,

$$f(\lambda A + (1-\lambda)B) \leq \lambda f(A) + (1-\lambda)f(B) \tag{12.103}$$

in the operator sense of Eq. (12.102). Intuitively, this condition means that the graph of $f(A)$ lies below the straight line joining $f(A)$ and $f(B)$, as in the scalar case, but evaluated through the operator order. Many important functions are operator convex on the positive semidefinite cone, including $f(x) = x\ln x$ and $f(x) = -\ln x$. Because $f(x) = -x\ln x$ is operator *concave*, it plays a central role in quantum entropy inequalities.

We now describe *completely positive trace-preserving (CPTP) maps.* A linear map Φ : $\mathcal{B}(\mathcal{H}) \to \mathcal{B}(\mathcal{H})$ is *positive* if it sends positive operators to positive operators: $A \geq 0 \Rightarrow \Phi(A) \geq 0$. It is *completely positive* if this property holds even when Φ acts on one part of a larger system:

$$\Phi \otimes \mathbb{I}_n : \mathcal{B}(\mathcal{H} \otimes \mathbb{C}^n) \to \mathcal{B}(\mathcal{H} \otimes \mathbb{C}^n)$$

remains positive for all $n \geq 1$. Complete positivity ensures physical consistency when the system is entangled with an external ancilla.

Every CPTP map admits a *Kraus representation,*

$$\Phi(\hat{\rho}) = \sum_{\alpha} \hat{K}_{\alpha}\hat{\rho}\hat{K}_{\alpha}^{\dagger}, \qquad \sum_{\alpha} \hat{K}_{\alpha}^{\dagger}\hat{K}_{\alpha} = \mathbb{I}, \tag{12.104}$$

where the $\hat{K}_{\alpha}$ are called *Kraus operators.* The condition $\sum_{\alpha} \hat{K}_{\alpha}^{\dagger}\hat{K}_{\alpha} = \mathbb{I}$ guarantees that $\mathrm{Tr}\,\Phi(\hat{\rho}) = \mathrm{Tr}\,\hat{\rho}$, i.e., that probability is conserved. Physically, CPTP maps describe the most general physically allowed evolution of a quantum state: they encompass unitary dynamics, measurements, decoherence, and environmental noise. Their Stinespring dilation theorem ensures that any such map can be realized as a unitary evolution on a larger Hilbert space followed by a partial trace over an environment.

At this point, we are in the position of proving the second law. Given the operator concavity of $f(x) = -x\ln x$, one can show that

$$\mathrm{Tr}[f(\Phi(\hat{\rho}))] \geq \mathrm{Tr}[f(\hat{\rho})] \tag{12.105}$$

for any CPTP map Φ; this is the operator version of Jensen's inequality. Since the von Neumann entropy is $S(\hat{\rho}) = -k_{\mathrm{B}}\mathrm{Tr}(\hat{\rho}\ln\hat{\rho}) = k_{\mathrm{B}}\mathrm{Tr}[f(\hat{\rho})]$, inequality (12.105) implies

$$S(\Phi[\hat{\rho}]) \geq S(\hat{\rho}), \tag{12.106}$$

with equality if and only if Φ is unitary or leaves $\hat{\rho}$ invariant. This result, rigorously proven by Herbert Spohn in 1978 [Spo78], is known as *Spohn's inequality* or the *monotonicity of the von Neumann entropy* under CPTP maps. It establishes that for any quantum dynamical semigroup of the Lindblad form, the von Neumann entropy can never decrease:

$$S(\Phi_t[\hat{\rho}]) \geq S(\hat{\rho}), \qquad t \geq 0,$$

with equality only for stationary states. In physical terms, Spohn's theorem expresses the second law of thermodynamics in quantum-mechanical language: no physically admissible CPTP evolution can reduce the total entropy of an open system. Equivalently, completely positive maps—representing all realizable quantum processes—can only decrease the distinguishability of states, providing an information-theoretic foundation for irreversibility.

To make this statement explicit, consider the entropy production rate derived from the Lindblad generator Eq. (12.98):

$$\frac{dS}{dt} = -k_{\mathrm{B}} \sum_{\alpha} \mathrm{Tr}\left[\left(\hat{L}_\alpha \hat{\rho} \hat{L}_\alpha^\dagger - \tfrac{1}{2}\{\hat{L}_\alpha^\dagger \hat{L}_\alpha, \hat{\rho}\}\right) \ln \hat{\rho}\right]. \tag{12.107}$$

Each Lindblad term $\mathcal{L}_\alpha[\hat{\rho}] = \hat{L}_\alpha \hat{\rho} \hat{L}_\alpha^\dagger - \frac{1}{2}\{\hat{L}_\alpha^\dagger \hat{L}_\alpha, \hat{\rho}\}$ defines a completely positive generator that contracts the quantum relative entropy

$$D(\hat{\rho}_1 \| \hat{\rho}_2) = \mathrm{Tr}(\hat{\rho}_1(\ln \hat{\rho}_1 - \ln \hat{\rho}_2)),$$

a property known as *Uhlmann's monotonicity*. Applying this to the steady state $\hat{\rho}_{\mathrm{ss}}$ of the dynamics leads to Spohn's result:

$$-\frac{d}{dt} D(\hat{\rho}(t) \| \hat{\rho}_{\mathrm{ss}}) = \frac{1}{k_{\mathrm{B}}} \frac{dS}{dt} - \sum_{\alpha} J_\alpha(t)\, \beta_\alpha \geq 0, \tag{12.108}$$

where J_α are the heat currents associated with each environmental channel at inverse temperature β_α. In the single-bath case, this inequality reduces to

$$\frac{dS}{dt} - \frac{\dot{Q}}{T} \geq 0, \tag{12.109}$$

which is the Clausius form of the second law. Consequently,

$$\frac{dS}{dt} \geq 0, \tag{12.110}$$

for any Lindblad evolution that is completely positive and trace preserving. Equality holds only for stationary states $\hat{\rho}_{\mathrm{ss}}$ that satisfy $\mathcal{L}[\hat{\rho}_{\mathrm{ss}}] = 0$, that is, when the density operator commutes with the Hamiltonian and the dissipative flows balance exactly:

$$[\hat{\mathcal{H}}, \hat{\rho}_{\mathrm{ss}}] = 0, \qquad \sum_{\alpha} \left(\hat{L}_\alpha \hat{\rho}_{\mathrm{ss}} \hat{L}_\alpha^\dagger - \tfrac{1}{2}\{\hat{L}_\alpha^\dagger \hat{L}_\alpha, \hat{\rho}_{\mathrm{ss}}\}\right) = 0.$$

These stationary states correspond to thermal or steady-state equilibria, depending on the specific form of the dissipators.

REMARK 12.8 Equation (12.110) encapsulates the *quantum $\mathcal{H}$-theorem*: under any completely positive, trace-preserving, and Markovian evolution, the von Neumann entropy of a system can never decrease. In information-theoretic terms, open quantum dynamics is a noisy channel that irreversibly degrades information about the system's initial state. Thermodynamically, it guarantees that entropy production $\sigma(t) = k_B^{-1}\, dS/dt$ is non-negative and vanishes only at equilibrium.

Equation (12.110) formalizes the intuitive statement that coupling to an environment leads to loss of phase coherence and to a progressive mixing of states. The entropy increase reflects the irreversible flow of information from the system to its surroundings. In microscopic terms, the Lindblad operators $\hat{L}_\alpha$ destroy off-diagonal coherences in the system's density matrix, effectively converting quantum uncertainty into classical statistical uncertainty. The long-time stationary state of Eq. (12.98) corresponds to thermal equilibrium if the dissipator satisfies the Kubo–Martin–Schwinger (KMS) detailed-balance condition at temperature T.

Entropy Increase for a Damped Qubit

Consider a two-level system with Hamiltonian $\hat{H} = \frac{\hbar\omega}{2}\hat{\sigma}_z$ subject to spontaneous emission with rate γ. The Lindblad equation is

$$\frac{d\hat{\rho}}{dt} = -\frac{i}{\hbar}[\hat{H}, \hat{\rho}] + \gamma\big(\hat{\sigma}_-\hat{\rho}\hat{\sigma}_+ - \tfrac{1}{2}\{\hat{\sigma}_+\hat{\sigma}_-, \hat{\rho}\}\big)\,.$$

Starting from a pure superposition $|\psi(0)\rangle = \frac{1}{\sqrt{2}}(|0\rangle + |1\rangle)$, the off-diagonal elements of $\hat{\rho}(t)$ decay exponentially: $\rho_{01}(t) = \rho_{01}(0)e^{-(i\omega+\gamma/2)t}$. The eigenvalues of $\hat{\rho}(t)$ become increasingly unequal in time, and the entropy

$$S(t) = -k_B \sum_{i=\pm} \lambda_i(t) \ln \lambda_i(t)$$

monotonically increases from 0 (pure state) to $k_B \ln 2$ (maximally mixed state), illustrating Eq. (12.110).

REMARK 12.9 For closed systems, the Liouville–von Neumann equation conserves entropy, reflecting microscopic reversibility. For open systems, the Lindblad equation introduces a controlled breaking of unitarity that makes entropy non-decreasing, embodying the quantum-mechanical origin of irreversibility and the arrow of time. The von Neumann entropy thus provides the rigorous bridge between the reversible dynamics of quantum mechanics and the irreversible laws of thermodynamics.

Quantum Canonical Ensemble

The canonical ensemble describes a quantum system in contact with a large thermal reservoir that fixes its temperature T, while allowing energy exchange. It is the equilibrium ensemble that maximizes entropy under the constraints of normalization and fixed mean energy. The resulting distribution is the quantum analogue of the Boltzmann factor, but now formulated in operator form through the density matrix.

Following Gibbs's and von Neumann's reasoning, the canonical density operator is obtained by minimizing the Helmholtz free energy at fixed temperature,

$$F[\hat{\rho}] = \mathrm{Tr}(\hat{\rho}\,\hat{H}) + k_B T\, \mathrm{Tr}(\hat{\rho} \ln \hat{\rho}), \tag{12.111}$$

subject to the constraint $\mathrm{Tr}(\hat{\rho}) = 1$. This functional combines the mean energy with the von Neumann entropy and reduces, for diagonal $\hat{\rho}$, to the classical expression $F = U - TS$. Its stationary point under variations $\hat{\rho} \to \hat{\rho} + \epsilon\hat{\eta}$, with $\mathrm{Tr}(\hat{\eta}) = 0$, determines the equilibrium density matrix.

We compute the variation

$$\delta F = \mathrm{Tr}\big(\hat{\eta}\,(\hat{H} + k_{\mathrm{B}}T(\ln\hat{\rho} + \mathbb{I}))\big)\,. \tag{12.112}$$

Setting $\delta F = 0$ for all Hermitian perturbations $\hat{\eta}$ gives the condition

$$\ln\hat{\rho} = -\beta\hat{H} + \alpha\mathbb{I}, \tag{12.113}$$

where the Lagrange multiplier α enforces normalization. Exponentiating both sides yields

$$\hat{\rho}_{\mathrm{can}} = e^{\alpha}\, e^{-\beta\hat{H}} = \frac{e^{-\beta\hat{H}}}{Z}, \tag{12.114}$$

where the partition function Z ensures $\mathrm{Tr}\,\hat{\rho}_{\mathrm{can}} = 1$,

$$Z = \mathrm{Tr}\, e^{-\beta\hat{H}}. \tag{12.115}$$

This is the quantum generalization of the Gibbs distribution. Because $e^{-\beta\hat{H}}$ is a function of $\hat{H}$, the ensemble automatically satisfies $[\hat{H}, \hat{\rho}] = 0$ and is stationary.

All thermodynamic quantities follow directly from the partition function:

$$U = \langle\hat{H}\rangle = \mathrm{Tr}(\hat{\rho}_{\mathrm{can}}\hat{H}) = -\frac{\partial\ln Z}{\partial\beta}, \tag{12.116}$$

$$F = -k_{\mathrm{B}}T\ln Z, \tag{12.117}$$

$$S = -\frac{\partial F}{\partial T} = k_{\mathrm{B}}\big(\beta U + \ln Z\big). \tag{12.118}$$

These relations are identical in form to those of classical statistical mechanics, but their meaning is intrinsically quantum: expectation values are computed through operator traces rather than phase-space integrals.

Both the partition function and expectation values can be expressed in terms of eigenvalues. Let $\{\epsilon_n, |n\rangle\}$ be the eigenvalues and eigenvectors of the Hamiltonian, $\hat{H}|n\rangle = \epsilon_n|n\rangle$. In this basis, Eq. (12.115) becomes

$$Z = \sum_n e^{-\beta\epsilon_n}, \tag{12.119}$$

and the ensemble average of any observable reduces to

$$\langle\hat{Q}\rangle = \frac{\sum_n e^{-\beta\epsilon_n}\langle n|\hat{Q}|n\rangle}{\sum_n e^{-\beta\epsilon_n}}. \tag{12.120}$$

Hence, the classical Boltzmann weighting appears as the diagonal form of the operator expression when the Hamiltonian is diagonalized.

In the classical ensemble, the probability density in phase space is $p(\Gamma) = e^{-\beta H(\Gamma)}/Z$, with

$$Z = \frac{1}{N!h^{3N}}\int e^{-\beta H(\Gamma)}\, d\Gamma.$$

In the quantum ensemble, the integral is replaced by a trace over discrete states, and h^{3N} is replaced by the intrinsic quantization of phase space. The difference is both conceptual and structural:

- *Continuity vs. discreteness:* In quantum mechanics, the energy spectrum is discrete (or quantized), and the trace sums over eigenstates rather than integrating over a continuum.
- *Basis invariance:* The operator form of $Z = \mathrm{Tr}\, e^{-\beta \hat{H}}$ is invariant under any unitary transformation, making it manifestly representation independent.
- *Indistinguishability and statistics:* Quantum particles obey Bose–Einstein or Fermi–Dirac statistics, which modify the counting of microstates in many-body systems; these appear automatically when constructing Z using symmetrized or anti-symmetrized states.

In the classical limit, when energy levels are dense and $\hbar \to 0$, the trace recovers the familiar phase-space integral through

$$\mathrm{Tr}\,(\cdot) \;\longrightarrow\; \frac{1}{N! h^{3N}} \int d\Gamma_N\,(\cdot),$$

and the two formulations coincide.

The quantum canonical ensemble represents the most fundamental equilibrium state in quantum statistical mechanics. It bridges dynamics and thermodynamics through a single operator equation,

$$\hat{\rho}_{\mathrm{can}} = \frac{e^{-\beta \hat{H}}}{Z},$$

which maximizes the von Neumann entropy while maintaining fixed mean energy. Its classical limit reproduces the Boltzmann distribution, while its quantum formulation naturally accommodates discrete spectra, degeneracies, and particle statistics. The trace formulation, being basis independent, unifies all representations—whether discrete, continuous, or field-theoretic—under a single compact expression.

Canonical Density Operator and Free Energy

The equilibrium density operator minimizes the free-energy functional

$$F[\hat{\rho}] = \mathrm{Tr}(\hat{\rho}\hat{H}) + k_{\mathrm{B}} T\, \mathrm{Tr}(\hat{\rho} \ln \hat{\rho}),$$

subject to $\mathrm{Tr}\,\hat{\rho} = 1$. For any trial density matrix $\hat{\rho}_t$, the Gibbs–Bogoliubov inequality holds:

$$F[\hat{\rho}_t] \geq F[\hat{\rho}_{\mathrm{can}}] = -k_{\mathrm{B}} T \ln Z,$$

with equality only for $\hat{\rho}_t = \hat{\rho}_{\mathrm{can}}$. This provides a variational characterization of equilibrium: among all admissible states, the canonical ensemble minimizes the free energy.

Second Quantization and the Quantum Grand-Canonical Ensemble

As in the classical case, when passing from the canonical to the quantum grand-canonical ensemble, we extend the system's freedom: the system may now exchange both energy and particles with a large reservoir. In the classical theory, this corresponds to allowing the number of particles N to fluctuate, with the chemical potential μ controlling the average particle number. In quantum mechanics, such a description requires a formulation that naturally accommodates variable particle number—this is the *second-quantized* or *Fock-space* formalism. We offer here only a brief recap; a more complete introduction will appear in Volume II.

For a system of N identical quantum particles, the total Hilbert space is the tensor product of the single-particle Hilbert spaces,

$$\mathcal{H}^{(N)} = \mathcal{H}_1^{\otimes N}. \tag{12.121}$$

The symmetry of the wavefunction under particle exchange depends on the particle statistics:

- *Bosons:* the wavefunction is symmetric under particle exchange.
- *Fermions:* the wavefunction is antisymmetric under particle exchange.

These symmetry properties are dictated by the *spin–statistics theorem*: particles with integer spin are bosons, while particles with half-integer spin are fermions. The Pauli exclusion principle for fermions follows immediately: two identical fermions cannot occupy the same quantum state, whereas any number of bosons may do so.

The Hilbert space of a system with an arbitrary (possibly variable) number of particles is constructed as the direct sum over all particle sectors:

$$\mathcal{F}_\pm(\mathcal{H}_1) = \bigoplus_{N=0}^{\infty} \mathbb{S}_\pm \mathcal{H}_1^{\otimes N}, \tag{12.122}$$

where the symbol $\mathbb{S}_+$ ($\mathbb{S}_-$) denotes the symmetrization (antisymmetrization) operator, ensuring the correct exchange symmetry. This space $\mathcal{F}_\pm$ is known as the *Fock space*. It provides the natural stage for second quantization, allowing the description of states with indefinite particle number.

For example, for two single-particle states $|n_1\rangle$ and $|n_2\rangle$, the normalized two-particle states are

$$\begin{aligned} |2\rangle_+ &= \frac{1}{\sqrt{2}}\big(|n_1, n_2\rangle + |n_2, n_1\rangle\big) && \text{(bosons)}, \\ |2\rangle_- &= \frac{1}{\sqrt{2}}\big(|n_1, n_2\rangle - |n_2, n_1\rangle\big) && \text{(fermions)}. \end{aligned} \tag{12.123}$$

On Fock space, the total particle-number operator takes the form

$$\hat{N} = \sum_{i=1}^{M} \mathbb{I} \otimes \cdots \otimes \underbrace{\hat{N}_i}_{i\text{th factor}} \otimes \cdots \otimes \mathbb{I}, \tag{12.124}$$

and similarly, the noninteracting Hamiltonian is

$$\hat{H}_0 = \sum_{i=1}^{M} \mathbb{I} \otimes \cdots \otimes \underbrace{\hat{\mathcal{H}}_i}_{i\text{th factor}} \otimes \cdots \otimes \mathbb{I}. \tag{12.125}$$

If pairwise interactions are present, the total Hamiltonian becomes $\hat{H} = \hat{H}_0 + \hat{H}_{\text{int}}$, where $\hat{H}_{\text{int}}$ acts on two-particle subspaces simultaneously.

Variational Principle and Grand-Canonical Distribution

In the grand-canonical ensemble, both energy and particle number fluctuate, while their averages are fixed by the temperature T and the chemical potential μ. The equilibrium density operator maximizes the generalized free-energy functional

$$\mathcal{F}[\hat{\rho}] = S(\hat{\rho}) - k_{\text{B}}\beta \left(\text{Tr}[\hat{\rho}\,\hat{H}] - \mu\,\text{Tr}[\hat{\rho}\,\hat{N}]\right), \tag{12.126}$$

under the constraint $\text{Tr}\,\hat{\rho} = 1$. Performing the variation $\delta\mathcal{F} = 0$ yields

$$\ln\hat{\rho} + \mathbb{I} - \beta(\hat{H} - \mu\hat{N}) = \lambda\,\mathbb{I}, \tag{12.127}$$

where the constant λ enforces normalization. Exponentiation gives the grand-canonical density operator,

$$\hat{\rho}_{\text{GC}} = \frac{e^{-\beta(\hat{H}-\mu\hat{N})}}{\mathcal{G}}, \qquad \mathcal{G} = \text{Tr}\,e^{-\beta(\hat{H}-\mu\hat{N})}. \tag{12.128}$$

The operator $\hat{H} - \mu\hat{N}$ is the *grand Hamiltonian*, and $\mathcal{G}$ is the *grand-canonical partition function.*

As in the quantum canonical ensemble, the grand-canonical ensemble can be expressed in terms of the spectrum of $\hat{H}$ and $\hat{N}$. Let $|n, \epsilon_n\rangle$ denote simultaneous eigenstates of particle number and energy:

$$\hat{N}|n, \epsilon_n\rangle = n|n, \epsilon_n\rangle, \qquad \hat{H}|n, \epsilon_n\rangle = \epsilon_n|n, \epsilon_n\rangle.$$

Using Eq. (12.128),

$$\begin{aligned} \mathcal{G} = \operatorname{Tr} e^{-\beta(\hat{H}-\mu\hat{N})} &= \sum_{n,\epsilon_n} \langle n, \epsilon_n | e^{-\beta(\hat{H}-\mu\hat{N})} | n, \epsilon_n\rangle \\ &= \sum_n e^{\beta\mu n} \operatorname{Tr} e^{-\beta\hat{H}(n)} = \sum_{n=0}^{\infty} \zeta^n Z_n, \end{aligned} \tag{12.129}$$

where $\zeta = e^{\beta\mu}$ is the fugacity, and Z_n is the canonical partition function with fixed particle number n. Thus the grand-canonical ensemble is a weighted sum of canonical ensembles with different n.

All macroscopic thermodynamic quantities follow directly from the grand partition function:

$$\Omega = -k_{\mathrm{B}}T \ln \mathcal{G} = -PV, \tag{12.130}$$

where Ω is the grand potential. The mean particle number and internal energy are

$$\langle\hat{N}\rangle = \frac{1}{\beta}\frac{\partial \ln \mathcal{G}}{\partial \mu}, \qquad \langle\hat{H}\rangle = -\frac{\partial \ln \mathcal{G}}{\partial \beta}. \tag{12.131}$$

These relations parallel the classical formulas; the quantum nature enters only through the operator structure and the symmetrization of the Fock space.

REMARK 12.10 The second-quantized formulation provides the most general and elegant framework for quantum statistical mechanics. It automatically incorporates particle indistinguishability and the spin-statistics connection, while permitting creation and annihilation processes. The grand-canonical density operator

$$\hat{\rho}_{\mathrm{GC}} = \frac{e^{-\beta(\hat{H}-\mu\hat{N})}}{\mathcal{G}}$$

is the foundation of the Bose–Einstein and Fermi–Dirac distributions, which will be discussed in the following sections.

12.7 Thermodynamic Consistency of Quantum Ensembles

The three density operators introduced above define probability measures over the energy eigenstates of the system. Thermodynamics emerges by associating to each ensemble a suitable entropy and thermodynamic potential, and then showing that the usual relations between U, S, T, P, μ, etc. hold.

In all cases, the appropriate entropy functional is the *Gibbs–von Neumann entropy*

$$S[\hat{\rho}] = -k_{\mathrm{B}} \operatorname{Tr}(\hat{\rho} \ln \hat{\rho}), \tag{12.132}$$

which reduces to the Shannon entropy of the eigenvalues of $\hat{\rho}$.

Micro-Canonical Ensemble

For an isolated system with fixed energy E, the micro-canonical density operator

$$\hat{\rho}_{\text{micro}} = \frac{\delta(E - \hat{H})}{\Omega(E)}, \qquad \Omega(E) = \operatorname{Tr} \delta(E - \hat{H}),$$

assigns equal weight to all energy eigenstates in the shell at energy E. Since all nonzero eigenvalues of $\hat{\rho}_{\text{micro}}$ are equal to $1/\Omega(E)$, the entropy is

$$S_{\text{micro}}(E) = -k_{\text{B}}\, \Omega(E)\, \frac{1}{\Omega(E)} \ln\left(\frac{1}{\Omega(E)}\right) = k_{\text{B}} \ln \Omega(E), \tag{12.133}$$

which is the familiar Boltzmann formula. Defining the temperature by

$$\frac{1}{T} = \left(\frac{\partial S_{\text{micro}}}{\partial E}\right)_{V,N}, \tag{12.134}$$

and similarly for pressure and chemical potential via derivatives with respect to V and N, one recovers the usual micro-canonical thermodynamics.

Canonical Ensemble

In the canonical case,

$$\hat{\rho}_{\text{can}} = \frac{e^{-\beta \hat{H}}}{Z}, \qquad Z = \operatorname{Tr} e^{-\beta \hat{H}},$$

we have

$$\ln \hat{\rho}_{\text{can}} = -\beta \hat{H} - \ln Z.$$

Hence

$$\begin{aligned} S_{\text{can}} &= -k_{\text{B}} \operatorname{Tr}(\hat{\rho}_{\text{can}} \ln \hat{\rho}_{\text{can}}) = -k_{\text{B}} \operatorname{Tr}\left[\hat{\rho}_{\text{can}}(-\beta \hat{H} - \ln Z)\right] \\ &= k_{\text{B}} \beta \operatorname{Tr}(\hat{\rho}_{\text{can}} \hat{H}) + k_{\text{B}} \ln Z = \frac{U}{T} + k_{\text{B}} \ln Z, \end{aligned} \tag{12.135}$$

where $U = \langle \hat{H} \rangle_{\text{can}}$ is the mean energy. Defining the Helmholtz free energy

$$F(T, V, N) = -k_{\text{B}} T \ln Z, \tag{12.136}$$

this becomes

$$F = U - TS, \tag{12.137}$$

the standard thermodynamic identity. Differentiating $\ln Z$ yields

$$U = -\frac{\partial}{\partial \beta} \ln Z, \qquad S = -\left(\frac{\partial F}{\partial T}\right)_{V,N}, \tag{12.138}$$

and using $F = U - TS$ together with $dF = -S\, dT - P\, dV + \mu\, dN$ recovers the familiar canonical relations between F, U, S, P, and μ.

Grand-Canonical Ensemble

When both energy and particle number fluctuate,

$$\hat{\rho}_{\text{grand}} = \frac{e^{-\beta(\hat{H} - \mu \hat{N})}}{\mathcal{G}}, \qquad \mathcal{G} = \operatorname{Tr} e^{-\beta(\hat{H} - \mu \hat{N})}.$$

Then

$$\ln \hat{\rho}_{\text{grand}} = -\beta(\hat{H} - \mu\hat{N}) - \ln \mathcal{G},$$

and the entropy becomes

$$\begin{aligned} S_{\text{grand}} &= -k_{\text{B}} \operatorname{Tr}(\hat{\rho}_{\text{grand}} \ln \hat{\rho}_{\text{grand}}) \\ &= k_{\text{B}}\beta \operatorname{Tr}[\hat{\rho}_{\text{grand}}(\hat{H} - \mu\hat{N})] + k_{\text{B}} \ln \mathcal{G} \\ &= \frac{U - \mu N}{T} + k_{\text{B}} \ln \mathcal{G}, \end{aligned} \tag{12.139}$$

with

$$U = \langle \hat{H} \rangle_{\text{grand}}, \qquad N = \langle \hat{N} \rangle_{\text{grand}}.$$

Introducing the grand potential

$$\Omega(T, V, \mu) = -k_{\text{B}}T \ln \mathcal{G}, \tag{12.140}$$

we obtain

$$\Omega = U - TS - \mu N, \tag{12.141}$$

so that

$$S = -\left(\frac{\partial \Omega}{\partial T}\right)_{V,\mu}, \qquad N = -\left(\frac{\partial \Omega}{\partial \mu}\right)_{T,V}, \qquad P = -\left(\frac{\partial \Omega}{\partial V}\right)_{T,\mu}. \tag{12.142}$$

In equilibrium, $\Omega = -PV$, and the differential identity

$$d\Omega = -S\,dT - P\,dV - N\,d\mu$$

is the quantum analogue of the classical grand-canonical thermodynamic relation.

The quantum micro-canonical, canonical, and grand-canonical density operators are thermodynamically consistent: when combined with the Gibbs–von Neumann entropy, they reproduce the standard thermodynamic potentials (S, F, Ω) and all associated equilibrium identities. The difference from the classical case lies only in the operator nature of $\hat{H}$ and $\hat{N}$ and in the quantum spectrum; the Legendre-transform structure of thermodynamics remains unchanged.

12.8 The Quantum Ideal Gas

The statistical mechanics of quantum gases can be elegantly formulated in the *occupation-number* or *second-quantized* representation. In this formalism, each single-particle state (labeled by momentum, energy level, or mode index j) is characterized by its occupation number n_j, which specifies how many indistinguishable particles populate that state. The occupation numbers are constrained by the statistics of the particles:

$$n_j = \begin{cases} 0, 1, 2, \ldots, & \text{for bosons,} \\ 0, 1, & \text{for fermions.} \end{cases} \tag{12.143}$$

The total energy and total particle number are then given by

$$E = \sum_j \epsilon_j n_j, \qquad N = \sum_j n_j, \tag{12.144}$$

where ϵ_j are the single-particle energy levels of the system.

Micro-Canonical Ensemble: Counting of States

To illustrate microstate counting in a simple quantum system, consider N identical harmonic oscillators, each with discrete energy levels

$$\epsilon_n = \hbar\omega\left(n + \tfrac{1}{2}\right), \qquad n = 0, 1, 2, \ldots$$

The total energy is therefore

$$E = \sum_{j=1}^{N} \hbar\omega\left(n_j + \tfrac{1}{2}\right), \tag{12.145}$$

and we wish to determine the number of distinct sets of non-negative integers $\{n_j\}$ compatible with a given total energy E.

Defining

$$M = \frac{E}{\hbar\omega} - \frac{N}{2}, \tag{12.146}$$

the problem reduces to counting the number of non-negative integer solutions of

$$\sum_{j=1}^{N} n_j = M. \tag{12.147}$$

This is a standard combinatorial problem: the number of ways of partitioning an integer M into N non-negative parts is

$$\Gamma(U) = \frac{(M + N - 1)!}{M!\,(N-1)!}, \tag{12.148}$$

obtained by the "pebbles and sticks" argument illustrated in Fig. 12.1. Here, M identical energy quanta (*pebbles*) are separated by $N-1$ dividers (*sticks*), each representing a partition boundary.

FIGURE 12.1 Illustration of the "pebbles and sticks" method for counting non-negative integer partitions of M into N parts.

The corresponding micro-canonical entropy is

$$S(E) = k_{\mathrm{B}} \ln \Gamma(E) = k_{\mathrm{B}} \ln \frac{\left(\frac{E}{\hbar\omega} + \frac{N}{2} - 1\right)!}{\left(\frac{E}{\hbar\omega} - \frac{N}{2}\right)!\,(N-1)!}. \tag{12.149}$$

For large N, applying Stirling's approximation and introducing the dimensionless energy per oscillator,

$$u = \frac{E}{N\hbar\omega}, \tag{12.150}$$

we obtain

$$S(u) \simeq N k_{\mathrm{B}} \Big[\left(u + \tfrac{1}{2}\right) \ln\left(u + \tfrac{1}{2}\right) - \left(u - \tfrac{1}{2}\right) \ln\left(u - \tfrac{1}{2}\right)\Big]. \tag{12.151}$$

Differentiating with respect to total energy yields the temperature:

$$\frac{1}{T} = \frac{\partial S}{\partial U} = \frac{k_{\mathrm{B}}}{\hbar\omega} \ln\left(\frac{u + \frac{1}{2}}{u - \frac{1}{2}}\right). \tag{12.152}$$

Solving for u gives

$$u = \frac{1}{2} \coth\left(\frac{\beta\hbar\omega}{2}\right), \tag{12.153}$$

which matches exactly the mean oscillator energy obtained in the canonical ensemble, confirming the equivalence of ensembles in the thermodynamic limit.

Entropy Maximization and the Maxwell–Boltzmann Distribution

The same reasoning extends to a general spectrum with level degeneracies d_i and occupancies n_i. Using Stirling's approximation for $n_i \gg 1$, the entropy reads

$$S = k_{\rm B} \ln \Omega = k_{\rm B} \sum_i n_i \big(\ln d_i - \ln n_i + 1 \big). \tag{12.154}$$

Maximize S under

$$N = \sum_i n_i, \qquad U = \sum_i n_i \epsilon_i,$$

with Lagrange multipliers λ, λ'. Setting $\delta S = 0$ gives

$$\ln d_i - \ln n_i + 1 + \lambda + \lambda' \epsilon_i = 0 \quad \Rightarrow \quad \frac{n_i}{d_i} = A\, e^{-\beta \epsilon_i}, \;\; A = e^{-\lambda}, \; \beta = -\lambda' = \frac{1}{k_{\rm B}T}. \tag{12.155}$$

Thus the equilibrium occupancy is

$$n_i = N p_i, \qquad p_i = \frac{d_i e^{-\beta\epsilon_i}}{Z_1}, \qquad Z_1 \equiv \sum_i d_i e^{-\beta \epsilon_i}. \tag{12.156}$$

Equation (12.156) is the *Maxwell–Boltzmann (MB) distribution*: in the dilute limit (sparse level occupancy), bosons and fermions behave classically.

Deriving the Gibbs Factor from Quantum Counting

Insert the equilibrium $n_i = N p_i$ into Eq. (12.154):

$$\ln \Omega = \sum_i N p_i \big(\ln d_i - \ln(N p_i) + 1 \big) = N \left[\sum_i p_i \ln \frac{d_i}{p_i} - \ln N + 1 \right]. \tag{12.157}$$

Using $p_i \propto d_i e^{-\beta \epsilon_i}$,

$$\sum_i p_i \ln \frac{d_i}{p_i} = \sum_i p_i \ln\big(Z_1 e^{\beta \epsilon_i} \big) = \ln Z_1 + \beta \sum_i p_i \epsilon_i = \ln Z_1 + \frac{\beta U}{N}.$$

Hence

$$\ln \Omega = N \ln Z_1 - N \ln N + N + \beta U, \quad \Rightarrow \quad S = k_{\rm B} \ln \Omega = k_{\rm B} (\beta U + N \ln Z_1 - N \ln N + N)\,. \tag{12.158}$$

Applying Stirling's formula $\ln N! = N \ln N - N + o(N)$,

$$S = k_{\rm B} \left(\beta U + \ln \frac{Z_1^N}{N!} \right) + o(N) = k_{\rm B} (\beta U + \ln Z_N) + o(N), \qquad Z_N \equiv \frac{Z_1^N}{N!}. \tag{12.159}$$

Thus the classical $N!$ *Gibbs correction* emerges naturally from the underlying quantum (indistinguishable) counting in the dilute limit: the MB N-particle partition function is

$$Z_N = \frac{Z_1^N}{N!}, \qquad F = -k_{\rm B} T \ln Z_N, \qquad S = \frac{U - F}{T}.$$

This shows in what sense classical statistics are a limiting case of the quantum ones.

Canonical Ensemble: The Quantum Harmonic Oscillator

For the canonical ensemble, the partition function of a single quantum oscillator is

$$Z_1 = \sum_{n=0}^{\infty} e^{-\beta\hbar\omega(n+\frac{1}{2})} = \frac{e^{-\beta\hbar\omega/2}}{1-e^{-\beta\hbar\omega}} = \frac{1}{2\sinh\left(\frac{\beta\hbar\omega}{2}\right)}. \tag{12.160}$$

The corresponding mean energy is

$$U_1 = -\frac{\partial \ln Z_1}{\partial \beta} = \frac{\hbar\omega}{2}\coth\left(\frac{\beta\hbar\omega}{2}\right), \tag{12.161}$$

and the Helmholtz free energy is

$$F_1 = -k_B T \ln Z_1 = k_B T \ln\left(2\sinh\frac{\beta\hbar\omega}{2}\right). \tag{12.162}$$

For N independent and distinguishable oscillators, the total partition function factorizes as

$$Z_N = Z_1^N, \tag{12.163}$$

and the thermodynamic quantities scale linearly:

$$F_N = NF_1, \qquad U_N = NU_1. \tag{12.164}$$

This example illustrates the power of quantum statistical mechanics. While in the classical case the energy spectrum is continuous and integrals over phase space must be evaluated, in the quantum case the discreteness of the spectrum simplifies the analysis. Moreover, the harmonic oscillator serves as the fundamental building block for the quantum ideal gas, where occupation numbers n_j describe the population of quantized single-particle states. The techniques developed here—counting of microstates, partition functions, and the link between micro-canonical and canonical ensembles—will be the foundation for deriving the Bose–Einstein and Fermi–Dirac distributions in the grand-canonical ensemble.

Grand-Canonical Ensemble

In the quantum grand-canonical ensemble, both energy and particle number fluctuate. Working in the occupation-number (Fock) basis, a many-body state is written as

$$|\psi\rangle = |n_1, n_2, \ldots\rangle, \qquad N = \sum_j n_j, \qquad E = \sum_j \epsilon_j n_j,$$

with $n_j \in \{0,1,2,\ldots\}$ for bosons and $n_j \in \{0,1\}$ for fermions.

The grand-canonical density operator is

$$\hat{\rho}_{GC} = \frac{e^{-\beta(\hat{H}-\mu\hat{N})}}{\mathcal{G}}, \qquad \mathcal{G}(\beta,\mu) = \mathrm{Tr}\, e^{-\beta(\hat{H}-\mu\hat{N})}.$$

Evaluating the trace in the occupation-number basis gives

$$\mathcal{G} = \sum_{\{n_j\}} \exp\left[-\beta\sum_j(\epsilon_j-\mu)n_j\right] = \prod_j \sum_{n_j} \left(e^{-\beta(\epsilon_j-\mu)}\right)^{n_j}. \tag{12.165}$$

The single-mode sums are

$$\mathcal{G} = \begin{cases} \prod_j \dfrac{1}{1-e^{-\beta(\epsilon_j-\mu)}} & \text{(bosons)}, \\ \prod_j \left(1+e^{-\beta(\epsilon_j-\mu)}\right) & \text{(fermions)}. \end{cases}$$

Thermal averages follow from $\ln \mathcal{G}$. The mean occupation of level j is

$$\langle n_j \rangle = -\frac{1}{\beta}\frac{\partial \ln \mathcal{G}}{\partial \epsilon_j} = \frac{1}{e^{\beta(\epsilon_j-\mu)} \mp 1} = \begin{cases} \dfrac{1}{e^{\beta(\epsilon_j-\mu)}-1} & \text{(bosons)}, \\ \dfrac{1}{e^{\beta(\epsilon_j-\mu)}+1} & \text{(fermions)}, \end{cases} \tag{12.166}$$

where the upper (lower) sign corresponds to bosons (fermions). Summing over j gives the mean particle number $\langle N \rangle = \sum_j \langle n_j \rangle$.

Classical (Boltzmann) Limit

The Maxwell–Boltzmann limit corresponds to a dilute or high-temperature regime,

$$e^{\beta(\epsilon_j-\mu)} \gg 1 \quad \Longleftrightarrow \quad z\, e^{-\beta\epsilon_j} \ll 1, \qquad z \equiv e^{\beta\mu} \text{ (fugacity)}.$$

Then quantum statistics are negligible. Expanding the denominators in Eq.(12.166),

$$\langle n_j \rangle_{\mathrm{B}} = \frac{1}{e^{\beta(\epsilon_j-\mu)}-1} = e^{-\beta(\epsilon_j-\mu)}\Big[1 + e^{-\beta(\epsilon_j-\mu)} + \cdots\Big] = z\, e^{-\beta\epsilon_j} + O((ze^{-\beta\epsilon_j})^2),$$

$$\langle n_j \rangle_{\mathrm{F}} = \frac{1}{e^{\beta(\epsilon_j-\mu)}+1} = e^{-\beta(\epsilon_j-\mu)}\Big[1 - e^{-\beta(\epsilon_j-\mu)} + \cdots\Big] = z\, e^{-\beta\epsilon_j} + O((ze^{-\beta\epsilon_j})^2). \tag{12.167}$$

Thus, to leading order,

$$\langle n_j \rangle \simeq z\, e^{-\beta\epsilon_j} \qquad \text{(Maxwell–Boltzmann limit)}.$$

Equivalently, the mean occupations across regimes are summarized as

$$\langle n_j \rangle = \begin{cases} \dfrac{1}{e^{\beta(\epsilon_j-\mu)}-1} & \text{Bose–Einstein}, \\ \dfrac{1}{e^{\beta(\epsilon_j-\mu)}+1} & \text{Fermi–Dirac}, \\ e^{-\beta(\epsilon_j-\mu)} & \text{Maxwell–Boltzmann (classical limit)}. \end{cases}$$

Thus, at high temperature or low density, quantum statistical effects disappear and both Bose and Fermi gases behave classically.

12.9 Closed-Box Formalism and Quantum Counting

A complementary way to describe a quantum ideal gas is to consider the quantization of momentum in a finite volume, the so-called *closed-box formalism*. Here, particles are confined to a cubic box of side L with periodic boundary conditions. This approach provides a direct link between the microscopic counting of states and the macroscopic thermodynamic quantities of the gas.

Quantized Energy Levels

For a single particle in a box, the allowed momenta are discrete:

$$\vec{p}_k = \frac{2\pi\hbar}{L}\,\vec{k}, \qquad \vec{k} = (k_x, k_y, k_z) \in \mathbb{Z}^3, \tag{12.168}$$

and the single-particle energies are

$$\epsilon_k = \frac{p_k^2}{2m} = \frac{(2\pi\hbar)^2}{2mL^2}(k_x^2 + k_y^2 + k_z^2). \tag{12.169}$$

Each energy level ϵ_k may be degenerate, with degeneracy d_k, corresponding to the number of different $\vec{k}$-vectors yielding the same energy. In the thermodynamic limit, sums over $\vec{k}$ become integrals over momentum space, according to

$$\frac{V}{h^3}\int d^3p \longleftrightarrow \sum_{\vec{k}}, \tag{12.170}$$

but the discrete representation is particularly useful for understanding the combinatorial origin of different quantum statistics.

Counting Microstates

A microstate of the system is specified by the set of occupation numbers $\{n_k\}$, where n_k denotes the number of particles occupying the energy level ϵ_k. The total energy and particle number are

$$E = \sum_k n_k \epsilon_k, \qquad N = \sum_k n_k. \tag{12.171}$$

For a given energy E, the number of possible microstates $\Gamma(E)$ depends on the statistical nature of the particles—Boltzmann, Bose, or Fermi. This can be analyzed by standard combinatorial arguments, such as the "pebbles and sticks" method used earlier for partition problems.

Boltzmann Statistics

For a classical gas of distinguishable particles, there are no restrictions on how many particles can occupy a given state. Let n_i denote the number of particles in the ith energy level of degeneracy d_i. The number of possible arrangements (microstates) corresponding to a given configuration $\{n_i\}$ is

$$\#_B(E) = \frac{N!}{\prod_i n_i!}. \tag{12.172}$$

Each of the n_i particles in level i can occupy any of the d_i degenerate substates, giving an additional factor $d_i^{n_i}$. Because the particles are indistinguishable, the overall number of accessible microstates is

$$\Gamma_B(E) = \frac{1}{N!}\prod_i \frac{d_i^{n_i}}{n_i!}. \tag{12.173}$$

This is the classical *Boltzmann counting rule*. If the spectrum were continuous, a factor of $1/h^3$ would appear from phase-space normalization.

Historical Note: Bose's Original Paper Sent to Einstein

In 1924 the physicist Satyendra Bose from Dhaka University in India (today Bangladesh) sent Einstein a paper, in which the Planck distribution law for photons was derived by entirely statistical arguments without resort to results from classical electrodynamics. Einstein realized the importance of the paper, translated it into German and submitted it for publication in Zeitschrift für Physik. He immediately started to work on the problem himself, and published two papers, developing the full picture of the quantum theory of bosonic particles. The concept of particles obeying Bose–Einstein statistics was born. Einstein noted that if the number of particles is conserved, even totally noninteracting particles will undergo a phase transition at low enough temperatures. This transition is termed Bose–Einstein condensation (BEC). Bose did not find this feature because he was discussing photons which, being massless, do not need to condense, because they can disappear instead when the energy of the system is decreased.

Bose–Einstein Statistics

For bosons, there is no restriction on the number of particles that can occupy a single quantum state. In the ith energy level of degeneracy d_i, we must distribute n_i identical particles among d_i identical boxes. This is a combination with repetitions, whose number is

$$\Gamma_{\text{Bose}}(n_i) = \frac{(n_i + d_i - 1)!}{n_i!\,(d_i - 1)!}. \tag{12.174}$$

Multiplying over all levels, the total number of microstates is

$$\Gamma_{\text{Bose}}(E) = \prod_i \frac{(n_i + d_i - 1)!}{n_i!\,(d_i - 1)!}. \tag{12.175}$$

Fermi–Dirac Statistics

For fermions, the Pauli exclusion principle limits each substate to at most one particle. Thus, in each level of degeneracy d_i, the n_i particles can occupy n_i of the d_i available states without repetition, yielding

$$\Gamma_{\text{Fermi}}(n_i) = \frac{d_i!}{n_i!\,(d_i - n_i)!}. \tag{12.176}$$

The total number of configurations is then

$$\Gamma_{\text{Fermi}}(E) = \prod_i \frac{d_i!}{n_i!\,(d_i - n_i)!}. \tag{12.177}$$

Combinatorics of Classical and Quantum Statistics

The three forms of counting are summarized in Tab. 12.1, which highlights how the fundamental distinction between distinguishable and indistinguishable particles—and the symmetry of their quantum states—determines the combinatorial structure of statistical mechanics.

TABLE 12.1 Comparison of combinatorial rules for Boltzmann, Bose, and Fermi statistics.

Statistics Type	Counting Formula	Combinatorial Interpretation
Boltzmann	$\dfrac{d_i^{n_i}}{n_i!}$	Permutations with repetitions
Bose–Einstein	$\dfrac{(d_i + n_i - 1)!}{n_i!\,(d_i - 1)!}$	Combinations with repetitions
Fermi–Dirac	$\dfrac{d_i!}{n_i!\,(d_i - n_i)!}$	Combinations without repetitions

REMARK 12.11 In the classical (Boltzmann) limit, the degeneracy d_i is large and the occupation numbers n_i are small, $n_i/d_i \ll 1$. In this regime, the Bose and Fermi counting rules both reduce to the Boltzmann formula, recovering classical statistical mechanics. Quantum statistics thus emerge as corrections to Boltzmann's theory that become significant only when particles are indistinguishable and their wavefunctions overlap in phase space.

Anyonic (Fractional) Statistics and Two-Dimensional Quantum Matter

In three spatial dimensions, the exchange of two identical particles can only change the sign of the wavefunction: $\psi(\vec{r}_1, \vec{r}_2) = \pm\psi(\vec{r}_2, \vec{r}_1)$, giving rise exclusively to *Bose* (+) or *Fermi* (−) statistics. However, in two dimensions the topology of configuration space allows for a continuous family of intermediate phases. When two particles are adiabatically exchanged in a plane, the wavefunction may acquire an arbitrary phase factor:

$$\psi(\vec{r}_1, \vec{r}_2) = e^{i\theta}\psi(\vec{r}_2, \vec{r}_1), \qquad 0 \le \theta \le \pi.$$

Particles obeying this generalized exchange rule are called *anyons*—a term introduced by Frank Wilczek in 1982 [Wil82]. They interpolate smoothly between bosons ($\theta = 0$) and fermions ($\theta = \pi$) and obey *fractional statistics.*

Topological Origin

The existence of anyons is a purely two-dimensional phenomenon. In a plane, continuous deformations of particle trajectories cannot remove the braiding of one particle around another: the fundamental group of configuration space is not the permutation group but the *braid group*. Each braid operation multiplies the many-body wavefunction by a phase $e^{i\theta}$, or more generally by a unitary matrix for non-Abelian anyons. This topological character makes anyons remarkably robust to local perturbations—a property that underlies their importance in modern topological materials and quantum computing.

Statistical Generalization

In terms of counting microstates, fractional or anyonic statistics can be expressed as a continuous interpolation between Bose and Fermi statistics. For a level of degeneracy d_i with n_i particles, the number of accessible states can be written in the generalized form introduced by Giovanni Gentile (1940) and reformulated by Frederick Haldane (1991) [Gen40, Hal91]:

$$\Omega_i^{(\alpha)} = \frac{\big[d_i + (n_i - 1)(1-\alpha)\big]!}{n_i!\,\big[d_i - 1 - \alpha(n_i - 1)\big]!}, \qquad 0 \le \alpha \le 1. \tag{12.178}$$

Here the parameter $\alpha = \theta/\pi$ quantifies the statistical phase: $\alpha = 0$ reproduces Bose–Einstein statistics, $\alpha = 1$ yields Fermi–Dirac statistics, and intermediate values $0 < \alpha < 1$ correspond to anyonic or semionic particles.

In the dilute or high-temperature limit, $d_i \gg 1$ and $n_i/d_i \ll 1$, Eq. (12.178) reduces to $\Omega_i^{(\alpha)} \approx d_i^{n_i}/n_i!$, recovering the classical Boltzmann statistics as the universal limit of all quantum statistics.

Anyonic quasiparticles arise naturally in two-dimensional electronic systems with strong correlations and topological order. The paradigmatic example is the *fractional quantum Hall effect*, where adiabatic quasiparticle exchanges generate fractional statistics and fractional electric charge [TSG82, Lau83, ASW84]. Other prominent realizations include *topological superconductors* and *quantum spin liquids*, whose emergent excitations can obey non-Abelian braiding statistics described by matrix-valued braid-group representations [Kit06, Wen04]. Engineered systems such as *Majorana zero modes* in one-dimensional topological superconductors [Kit01, Ali12] and proposed *quantum Hall lattice* platforms are likewise predicted to host anyonic quasiparticles, with non-Abelian species forming the basis for fault-tolerant *topological quantum computation* [NSS+08].

In all these settings, the macroscopic properties—thermal, magnetic, or transport—reflect modified distribution functions that interpolate between Bose and Fermi limits. The statistical angle θ (or equivalently $\alpha = \theta/\pi$) encodes the topological phase accumulated under particle exchange, linking microscopic braiding statistics to measurable thermodynamic behavior.

From Bosons and Fermions to Anyons

Thus, Bose and Fermi statistics appear as the two discrete endpoints of a continuous spectrum of possible quantum statistics permitted in two dimensions. Anyonic statistics unify and generalize them:

$$\begin{cases} \text{Bose limit:} & \theta = 0,\ \alpha = 0,\ \psi \text{ symmetric}, \\ \text{Fermi limit:} & \theta = \pi,\ \alpha = 1,\ \psi \text{ antisymmetric}, \\ \text{Anyonic (fractional):} & 0 < \theta < \pi,\ \alpha \in (0,1),\ \psi \text{ multivalued under exchange.} \end{cases}$$

The study of such intermediate statistics has become central to understanding quantum Hall states, topological insulators, and other strongly correlated systems where topology, rather than simple symmetry, dictates the fundamental statistical behavior of quasiparticles.

13

Properties of Quantum Gases

13.1 Properties of Ideal Quantum Gases

The thermodynamic theory of quantum gases constitutes one of the cornerstones of modern statistical mechanics and is presented in many classic references, including the textbooks of Kerson Huang [Hua87], Raj K. Pathria and Paul Beale [PB11], and Charles Kittel and Herbert Kroemer [KK80]. The formalism developed in this and the following chapters follows the same general approach: starting from the combinatorial foundations of quantum statistics, we derive the macroscopic properties of gases composed of indistinguishable particles.

In the previous chapter, we examined how classical (Boltzmann) statistics emerge as a limiting case of quantum mechanics and how the exchange symmetry of identical particles leads to distinct Bose and Fermi distributions. We also introduced the notion of fractional, or semionic, statistics, which interpolate continuously between these two extremes.

Here we focus on the *ideal quantum gases*—ensembles of non-interacting particles obeying Bose–Einstein, Fermi–Dirac, or more general fractional statistics. Even in the absence of interactions, such systems exhibit collective behaviors of purely quantum origin, such as Fermi degeneracy pressure in metals and neutron stars, or Bose–Einstein condensation in ultracold atomic gases.

For convenience and generality, we begin with the most inclusive statistical formulation, based on the generalized counting rule for semionic particles. This unified framework allows all limiting cases—Bose, Fermi, and classical—to be obtained by setting the statistical parameter $\alpha = 0$, 1, or by taking the high-temperature limit. Subsequent sections specialize these general results to describe the thermodynamics of Bose and Fermi gases in detail.

In the previous discussion of fractional statistics, the number of accessible microstates for a level i with degeneracy d_i and occupation n_i was written as

$$\Omega_i(\alpha) = \frac{[d_i + (n_i - 1)(1 - \alpha)]!}{n_i!\,[d_i - 1 - \alpha(n_i - 1)]!}, \tag{13.1}$$

where the parameter α determines the statistics: $\alpha = 0$ for bosons, $\alpha = 1$ for fermions, and $0 < \alpha < 1$ for fractional (semionic) particles.

DOI: 10.1201/9781003603528-13

We now derive the mean occupation numbers n_i in the micro-canonical ensemble by maximizing the entropy under appropriate constraints.

Entropy and Constraints

The total entropy is

$$S = k_{\mathrm{B}} \ln \Omega = k_{\mathrm{B}} \sum_i \ln \Omega_i(\alpha), \tag{13.2}$$

subject to conservation of total particle number and total energy:

$$\sum_i n_i = N, \qquad \sum_i n_i \epsilon_i = U. \tag{13.3}$$

Introducing Lagrange multipliers β and $\mu\beta$ for these constraints, we define the functional

$$\tilde{S} = k_{\mathrm{B}} \sum_i \ln \Omega_i(\alpha) - \beta\left(\sum_i n_i \epsilon_i - U\right) - \mu\beta\left(\sum_i n_i - N\right), \tag{13.4}$$

and impose $\delta\tilde{S} = 0$ to find the equilibrium configuration.

Using Stirling's approximation $\ln n! \simeq n \ln n - n$ and differentiating Eq. (13.4) with respect to n_i, we obtain

$$0 = \frac{\partial \tilde{S}}{\partial n_i} = k_{\mathrm{B}}\Big[(1-\alpha)\ln\big(d_i + (n_i - 1)(1-\alpha)\big) - \ln n_i - \alpha \ln\big(d_i - 1 - \alpha(n_i - 1)\big)\Big] - k_{\mathrm{B}}\beta(\epsilon_i - \mu). \tag{13.5}$$

Rearranging, we find the general equilibrium condition

$$(1-\alpha)\ln\big(d_i + (n_i - 1)(1-\alpha)\big) - \alpha \ln\big(d_i - 1 - \alpha(n_i - 1)\big) - \ln n_i + \beta(\epsilon_i - \mu) = 0. \tag{13.6}$$

Fermi and Bose Limits

For fermions ($\alpha = 1$), Eq. (13.6) reduces to

$$\ln(d_i - n_i) - \ln n_i + \beta(\epsilon_i - \mu) = 0, \tag{13.7}$$

yielding

$$n_i = \frac{d_i}{e^{\beta(\epsilon_i - \mu)} + 1}. \tag{13.8}$$

For bosons ($\alpha = 0$), we obtain

$$\ln(d_i + n_i - 1) - \ln n_i + \beta(\epsilon_i - \mu) = 0, \tag{13.9}$$

which gives

$$n_i = \frac{d_i}{e^{\beta(\epsilon_i - \mu)} - 1}. \tag{13.10}$$

These are the well-known *Fermi–Dirac* and *Bose–Einstein* distributions, respectively.

In the high-temperature or low-density limit where $e^{\beta(\epsilon_i - \mu)} \gg 1$, both reduce to the classical Maxwell–Boltzmann form:

$$n_i \simeq d_i \, e^{-\beta(\epsilon_i - \mu)}. \tag{13.11}$$

Starting from the most general fractional (semionic) statistics allows us to recover all familiar cases as specific limits: bosons for $\alpha = 0$, fermions for $\alpha = 1$, and classical particles in the limit of large degeneracy and small occupancy. This unified approach is not only mathematically convenient but also physically motivated: in modern two-dimensional systems such as quantum Hall states and topological insulators, fractional statistics and anyonic excitations naturally appear, making this generalized formulation essential for contemporary quantum statistical mechanics.

13.2 Macroscopic Properties of Ideal Quantum Gases

The thermodynamic quantities of an ideal quantum gas are obtained from summations over discrete energy levels of the form

$$\sum_n e^{-\beta\epsilon_n}.$$

For macroscopic systems, where energy levels are extremely dense, these sums can be replaced by integrals over energy or momentum space weighted by the density of states.

For a free particle of mass m confined in a cubic box of side L (volume $V = L^3$), the single-particle energies are

$$\epsilon_{\vec{p}} = \frac{p^2}{2m}, \qquad \vec{p} = \frac{2\pi\hbar}{L}\vec{k}, \qquad \vec{k} \in \mathbb{Z}^3.$$

A sum over all allowed momentum states may thus be written as

$$\sum_{\vec{p}} e^{-\beta\epsilon_{\vec{p}}} \simeq \sum_{k_x,k_y,k_z} e^{-\frac{\beta(2\pi\hbar)^2}{2mL^2}(k_x^2+k_y^2+k_z^2)} \approx \int d^3k\, e^{-\frac{\beta(2\pi\hbar)^2k^2}{2mL^2}}. \tag{13.12}$$

After changing variables from $\vec{k}$ to $\vec{p} = (2\pi\hbar/L)\vec{k}$, we obtain

$$\sum_{\vec{p}} \longrightarrow \frac{V}{h^3}\int d^3p = \frac{4\pi V}{h^3}\int_0^\infty p^2\,dp, \tag{13.13}$$

a relation that is central to the study of ideal quantum gases. The factor V/h^3 represents the number of quantum states per unit volume in phase space.

For noninteracting particles, the grand-canonical partition function factorizes over single-particle states:

$$\ln Z = \begin{cases} -\sum_j \ln\left(1-\zeta e^{-\beta\epsilon_j}\right) & \text{(bosons)}, \\ \sum_j \ln\left(1+\zeta e^{-\beta\epsilon_j}\right) & \text{(fermions)}, \end{cases} \tag{13.14}$$

where $\zeta = e^{\beta\mu}$ is the fugacity. Using Eq. (13.13), the sum becomes an integral over momentum space, which for fermions reads

$$\ln Z = \frac{4\pi V}{h^3}\int_0^\infty p^2 \ln\left(1+\zeta e^{-\beta p^2/(2m)}\right) dp. \tag{13.15}$$

A completely analogous expression with a minus sign in the logarithm holds for bosons.

Fermi Systems: Low- and High-Density Regimes

For a degenerate Fermi gas, the logarithm of the grand partition function,

$$\ln Z = \frac{PV}{k_\mathrm{B}T} = \sum_{\mathbf{p}} \ln\left(1+\zeta e^{-\beta\epsilon_{\mathbf{p}}}\right), \tag{13.16}$$

is approximated by converting the sum into the integral in Eq. (13.15):

$$\ln Z = \frac{4\pi V}{h^3}\int_0^\infty p^2 \ln\left(1+\zeta e^{-\beta p^2/(2m)}\right) dp. \tag{13.17}$$

Expanding the logarithm as a power series, $\ln(1+x) = \sum_{l=1}^{\infty}(-1)^{l+1}x^l/l$, we obtain

$$\ln Z = \frac{4\pi V}{h^3}\sum_{l=1}^{\infty}\frac{(-1)^{l+1}\zeta^l}{l}\int_0^{\infty} p^2 e^{-l\beta p^2/(2m)}\, dp. \tag{13.18}$$

Evaluating the Gaussian integral,

$$\int_0^{\infty} p^2 e^{-l\beta p^2/(2m)}\, dp = \frac{\sqrt{\pi}}{4}\left(\frac{2m}{\beta l}\right)^{3/2},$$

leads to

$$\ln Z = \frac{V}{\lambda^3}\sum_{l=1}^{\infty}\frac{(-1)^{l+1}\zeta^l}{l^{5/2}}, \tag{13.19}$$

where

$$\lambda = \frac{h}{\sqrt{2\pi m k_{\mathrm{B}} T}} \tag{13.20}$$

is the thermal de Broglie wavelength, setting the scale of quantum effects.

It is convenient to introduce the *Fermi functions*

$$f_\nu(\zeta) = \sum_{l=1}^{\infty}\frac{(-1)^{l+1}\zeta^l}{l^\nu}, \tag{13.21}$$

which generalize the polylogarithm function $\mathrm{Li}_\nu(\zeta)$. In terms of these, the pressure and number density take the compact forms

$$\frac{P}{k_{\mathrm{B}}T} = \frac{1}{\lambda^3} f_{5/2}(\zeta), \tag{13.22}$$

$$\frac{N}{V} = \frac{1}{\lambda^3} f_{3/2}(\zeta). \tag{13.23}$$

Equations (13.22) and (13.23) summarize the equation of state of a free Fermi gas. In the low-fugacity limit ($\zeta \ll 1$), the functions $f_\nu(\zeta)$ approach ζ, recovering the classical Maxwell–Boltzmann results. At low temperatures or high densities ($\zeta \gtrsim 1$), quantum effects become dominant: the occupation of momentum states forms a sharp boundary at the *Fermi surface*, characterized by the Fermi momentum p_{F} and Fermi energy $\epsilon_{\mathrm{F}} = p_{\mathrm{F}}^2/(2m)$.

These functions, shown schematically in Fig. 13.1, interpolate smoothly between classical and degenerate regimes and form the basis for analyzing low- and high-temperature properties of Fermi systems.

Equation of State and Quantum Corrections: The Virial Expansion

The relations derived previously,

$$\frac{P}{k_{\mathrm{B}}T} = \frac{1}{\lambda^3} f_{5/2}(\zeta), \tag{13.24}$$

$$\frac{N}{V} = \frac{1}{\lambda^3} f_{3/2}(\zeta), \tag{13.25}$$

provide a parametric representation of the equation of state of an ideal Fermi (or Bose) gas. Here $\lambda = h/\sqrt{2\pi m k_{\mathrm{B}}T}$ is the thermal de Broglie wavelength and $\zeta = e^{\beta\mu}$ the fugacity. In contrast to the classical case, one must first determine the effective fugacity ζ^* from the number equation (13.25), and then substitute it into the pressure relation (13.24) to obtain the full equation of state:

$$\frac{P}{k_{\mathrm{B}}T} = \frac{1}{\lambda^3} f_{5/2}(\zeta^*), \qquad \frac{N}{V} = \frac{1}{\lambda^3} f_{3/2}(\zeta^*). \tag{13.26}$$

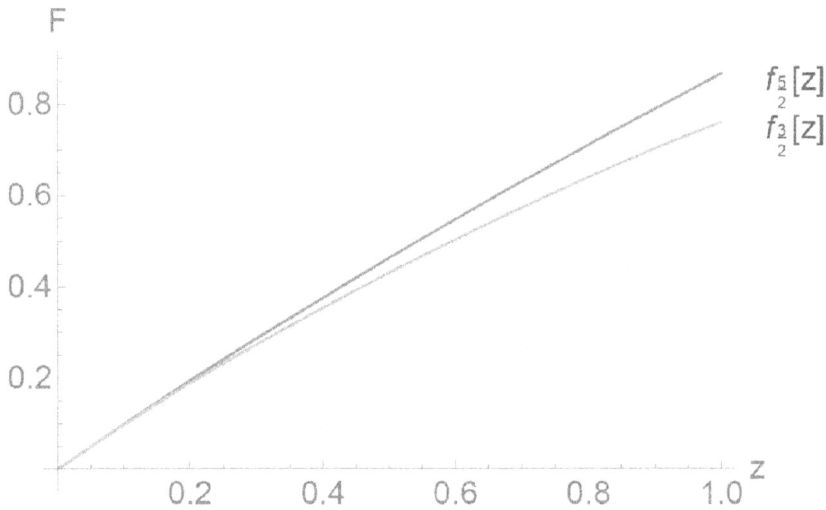

FIGURE 13.1 The behavior of the F-functions as a function of the input.

Classical Limit

For small fugacity ($\zeta \ll 1$), corresponding to high temperatures or low densities, the Fermi functions can be expanded as power series:

$$f_{5/2}(\zeta) = \zeta - \frac{\zeta^2}{2^{5/2}} + O(\zeta^3), \qquad f_{3/2}(\zeta) = \zeta - \frac{\zeta^2}{2^{3/2}} + O(\zeta^3).$$

Substituting the leading term into Eqs. (13.26), we find

$$\frac{P}{k_B T} \simeq \frac{N}{V}, \tag{13.27}$$

which is simply the ideal-gas law. The next term in the series, however, introduces a quantum correction that persists even for noninteracting particles.

Second Virial Coefficient

Retaining terms up to $O(\zeta^2)$ in the expansion of $f_{3/2}(\zeta)$,

$$\frac{N}{V} = \frac{1}{\lambda^3}\left(\zeta - \frac{\zeta^2}{2^{3/2}}\right),$$

we can invert this relation to express the fugacity in powers of the density:

$$\zeta = \frac{\lambda^3 N}{V} + \frac{1}{2^{3/2}}\left(\frac{\lambda^3 N}{V}\right)^2 + O\left[\left(\frac{N}{V}\right)^3\right]. \tag{13.28}$$

Substituting Eq. (13.28) into the expansion of $f_{5/2}(\zeta)$,

$$f_{5/2}(\zeta) = \zeta - \frac{\zeta^2}{2^{5/2}} + O(\zeta^3),$$

and keeping terms up to second order, we obtain

$$\frac{P}{k_B T} = \frac{1}{\lambda^3}\left[\frac{\lambda^3 N}{V} + \frac{1}{4\sqrt{2}}\left(\frac{\lambda^3 N}{V}\right)^2 + O\left[\left(\frac{N}{V}\right)^3\right]\right]. \tag{13.29}$$

Equivalently, in virial form,

$$\frac{PV}{Nk_{\mathrm{B}}T} = 1 + \frac{1}{4\sqrt{2}}\frac{\lambda^3 N}{V} + O\left[\left(\frac{N}{V}\right)^2\right]. \tag{13.30}$$

The second term represents the *second virial coefficient* of the quantum gas:

$$B_2 = \frac{\lambda^3}{4\sqrt{2}}. \tag{13.31}$$

It arises purely from Fermi–Dirac statistics, without any interparticle interactions. For bosons, the sign of the second term is reversed, reflecting the effective "attraction" due to the tendency of bosons to occupy the same quantum state.

This correction is formally analogous to a van der Waals gas with parameters $b = 0$ and $a = \lambda^3/(4\sqrt{2})$: a nonclassical pressure term originating solely from quantum statistics.

Occupation Numbers and the Classical Limit

The mean occupation number of a single-particle state of energy ϵ_j is

$$\langle n_j \rangle_\pm = \frac{\zeta e^{-\beta\epsilon_j}}{1 \pm \zeta e^{-\beta\epsilon_j}}, \tag{13.32}$$

where the upper sign corresponds to fermions and the lower to bosons. In the classical limit ($\zeta \ll 1$ or equivalently $\lambda \to 0$), Eq. (13.32) reduces to

$$\langle n_j \rangle \simeq \zeta e^{-\beta\epsilon_j} = \frac{\lambda^3 N}{V} e^{-\beta\epsilon_j}, \tag{13.33}$$

which is precisely the Maxwell–Boltzmann distribution. Thus, the classical law of equipartition emerges from quantum statistics in the limit where the thermal wavelength is much smaller than the mean interparticle spacing.

The virial expansion demonstrates how quantum effects appear as corrections to the ideal-gas law even in the absence of interactions. For fermions, these corrections increase the pressure (reflecting the Pauli exclusion principle). For bosons, instead, they reduce it (reflecting the bosonic enhancement of low-energy states). Both limits merge smoothly into the classical Boltzmann regime as $\lambda^3 N/V \to 0$.

The Fermi Surface and Degenerate Quantum Gases

In the limit of low density or high temperature—when the thermal de Broglie wavelength

$$\lambda = \frac{h}{\sqrt{2\pi m k_{\mathrm{B}} T}}$$

is much smaller than the mean interparticle spacing—the ideal Fermi gas reduces to the classical Maxwell–Boltzmann gas. We now turn to the opposite regime,

$$\frac{\lambda^3 N}{V} \gg 1,$$

where the particle density is high enough that quantum effects dominate. In this regime, the Pauli exclusion principle prevents multiple fermions from occupying the same quantum state, producing a finite pressure even at zero temperature.

For fermions, the mean occupation number of a single-particle energy level ϵ_j is

$$\langle n_j \rangle_+ = \frac{\zeta e^{-\beta\epsilon_j}}{1 + \zeta e^{-\beta\epsilon_j}} = \frac{1}{1 + \zeta^{-1} e^{\beta\epsilon_j}}, \tag{13.34}$$

where $\zeta = e^{\beta\mu}$ is the fugacity. At high densities (large ζ), the Fermi functions $f_\nu(\zeta)$ must be evaluated for $\zeta \gg 1$. Sommerfeld showed that for large ζ,

$$f_{3/2}(\zeta \gg 1) \simeq \frac{4}{3\sqrt{\pi}} (\ln \zeta)^{3/2}. \tag{13.35}$$

Substituting Eq. (13.35) into the particle-number relation,

$$\frac{N}{V} = \frac{1}{\lambda^3} f_{3/2}(\zeta),$$

and solving for ζ in the degenerate limit, we find

$$\zeta \simeq e^{\beta\epsilon_{\mathrm{F}}},$$

where ϵ_{F} is defined as the *Fermi energy*:

$$\epsilon_{\mathrm{F}} = \frac{\hbar^2}{2m} \left(\frac{6\pi^2 N}{V}\right)^{2/3}. \tag{13.36}$$

The corresponding *Fermi temperature* is given by

$$k_{\mathrm{B}} T_{\mathrm{F}} = \epsilon_{\mathrm{F}}.$$

This temperature sets the energy scale separating the quantum and classical regimes of a Fermi gas.

At $T \to 0$, Eq. (13.34) becomes a step function:

$$\langle n_j \rangle_+ (T = 0) = \begin{cases} 1, & \epsilon_j < \epsilon_{\mathrm{F}}, \\ 0, & \epsilon_j > \epsilon_{\mathrm{F}}. \end{cases} \tag{13.37}$$

All states with energies below ϵ_{F} are occupied, while those above are empty. This defines the *Fermi surface*: the set of states in momentum space satisfying $\epsilon_p = \epsilon_{\mathrm{F}}$.

The Fermi Sphere

For a free electron gas, where $\epsilon_p = p^2/(2m)$, the total number of occupied states at $T = 0$ is obtained by integrating over momentum space. In three dimensions, we have

$$N = 4\pi \int_0^{p_{\mathrm{F}}} \frac{p^2\, dp}{h^3/(2s+1)} = \frac{4\pi p_{\mathrm{F}}^3}{3h^3} (2s+1) V, \tag{13.38}$$

where $s = 1/2$ for electrons and p_{F} is the Fermi momentum defined by $\epsilon_{\mathrm{F}} = p_{\mathrm{F}}^2/(2m)$. The occupied states form a sphere of radius p_{F} in momentum space—the *Fermi sphere*—which contains all N particles.

At zero temperature the system therefore consists of a completely filled Fermi sphere, with each quantum state occupied by a single fermion (or two when spin degeneracy is included). The Fermi surface forms the boundary between occupied and unoccupied states, and its properties determine the low-temperature behavior of metals and other degenerate Fermi systems.

The presence of a Fermi surface is a direct manifestation of the Pauli exclusion principle. Because each momentum state can contain at most one fermion per spin orientation, the particles fill successive energy levels up to ϵ_{F}. Even at absolute zero, the system possesses a finite internal energy and exerts a pressure known as the *degeneracy pressure*. This quantum mechanical pressure is crucial in stabilizing white dwarfs and neutron stars against gravitational collapse, and governs the electronic properties of solids at low temperatures.

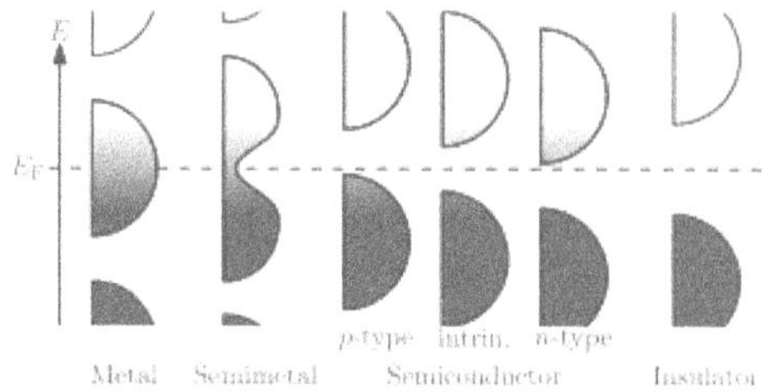

FIGURE 13.2 When the temperature is lower than the Fermi temperature, fermions perform a kind of "musical-chair ballet"—filling all available low-energy states until the Fermi surface is reached. Because no two fermions can occupy the same quantum state, additional particles must occupy higher-energy levels. This mechanism explains why materials exhibit very different conductive properties: in metals, the Fermi energy lies within a conduction band, so no energy gap separates occupied and unoccupied states, allowing electrons to move freely. (Source: Wikipedia)

Massless Bose Systems: The Photon Gas

We now turn to the statistical mechanics of a gas of photons—massless bosons confined in a cavity. Photons obey Bose–Einstein statistics but differ from massive particles in one crucial respect: their number is not conserved. As a result, the chemical potential vanishes, $\mu = 0$, and hence the fugacity is fixed at $\zeta = 1$.

For a photon of frequency ω and wavevector $\vec{k}$, we have

$$\epsilon = \hbar\omega, \qquad \vec{p} = \hbar\vec{k}, \qquad |\vec{k}| = \frac{\omega}{c}. \tag{13.39}$$

Because electromagnetic waves are transverse, the polarization vector $\boldsymbol{\epsilon}$ satisfies $\vec{\epsilon}\cdot\vec{k} = 0$. This yields two independent polarization states, giving a degeneracy factor of $g = 2$.

In a cubic box of side L, the allowed wavevectors are quantized as

$$\vec{k} = \frac{2\pi}{L}\vec{n}, \qquad \vec{n} \in \mathbb{N}^3.$$

The mean occupation number of each mode is then

$$n_{\vec{k}} = \frac{g}{e^{\beta\hbar\omega_{\vec{k}}} - 1} = \frac{2}{e^{\beta\hbar\omega_{\vec{k}}} - 1}. \tag{13.40}$$

The total number of photons in the cavity is obtained by integrating over all frequencies:

$$N = \sum_{\mathbf{k}} n_{\mathbf{k}} \simeq \frac{V}{\pi^2 c^3}\int_0^\infty \frac{\omega^2\, d\omega}{e^{\beta\hbar\omega} - 1} = \frac{2V(k_B T)^3}{\pi^2 c^3 \hbar^3}\int_0^\infty \frac{x^2\, dx}{e^x - 1}, \tag{13.41}$$

where we used the substitution $x = \beta\hbar\omega$. The remaining integral defines the Riemann zeta function:

$$\int_0^\infty \frac{x^k\, dx}{e^x - 1} = \Gamma(k+1)\,\zeta(k+1), \tag{13.42}$$

and for $k = 2$, $\zeta(3) \approx 1.202$. Thus,

$$N = \frac{2\,\zeta(3)}{\pi^2 c^3 \hbar^3} V (k_B T)^3 \equiv b\, V T^3, \qquad b = \frac{2\,\zeta(3) k_B^3}{\pi^2 c^3 \hbar^3}. \tag{13.43}$$

The number of photons in thermal equilibrium scales as T^3.

Internal Energy and Equation of State

The total internal energy is obtained similarly:

$$U = \sum_{\vec{k}} \epsilon_{\vec{k}} n_{\vec{k}} = \frac{2V}{\pi^2 c^3} \int_0^\infty \frac{\hbar\omega^3\, d\omega}{e^{\beta\hbar\omega} - 1}. \tag{13.44}$$

Changing variables again to $x = \beta\hbar\omega$, we find

$$\frac{U}{V} = \frac{2(k_B T)^4}{\pi^2 c^3 \hbar^3} \int_0^\infty \frac{x^3\, dx}{e^x - 1} = \frac{2(k_B T)^4}{\pi^2 c^3 \hbar^3} \Gamma(4)\, \zeta(4). \tag{13.45}$$

Since $\Gamma(4) = 6$ and $\zeta(4) = \pi^4/90$,

$$\frac{U}{V} = \frac{\pi^2 k_B^4 T^4}{15\hbar^3 c^3}. \tag{13.46}$$

The energy density therefore scales as T^4. From the thermodynamic relation $P = (1/3)U/V$, valid for massless relativistic particles, the equation of state of the photon gas is

$$P = \frac{1}{3}\frac{U}{V}. \tag{13.47}$$

Equation (13.46) represents the celebrated *Planck radiation law*, which solved the classical ultraviolet catastrophe and marked the birth of quantum mechanics.

Differentiating Eq. (13.46) with respect to T gives the specific heat at constant volume:

$$c_V = \left(\frac{\partial U}{\partial T}\right)_V = \frac{4\pi^2 k_B^4 T^3}{15\hbar^3 c^3}. \tag{13.48}$$

Both the internal energy and specific heat grow rapidly with T^4 and T^3, respectively, reflecting the unbounded increase in photon number as temperature rises ($\mu = 0$ implies no saturation of photon density).

If the cavity has a small aperture, the radiation escaping per unit area and time—the *emitted intensity*—is given by

$$I(\omega, T) = \frac{c}{4\pi} \int u(\omega, T) \cos\theta\, d\Omega = \frac{c}{4} u(\omega, T), \tag{13.49}$$

where $u(\omega, T)$ is the spectral energy density. Integrating over all frequencies yields the total radiative flux:

$$I(T) = \frac{c}{4}\frac{U}{V} = \sigma T^4, \tag{13.50}$$

with

$$\sigma = \frac{\pi^2 k_B^4}{60\hbar^3 c^2} \tag{13.51}$$

the *Stefan–Boltzmann constant*. Equation (13.51) is the theoretical foundation of the Stefan–Boltzmann law, which accurately describes the total power radiated by blackbodies—from laboratory cavities to stellar surfaces.

Blackbody Radiation Measurements and the Discovery of the Photon

The *ultraviolet catastrophe*, also known as the *Rayleigh–Jeans catastrophe*, refers to the classical prediction that an ideal blackbody at thermal equilibrium should emit an arbitrarily large amount of energy at short wavelengths. According to classical electrodynamics and equipartition, the radiated energy diverges in the ultraviolet region,

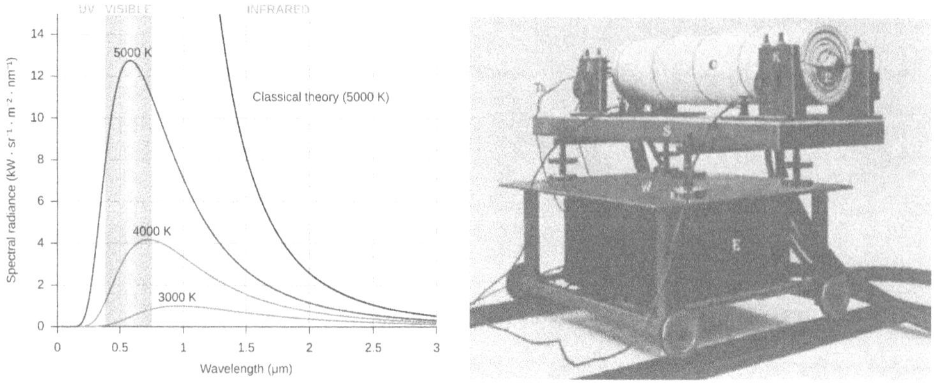

FIGURE 13.3 (a) Illustration of black-body radiation and the ultraviolet catastrophe (source: Wikipedia). **(b) Lummer and Kurlbaum's electrically heated black-body cavity**, reproduced from [LK98].

in stark contradiction with high-precision cavity-radiation measurements at the end of the nineteenth century. This failure of classical theory signaled that a crucial physical ingredient was missing.

Between 1897 and 1900, Otto Lummer and Ernst Pringsheim performed high-precision measurements of blackbody radiation up to wavelengths of about 6 μm and temperatures between 800 °C and 1400 °C. In 1900, Rubens and Kurlbaum employed the *Reststrahlenmethode* (residual-ray method) to probe even longer wavelengths. Their measurements clearly showed that Wien's radiation law, although accurate at short wavelengths, deviated systematically at long wavelengths, whereas the Rayleigh–Jeans formula diverged. These experiments revealed, beyond doubt, the inconsistency of classical radiation theory. The *law of blackbody radiation*, often called the *Wien-Planck law*, describes how a body emits electromagnetic radiation as a function of temperature. In 1893, Wilhelm Wien formulated his displacement law, identifying the wavelength of maximum emission at a given temperature, and proposed a spectral law valid at short wavelengths. However, his formula failed in the long-wavelength limit. In 1900, Max Planck introduced a new radiation law that fit the full spectrum by assuming that electromagnetic energy is emitted in discrete quanta. In 1905, Albert Einstein took Planck's hypothesis a decisive step further: he argued that Planck's energy quanta are not merely a mathematical device but represent real physical particles, later called *photons.* By modifying Boltzmann's statistical mechanics to describe an ensemble of such quanta, Einstein explained the photoelectric effect, a phenomenon that had defied classical theory, as well as an unpublished empirical law of Stokes concerning fluorescence.

A sequence of Nobel Prizes recognized the profound significance of blackbody radiation and the photon concept. The **1911 Nobel Prize in Physics** was awarded to Wilhelm Wien for *his discoveries regarding the laws governing the radiation of heat.* The **1918 Nobel Prize in Physics** went to Max Planck *in recognition of the services he rendered to the advancement of physics by his discovery of energy quanta.* The **1921 Nobel Prize in Physics** was awarded to Albert Einstein for *his services to theoretical physics, and especially for his discovery of the law of the photoelectric effect.*

Massive Bose Systems: Bose–Einstein Condensation

We now consider a system composed of noninteracting massive bosons. As in the Fermi case, it is convenient to work in the grand-canonical ensemble, where the particle number is allowed to fluctuate. The grand partition function for an ideal Bose gas is

$$\ln Z = -\sum_{\mathbf{p}} \ln\left(1 - \zeta e^{-\beta\epsilon_{\vec{p}}}\right), \tag{13.52}$$

where $\epsilon_{\vec{p}} = p^2/(2m)$ and $\zeta = e^{\beta\mu}$ is the fugacity.

The sum in Eq. (13.52) diverges as $\zeta \to 1$ (or $\mu \to 0$) due to the zero-momentum state $\vec{p} = 0$. It is therefore convenient to separate this singular term:

$$\ln Z = \ln(1-\zeta)^{-1} + \sum_{\vec{p}\neq 0} \ln\left(1 - \zeta e^{-\beta\epsilon_{\vec{p}}}\right)^{-1}. \tag{13.53}$$

Replacing the sum over nonzero $\vec{p}$ with an integral over momentum space yields

$$\ln Z = \ln(1-\zeta)^{-1} + \frac{4\pi V}{h^3}\int_0^\infty p^2 \ln\left(1 - \zeta e^{-\beta p^2/(2m)}\right)^{-1} dp. \tag{13.54}$$

The first term accounts for the occupancy of the ground state ($p = 0$), which becomes macroscopically large below the condensation temperature, while the integral describes the contribution of all excited states.

Expanding the logarithm as $\ln(1-x)^{-1} = \sum_{l=1}^{\infty} x^l/l$, we obtain

$$\begin{aligned}\frac{4\pi V}{h^3}\int_0^\infty p^2 \ln\left(1 - \zeta e^{-\beta p^2/(2m)}\right)^{-1} dp &= \frac{4\pi V}{h^3}\sum_{l=1}^{\infty}\frac{\zeta^l}{l}\int_0^\infty p^2 e^{-l\beta p^2/(2m)}\, dp \\ &= \frac{V}{\lambda^3}\, g_{5/2}(\zeta),\end{aligned} \tag{13.55}$$

where

$$g_\nu(\zeta) = \sum_{l=1}^{\infty}\frac{\zeta^l}{l^\nu}$$

is the *Bose function*, analogous to the Fermi function $f_\nu(\zeta)$, and $\lambda = h/\sqrt{2\pi m k_B T}$ is again the thermal de Broglie wavelength.

From Eq. (13.52), the pressure and number density are given by

$$\frac{P}{k_B T} = \frac{1}{V}\ln Z = \frac{1}{V}\ln(1-\zeta)^{-1} + \frac{1}{\lambda^3} g_{5/2}(\zeta), \tag{13.56}$$

$$\frac{N}{V} = \zeta\frac{\partial}{\partial\zeta}\left(\frac{1}{V}\ln Z\right) = \frac{\zeta}{1-\zeta} + \frac{1}{\lambda^3} g_{3/2}(\zeta). \tag{13.57}$$

The first term in each expression arises from the ground state and diverges as $\zeta \to 1$, signaling the onset of *Bose–Einstein condensation (BEC)*.

In the dilute regime ($N/V \ll \lambda^{-3}$), the fugacity ζ is small, and Eqs. (13.56)–(13.57) reduce to the classical ideal-gas law. Expanding $g_{3/2}(\zeta)$ and $g_{5/2}(\zeta)$ to second order,

$$g_{3/2}(\zeta) = \zeta + \frac{\zeta^2}{2^{3/2}} + O(\zeta^3), \qquad g_{5/2}(\zeta) = \zeta + \frac{\zeta^2}{2^{5/2}} + O(\zeta^3),$$

we can express ζ in terms of the density:

$$\zeta = \frac{\lambda^3 N}{V} - \frac{1}{2^{3/2}}\left(\frac{\lambda^3 N}{V}\right)^2 + O\left[\left(\frac{N}{V}\right)^3\right]. \tag{13.58}$$

Substituting Eq. (13.58) into $g_{5/2}(\zeta)$ yields

$$\frac{P}{k_B T} = \frac{1}{\lambda^3}\left[\frac{\lambda^3 N}{V} - \frac{1}{4\sqrt{2}}\left(\frac{\lambda^3 N}{V}\right)^2 + O\left[\left(\frac{N}{V}\right)^3\right]\right], \tag{13.59}$$

or equivalently,

$$\frac{PV}{Nk_B T} = 1 - \frac{1}{4\sqrt{2}}\frac{\lambda^3 N}{V} + O\left[\left(\frac{N}{V}\right)^2\right]. \tag{13.60}$$

The second virial coefficient for the Bose gas is therefore

$$B_2 = -\frac{\lambda^3}{4\sqrt{2}}, \tag{13.61}$$

opposite in sign to that of the Fermi gas. The negative correction reflects the bosonic tendency toward occupation of the same quantum state—an effective "attraction" of purely statistical origin.

At higher densities, as $\zeta \to 1$, the contribution from excited states saturates at a finite value

$$\frac{N_{\text{exc}}}{V} = \frac{1}{\lambda^3} g_{3/2}(1) = \frac{1}{\lambda^3}\zeta(3/2) \approx 2.612\,\lambda^{-3},$$

and any excess particles accumulate in the ground state. The critical temperature for the onset of condensation is determined by the condition $N = N_{\text{exc}}$:

$$T_c = \frac{2\pi\hbar^2}{mk_B}\left[\frac{n}{\zeta(3/2)}\right]^{2/3}, \tag{13.62}$$

where $n = N/V$ is the number density. Below T_c, a macroscopic fraction of the bosons occupies the ground state, forming a Bose–Einstein condensate.

The ideal Bose gas thus exhibits quantum corrections opposite in sign to those of the Fermi gas. At low densities, both reduce to the Maxwell–Boltzmann limit, but as the temperature decreases or the density increases, bosons begin to "condense" into the lowest energy state. The emergence of a macroscopic ground-state population below T_c is one of the most striking consequences of quantum statistics, confirmed experimentally in ultracold atomic gases and superfluid helium.

The Bose–Einstein Phase Transition

We have seen that in the grand-canonical partition function of a Bose gas, one particular term diverges as the fugacity approaches unity ($\zeta \to 1$). This divergence originates from the *zero-momentum mode* ($p = 0$), corresponding to a macroscopic occupation of the ground state. Physically, this signals the onset of a new phase—the *Bose–Einstein condensate*—in which a finite fraction of the particles collapses into the lowest-energy quantum state.

The total number density can be written as

$$n = \frac{N}{V} = n_0 + n^{(h)} = \frac{1}{V}\frac{\zeta}{1-\zeta} + \frac{1}{\lambda^3} g_{3/2}(\zeta), \tag{13.63}$$

where n_0 is the density of particles in the ground state and $n^{(h)}$ that of excited (thermal) states. From Eq. (13.63), the occupation of the zero mode is

$$n_0 = \frac{\zeta}{V(1-\zeta)} = \frac{1}{V}\frac{1}{e^{-\beta\mu} - 1}. \tag{13.64}$$

As the temperature decreases, the chemical potential μ increases from large negative values toward zero, but never exceeds it ($\mu < 0$). Near the transition,

$$n - n^{(h)} \simeq -\frac{1}{\beta\mu}, \tag{13.65}$$

implying

$$\mu \simeq -\frac{k_B T}{n - n^{(h)}} < 0, \tag{13.66}$$

so that $\mu \to 0^-$ as $T \to 0$. This shows that, at sufficiently low temperature, the ground state begins to accumulate a macroscopic number of particles.

The total density can be rewritten as

$$n = n_0 + \frac{1}{\lambda^3} g_{3/2}(\zeta). \tag{13.67}$$

The critical temperature T_c is defined by the condition $\zeta = 1$, at which the number of excited particles reaches its maximum value:

$$n_{\text{max}}^{(h)} = \frac{1}{\lambda_c^3} g_{3/2}(1) = \frac{1}{\lambda_c^3} \zeta(3/2) \simeq \frac{2.612}{\lambda_c^3}. \tag{13.68}$$

Here $\lambda_c = h/\sqrt{2\pi m k_B T_c}$ is the thermal wavelength at $T = T_c$. The critical temperature is therefore

$$T_c = \frac{2\pi\hbar^2}{m k_B} \left[\frac{n}{\zeta(3/2)}\right]^{2/3}. \tag{13.69}$$

For $T < T_c$, the fugacity remains effectively fixed at $\zeta \simeq 1$, and the fraction of condensed particles becomes

$$\frac{n_0}{n} = 1 - \frac{n^{(h)}}{n} = 1 - \frac{1}{n\lambda^3} g_{3/2}(1) = 1 - \left(\frac{T}{T_c}\right)^{3/2}. \tag{13.70}$$

Thus, below the critical temperature, a macroscopic fraction of the total particle number resides in the ground state. The function $n_0(T)$ is continuous across the transition, but its temperature derivative is not—indicating that the Bose–Einstein condensation is a *second-order phase transition.*

Chemical Potential and the Approach to Condensation

For $T > T_c$, the chemical potential remains negative and is determined by Eq. (13.67). As the temperature approaches T_c from above, $\mu \to 0^-$ and the population of the ground state diverges according to Eq. (13.64). Below T_c, μ remains pinned at zero, while n_0 grows rapidly with decreasing temperature as in Eq. (13.70).

At high temperature ($T > T_c$), the system behaves as a normal Bose gas, with particles distributed over many excited states. As the temperature decreases to T_c, the thermal wavelength λ becomes comparable to the interparticle spacing, and quantum coherence extends across the system. Below T_c, the particles collectively occupy the same quantum state, forming a macroscopic wavefunction—the *Bose–Einstein condensate.*

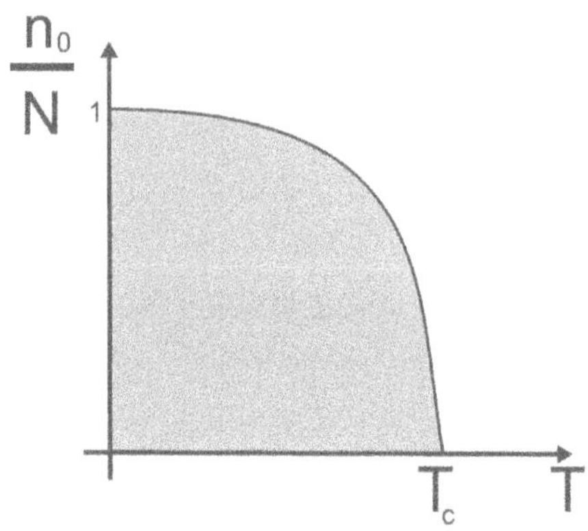

FIGURE 13.4 Temperature dependence of the condensate fraction n_0/N. Below the critical temperature T_c, a macroscopic fraction of particles occupies the ground state, while above T_c the occupation number rapidly decreases. The derivative $\partial_T n_0(T)$ is discontinuous at T_c, reflecting the non-analytic behavior of the Bose function $g_{3/2}(\zeta)$ at $\zeta = 1$.

Experimental Demonstration of Bose–Einstein Condensation: A Tour de Force

The possibility of realizing Bose–Einstein condensation (BEC) in a dilute gas of atomic particles was understood early on to be of profound physical interest. The requirements, however, are formidable. *For atoms to "see" one another, their wave functions must extend sufficiently to overlap appreciably.* The minimal spatial extent of an atomic wave packet is set by the *thermal de Broglie wavelength*, which must exceed the interparticle spacing in order for quantum statistics to play a significant role. Achieving this regime demands both very low energies (i.e., low temperatures) and high densities. Under such conditions, most atomic species tend to form molecules or condense into a liquid, so the challenge was to reach the required densities while maintaining a stable atomic gas.

A major conceptual breakthrough occurred in 1975, when T. Hänsch and A. Schawlow realized that laser light could be used to cool free atoms. Absorption of a photon, followed by spontaneous emission, imparts a net momentum change that—in aggregate—reduces the atomic velocity. With an appropriate three-dimensional configuration of laser beams, all motional degrees of freedom can be cooled. This mechanism, known as *Doppler cooling*, reaches a limit set by the randomness of spontaneous emission.

In 1986, J. Dalibard proposed combining magnetic-field gradients with Zeeman tuning to enhance trapping efficiency. This led to the development of the *magneto-optical trap (MOT)*, which simultaneously cools and confines atoms, has a large capture range, and became a cornerstone of all subsequent progress. Soon after, S. Chu, C. Cohen-Tannoudji, and W. D. Phillips refined laser cooling and trapping to unprecedented precision.

Reaching the ultralow temperatures required for BEC demanded one further step: *evaporative cooling*. Suggested by H. Hess in 1986, this method selectively removes the most energetic atoms from the trap, leaving the remaining ones at a lower effective temperature—analogous to how a cup of hot tea cools as steam escapes.

At JILA in Boulder, Colorado, Wieman proposed using alkali atoms, which are particularly amenable to laser cooling. His strategy was to cool atoms in a MOT, transfer them to a purely magnetic trap, and then apply evaporative cooling to reach the requisite low temperatures. Working with E. Cornell, this approach culminated in the first observation of BEC in ^{87}Rb in June 1995.

In parallel, W. Ketterle at MIT conducted experiments with ^{23}Na, achieving BEC only months after the JILA breakthrough. The MIT condensates contained more than two orders of magnitude more atoms, demonstrating both robustness and scalability of the phenomenon.

These achievements were recognized with two Nobel Prizes. The **1997 Nobel Prize in Physics** was awarded to Steven Chu, Claude Cohen-Tannoudji, and William Phillips *for development of methods to cool and trap atoms with laser light.* The **2001 Nobel Prize in Physics** was awarded jointly to Eric Cornell, Wolfgang Ketterle, and Carl Wieman *for the achievement of Bose–Einstein condensation in dilute gases of alkali atoms, and for early fundamental studies of the properties of the condensates.*

Further Reading. A comprehensive treatment of the underlying principles and experimental methods that enabled the observation of Bose–Einstein condensation are presented in the *Proceedings of the International School of Physics "Enrico Fermi", Course CXL: Bose–Einstein Condensation in Atomic Gases.* In particular, see the chapters by E. A. Cornell, J. R. Ensher, and C. E. Wieman on *Experiments in Dilute Atomic Bose–Einstein Condensation*; by W. Ketterle, D. S. Durfee, and D. M. Stamper-Kurn on *Making, Probing, and Understanding Bose–Einstein Condensation*; and by K. Helmerson and W. D. Phillips on *Cooling, Trapping and Manipulation of Neutral Atoms and BEC by Electromagnetic Fields.*

13.2.1 Energy and the Equation of State for Quantum Gases

The internal energy of an ideal quantum gas can be obtained directly from the grand potential. In the grand-canonical ensemble,

$$U = -\left(\frac{\partial \ln Z}{\partial \beta}\right)_\zeta = -\frac{\partial}{\partial \beta}\left(\frac{PV}{k_\mathrm{B}T}\right)_\zeta, \tag{13.71}$$

where the derivative is taken at constant fugacity $\zeta = e^{\beta\mu}$.

Carrying out the derivative, one finds

$$U = \begin{cases} \dfrac{3}{2}\dfrac{k_\mathrm{B}T}{\lambda^3}\, f_{5/2}(\zeta), & \text{Fermi gas,} \\[2ex] \dfrac{3}{2}\dfrac{k_\mathrm{B}T}{\lambda^3}\, g_{5/2}(\zeta), & \text{Bose gas,} \end{cases} \tag{13.72}$$

where $f_\nu(\zeta)$ and $g_\nu(\zeta)$ are the Fermi and Bose functions defined earlier, and $\lambda = h/\sqrt{2\pi m k_\mathrm{B}T}$ is the thermal de Broglie wavelength.

Neglecting the small ground-state term $(1/V)\ln(1-\zeta)$ in the Bose case (valid away from $T \simeq 0$), we find a simple relation between pressure and internal energy:

$$U = \frac{3}{2}PV, \tag{13.73}$$

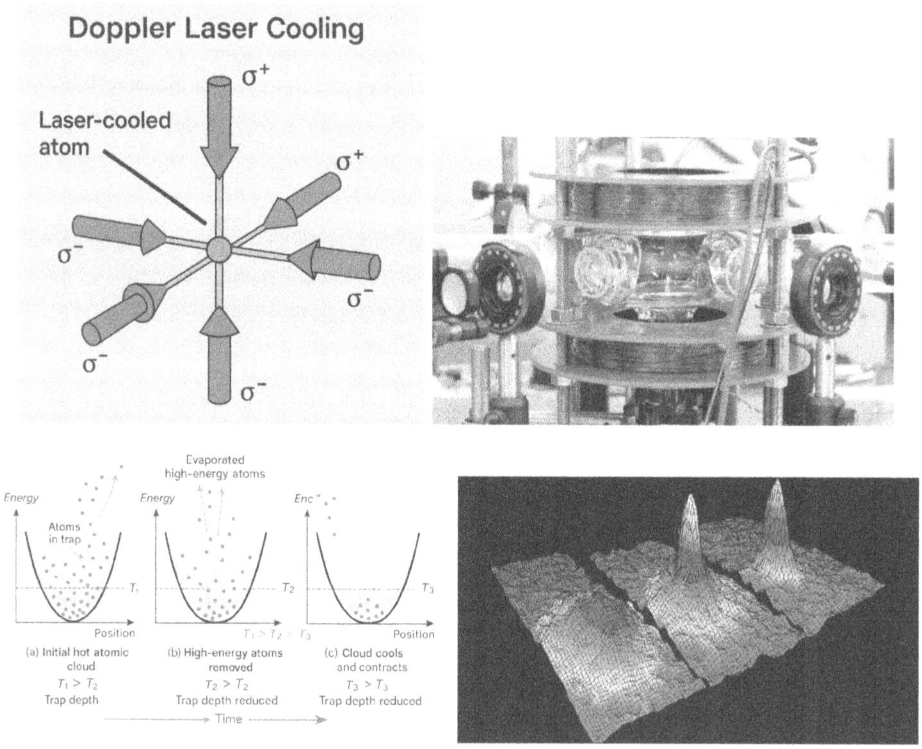

FIGURE 13.5 (Top left) Concept of laser cooling; (Top right) A magneto-optical trap; (Bottom left) Concept of evaporative cooling; (Bottom right) Observation of BEC in ^{87}Rb. (a) Corresponds to the situation just before the condensation sets in; (b) a condensate peak with a thermal background is observed; and (c) the situation where almost all atoms participate in the condensate, from [AEM$^+$95]. Figure reproduced from NIST/JILA/CU-Boulder [Geo20].

identical in form to the classical ideal-gas result. This proportionality holds for both Fermi and Bose gases, provided the temperature is above the degeneracy or condensation regime.

Starting from the grand-canonical expression, the pressure can also be written explicitly as an integral over momentum:

$$P = \pm \frac{k_B T}{2\pi^2} \int_0^\infty p^2 \ln\left(1 \pm e^{-\beta(\epsilon_p - \mu)}\right) dp, \tag{13.74}$$

where the upper sign refers to fermions and the lower to bosons. Integrating by parts, and noting that the surface terms vanish, gives

$$P = \frac{1}{3\pi^2} \int_0^\infty \frac{p^3}{m} \frac{dp}{e^{\beta(\epsilon_p - \mu)} \pm 1} = \frac{1}{2\pi^2} \int_0^\infty \frac{\hbar^2 p^2}{3m} n_p \, dp, \tag{13.75}$$

where $n_p = [e^{\beta(\epsilon_p - \mu)} \pm 1]^{-1}$ is the mean occupation number of the state with momentum p. Equation (13.75) highlights the microscopic origin of the pressure as the momentum flux of occupied states.

Expressed in dimensionless form using $x = p\sqrt{\beta/(2mk_B T)}$,

$$P = \frac{k_B T}{3\pi^2} \left(\frac{2mk_B T}{\hbar^2}\right)^{3/2} \int_0^\infty \frac{x^4 \, dx}{e^{x^2 - \beta\mu} \pm 1}. \tag{13.76}$$

This integral cannot be evaluated analytically in general, but its limiting forms reproduce the classical and degenerate-gas results derived earlier.

Experimental Demonstration of Superfluidity in ^{4}He

In 1908, Heike Kamerlingh Onnes achieved the first liquefaction of helium. Employing multiple pre-cooling stages and the Hampson–Linde cycle, which relies on the *Joule–Thomson effect*, he succeeded in lowering the temperature to helium's boiling point (−269 °C, or 4.2 K). By subsequently reducing the pressure of the liquid helium, Onnes reached temperatures close to 1.5 K—at that time, the lowest temperature ever attained on Earth.

In 1934, Pyotr Kapitsa constructed a new device for producing liquid helium, which cooled the gas by periodic expansions. For the first time, a machine had been made which could produce liquid helium in large quantities without previous cooling with liquid hydrogen. This heralded a new epoch in the field of low-temperature physics.

In the 1920s, it had been found that when liquid helium was exposed to a temperature of less than 2.3 degrees above absolute zero, it was changed into an unusual form, which was named He II, or *helium two*.

The superfluidity effect was discovered by Kapitsa and independently by John Allen and Don Misener in 1937. They showed that He II had such great internal mobility and *negligible or vanishing viscosity*, that it could better be characterized as a *superfluid*. During the next few years, Kapitsa's experiments on the properties of He II indicated that it is in a macroscopic quantum state, and that He II is therefore a *quantum fluid with zero entropy*, i.e., that it has a perfect atomic order. Later experiments by G. Rayfield and F. Reif demonstrated the existence of *quantized vortex lines* and O. Avenel and E. Varoquaux showed *macroscopic quantum tunneling (Josephson effect)* in ^{4}He.

Three Nobel Prizes were awarded related to superfluity in ^{4}He. The **1913 Nobel Prize in Physics** was awarded to Heike Kamerlingh Onnes *for his investigations on the properties of matter at low temperatures which led, inter alia, to the production of liquid helium.* The **1962 Nobel Prize in Physics** was awared to Lev Landau *for his pioneering theories for condensed matter, especially liquid helium.* The **1978 Nobel Prize in Physics** was awarded to Pyotr Kapitsa *for his basic inventions and discoveries in the area of low-temperature physics.*

Degenerate Fermi Gas

At very low temperatures, the Fermi gas remains strongly degenerate and the internal energy approaches

$$U = \frac{3}{5} N\epsilon_{\mathrm{F}} \left[1 + \frac{5\pi^2}{12} \left(\frac{T}{T_{\mathrm{F}}} \right)^2 + \cdots \right], \tag{13.77}$$

where ϵ_{F} and $T_{\mathrm{F}} = \epsilon_{\mathrm{F}}/k_{\mathrm{B}}$ are the Fermi energy and Fermi temperature, respectively. Even at $T = 0$, the gas exerts a finite *degeneracy pressure*

$$P_0 = \frac{2}{3}\frac{U_0}{V} = \frac{2}{5} n\epsilon_{\mathrm{F}}, \tag{13.78}$$

arising solely from the Pauli exclusion principle.

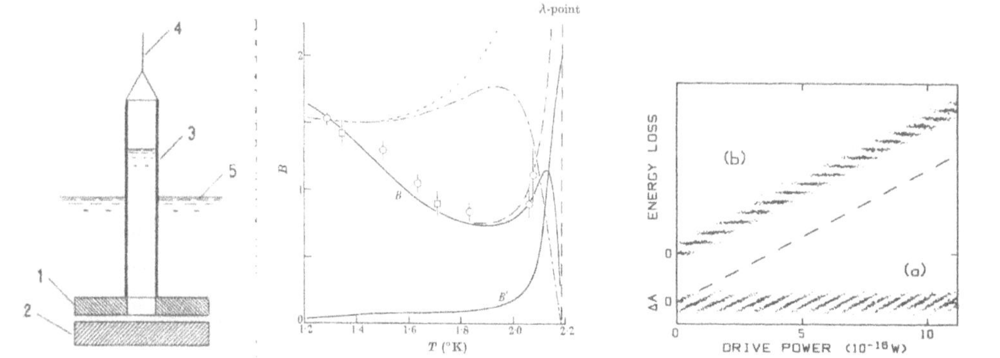

FIGURE 13.6 **(Left)** First apparatus used to measure the ultralow viscosity of liquid[4] He below the λ-point, reproduced from [Kap38]; **(Center)** Comparison between theory and experiment for the mutual-friction coefficient between the normal and superfluid components of uniformly rotating[4] He as a function of temperature, reproduced from Fig. 1 of [HV56]; **(Right)** Observation of quantized dissipation consistent with Josephson physics in the critical flow of superfluid ^{4}He through an aperture, reproduced from Fig. 3 of [AV85].

Experimental Verification of the Fermi-Dirac Statistics

The Fermi–Dirac statistics was first published in 1926 by Enrico Fermi and Paul Dirac. According to Dirac, it was first studied by Fermi, and Dirac called it *Fermi statistics* and the corresponding particles *fermions.*

Walter Jaeger and Heinrich Diesselhorst made in 1900 the most precise measurements of the relationship between the thermal and electrical conductivities of metals. They developed and applied a precise method for simultaneously measuring thermal conductivity, electrical conductivity, specific heat, and thermal electromotive force (thermoelectric power). Their methods were later refined and used by other physicists, such as Charles H. Lees and H. Kamerlingh Onnes, to extend measurements to lower temperatures.

In 1927 Arnold Sommerfeld applied the Fermi–Dirac statistics to electrons in metals, developed the free electron model, and compared classical and quantum predictions against Jaeger and Diesselhorst measurements of the so-called *Wiedemann–Franz law*: the ratio of thermal conductivity to electric conductivity in metals is proportional to temperature, $\kappa/\sigma = LT$, where L is the *Lorentz factor*. Classical theory predicted $L = (\pi^2/2)(k_\mathrm{B}/e)^2$ (Lorentz model), while Fermi–Dirac statistics predicted $L = (\pi^2/3)(k_\mathrm{B}/e)^2$, much closer to the experimental data.

Equations (13.72)–(13.76) encapsulate the full thermodynamics of ideal Fermi and Bose gases. At high temperatures, both reduce to the classical ideal-gas law $PV = Nk_\mathrm{B}T$. At low temperatures, quantum corrections become significant: for fermions, they produce a residual pressure and energy even at zero temperature, while for bosons they drive the system toward Bose–Einstein condensation.

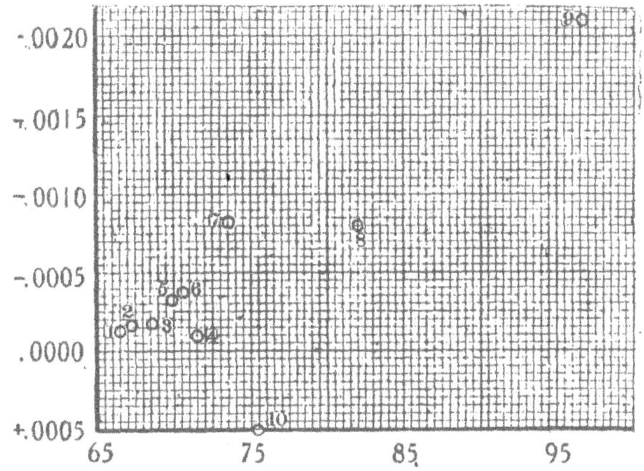

FIGURE 13.7 Temperature coefficient of thermal conductivity against Wiedemann–Franz ratio times 10^{-9}. Taken from Fig. 1 of [Bri22].

Discovery of Superfluidity in ^{3}He

In the early 1970s, David M. Lee, Douglas D. Osheroff, and Robert C. Richardson conducted a series of pioneering experiments at Cornell University that led to the discovery of *superfluidity in fermionic helium-3*. Their initial report of the new phases appeared in 1972 [ORL72] and quickly reshaped the landscape of low-temperature physics.

Unlike bosonic ^{4}He, which undergoes Bose–Einstein condensation and becomes superfluid at $T \approx 2.17$ K, ^{3}He is a *fermion* (nuclear spin 1/2) and cannot condense without first forming bosonic pairs. The idea that fermions can pair to form effective bosons was established in the Bardeen–Cooper–Schrieffer theory of superconductivity [BCS57]. In metals, this pairing is mediated by lattice vibrations and produces singlet Cooper pairs with $L = S = 0$. In ^{3}He, however, the atoms themselves provide an effective magnetic interaction that allows pairing in the *triplet*, *p*-wave channel. The resulting Cooper pairs carry internal orbital angular momentum $L = 1$ and spin $S = 1$, and possess a highly nontrivial, multi-component order parameter described in detail by Leggett's theory [Leg06]. In total, the ^{3}He order parameter has $2(2L + 1)(2S + 1) = 18$ real components, giving rise to a remarkably rich pattern of broken symmetries.

Experimental Signatures

The Cornell group was not initially searching for superfluidity, but for magnetic ordering in solid ^{3}He (see Fig. 13.8). Instead, they observed subtle anomalies in the melting-curve pressure of a liquid–solid ^{3}He mixture, showing distinct kinks at 2.7 mK and 1.8 mK. These turned out to be transitions into the now-famous **A** and **B** superfluid phases. **A** magnetic field favors the ***A*** phase and produces an additional spin-polarized phase, $\boldsymbol{A}_1$.

The discovery relied on a remarkable property of ^{3}He first recognized by Pomeranchuk [Pom50]: at millikelvin temperatures, *the liquid has lower entropy than*

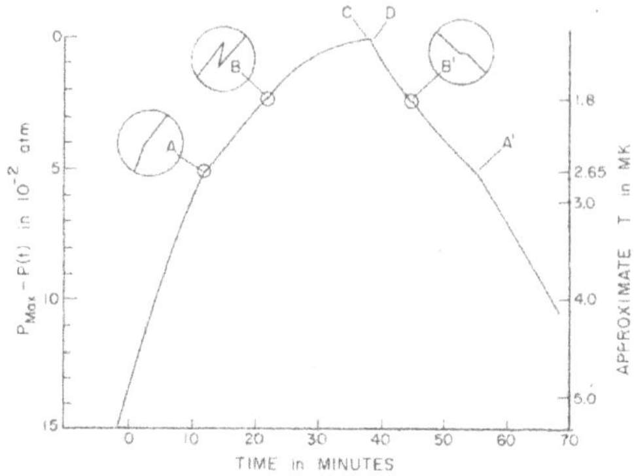

FIGURE 13.8 Internal pressure in a Pomeranchuk cell containing a liquid–solid^3 He mixture under uniform compression and decompression. Slope discontinuities at A and B mark the two superfluid transitions. Adapted from Osheroff, Richardson, and Lee [ORL72].

the solid. Compressing the liquid freezes a portion of it, requiring heat that is extracted from the remaining liquid, thereby cooling it. This *Pomeranchuk effect*, developed experimentally by Anufriev and later refined at Cornell, enabled temperatures below 2 mK.

Soon after the discovery, two independent groups confirmed the hydrodynamic signatures of superfluid ^{3}He. Lounasmaa's group in Helsinki measured the damping of a vibrating wire and observed a thousand-fold decrease in friction below 2 mK [Lou74]. Wheatley's group at La Jolla, California, detected fourth sound, a hallmark of superfluid flow in porous media [Whe75].

The significance of these breakthroughs was underscored by two Nobel Prizes: the **1996 Nobel Prize in Physics** was awarded to Lee, Osheroff, and Richardson *for their discovery of superfluidity in helium-3*, and the **2003 Nobel Prize in Physics** was awarded to Abrikosov, Ginzburg, and Leggett *for fundamental theoretical contributions to superconductors and superfluids* [LOR97, AGL04].

Bose–Einstein Condensation of Photons

A natural question in statistical mechanics is whether *photons*, the most ubiquitous bosons in nature, can undergo Bose–Einstein condensation (BEC). In thermal equilibrium, the answer is ordinarily *no*. In blackbody radiation, the photon number is *not conserved*: as the temperature decreases, photons are absorbed by the cavity walls, maintaining a chemical potential $\mu = 0$. As a result, the ground state is never macroscopically populated, and the usual BEC transition does not occur.

However, several theoretical works showed that if photons can undergo a number-conserving thermalization process—resulting in an effective nonzero chemical potential—then a genuine photonic BEC should be possible. This requires two ingredients: (i) a mechanism that thermalizes photon energies while conserving photon number, and (ii) a medium that acts as a reservoir for the photon gas without absorbing photons at low temperature.

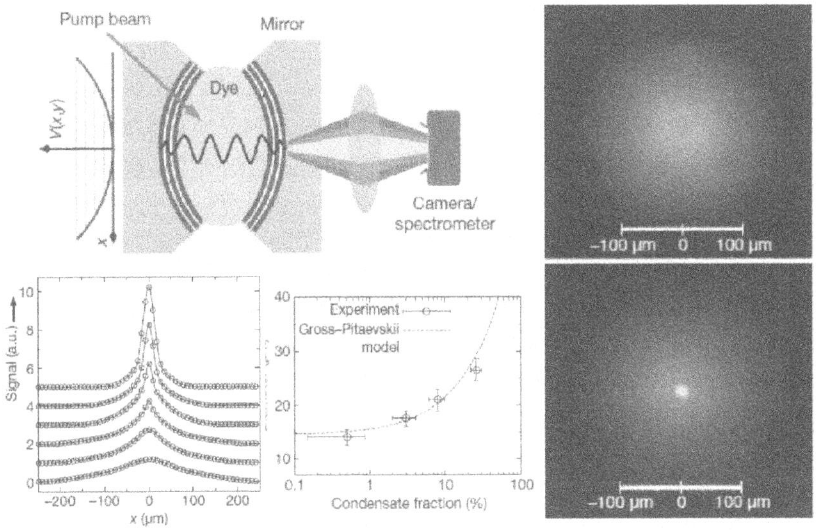

FIGURE 13.9 BEC photonic condensation reproduced from Klaers, Schmitt, Vewinger and Weitz [KSVW10]. **Top Left:** Schematic of the optical microcavity. The curvature of the mirrors provides a harmonic trapping potential for photons. **Right top and bottom:** Spatial intensity distribution through one cavity mirror. Below threshold (upper panel) the distribution is thermal; above threshold (lower panel) a bright, sharply peaked TEM_{00} condensate emerges. **Bottom left:** Cuts of the central intensity profile as the pump power increases, illustrating the rapid growth of the condensate peak. **Bottom center:** Condensate width versus condensate fraction, compared with a theoretical model based on the Gross–Pitaevskii equation.

Experimental Realization

In 2010, photonic BEC was demonstrated experimentally for the first time by Klaers, Schmitt, Vewinger, and Weitz [KSVW10]. The experiment used a dye-filled optical microcavity, acting as a "white-wall" box where photons repeatedly scatter off dye molecules without being lost. The curved cavity mirrors impose a harmonic trapping potential in the transverse direction and create an effective photon mass. Multiple absorption and re-emission events by the dye establish a quasi-thermal equilibrium at the temperature of the solution (approximately room temperature), while the cavity cutoff frequency fixes the photon number, giving rise to an effective nonzero chemical potential.

In this geometry, the photon gas behaves formally like a two-dimensional trapped gas of massive bosons. When the pump power is increased, the system crosses the BEC threshold and large numbers of photons accumulate in the lowest transverse mode (TEM_{00}), forming a macroscopic coherent condensate.

Bose–Einstein Condensation of Photons

Experimental Signatures

Figure 13.9 (from [KSVW10]) illustrates the key observations. The experiment demonstrated that BEC is not limited to massive particles such as atoms or excitons, but can arise in a *photon gas* when number conservation and appropriate thermalization are engineered. This result opened the door to new research directions, including photon fluids, non-equilibrium condensates, polariton BECs, and quantum simulators based on driven optical systems.

14

Classical Many-Body Physics: Introduction to Ferromagnets

14.1 Ferromagnets

Ferromagnetism is one of the most striking examples of collective behavior in matter: the emergence of macroscopic magnetization from microscopic magnetic moments. Its origin lies in the quantum mechanical property of particles known as *spin*. At high temperature, thermal agitation dominates and spins are randomly oriented, producing no net magnetic moment. Below a certain critical temperature, however, interactions between atomic moments lead to spontaneous alignment of spins, giving rise to a permanent magnetization. This spontaneous ordering represents a fundamental example of a *phase transition*.

The word "magnet" itself dates back to ancient Greece: the mineral *magnetite* (Fe_3O_4) was known for its mysterious ability to attract iron. Early references to magnetic materials also appear in Chinese and Indian sources, often in connection with early navigation. Modern understanding of ferromagnetism, however, began in the nineteenth and twentieth centuries with the work of Pierre Curie, Pierre Weiss, and later quantum theorists such as Heisenberg, who related magnetic ordering to exchange interactions between electron spins.

Ferromagnets are characterized by domains of uniformly aligned magnetic moments, while other forms of magnetism arise from different patterns of ordering:

- *Ferromagnetism:* parallel alignment of spins leading to a large net magnetization.
- *Ferrimagnetism:* antiparallel alignment of unequal spins, giving a smaller net moment.
- *Antiferromagnetism:* antiparallel alignment of equal spins, producing zero net magnetization below the Néel temperature.
- *Paramagnetism:* random spin orientations above the ordering temperature.

Let us now focus on the classical statistical mechanics of ferromagnets. The simplest microscopic models of magnetism are based on the behavior of the *magnetic moments* of atoms or

DOI: 10.1201/9781003603528-14

FIGURE 14.1 Schematic representation of magnetic ordering and experimental realization. A macroscopic magnetized sample.

TABLE 14.1 Examples of ferromagnetic and antiferromagnetic materials and their corresponding critical temperatures. The Curie temperature marks the onset of ferromagnetic order, while the Néel temperature marks the antiferromagnetic transition.

Material	Curie temperature T_c (K)	Néel temperature T_N (K)
Fe	1043	0
Co	1388	0
Ni	627	0
Gd	293	0
Dy	85	0
MnO	0	116
MnS	0	160
MnTe	0	307
MnF_2	0	67

ions in a crystalline solid. Each atom carries a magnetic moment proportional to its spin:

$$\vec{\mu}_i = \mu \vec{S}_i, \tag{14.1}$$

where $\vec{S}_i = (S_i^x, S_i^y, S_i^z)$ is a unit vector representing the direction of the spin, with $\vec{S}_i \cdot \vec{S}_i = 1$, and μ is the magnetic moment per particle. The moments are typically localized: we assume that the atoms are fixed at the sites of a crystal lattice and cannot move.

Magnetic Units and Scales

The magnetic moment μ is measured in ampere–square meters ($\mathrm{A} \cdot \mathrm{m}^2$) or equivalently in joules per tesla (J/T). The magnetic field B is measured in teslas ($1\,\mathrm{T} = 1\,\mathrm{Wb/m^2} = 1\,\mathrm{V \cdot s/m^2}$). For reference, the Earth's magnetic field varies between 20 and 60 μT, while the strongest continuous laboratory fields reach about 45 T (at the MagLab, Tallahassee, Florida), and pulsed fields can exceed 100 T (at the MagLab in Los Alamos, New Mexico).

In ferromagnetic materials, the collective magnetization is characterized by the *saturation magnetization* M_s, the maximum magnetization attainable when all spins are aligned. For example, in permalloy ($Fe_{80}Ni_{20}$), M_s typically ranges between 0.7 and 1.2 T, depending on purity and composition.

At equilibrium, various models have been developed to describe the temperature dependence of the magnetization. In the simplest case of a *non-interacting Pauli paramagnet*, each atomic spin can take only two orientations relative to an external magnetic field. This leads naturally to the introduction of discrete spin variables $s_i = \pm 1$, which serve as the basis of the *Ising model*. In this context, the system's macroscopic properties are governed by two key external parameters:

- the temperature T, which controls the strength of thermal fluctuations;
- the external magnetic field h, which couples linearly to the total magnetization.

The competition between the ordering tendency of the interactions and the disordering influence of thermal fluctuations lies at the heart of mean-field theory and of our statistical description of ferromagnets.

14.2 Basics of Magnetism: The $O(n)$ Interaction Models

The microscopic description of magnetism can be formulated in terms of interacting spins located on a lattice. Depending on the dimensionality of the spins, we obtain the general class of $O(n)$ *models*, where each spin $\vec{S}_i$ is an n-component unit vector. These include several important special cases:

- $O(1)$: the *Ising model*, where spins take only two values $S_i = \pm 1$. In this case $\mathcal{Z}_2$ symmetry is invariance under the transformation $S_i \to -S_i$. In this Volume, we use the notation the Ising spins are called σ_i, although more broadly in the literature they are also notationally represented as s_i.
- $O(2)$: the *XY model*, describing planar rotators with continuous symmetry in two dimensions;
- $O(3)$: the *Heisenberg model*, where spins can point freely in three-dimensional space.

Such models serve as idealized representations of real magnetic materials and, more generally, as paradigms for continuous symmetries and phase transitions in statistical mechanics.

Experimental Realizations of $O(N)$ Spin Models

Idealized spin models such as the Ising, XY, and Heisenberg models are central to statistical mechanics because they isolate the essential symmetries and collective behavior of many-body systems. Despite their simplicity, these models arise naturally in a wide range of real materials—and even in systems far outside of traditional physics.

The Ising model applies to systems whose microscopic variables take only two states (e.g., up/down, open/closed). Real physical realizations typically arise when the underlying Heisenberg Hamiltonian acquires a strong anisotropy that selects a preferred axis. For example, layered ferromagnets such as $CoNb_2O_6$ are celebrated realizations of a quasi-one-dimensional Ising chain, exhibiting both quantum and classical critical behavior [CTW+].

Strikingly, Ising-like behavior also appears in biological and cognitive systems. Cooperative base-pair opening in DNA and two-state protein folding/unfolding transitions behave as effective Ising variables, allowing statistical-mechanical descriptions of biological regulation [PS70, ME99]. In neuroscience and collective decision-making, the pairwise Ising model successfully predicts global neural network activity from recorded spike trains, demonstrating Ising-like correlations in cortical circuits [SBSB06].

These examples illustrate why the simplest $O(1)$ model is so ubiquitous, appearing across magnetism, biophysics, and collective behavior in complex systems.

Systems with a continuous $U(1)$ planar rotational symmetry correspond to the XY or $O(2)$ model. Its experimental relevance is tightly connected to the Berezinskii–Kosterlitz–Thouless (BKT) transition, in which topological vortices unbind.

Ultrathin superconducting films (e.g., Nb, Al) display BKT behavior in their resistivity curves, with vortex–antivortex unbinding driving the transition, as predicted by BKT theory [HN79, KT73]. Similarly, two-dimensional ^{4}He films on graphite exhibit the BKT superfluid transition with the universal jump in superfluid density observed experimentally [BR78]. In ultracold atoms, Bose gases confined in effectively two-dimensional traps realize XY physics, and direct imaging experiments have resolved the vortex proliferation at the BKT transition [HKC$^+$06].

These systems highlight the deep connection between $O(2)$ symmetry, topology, and the emergence of superfluid or superconducting phases in two dimensions.

The Heisenberg model describes systems with full three-dimensional rotational symmetry. Many classical magnets fall into this universality class. Simple ferromagnets such as EuO and antiferromagnets such as MnF_2 display critical behavior consistent with the $O(3)$ universality class, as measured in neutron-scattering and bulk magnetic experiments [BG70].

Liquid crystals also feature vector order parameters. The nematic–isotropic transition involves fluctuations of a director field that are well captured by $O(3)$-symmetric continuum theories [dGP93]. Spinor Bose–Einstein condensates (spin-1 or spin-2) realize nearly perfect $O(3)$ symmetry in a highly tunable platform, enabling precise measurements of Heisenberg-like criticality [SKU13].

These realizations show how the $O(3)$ model governs the collective behavior of three-dimensional spin systems, from traditional magnetic materials to quantum gases.

Taken together, the $O(N)$ models serve not only as theoretical paradigms but also as experimentally realizable frameworks across condensed matter physics, biophysics, and complex systems.

The Heisenberg Model: An $O(3)$ System

To construct a general energy functional for a magnetic system, let us assume that the free energy $f(\{\vec{S}\}, \vec{H})$ is a continuous and analytic function of the local spin directions $\vec{S}_i$ and of an external magnetic field $\vec{H}$. Expanding f around a reference configuration, for example, where all spins are aligned along the z-axis ($\vec{S}_i = \hat{z}$), we can write

$$\begin{aligned} f(\vec{S} \approx \hat{z}, \vec{H}) \approx f(\hat{z}, \vec{H}) &+ \sum_i (\vec{S}_i - \hat{z}) \cdot \nabla_{\vec{S}_i} f \big|_{\vec{S}=\hat{z}} \\ &+ \frac{1}{2} \sum_{i,j} (\vec{S}_i - \hat{z}) \cdot \left[\nabla_{\vec{S}_i} \nabla_{\vec{S}_j} f \big|_{\vec{S}=\hat{z}} \right] (\vec{S}_j - \hat{z}) + \cdots . \end{aligned} \tag{14.2}$$

Neglecting constant terms and higher-order corrections, this expansion can be rewritten as

$$f(\vec{S}, \vec{H}) \simeq \sum_i \vec{S}_i \cdot \tilde{\vec{h}}_i + \frac{1}{2} \sum_{i,j} \sum_{\mu,\nu} S_i^\mu S_j^\nu \, J_{ij}^{\mu\nu} , \tag{14.3}$$

where the coefficients $J_{ij}^{\mu\nu}$ are the *exchange couplings* between spins i and j. The first term represents the coupling of each spin to an effective local magnetic field $\tilde{\vec{h}}_i$.

Rotational Invariance

A realistic magnetic system in the absence of an external field must be rotationally invariant: if all spins are rotated by the same orthogonal transformation $O_{\mu\nu}$, the energy should remain unchanged. This condition requires

$$J_{ij}^{\mu\nu} = \sum_{\mu',\nu'} O_{\mu\mu'} O_{\nu\nu'} J_{ij}^{\mu'\nu'},$$

which is satisfied if

$$J_{ij}^{\mu\nu} = J_{ij}\,\delta_{\mu\nu}. \tag{14.4}$$

In that case, the interaction term reduces to the scalar product

$$\sum_{\mu,\nu} S_i^{\mu} S_j^{\nu} J_{ij}^{\mu\nu} = J_{ij}\,\vec{S}_i \cdot \vec{S}_j, \tag{14.5}$$

which is invariant under all orthogonal rotations. The Hamiltonian can thus be written compactly as

$$H = -\sum_i \vec{S}_i \cdot \vec{h} - \sum_{i<j} J_{ij}\,\vec{S}_i \cdot \vec{S}_j, \tag{14.6}$$

known as the *Heisenberg Hamiltonian.* This model describes spins with continuous rotational $(O(3))$ symmetry, and the constants J_{ij} quantify the strength of the exchange interaction.

REMARK 14.1 The exchange interaction originates from quantum mechanics: it is a consequence of the Pauli exclusion principle and the antisymmetry of the many-electron wavefunction. A positive $J_{ij} > 0$ favors parallel alignment (ferromagnetism), while $J_{ij} < 0$ favors antiparallel alignment (antiferromagnetism).

Historical Note: Curie, Weiss, Heisenberg, and the Ising Model

The scientific study of magnetism began with the pioneering experiments of Pierre Curie, who established that the magnetic susceptibility χ of many paramagnetic materials varies inversely with temperature:

$$\chi = \frac{C}{T},$$

now known as the *Curie law.* This simple empirical observation demonstrated that thermal agitation acts to randomize magnetic moments, while lower temperatures favor alignment.

Building on Curie's results, Pierre Weiss introduced the concept of a *molecular field*—an internal effective magnetic field proportional to the overall magnetization M. This idea led to the *Curie–Weiss law,*

$$\chi = \frac{C}{T - T_c},$$

which successfully describes the divergence of magnetic susceptibility at the critical temperature T_c, where spontaneous magnetization sets in. Weiss's mean-field hypothesis provided the first theoretical link between microscopic spin interactions and macroscopic magnetic order.

In 1928, Werner Heisenberg gave the first microscopic quantum-mechanical foundation of ferromagnetism. He demonstrated that the alignment of neighboring spins arises from the *exchange interaction,* a direct consequence of the antisymmetry of the electronic wavefunction under particle exchange. This result led to the formulation of the *Heisenberg Hamiltonian,* the prototype for describing interacting spin systems with continuous rotational $(O(3))$ symmetry.

A few years earlier, in 1920, Wilhelm Lenz had proposed to his student Ernst Ising a simplified, one-dimensional version of Weiss's mean-field idea, where each spin variable $s_i = \pm 1$ interacts only with its nearest neighbors. Ising solved the model exactly in his 1925 doctoral thesis [Isi25] and showed that in one dimension no phase transition occurs at finite temperature, a result that initially suggested the model's limitations.

However, the two-dimensional version of the model became a cornerstone of statistical mechanics when Lars Onsager obtained its exact solution in 1944 [Ons44], demonstrating the existence of a finite-temperature phase transition and providing the first exact calculation of a critical point in a lattice model. Onsager's work inaugurated the modern study of critical phenomena, leading eventually to scaling theory and the renormalization group.

Together, the contributions of Curie, Weiss, Heisenberg, Lenz, Ising, and Onsager form the historical and conceptual foundation of our present understanding of magnetic ordering and phase transitions.

The Heisenberg model represents the starting point for the statistical mechanics of magnetic systems and, more broadly, for the study of systems with continuous symmetries. In subsequent sections we will see how simplifying this model (for example by reducing spin dimensionality to one component) leads to tractable systems such as the Ising and XY models, which remain paradigms in the study of phase transitions and critical phenomena.

Quantum Origin of the Exchange Interaction

The microscopic origin of ferromagnetism lies in the *exchange interaction*—a purely quantum-mechanical effect arising from the indistinguishability of electrons and the symmetry of their wavefunctions. This mechanism was first identified by Werner Heisenberg in 1928 as the essential source of magnetic ordering.

Consider two atoms with neighboring electrons, each carrying spin 1/2. As a simplification, let each electron be described by a localized orbital $\phi_i(\vec{r})$. For identical fermions, the total wavefunction must be antisymmetric under exchange of both spatial and spin coordinates:

$$\Psi(1,2) = -\Psi(2,1).$$

We therefore construct symmetric and antisymmetric spatial wavefunctions:

$$\Phi_\pm(\vec{r}_1, \vec{r}_2) = \frac{1}{\sqrt{2}} \left[\phi_1(\vec{r}_1)\phi_2(\vec{r}_2) \pm \phi_2(\vec{r}_1)\phi_1(\vec{r}_2)\right], \tag{14.7}$$

together with the corresponding spin states:

- *Triplet states* (symmetric in spin):

$$|S=1, m_s=1\rangle = |++\rangle, \qquad |S=1, m_s=-1\rangle = |--\rangle,$$
$$|S=1, m_s=0\rangle = \tfrac{1}{\sqrt{2}}(|+-\rangle + |-+\rangle).$$

- *Singlet state* (antisymmetric in spin):

$$|S=0, m_s=0\rangle = \tfrac{1}{\sqrt{2}}(|+-\rangle - |-+\rangle).$$

Because fermions require the *total* wavefunction to be antisymmetric, the allowed combinations are:

$$\Psi(1,2) = \begin{cases} \Phi_+(\vec{r}_1,\vec{r}_2)\,|S=0\rangle, & \text{(singlet)}, \\ \Phi_-(\vec{r}_1,\vec{r}_2)\,|S=1\rangle, & \text{(triplet)}. \end{cases} \tag{14.8}$$

The two-electron Hamiltonian is

$$\hat{H} = \hat{H}_1 + \hat{H}_2 + V(|\vec{r}_1 - \vec{r}_2|), \tag{14.9}$$

where V is the Coulomb repulsion between electrons. Evaluating its expectation value in the states of Eq. (14.8) gives, to first order in perturbation theory,

$$E_{\text{singlet}} = \Delta + J_{12}, \qquad E_{\text{triplet}} = \Delta - J_{12}, \tag{14.10}$$

where Δ depends only on the radial shape of the orbitals, and

$$J_{12} = \int d^3r_1\, d^3r_2\, \phi_1^*(\vec{r}_1)\phi_2^*(\vec{r}_2)\, V(|\vec{r}_1 - \vec{r}_2|)\, \phi_2(\vec{r}_1)\phi_1(\vec{r}_2) \tag{14.11}$$

is the *exchange integral.* If $J_{12} > 0$, the triplet state has lower energy—implying parallel spin alignment and thus the microscopic origin of *ferromagnetism.* Paul Dirac later showed that the exchange effect can be represented by an effective spin–spin interaction [Dir29] of the form

$$H_{\text{exch}} = -J(|\vec{R}_1 - \vec{R}_2|)\left(\frac{1}{2} + 2\,\vec{S}_1 \cdot \vec{S}_2\right), \tag{14.12}$$

where J is the exchange coupling. This effective interaction is precisely the one appearing in the Heisenberg Hamiltonian [Eq. (14.6)], derived earlier from symmetry considerations. Equation (14.12) embodies the fundamental principle that the antisymmetry of fermionic wavefunctions gives rise to an effective spin-dependent force.

In real materials, several mechanisms can mediate exchange interactions. Direct exchange arises from the direct overlap of electronic orbitals on neighboring atoms, as discussed above. In the case of *superexchange,* it arises from virtual hopping of electrons via an intermediate non-magnetic ion (important in oxides). For *indirect exchange (RKKY),* it is mediated by conduction electrons in metals, leading to oscillatory couplings [RK54]. Instead, the *Dzyaloshinskii–Moriya interaction* is an antisymmetric exchange interaction due to spin–orbit coupling and lack of inversion symmetry [Dzy58, Mor60]. All these mechanisms can be viewed as manifestations of the same underlying quantum principle, the correlation between spin alignment and spatial overlap of electronic wavefunctions.

Historical Note: The Bohr–van Leeuwen Theorem

One of the most surprising results in early twentieth-century statistical physics was the *Bohr–van Leeuwen theorem,* discovered in Bohr's 1911 doctoral thesis and independently by van Leeuwen in 1919 [Boh72, vL21]. It states that *no classical system of charged particles in thermal equilibrium can exhibit magnetism*: both diamagnetism and paramagnetism vanish identically. At face value, the theorem appears to make all magnetic phenomena impossible in classical physics—a profound contradiction with experiment, which ultimately helped motivate the emergence of quantum mechanics and the quantum theory of magnetism, as famously reviewed by Van Vleck [Vle32].

Classical Derivation

Consider a classical system of N charged particles in an external magnetic field $\vec{B} = \nabla \times \vec{A}$. The Hamiltonian is

$$H(\{\vec{r}_i, \vec{p}_i\}) = \sum_{i=1}^{N} \frac{1}{2m_i}\left[\vec{p}_i - q_i\vec{A}(\vec{r}_i)\right]^2 + V(\{\vec{r}_i\}).$$

The canonical partition function in phase space is

$$Z(\vec{B}) = \frac{1}{h^{3N}} \int d^{3N}r \, d^{3N}p \, \exp\Big[-\beta H(\vec{r}, \vec{p}; \vec{B})\Big] .$$

Perform in the momentum variables the shift $\vec{p}_i \mapsto \vec{p}_i' = \vec{p}_i - q_i \vec{A}(\vec{r}_i)$. Since the Jacobian is unity,

$$\int d^3p_i \, e^{-\beta[\vec{p}_i - q_i\vec{A}]^2/2m_i} = \int d^3p_i' \, e^{-\beta(p_i')^2/2m_i},$$

and the vector potential disappears completely from the partition function. Thus Z is *independent of* $\vec{B}$,

$$\frac{\partial Z}{\partial \vec{B}} = 0, \qquad \Rightarrow \qquad \langle \vec{M} \rangle = -\frac{1}{\beta}\frac{\partial \ln Z}{\partial \vec{B}} = 0.$$

The average magnetization is therefore *exactly zero* in any classical system described by a Gibbs ensemble.

Although often presented as an absolute no-go result, the classical Bohr–van Leeuwen theorem relies on several implicit assumptions in its derivation:

1. translational invariance of the Hamiltonian,
2. absence of additional conserved quantities beyond energy,
3. a simply connected geometry without topological constraints, and
4. boundary conditions that do not impose restrictions on phase-space trajectories.

When any of these assumptions is violated, the conclusion of vanishing magnetization need not hold. In particular, the geometry and symmetry of the confining boundary can impose additional conservation laws—such as angular momentum conservation in rotationally symmetric domains—leading to non–Gibbsian steady states and persistent chiral motion. Recent work on classical hard–disk gases in circular cavities has demonstrated precisely this effect: the circular boundary enforces conservation of angular momentum and drives the system toward a *Generalized Gibbs Ensemble (GGE)* for a large number of particles, an exponential family depending on all conserved quantities, including L_z [CV24, Inn02, KS18]. Because the resulting distribution explicitly breaks time-reversal symmetry, a finite average angular momentum—and therefore a finite magnetic moment—emerges, even in purely classical dynamics. Earlier analyses also emphasized that additional integrals of motion or boundary-induced constraints can invalidate the classical Gibbs measure [Dub11, Dub08, FG39].

Thus, the Bohr–van Leeuwen theorem is best understood not as a fundamental prohibition of classical magnetism, but as a consequence of the specific symmetry assumptions entering the Gibbs ensemble. It is intended for real physical systems in which imperfections are natural phenomena, as it is clear, for instance, in Tolman [Tol38]. Modern studies make clear that geometry, boundary conditions, and conservation laws can induce equilibrium distributions that fall outside the canonical ensemble, allowing finite classical magnetization to appear under well-defined conditions.

14.2.1 The XY Model: $O(2)$

The $O(2)$ or *XY model* is a dimensionally reduced version of the Heisenberg model in which the spins are constrained to lie in a plane. Each spin $\mathbf{S}_i$ is a two-dimensional unit vector characterized by an angle θ_i:

$$\vec{S}_i = \big(\cos\theta_i,\, \sin\theta_i\big). \tag{14.13}$$

The Hamiltonian takes the form

$$H = -\sum_{i<j} J_{ij}\, \vec{S}_i \cdot \vec{S}_j = -\sum_{i<j} J_{ij} \cos(\theta_i - \theta_j), \tag{14.14}$$

where the coupling J_{ij} favors alignment of neighboring spins.

The XY model exhibits rich and subtle behavior. In two dimensions, it does not undergo a conventional symmetry-breaking phase transition at finite temperature. Instead, it supports a topological transition, the *Berezinskii–Kosterlitz–Thouless (BKT) transition*, in which bound pairs of vortex and antivortex excitations unbind above a critical temperature [KT73, Kos74]. Below the BKT temperature, long-range order is replaced by a state with algebraically decaying correlations, characteristic of a topological phase. The XY model is thus a paradigmatic example of how continuous symmetries behave in low-dimensional systems. This will be analyzed in detail in Volume II.

14.2.2 The Ising Model: $O(1)$

The *Ising model* represents the discrete ($O(1)$) limit of the Heisenberg model, where each spin can only point "up" or "down" along a fixed axis. It was introduced by Wilhelm Lenz in 1920 and solved in one dimension by his student Ernst Ising in 1925, who found no finite-temperature transition. Nevertheless, its two-dimensional version later became one of the most important exactly solvable models in statistical mechanics.

The Hamiltonian of the Ising model is

$$H = -\sum_{i<j} J_{ij}\, S_i^z S_j^z - h\sum_i S_i^z \equiv -\sum_{i<j} J_{ij}\, \sigma_i\sigma_j - h\sum_i \sigma_i \tag{14.15}$$

where $S_i^z = \sigma_i = \pm 1$ represents the orientation of the spin along the z-axis, J_{ij} is the coupling constant, and h is a uniform external magnetic field. For $h \neq 0$, this model is *not* $\mathcal{Z}_2$ invariant. The model captures the competition between interaction-induced order (favoring parallel alignment) and thermal disorder (favoring random orientations). Notationwise, the Ising varibales are often written as s_i or σ_i in the literature, dropping the components.

A remarkable feature of the Ising model is its equivalence to a simple model of a classical fluid—the *lattice gas*. This correspondence provides a direct statistical-mechanical link between magnetism and the liquid–gas transition.

Consider a classical system of particles interacting through a short-range potential approximated by a square-well form:

$$U(r) = \begin{cases} \infty, & r < r_0, \\ -\epsilon, & r_0 \le r < r_0(1+\delta), \\ 0, & r \ge r_0(1+\delta). \end{cases} \tag{14.16}$$

Because the potential is strongly repulsive at short distance, no two particles can occupy the same site. We can therefore consider a regular lattice where each site i may either be occupied ($s_i = 1$) or empty ($s_i = 0$). The total energy of a configuration $\{s_i\}$ is

$$H = \sum_{i<j} \phi_{ij}\, s_i s_j, \tag{14.17}$$

where $\phi_{ij} = U(|\vec{r}_i - \vec{r}_j|)$ encodes the pairwise interaction.

Introducing spin variables

$$s_i = \frac{1-\sigma_i}{2}, \qquad \sigma_i = \pm 1, \tag{14.18}$$

the Hamiltonian becomes

$$H = \frac{1}{4}\sum_{i<j} \phi_{ij} - \sum_i \Phi_i \sigma_i + \frac{1}{4}\sum_{i<j} \phi_{ij}\sigma_i\sigma_j, \tag{14.19}$$

$$\Phi_i = \frac{1}{2}\sum_j \phi_{ij}. \tag{14.20}$$

Neglecting constant terms, we see that the lattice gas maps exactly onto an Ising model in an *effective external field* proportional to the chemical potential. Consequently, the Ising model and the lattice gas share the same critical behavior, providing a unifying statistical-mechanical description of both ferromagnetism and the liquid–gas transition.

REMARK 14.2 In this sense, the $O(1)$, $O(2)$, and $O(3)$ models can be viewed as successive classical limits of spin systems with decreasing degrees of rotational freedom. The Ising model describes discrete order parameters, the XY model continuous planar symmetry and topological order, and the Heisenberg model full rotational symmetry in three dimensions. Together they form a hierarchy of models that capture the essential statistical features of magnetism and classical many-body physics.

14.2.3 The One-Dimensional Ferromagnetic Ising Model

The one-dimensional Ising model provides an exact example of how microscopic interactions can be solved analytically, and it introduces one of the most powerful techniques in statistical mechanics—the *transfer matrix method.* Although the model does not exhibit a finite-temperature phase transition, it remains a paradigmatic case for understanding spin correlations and collective behavior.

Consider a chain of N spins $\sigma_i = \pm 1$ with nearest-neighbor ferromagnetic coupling $J > 0$ and periodic boundary conditions ($\sigma_{N+1} = \sigma_1$). In the presence of a magnetic field h, the Hamiltonian is

$$H = -J\sum_{i=1}^{N} \sigma_i\sigma_{i+1} - h\sum_{i=1}^{N} \sigma_i. \tag{14.21}$$

The partition function is then

$$Z = \sum_{\{\sigma_i = \pm 1\}} e^{\beta J \sum_i \sigma_i\sigma_{i+1} + \beta h \sum_i \sigma_i}. \tag{14.22}$$

The key observation is that each pair interaction in Eq. (14.21) involves only two neighboring spins. This allows the Boltzmann factor to be written as a product of local terms:

$$\begin{aligned} Z &= \sum_{\{\sigma\}} \prod_{i=1}^{N} e^{\beta J\sigma_i\sigma_{i+1} + \frac{1}{2}\beta h(\sigma_i + \sigma_{i+1})} = \sum_{\{\sigma\}} \mathcal{T}_{12}\mathcal{T}_{23}\cdots\mathcal{T}_{N1} \\ &= \mathrm{Tr}(\mathcal{T}^N), \end{aligned} \tag{14.23}$$

where $\mathcal{T}$ is the *transfer matrix* defined by

$$\mathcal{T} = \begin{pmatrix} e^{\beta(J+h)} & e^{-\beta J} \\ e^{-\beta J} & e^{\beta(J-h)} \end{pmatrix}. \tag{14.24}$$

The trace arises from the periodic boundary condition $\sigma_{N+1} = \sigma_1$.

Because $\mathcal{T}$ is real and symmetric, it can be diagonalized. Let $\lambda_1 > \lambda_2$ denote its eigenvalues. Then,

$$Z = \lambda_1^N + \lambda_2^N. \tag{14.25}$$

In the thermodynamic limit ($N \to \infty$), the free energy per spin is dominated by the largest eigenvalue:

$$f = -\frac{1}{\beta} \ln \lambda_1, \qquad F = Nf. \tag{14.26}$$

The eigenvalues of the transfer matrix are obtained directly from Eq. (14.24):

$$\lambda_{1,2} = e^{\beta J} \cosh(\beta h) \pm \sqrt{e^{2\beta J} \sinh^2(\beta h) + e^{-2\beta J}}. \tag{14.27}$$

For zero field ($h = 0$),

$$\lambda_1 = 2\cosh(\beta J), \qquad \lambda_2 = 2\sinh(\beta J).$$

Hence, the free energy per spin becomes

$$f = -\frac{1}{\beta} \ln[2\cosh(\beta J)]. \tag{14.28}$$

All thermodynamic quantities can be derived from f by differentiation. In particular, since f is analytic for all finite T, the one-dimensional Ising model exhibits *no phase transition* at nonzero temperature.

The thermal average of the magnetization per spin is

$$m = \frac{1}{N} \sum_i \langle \sigma_i \rangle = \frac{\mathrm{Tr}(\sigma_1^z \mathcal{T}^N)}{Z}. \tag{14.29}$$

In the large-N limit, only the largest eigenvalue contributes, and using the normalized eigenvector $|u_1\rangle$ associated with λ_1, one obtains

$$m = \langle u_1 | \sigma^z | u_1 \rangle = \frac{e^{\beta J} \sinh(\beta h)}{\sqrt{e^{2\beta J} \sinh^2(\beta h) + e^{-2\beta J}}}. \tag{14.30}$$

At zero field, $m = 0$ for all $T > 0$, confirming the absence of spontaneous magnetization in one dimension. The susceptibility $\chi = \partial m / \partial h$ and specific heat $C_V = -T\, \partial^2 f / \partial T^2$ can be obtained directly from Eqs. (14.27) and (14.30).

The correlation function between spins at sites k and $k + r$ is defined as

$$g(r) = \langle \sigma_k \sigma_{k+r} \rangle - \langle \sigma_k \rangle \langle \sigma_{k+r} \rangle. \tag{14.31}$$

In zero field, $\langle \sigma_i \rangle = 0$, and one finds

$$g(r) = \langle \sigma_k \sigma_{k+r} \rangle = \left(\frac{\lambda_2}{\lambda_1} \right)^r = e^{-r/\xi}, \tag{14.32}$$

where the correlation length is given by

$$\xi^{-1} = \ln\left(\frac{\lambda_1}{\lambda_2} \right). \tag{14.33}$$

Thus, correlations decay exponentially with distance, and the correlation length remains finite for all $T > 0$. Only at $T = 0$ ($\lambda_1 = \lambda_2$) does the system become fully ordered.

Two-Point Correlations and the Correlation Length

We now compute the two-point correlation function of the one-dimensional Ising chain to make explicit the absence of long-range order at any finite temperature.

The correlation between neighboring spins can be expressed as a derivative of the free energy with respect to the coupling constant $J_{i,i+1}$:

$$\langle \sigma_i \sigma_{i+1} \rangle = -\frac{1}{\beta}\frac{\partial \ln Z}{\partial J_{i,i+1}} = -\frac{1}{\beta Z}\frac{\partial Z}{\partial J_{i,i+1}}. \tag{14.34}$$

Equivalently, in terms of the magnetic field variables, one can write

$$\langle \sigma_i \sigma_{i+1} \rangle = \frac{1}{\beta^2}\frac{\partial^2 \ln Z}{\partial h_i\, \partial h_{i+1}}. \tag{14.35}$$

From the transfer matrix $\mathcal{T}$ defined in Eq. (14.24), it is convenient to introduce the operator

$$\mathcal{T}_2 = \begin{pmatrix} -e^{\beta J} & e^{-\beta J} \\ e^{-\beta J} & -e^{\beta J} \end{pmatrix}, \tag{14.36}$$

which acts as a "defect insertion" in the transfer-matrix product when differentiating with respect to $J_{i,i+1}$. Then Eq. (14.34) can be written as

$$\langle \sigma_i \sigma_{i+1} \rangle = \left.\frac{\mathrm{Tr}(\mathcal{T}_2\, \mathcal{T}^{N-1})}{\mathrm{Tr}(\mathcal{T}^N)}\right|_{J_{i,i+1}=J}. \tag{14.37}$$

By successive differentiation, the correlation between spins separated by L sites becomes

$$\langle \sigma_i \sigma_{i+L} \rangle = \left.\frac{\mathrm{Tr}(\mathcal{T}_2^L\, \mathcal{T}^{N-L})}{\mathrm{Tr}(\mathcal{T}^N)}\right|_{J_{i,i+1}=J}. \tag{14.38}$$

For matrices of the symmetric form

$$A = \begin{pmatrix} a & b \\ b & a \end{pmatrix}, \qquad B = \begin{pmatrix} -a & b \\ b & -a \end{pmatrix},$$

one easily verifies that

$$AB = (b^2 - a^2)\, I, \tag{14.39}$$

where I is the 2×2 identity matrix. Applying this relation to $\mathcal{T}$ and $\mathcal{T}_2$, we find

$$\langle \sigma_i \sigma_{i+L} \rangle = (b^2 - a^2)^L \frac{\mathrm{Tr}(\mathcal{T}^{N-2L})}{\mathrm{Tr}(\mathcal{T}^N)}, \tag{14.40}$$

where a and b denote the matrix elements of $\mathcal{T}$:

$$a = e^{\beta J}\cosh(\beta h), \qquad b = e^{-\beta J}.$$

Since $b^2 - a^2 = (b-a)(b+a)$, and recalling that $b + a = \lambda_1$ and $b - a = \lambda_2$ are the eigenvalues of $\mathcal{T}$, the expression simplifies to

$$\langle \sigma_i \sigma_{i+L} \rangle = \frac{\lambda_2^{N-L}\lambda_1^L + \lambda_2^L \lambda_1^{N-L}}{\lambda_1^N + \lambda_2^N}. \tag{14.41}$$

In the thermodynamic limit ($N \to \infty$), this reduces to

$$\langle \sigma_i \sigma_{i+L} \rangle \simeq \left(\frac{\lambda_2}{\lambda_1}\right)^L = [\tanh(\beta J)]^L. \tag{14.42}$$

Defining the correlation function as $g(L) = e^{-L/\xi}$, Eq. (14.42) gives

$$\xi^{-1} = \ln\left(\frac{\lambda_1}{\lambda_2}\right) = |\ln[\tanh(\beta J)]|. \tag{14.43}$$

Hence, the correlation length diverges only as $T \to 0$:

$$\xi \xrightarrow[T\to 0]{} \infty,$$

while remaining finite at any $T > 0$. This means that spins become correlated only over short distances, and there is no spontaneous magnetization at finite temperature.

Equation (14.43) confirms that the one-dimensional Ising model is *critical only at zero temperature.* Long-range order—and thus spontaneous magnetization—requires an infinite correlation length, which in turn occurs only as $T \to 0$. This result illustrates a fundamental property of low-dimensional systems: short-range interactions are insufficient to sustain a finite-temperature phase transition. The emergence of true long-range order requires higher dimensionality or long-range coupling between spins.

Alternative Solution: Ising's Original Combinatorial Derivation

Lenz originally proposed to his student Ernst Ising the study of what is now called the *Ising model* as an attempt to go beyond the Curie–Weiss mean-field theory of ferromagnets. In the main text we solved the one-dimensional Ising model using the modern transfer-matrix formalism. Here we present *Ising's original method,* based on a direct combinatorial enumeration of microstates. While today regarded as elementary, Ising's approach was innovative and technically delicate for the time [Isi25]. For historical context, see R. Folk [Fol22].

Counting Microstates

Ising considered configurations of ν_1 spins aligned with the external field and ν_2 spins aligned against it, with $\nu_1 + \nu_2 = n$. A key role is played by the positions where neighboring spins are opposite, which Ising called *energy places* (today: domain walls). If σ denotes the number of such places, then $\sigma = 2s + \delta$ where s counts internal domain walls and $\delta = 0, 1$ accounts for whether the first and last spins are parallel or antiparallel.
The number of ways to distribute r aligned spins among m slots is

$$\nu_r(m) = \binom{m-1}{r-1},$$

reflecting the number of partitions of m objects into r positive groups. Using this structure, the number of spin configurations with given (ν_1, ν_2, σ) becomes

$$\mathcal{N}(\nu_1, \nu_2, \sigma = 2s + \delta) = \binom{\nu_1 - 1}{s}\binom{\nu_2 - 1}{s + \delta - 1} + \binom{\nu_2 - 1}{s}\binom{\nu_1 - 1}{s + \delta - 1}.$$

Constructing the Partition Function

The energy of such a configuration depends on (i) the imbalance $\nu_1 - \nu_2$, which couples to the external field, and (ii) the number of domain walls $2s+\delta$, each contributing energy ϵ. Ising's partition function can therefore be written as

$$Z(n) = \sum_{\nu_1+\nu_2=n} \sum_{s=0}^{\infty} \sum_{\delta=0}^{1} \mathcal{N}(\nu_1, \nu_2, 2s + \delta)\, e^{(\nu_1-\nu_2)\alpha}\, e^{-(2s+\delta)\beta},$$

where $\alpha = mH/k_BT$ and $\beta = \epsilon/k_BT$.

Generating Function and Magnetization

Ising's crucial step was the introduction of the generating series

$$F(x) = \sum_{n=0}^{\infty} Z(n)\, x^n,$$

which reorganizes the double combinatorial sum into a closed rational function:

$$F(x) = \frac{2x\left[\cosh\alpha - (1 - e^{-\epsilon/k_BT})x\right]}{1 - 2\cosh\alpha\, x + (1 - e^{-2\epsilon/k_BT})x^2}.$$

For large n, the behavior of $Z(n)$ follows from the dominant pole of $F(x)$, permitting a closed expression for the magnetization:

$$M = mn\, \frac{\sinh\alpha}{\sqrt{\sinh^2\alpha + e^{-2\beta}}}.$$

Setting the external field to zero ($\alpha = 0$) yields $M = 0$ at any $T > 0$, demonstrating that the one-dimensional Ising chain has no finite-temperature ferromagnetic phase transition. This was the central conclusion of Ising's original paper—correct for one dimension, but unfortunately interpreted by him as implying the absence of ferromagnetism in higher dimensions as well.

Domain-Wall Argument

An intuitive argument for the absence of spontaneous magnetization in the one-dimensional Ising model is provided by the concept of *domain walls*. Consider the ground state in which all spins are aligned. If we flip all the spins to the right of a certain site, we create a single domain wall separating regions of opposite magnetization. The energy cost of such a defect is finite—specifically $2J$—since only the two neighboring spins at the boundary are misaligned, and this cost does not depend on the size of the reversed domain.

However, there are $N - 1$ possible positions for the domain wall along a chain of N spins, and hence the free energy associated with a single wall is

$$F = 2J - k_\mathrm{B}T\ln(N-1). \tag{14.44}$$

At any nonzero temperature, the entropic contribution eventually dominates as N increases, making the creation of domain walls thermodynamically favorable. Once domain walls proliferate, large regions of opposite magnetization appear, and long-range order is destroyed.

Moreover, inserting additional domain walls further reduces the free energy, ultimately leading to a completely disordered state at finite temperature. This argument explains why spontaneous ordering cannot occur in one dimension, while in two and higher dimensions, where the energy cost of a domain wall grows with its length, long-range order can be sustained.

REMARK 14.3 This line of reasoning, originally formulated by Rudolf Peierls in 1936 [Pei36], is now known as the *Peierls argument*. It provides a simple yet powerful geometrical explanation of why the one-dimensional Ising model lacks a finite-temperature phase transition, while the two-dimensional model exhibits one.

Combinatorial Methods for Solving Ising Models

In the preceding section, we solved the one-dimensional Ising model exactly using the transfer-matrix technique, a standard approach in most textbooks. Here, we present an alternative *combinatorial* or *graphical* method that emphasizes the underlying structure of the problem.

Consider a linear chain of N spins, each taking values $\sigma_i = \pm 1$. For nearest-neighbor interactions, the Hamiltonian is

$$H(s) = -J \sum_{i=1}^{N} \sigma_i \sigma_{i+1}, \tag{14.45}$$

with periodic boundary conditions ($\sigma_{N+1} = \sigma_1$). For open boundary conditions, the summation extends only to $N - 1$.

The canonical partition function is

$$Z(\beta) = \sum_{\{\sigma_i = \pm 1\}} e^{-\beta H(\sigma)} = \sum_{\{\sigma_i = \pm 1\}} \prod_{i=1}^{M} e^{\beta J \sigma_i \sigma_{i+1}}, \tag{14.46}$$

where $M = N$ for periodic boundaries and $M = N - 1$ for open ones.

Since each bond variable $\sigma_i \sigma_{i+1}$ can only take values ± 1, we can use the identity

$$e^{\beta J \sigma_i \sigma_{i+1}} = \cosh(\beta J) + \sigma_i \sigma_{i+1} \sinh(\beta J), \tag{14.47}$$

to rewrite the Boltzmann factor in terms of hyperbolic functions. Substituting Eq. (14.47) into the partition function gives

$$Z = \sum_{\{\sigma_i = \pm 1\}} \prod_{i=1}^{M} \left[\cosh(\beta J) + \sigma_i \sigma_{i+1} \sinh(\beta J) \right]. \tag{14.48}$$

Introducing the variable $u = \tanh(\beta J)$ and noting that $\sinh(\beta J) = \cosh(\beta J) u$, we can factor out the constant prefactor:

$$Z = (\cosh \beta J)^M \sum_{\{\sigma_i = \pm 1\}} \prod_{i=1}^{M} (1 + u\, \sigma_i \sigma_{i+1}). \tag{14.49}$$

The variable $u = \tanh(\beta J)$ acts as a small expansion parameter at high temperature ($u \to 0$ as $T \to \infty$).

Expanding the product in Eq. (14.49) yields a sum over all possible subsets of bonds:

$$\begin{aligned} \prod_{i=1}^{M} (1 + u\, \sigma_i \sigma_{i+1}) &= 1 + u \sum_{i} \sigma_i \sigma_{i+1} + u^2 \sum_{i<j} \sigma_i \sigma_{i+1} \sigma_j \sigma_{j+1} + \cdots \\ &\quad + u^{M-1} (\sigma_1 \sigma_2 \cdots \sigma_{M-1} \sigma_M) + u^M (\sigma_1 \sigma_2 \cdots \sigma_M \sigma_1). \end{aligned} \tag{14.50}$$

Each term corresponds to selecting a particular subset of bonds connecting nearest-neighbor spins. In a *graphical interpretation*, the spins σ_i correspond to vertices, and each factor $u\, \sigma_i \sigma_{i+1}$ corresponds to an occupied bond (edge) between sites i and $i + 1$. The expansion of Z therefore represents a sum over all possible bond configurations, weighted by the product of their corresponding spin factors.

For open boundary conditions, the final term in Eq. (14.50) is absent, while for periodic boundaries it represents a closed loop over the chain.

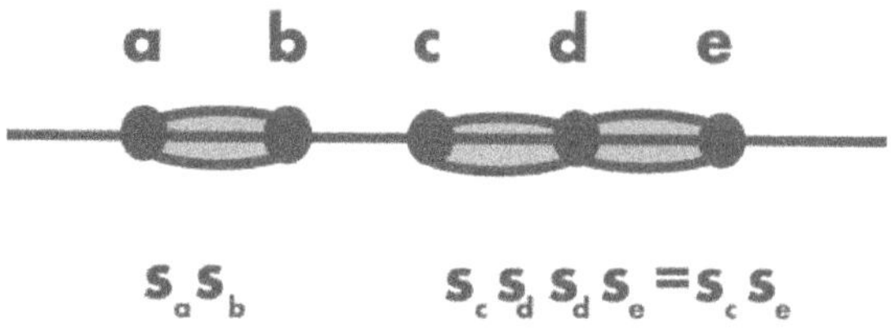

FIGURE 14.2 An example of a subgraph expansion at the order u^3. This is equivalent to the product of the bonds $(a,b),(c,d)$, and (d,e). The term in the expansion is then $\sigma_a,\sigma_b,\sigma_c,\sigma_e$ as σ_d appears an even number of times. This is not an admissible graph, as the sum over the spin configuration will lead this term to be zero.

This combinatorial representation offers a natural graphical interpretation of the partition function as a sum over bond configurations—a perspective that generalizes elegantly to higher-dimensional lattices and forms the basis of the high-temperature expansion methods used in modern statistical mechanics.

Each selected bond between spins i and $i+1$ contributes a factor $\sigma_i\sigma_{i+1}$. Thus, every term in the polynomial expansion of Eq. (14.50) corresponds uniquely to a particular subset of bonds—and therefore to a distinct *graph configuration* of the spin chain.

When performing the summation over all spin configurations $\{\sigma_i = \pm 1\}$, a crucial simplification arises. At any vertex (spin site), if an odd number of bonds are selected, the contribution vanishes because

$$\sum_{\sigma_i=\pm 1} \sigma_i = (+1) + (-1) = 0. \tag{14.51}$$

Therefore, only those subgraphs in which every vertex is incident to an even number of selected bonds contribute to the partition function. Such configurations are called *admissible graphs*, and we denote their set by $\mathcal{A}$.

In one dimension, the admissible graphs are particularly simple: they consist either of no bonds (all vertices of degree 0) or of disjoint pairs of bonds forming closed loops around the chain due to the periodic boundary conditions. Because $\sigma_i^2 = 1$, the spin variables completely factor out in these graphs, leaving only numerical factors of 2^N (from the sum over all spin states) and powers of $u = \tanh(\beta J)$ corresponding to the number of selected bonds. Hence, each admissible graph $\mathcal{G} \in \mathcal{A}$ contributes a factor $u^{m(\mathcal{G})}$, where $m(\mathcal{G})$ is the number of bonds in $\mathcal{G}$.

Using the identity $\cosh(\beta J) = (1-u^2)^{-1/2}$, and noting that the sum over spins has been performed, we obtain the general expression

$$Z = 2^N (1-u^2)^{-M/2} \Big[1 + \sum_{\mathcal{G}\in\mathcal{A}} u^{m(\mathcal{G})} \Big], \tag{14.52}$$

where M is the number of bonds ($M = N$ for periodic boundaries and $M = N-1$ for open boundaries).

For the one-dimensional Ising model (the “1D” subscript in Eq. (14.53)), the counting of admissible graphs is straightforward:

- For *open boundary conditions*, there are no closed loops and hence no nontrivial admissible graphs.
- For *periodic boundary conditions*, there is a single admissible graph—the one containing all N bonds of the ring.

Substituting these cases into Eq. (14.52) gives

$$Z_{1\mathrm{D}} = \begin{cases} 2^N (1-u^2)^{-(N-1)/2}, & \text{open boundaries,} \\ 2^N (1-u^2)^{-N/2} \big(1+u^N\big), & \text{periodic boundaries.} \end{cases} \tag{14.53}$$

This *combinatorial approach* elegantly reproduces the exact solution of the one-dimensional Ising model, previously derived via the transfer-matrix method. It also reveals the underlying graphical structure of the partition function—a perspective that generalizes naturally to higher dimensions and forms the conceptual basis of high-temperature expansions and diagrammatic methods in statistical mechanics.

15

Mean-Field Ferromagnets

15.1 Motivation for Mean-Field Theory

In the previous chapters, we encountered systems—such as the Ising and Heisenberg models—whose microscopic definitions are simple but whose exact solutions are possible only in a few special cases. The one-dimensional Ising model can be solved exactly, but its higher-dimensional counterparts, or more general spin systems, resist any straightforward analytical treatment. The reason is that the *interactions among many degrees of freedom create correlations* that make the partition function intractable beyond very small systems.

The essential idea of *mean-field theory (MFT)* is to replace these complex many-body correlations by an *effective average interaction.* Each spin or particle is assumed to feel the influence of all others only through an average, or "mean," field that represents the collective behavior of the system. In this approximation, one neglects fluctuations around the mean value, reducing the full interacting problem to an effective single-particle problem.

This approach is conceptually similar to *Boltzmann's Stosszahlansatz* (assumption of molecular chaos) in kinetic theory, where correlations between colliding particles are ignored, allowing the derivation of the Boltzmann equation from deterministic microscopic dynamics. In both cases, we replace a system of strongly coupled entities by a statistically independent ensemble subject to an averaged influence. The resulting approximation captures essential macroscopic behavior—such as phase transitions and spontaneous magnetization—while dramatically simplifying the mathematics.

From a physical point of view, mean-field theory provides a first self-consistent description of cooperative phenomena: each spin interacts with an average magnetization produced by all the others, and the system's collective behavior emerges from the self-consistency of this assumption.

DOI: 10.1201/9781003603528-15

15.2 Some Definitions about Graphs

To describe systems with many interacting degrees of freedom, it is convenient to introduce a language that encodes *who interacts with whom.* This is most naturally done using the formalism of graphs [Bol98]. In what follows, we recall only the minimal definitions necessary for the mean-field treatment of Ising models on arbitrary interaction networks.

Graphs and Adjacency Matrices

A *graph* $G = (V, E)$ consists of a set of vertices (or nodes) $V = \{1, \dots, N\}$ and a set of edges $E \subseteq V \times V$ representing interactions between them (Fig. 15.1). Edges may be directed or undirected, and can carry weights $w_{ij} \geq 0$. In statistical mechanics, each node will represent a spin or a particle, while edges encode the couplings between them.

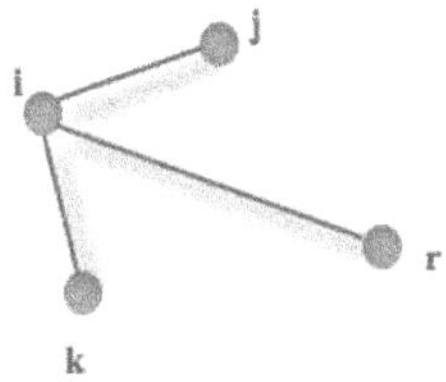

FIGURE 15.1 Example of a simple undirected graph with four vertices i, j, k, r and edges (i, j), (i, r), and (i, k). Reproduced from [Car25].

The connectivity of a graph is encoded in its *adjacency matrix* $A = (A_{ij})$, defined by

$$A_{ij} = \begin{cases} w_{ij}, & \text{if } (i,j) \in E, \\ 0, & \text{otherwise.} \end{cases} \tag{15.1}$$

For unweighted undirected graphs, $A_{ij} = A_{ji} \in \{0, 1\}$. The *degree* (or valency) of a node i is

$$d_i = \sum_j A_{ij}, \tag{15.2}$$

that is, the number of edges connected to node i. The total number of edges is $E = (1/2)\sum_{ij} A_{ij}$, a result sometimes referred to as the *handshaking lemma.*

Matrix Properties

The adjacency matrix provides a compact algebraic representation of the network:

- The matrix A is symmetric for undirected graphs and asymmetric for directed ones.
- The entries of A^k count the number of distinct paths of length k connecting two vertices: $(A^k)_{ij} \neq 0$ if at least one such path exists.
- If $A^r = 0$ for some integer r, the graph has no closed paths (i.e., it is acyclic).

A graph that can be decomposed into two disjoint vertex sets V_1 and V_2, such that all edges connect nodes in different sets, is called *bipartite.* After a suitable permutation of indices, its adjacency matrix takes the block form

$$A = \begin{pmatrix} 0 & B \\ B^{\mathrm{T}} & 0 \end{pmatrix}, \tag{15.3}$$

where B encodes the cross-links between the two partitions.

FIGURE 15.2 A graph whose adjacency matrix is banded. Reproduced from [Car25].

Incidence and Laplacian Matrices

Alongside the adjacency matrix, two other matrix representations play an important role in describing the structure and dynamics of graphs: the *incidence matrix* and the *Laplacian matrix*. They provide alternative ways to encode the connectivity of a network and serve as discrete analogues of gradient and Laplace operators on continuous spaces.

Incidence Matrix

For a graph G with n vertices and m edges, the *incidence matrix* $B(G)$ is an $n \times m$ matrix that relates nodes to edges. Each column of B corresponds to an edge, and each row to a vertex:

$$B_{i\beta} = \begin{cases} +1, & \text{if vertex } i \text{ is the source of edge } \beta, \\ -1, & \text{if vertex } i \text{ is the target of edge } \beta, \\ 0, & \text{otherwise.} \end{cases} \tag{15.4}$$

For undirected graphs, the signs are arbitrary but opposite at the two endpoints, while for directed graphs they encode edge orientation. The incidence matrix is closely related to the adjacency matrix of the *line graph* $\mathscr{L}(G)$, whose vertices correspond to the edges of G. Two edges of G are connected in $\mathscr{L}(G)$ if they share a common vertex, leading to the relation

$$A(\mathscr{L}(G)) = B^{\mathrm{T}}(G)B(G) - 2I_m, \tag{15.5}$$

where I_m is the identity matrix of dimension m.

The Graph Laplacian

For an undirected graph, the *Laplacian matrix* is defined as

$$L = D - A = B\,B^{\mathrm{T}}, \tag{15.6}$$

where A is the adjacency matrix and $D = \mathrm{diag}(d_1, \ldots, d_n)$ the diagonal matrix of vertex degrees. In weighted graphs, each edge (i, j) carries a weight $a_{ij} > 0$.

Properties

- The Laplacian is *positive semi-definite*: for any vector $\mathbf{x}$,

$$\vec{x}^{\mathrm{T}} L \vec{x} = \sum_{(i,j)\in E(G)} a_{ij}(x_i - x_j)^2 \geq 0. \tag{15.7}$$

- The smallest eigenvalue of L is always zero, corresponding to the uniform eigenvector $\vec{1}$, since $L\vec{1} = 0$.
- The Laplacian acts as the discrete analogue of the continuum Laplace operator ∇^2:

$$\nabla^2 u = \sum_i \partial^2_{x_i} u. \tag{15.8}$$

  In this analogy, L governs discrete diffusion or relaxation processes on networks.

The Laplacian determines how perturbations or excitations spread through a network. Its eigenvalues set the characteristic time scales of diffusion, synchronization, or spin relaxation on the underlying graph. In the mean-field theory of ferromagnets, the Laplacian spectrum reflects how correlations propagate across the interaction network and how collective modes emerge near criticality. The reader interested in these spectral methods is invited to read [Car25].

Representing interactions through adjacency and Laplacian matrices will allow us to formulate mean-field approximations on arbitrary graphs. In particular, the Laplacian encodes how deviations from average magnetization diffuse through the system—an idea that will reappear when analyzing fluctuations and stability in network-based mean-field models.

In particular, representing interactions through adjacency matrices provides a natural way to generalize the mean-field description of ferromagnets. In the Ising model with arbitrary interaction patterns, the energy function

$$H = -\sum_{i<j} J_{ij}\, \sigma_i \sigma_j \tag{15.9}$$

can be written compactly as $H = -(1/2)\, \vec{\sigma}^{\mathrm{T}} J\, \vec{\sigma}$, where J plays the role of a weighted adjacency matrix describing the interaction network. Fully connected graphs correspond to the traditional Curie–Weiss mean-field model, whereas sparse or structured graphs represent finite-connectivity networks such as random graphs, lattices, or small-world systems. The graph formalism thus allows us to treat mean-field ferromagnets as systems defined on arbitrary interaction topologies—a point of view that unifies many classical and modern models of collective behavior.

15.3 Curie–Weiss Theory

The Curie–Weiss model represents the first and simplest *mean-field approximation* of the Ising model. Historically, it was developed by Pierre Curie [Cur95] and later generalized by Pierre Weiss [Wei07], who introduced the concept of a *molecular field*—an internal effective magnetic field proportional to the overall magnetization. While Curie's law was purely empirical, Weiss's modification provided a simple microscopic picture that captures spontaneous magnetization below a critical temperature, now called the *Curie temperature.*

In the mean-field approximation, each spin interacts equally with all others, so spatial correlations are neglected. Replacing the adjacency matrix A_{ij} by 1 for all pairs, the Hamiltonian becomes

$$H = -\frac{J}{2}\sum_{i,j} \sigma_i \sigma_j - h \sum_{i=1}^{N} \sigma_i, \tag{15.10}$$

where $J > 0$ is the ferromagnetic coupling and H is an external magnetic field. The corresponding canonical partition function is

$$Z = \sum_{\{\sigma_i = \pm 1\}} \exp\left[\frac{\beta J}{2}\Big(\sum_i \sigma_i\Big)^2 + \beta h \sum_i \sigma_i\right]. \tag{15.11}$$

Introducing the *magnetization per spin,*

$$M = \frac{1}{N}\sum_i \sigma_i, \tag{15.12}$$

we can rewrite Eq. (15.11) as

$$Z = \sum_{\{\sigma_i = \pm 1\}} \exp\left[\frac{\beta J N^2}{2} M^2 + \beta h N M\right]. \tag{15.13}$$

Now, it is interesting to observe that for $h = 0$, the Hamiltonian is invariant under the transformation $\sigma_i \to -\sigma_i$, e.g., a $\mathcal{Z}_2$ symmetry. Thus, we would expect that in this case the system preserves this symmetry. The Curie–Weiss model is an example that explicitly shows that this is not the case at a particular critical temperature T_c. This is called *symmetry breaking.*

To linearize the quadratic term in M, we use the Gaussian integral identity

$$\int_{-\infty}^{\infty} e^{-x^2+2ax}\, dx = \sqrt{\pi}\, e^{a^2}. \tag{15.14}$$

This is often called *Hubbard–Stratonovich transformation.* Applying this to Eq. (15.13) yields

$$\begin{aligned} Z &= \frac{1}{\sqrt{\pi}} \sum_{\{\sigma_i = \pm 1\}} \int_{-\infty}^{\infty} e^{-x^2 + \left(2\sqrt{\frac{\beta J N^2}{2}}\, x + \beta h N\right) M}\, dx \\ &= \frac{1}{\sqrt{\pi}} \int_{-\infty}^{\infty} e^{-x^2} \sum_{\{\sigma_i = \pm 1\}} \prod_i e^{\frac{1}{N}(2\sqrt{\frac{\beta J N^2}{2}}\, x + \beta h N)\sigma_i}\, dx. \end{aligned} \tag{15.15}$$

The spin sum can now be performed independently at each site:

$$\sum_{\sigma_i = \pm 1} e^{a\sigma_i} = 2\cosh(a). \tag{15.16}$$

Hence,

$$\begin{aligned} Z &= \frac{1}{\sqrt{\pi}} \int_{-\infty}^{\infty} e^{-x^2} \left[2\cosh\left(\sqrt{2\beta J}\, x + \beta h\right)\right]^N dx \\ &= \frac{1}{\sqrt{\pi}} \int_{-\infty}^{\infty} \exp\left[-x^2 + N \ln\left(2\cosh\left(\sqrt{2\beta J}\, x + \beta h\right)\right)\right] dx. \end{aligned} \tag{15.17}$$

To ensure that the total energy scales linearly with N, we rescale the coupling as $J \to J/N$. With this substitution, Eq. (15.17) becomes

$$Z = \frac{\sqrt{N}}{\sqrt{2\pi\beta J}} \int_{-\infty}^{\infty} e^{-Nf(\beta, M, h)}\, dM, \tag{15.18}$$

where

$$f(\beta, M, H) = -\frac{\beta J}{2} M^2 + \ln\left[2\cosh\left(\beta J M + \beta h\right)\right]. \tag{15.19}$$

In the thermodynamic limit ($N \to \infty$), the integral in Eq. (15.18) is dominated by the stationary point of $g(\beta, M, H)$:

$$\frac{\partial g}{\partial M} = 0 \quad \Rightarrow \quad M = \tanh\left[\beta(JM + h)\right]. \tag{15.20}$$

This is the first example of a *saddle-point approximation* that we see in this book. Equation (15.20) is the celebrated *Curie–Weiss self-consistency equation,* which defines the mean magnetization as a function of temperature and external field. At high temperature ($\beta J \ll 1$), $M \simeq \beta(JM + H)$, leading to the linear relation

$$\chi = \frac{M}{h} = \frac{C}{T - T_c}, \qquad T_c = J/k_{\mathrm{B}}, \tag{15.21}$$

which reproduces the *Curie–Weiss law* for the magnetic susceptibility. In general, the magnetic susceptibility is written in terms

$$\chi \propto \left(\frac{1}{T - T_c}\right)^{\beta}, \qquad T_c = J/k_{\mathrm{B}}, \tag{15.22}$$

of generic exponents β, which vary for materials. Mean-field theory predicts $\beta = 1/2$ which is not often observed experimentally. This is one of the limitations of this approach.

In this model, the interaction of each spin with all others is approximated by an average internal field JM. The self-consistent Eq. (15.20) expresses the balance between this effective field and thermal agitation. Below the Curie temperature T_c, a nonzero solution $M \neq 0$ appears spontaneously, representing the onset of ferromagnetic order. Above T_c, only the paramagnetic solution $M = 0$ remains stable.

Experimental Verification of the Curie–Weiss Law in Real Ferromagnets

The Curie–Weiss theory provides a remarkably accurate phenomenological description of many three-dimensional ferromagnets, particularly near and above their critical (Curie) temperature T_c. Experimentally, the *magnetic susceptibility* χ of such materials follows the Curie–Weiss law,

$$\chi = \frac{C}{T - T_c}, \tag{15.23}$$

where C is the Curie constant and T_c is the Curie temperature determined from the extrapolation of the high-temperature linear regime.

Example: Permalloy ($Fe_{80}Ni_{20}$) Permalloy is a three-dimensional metallic ferromagnet composed primarily of iron and nickel. Its Curie temperature is approximately $T_c \simeq 850$ K, and its saturation magnetization at low temperature is $M_s \simeq 0.8–1.0$ T. Measurements of its magnetic susceptibility show that:

- For $T > T_c$, the inverse susceptibility $1/\chi$ varies linearly with temperature, confirming the Curie–Weiss law.
- The extrapolated line intersects the temperature axis at $T = T_c$, consistent with a second-order phase transition.
- For $T < T_c$, the magnetization $M(T)$ follows an empirical power law $M(T) \sim (T_c - T)^{\beta}$, with an experimentally measured critical exponent $\beta \approx 0.33$, slightly below the mean-field value $\beta_{\mathrm{MF}} = 1/2$.

An explanation for the difference between the mean-field and theoretical critical exponents will be provided later in this Volume. These measurements confirm the essential predictions of the Curie–Weiss theory: thermal agitation suppresses spontaneous magnetization above T_c, while below T_c the system exhibits collective alignment of atomic moments. The deviations of critical exponents from their mean-field values arise from short-range correlations and critical fluctuations that the mean-field approximation neglects.

A detailed discussion of experimental ferromagnetism and Curie–Weiss behavior in real materials can be found in standard solid-state physics and magnetism textbooks, see e.g., [Kit05, Boz93, Blu01].

15.3.1 Phenomenology of the Curie–Weiss Equation

The mean-field equation for the magnetization,

$$M = \tanh\big[\beta(JM + h)\big], \tag{15.24}$$

captures the essential phenomenology of the ferromagnetic transition. To analyze its behavior near the critical temperature, we expand the hyperbolic tangent function for small arguments. Keeping terms up to cubic order, we obtain

$$M \simeq \beta JM + \beta h - \frac{\beta^3}{3}(JM + h)^3, \tag{15.25}$$

which provides a good approximation in the high-temperature regime ($\beta J \lesssim 1$).

Case $h = 0$: Spontaneous Magnetization

For zero external field, Eq. (15.25) reduces to

$$M\big(1 - \beta J\big) + \frac{\beta^3 J^3}{3} M^3 = 0. \tag{15.26}$$

The stationary solutions are

$$M = 0, \qquad M = \pm \frac{\sqrt{3}\,\sqrt{\beta J - 1}}{\beta^{3/2} J^{3/2}}. \tag{15.27}$$

For $\beta J < 1$ (i.e., $T > T_c = J/k_{\mathrm{B}}$), the only real solution is $M = 0$, corresponding to the *paramagnetic phase.* When $\beta J > 1$ (i.e., $T < T_c$), two additional real solutions appear, symmetric about zero, corresponding to nonzero spontaneous magnetization. At $T = T_c$, the magnetization vanishes continuously, signaling a *second-order phase transition* driven by spontaneous symmetry breaking.

Case $h \neq 0$: External Field

When a finite external magnetic field h is applied, Eq. (15.24) must be solved numerically. The result is illustrated schematically in Fig. 15.3. Above the critical temperature ($T > T_c$), the magnetization is analytic and proportional to the field:

$$M \simeq \chi h, \qquad \chi = \frac{C}{T - T_c}, \tag{15.28}$$

where χ is the magnetic susceptibility. Below T_c, the system develops a finite spontaneous magnetization even as $h \to 0$:

$$\lim_{h \to 0^{\pm}} M(T < T_c, h) = \pm M_0(T) \neq 0. \tag{15.29}$$

The discontinuity between the limits $h \to 0^+$ and $h \to 0^-$ reflects the breaking of the $\mathbb{Z}_2$ spin-reversal symmetry of the Hamiltonian and defines the onset of magnetic order.

The emergence of two symmetric nonzero solutions in Eq. (15.27) below T_c signifies the phenomenon of *spontaneous symmetry breaking.* Although the underlying Hamiltonian is symmetric under $\sigma_i \to -\sigma_i$, the system in the thermodynamic limit chooses one of the two magnetized states, resulting in an ordered phase. This mean-field picture underlies much of the modern understanding of continuous phase transitions and critical phenomena.

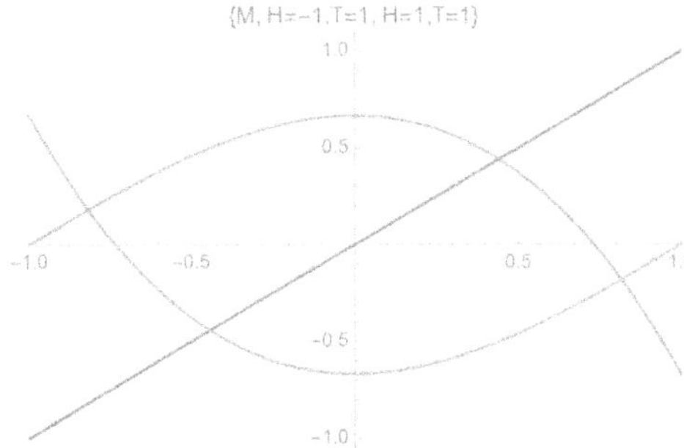

FIGURE 15.3 Plot of M versus the right-hand side of the mean-field equation for $H = \pm 1$. In this case we have a broken symmetry due to the external field, and the magnetization is nonzero.

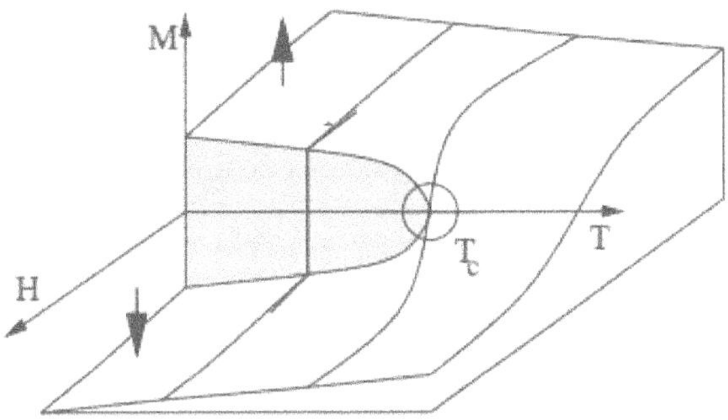

FIGURE 15.4 Magnetization of the mean-field Ising model in an external field. As we can see the magnetization becomes discontinuous at small temperature, below T_c, as we change the external magnetic field.

Ising on a Generic Graph

The Curie–Weiss model corresponds to an Ising ferromagnet defined on a *complete graph*, where each spin interacts equally with all others. More generally, a system of interacting spins can be represented by a *network* or *interaction graph* $G = (V, E)$, whose adjacency matrix A_{ij} encodes which pairs of spins interact and with what strength. This viewpoint naturally connects the statistical mechanics of many-body systems with the framework of *network science*, where nodes represent interacting units and edges encode pairwise couplings [New10].

For a graph with N vertices and adjacency matrix A_{ij}, the Ising Hamiltonian is

$$H = -\frac{J}{2}\sum_{i,j} A_{ij}\sigma_i\sigma_j - h\sum_{i=1}^{N}\sigma_i, \tag{15.30}$$

where $J > 0$ is the ferromagnetic coupling constant and h the external magnetic field. The adjacency matrix defines the connectivity pattern: $A_{ij} = 1$ if spins i and j interact, and 0 otherwise.

Each spin feels an effective local field

$$h_i^{\text{eff}} = J\sum_j A_{ij}\langle\sigma_j\rangle + h, \tag{15.31}$$

where correlations between spins are neglected. Defining $m_i = \langle \sigma_i \rangle$, we can express the mean-field free energy as a function of the magnetizations,

$$f(\{m_i\}) = -\frac{J}{2}\sum_{i,j} A_{ij} m_i m_j - h\sum_i m_i + \frac{1}{\beta}\sum_i s(m_i), \tag{15.32}$$

where $s(m_i)$ is the local spin entropy. The derivation of this free energy is reserved for later chapters in which we will introduce other mean-field techniques. The saddle-point approximation can be in fact interpreted as a minimization of a macroscopic free energy f. Minimizing f with respect to m_i gives the self-consistent mean-field equation

$$m_i = \tanh\left[\beta\left(J\sum_j A_{ij} m_j + h\right)\right]. \tag{15.33}$$

A detailed derivation of Eq. (15.32) will be presented later in this Volume, when we introduce the *variational formulation* of mean-field theory.

For approximately uniform degree $d_i \simeq d$ and magnetization $m_i \simeq m$, Eq. (15.33) reduces to

$$m = \tanh[\beta(Jdm + h)], \tag{15.34}$$

recovering the Curie–Weiss equation when $d = N - 1$. Linearizing Eq. (15.33) for small m_i gives

$$\vec{m} = \beta J A \vec{m}, \tag{15.35}$$

so that the critical temperature depends on the largest eigenvalue $\lambda_{\max}$ of A:

$$T_c = \frac{J\lambda_{\max}}{k_B}. \tag{15.36}$$

The network's spectral properties thus control the onset of collective order: highly connected or dense networks (large $\lambda_{\max}$) enhance cooperative alignment, while sparse or irregular ones suppress it. This graph-based formulation extends the Curie–Weiss picture to systems with arbitrary connectivity, providing a unifying language that links ferromagnetism, statistical mechanics, and the modern physics of complex networks.

15.4 The Spherical Model

The *spherical model*, introduced by Theodor Berlin and Mark Kac in 1952 [BK52], is an exactly solvable model of ferromagnetism that provides a continuous analogue of the Ising model. It replaces discrete spin variables with continuous ones subject to a global constraint, allowing the full partition function to be evaluated analytically. Beyond its intrinsic theoretical value, the spherical model serves as a pedagogical prototype for techniques—such as integral representations and saddle-point analysis—that will reappear throughout this Volume (see also [Bax82]).

Definition

Each site $i = 1, \ldots, N$ carries a real spin variable $\sigma_i \in \mathbb{R}$ constrained by

$$\sum_{i=1}^{N} \sigma_i^2 = N. \tag{15.37}$$

The Hamiltonian on an interaction graph with adjacency matrix A_{ij} is

$$H = -J\sum_{i,j} A_{ij}\sigma_i\sigma_j - \sum_i h_i\sigma_i, \tag{15.38}$$

with $J > 0$ denoting the ferromagnetic coupling. The canonical partition function reads

$$Z_N = \int_{-\infty}^{\infty} d\sigma_1 \cdots d\sigma_N \exp\left[\beta J\sum_{i,j} A_{ij}\sigma_i\sigma_j + \beta\sum_i h_i\sigma_i\right] \delta\left(N - \sum_i \sigma_i^2\right). \tag{15.39}$$

Integral Representation of the Constraint

The delta function enforcing Eq. (15.37) can be expressed as

$$\delta(x) = \frac{1}{2\pi}\int_{-\infty}^{\infty} e^{i\eta x}\, d\eta. \tag{15.40}$$

Substituting this into Eq. (15.39) and introducing a convergence factor $a > 0$, one obtains

$$Z_N = \frac{1}{2\pi}\int d^N\sigma\, d\eta \exp\Big[\beta J\sum_{i,j} A_{ij}\sigma_i\sigma_j + \beta\sum_i h_i\sigma_i + (i\eta + a)\big(N - \sum_i \sigma_i^2\big)\Big]. \tag{15.41}$$

Define

$$V_{ij} = (i\eta + a)\delta_{ij} - \beta J A_{ij}. \tag{15.42}$$

The exponent is now quadratic in $\vec{\sigma}$, allowing us to perform the Gaussian integral provided a is large enough for convergence. Integrating over the spins gives

$$Z_N = \frac{1}{2\pi}\int_{-\infty}^{\infty} d\eta\, \frac{\pi^{N/2}}{\sqrt{\det V}} \exp\left[\frac{\beta^2}{4}\mathbf{h}^{\mathrm{T}} V^{-1}\vec{h} + N(i\eta + a)\right], \tag{15.43}$$

where $\vec{h} = (h_1, \ldots, h_N)$.

Spectral Decomposition

Since V depends on A only through a diagonal shift, its eigenvalues are

$$\lambda_V(\vec{p}) = (i\eta + a) - \beta J\lambda_A(\vec{p}), \tag{15.44}$$

where $\lambda_A(\vec{p})$ are the eigenvalues of the adjacency matrix. For a d-dimensional cubic lattice with $N = L^d$, these are

$$\lambda_A(n_1, \ldots, n_d) = \sum_{r=1}^{d} \cos\big(\omega_{n_r}^r\big), \qquad \omega_{n_r}^r = \frac{2\pi n_r}{L}. \tag{15.45}$$

Replacing the discrete sum over momenta by an integral in the thermodynamic limit, we can express

$$\log\det V = \frac{N}{(2\pi)^d}\int_0^{2\pi} \cdots \int_0^{2\pi} \log\big[(i\eta + a) - \beta J\sum_{k=1}^{d} \cos(p_k)\big]\, dp_1 \cdots dp_d. \tag{15.46}$$

Reduced Form

Collecting terms, the partition function can be written compactly as

$$Z_N = \int_{-\infty}^{\infty} d\eta \, \exp\Big[\tfrac{\beta^2}{4}\mathbf{h}^{\mathrm{T}}V^{-1}\mathbf{h} + N\phi(\eta)\Big], \tag{15.47}$$

where

$$\phi(\eta) = (a+i\eta) - \frac{1}{2(2\pi)^d}\int_0^{2\pi}\cdots\int_0^{2\pi} \log\Big[(i\eta+a) - \beta J\sum_{k=1}^{d}\cos(p_k)\Big]\, dp_1\cdots dp_d. \tag{15.48}$$

After shifting variables

$$z = \frac{(i\eta+a) - d\beta J}{\beta J}, \qquad K = \beta J, \tag{15.49}$$

Eq. (15.48) becomes

$$\phi(z) = \beta J(z+d) - \frac{1}{2}\log(\beta J) - \frac{1}{2(2\pi)^d}\int_0^{2\pi}\cdots\int_0^{2\pi} \log\Big[z+d-\sum_{k=1}^{d}\cos(p_k)\Big]\, dp_1\cdots dp_d. \tag{15.50}$$

The partition function then takes the formal integral form

$$Z_N = \frac{K\, i\, \pi^{N/2}}{2\pi K^{N/4}} \int_{c-i\infty}^{c+i\infty} e^{N\phi(z)}\, dz, \tag{15.51}$$

where $c = (a - d\beta J)/(\beta J)$. The thermodynamic properties follow from evaluating this integral in the large-N limit using the saddle-point method.

Before performing the saddle-point evaluation of the partition function, we assume that the external magnetic field is homogeneous, $h_i = h$. In this case, only the lowest eigenmode of V^{-1} contributes to the field-dependent term, and

$$\vec{h}^{\mathrm{T}}V^{-1}\vec{h} = \frac{Nh^2}{Kz}, \qquad K = \beta J. \tag{15.52}$$

The partition function can therefore be written in the compact form

$$Z_N = \frac{K\, i\, \pi^{N/2}}{2\pi K^{N/2}} \int_{c-i\infty}^{c+i\infty} e^{N\Phi(z)}\, dz, \tag{15.53}$$

where

$$\Phi(z) = \frac{\beta^2 h^2}{4Kz} + Kz + Kd - \frac{1}{2(2\pi)^d}\int_0^{2\pi}\cdots\int_0^{2\pi} \log\Big[z+d-\sum_{k=1}^{d}\cos p_k\Big]\, dp_1\cdots dp_d. \tag{15.54}$$

Free Energy and Saddle-Point Equation

In the thermodynamic limit ($N \to \infty$), the integral in Eq. (15.53) can be evaluated by the saddle-point method. The free energy per spin is given by

$$\frac{F}{Nk_{\mathrm{B}}T} = -\frac{1}{N}\ln Z_N = \frac{1}{2}\ln\frac{K}{\pi} + \Phi(z_0), \tag{15.55}$$

where z_0 satisfies the saddle-point condition

$$\left.\frac{d\Phi}{dz}\right|_{z=z_0} = 0. \tag{15.56}$$

Differentiating Eq. (15.54) gives

$$-\frac{\beta^2 h^2}{4Kz_0^2} + K - \frac{1}{2(2\pi)^d}\int_0^{2\pi}\cdots\int_0^{2\pi} \frac{dp_1\cdots dp_d}{z_0+d-\sum_{k=1}^{d}\cos p_k} = 0. \tag{15.57}$$

Equation of State

The magnetization follows from

$$M = -\frac{1}{N}\frac{\partial F}{\partial h} = \frac{\beta h}{2Kz_0}, \tag{15.58}$$

so that $z_0 = h/(2JM)$. Substituting this relation into Eq. (15.57) gives the *equation of state*:

$$\phi'(z_0) - \frac{h^2}{4Kz_0^2} = \phi'\left(\frac{h}{2JM}\right) - \frac{J^2M^2}{K} = 0, \tag{15.59}$$

where

$$\phi'(z) = -K + \frac{1}{2(2\pi)^d}\int_0^{2\pi}\cdots\int_0^{2\pi}\frac{dp_1\cdots dp_d}{z + d - \sum_{k=1}^{d}\cos p_k}. \tag{15.60}$$

Integral Representation

Using the identity $1/\lambda = \int_0^\infty e^{-t\lambda}\,dt$ and the Bessel function relation

$$\frac{1}{2\pi}\int_0^{2\pi} e^{-t\cos p}\,dp = J_0(it), \tag{15.61}$$

we rewrite Eq. (15.60) as

$$\phi'(z) = -K + \frac{1}{2}\int_0^\infty e^{-t(z+d)}\,J_0(it)^d\,dt. \tag{15.62}$$

For large t, $J_0(it) \sim e^t/\sqrt{2\pi t}$, so that

$$\phi'(z) \simeq -K + \frac{1}{2}\int_0^\infty e^{-t(z+d)+dt}\frac{dt}{(2\pi t)^{d/2}}. \tag{15.63}$$

The integral converges only if $z > 0$ and reveals that

$$\phi'(0) = \begin{cases} \infty, & d \le 2, \\ \text{finite}, & d > 2. \end{cases} \tag{15.64}$$

Thus, the spherical model exhibits a finite critical temperature only for dimensions $d > 2$; for $d \le 2$, fluctuations destroy long-range order.

Spontaneous Magnetization

For small z and $d > 2$, one can expand

$$\phi'(z) \approx \phi'(0) + \tfrac{1}{2}R_d z, \tag{15.65}$$

where R_d is a dimension-dependent constant. Equation (15.59) then gives

$$-K + \frac{R_d h}{2JM} - \frac{J^2M^2}{K} = 0. \tag{15.66}$$

At zero field, the magnetization satisfies

$$\lim_{h\to 0} M(h) = \begin{cases} 0, & T > T_c, \\ \neq 0, & T < T_c, \end{cases} \tag{15.67}$$

signaling a phase transition and *spontaneous symmetry breaking* for $d > 2$. For $d \leq 2$, no such transition occurs, in agreement with general results for continuous symmetries in low dimensions.

The spherical model reproduces the Curie–Weiss behavior in high dimensions while remaining exactly solvable. It thus provides a clear illustration of how dimensionality controls the existence of spontaneous magnetization. A detailed discussion and comparison with other exactly solved models can be found in Baxter's textbook [Bax82].

15.5 General Features of the Ising Model

The general partition function of the ferromagnetic Ising model on an arbitrary lattice or interaction graph is

$$Z(\beta, h) = \sum_{\{\sigma_i = \pm 1\}} \exp\left[-\beta\left(\sum_{i,j} J_{ij}\sigma_i\sigma_j + \sum_i h_i\sigma_i\right)\right], \tag{15.68}$$

where J_{ij} encodes the interaction strength between spins at sites i and j, and h_i is an external magnetic field.

Order Parameters and Correlations

The order parameter of interest is the average magnetization,

$$M = \frac{1}{N}\sum_{i=1}^{N}\langle\sigma_i\rangle = -\frac{1}{N\beta}\sum_i \frac{\partial \ln Z}{\partial h_i}. \tag{15.69}$$

A discontinuity in M as a function of temperature or field signals a *first-order* phase transition according to the Ehrenfest classification.

A second quantity of interest is the two-point correlation function,

$$\langle\sigma_i\sigma_j\rangle = \frac{1}{\beta^2 Z}\frac{\partial^2 Z}{\partial h_i\,\partial h_j}, \tag{15.70}$$

whose discontinuity or divergence indicates a *second-order* (continuous) phase transition.

Graph Representation and Laplacian Form

Because $\sigma_i^2 = 1$, one can always add an arbitrary diagonal term to J_{ij} without changing the thermodynamics. Hence,

$$Z = \sum_{\{\sigma_i\}} \exp\left[-\beta\left(\sum_{i,j}(J_{ij} - D_i\delta_{ij})\sigma_i\sigma_j + \sum_i h_i\sigma_i\right)\right], \tag{15.71}$$

for any choice of D_i. If $J_{ij} = JA_{ij}$, where A_{ij} is the adjacency matrix of the underlying lattice and $J < 0$ (ferromagnetic), one can take $D_i = \sum_j A_{ij}$ (the node degree). Then $D_i\delta_{ij} - A_{ij} = L_{ij}$, the graph Laplacian, showing that the Laplacian naturally arises in the Ising model as the operator governing spin couplings.

Ferromagnetic and Antiferromagnetic Interactions

Consider the two-body interaction term

$$J_{ij}\sigma_i\sigma_j, \tag{15.72}$$

TABLE 15.1 Possible states and energies of the two-body interaction term in Eq. (15.72).

σ_i	σ_j	$E = J\sigma_i\sigma_j$
-1	-1	$-J$
-1	$+1$	$+J$
$+1$	-1	$+J$
$+1$	$+1$	$-J$

with $\sigma_i, \sigma_j = \pm 1$ and uniform coupling $J_{ij} = J$. The four possible states and their corresponding energies are shown in Tab. 15.1.

If $J < 0$, parallel spins ($\sigma_i = \sigma_j$) minimize the energy, leading to a *ferromagnetic* ground state with all spins aligned. If $J > 0$, antiparallel alignment ($\sigma_i = -\sigma_j$) is energetically favored, resulting in an *antiferromagnetic* configuration.

Examples and Symmetry

Figure 15.5 illustrates the ground states for a triangular plaquette with ferromagnetic and antiferromagnetic couplings. For $J < 0$, all spins align in either of two degenerate configurations related by the global $\mathbb{Z}_2$ symmetry $\sigma_i \to -\sigma_i$. If an external magnetic field h is applied, this degeneracy is lifted: the state aligned with h has the lower free energy, and the symmetry is said to be *spontaneously broken.*

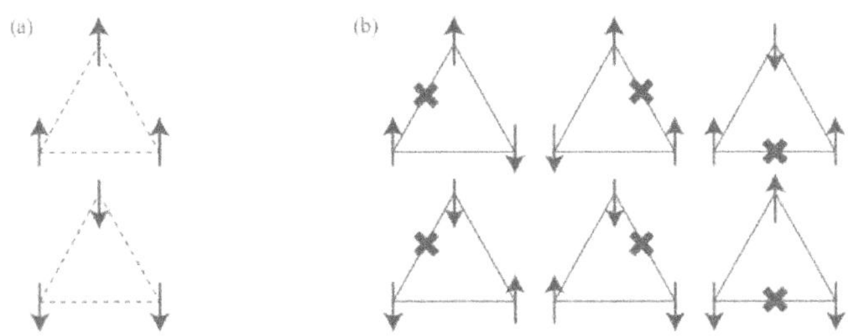

FIGURE 15.5 Ground states for the (a) ferromagnetic and (b) antiferromagnetic Ising model on a triangular plaquette.

The competition between energy and entropy determines the thermodynamic state:

$$F = U - TS. \tag{15.73}$$

At zero temperature, the ground state minimizes U; at finite temperature, entropy lifts degeneracy and can stabilize disordered phases.

Frustration and Antiferromagnetic Order

When couplings of mixed sign are present, some spins cannot simultaneously minimize all pairwise interactions—a situation known as *frustration.* Consider a simple triangular system:

$$H = J_1\sigma_1\sigma_2 + J_2\sigma_2\sigma_3 + J_3\sigma_3\sigma_1 + h(\sigma_1 + \sigma_2 + \sigma_3). \tag{15.74}$$

If $J_1 = J_2 = J_3 < 0$ (ferromagnetic), the ground state has all spins aligned and energy $E_0 = -3J$. For $h \neq 0$, this degeneracy is lifted, and the ground state aligns with the external field.

If instead $J_1 = J_2 > 0$ and $J_3 < 0$, the interactions compete: two bonds favor antiparallel spins, while one favors alignment. The system cannot satisfy all simultaneously, resulting in multiple degenerate ground states. This is the hallmark of frustration—a local rather than global degeneracy. A useful rule of thumb: a plaquette is frustrated if the product of the coupling signs around it is negative.

When all couplings are antiferromagnetic ($J > 0$), spins alternate directions on bipartite lattices, producing the *Néel state*, shown in Fig. 15.6. Such ordering minimizes the energy by aligning each spin opposite to its neighbors. If sublattices carry different spin magnitudes, the system retains a net magnetization and is called a *ferrimagnet.*

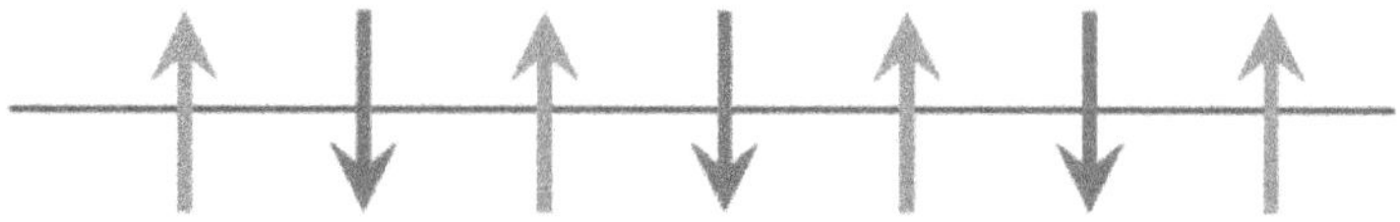

FIGURE 15.6 Néel order in a one-dimensional antiferromagnetic chain: neighboring spins point in opposite directions.

Boltzmann Weight and Effective Temperature

For ferromagnetic interactions, low-energy configurations are favored by the Boltzmann factor $e^{-\beta H}$. In the antiferromagnetic case, since flipping the sign of all J_{ij} corresponds to $H \to -H$, the statistical weight becomes $e^{+\beta H}$. This is equivalent to describing the system at an *effective negative temperature*, emphasizing the opposite tendency toward alignment.

REMARK 15.1 Frustration and Néel ordering are key ingredients in the physics of magnetic materials, spin glasses, and frustrated lattices such as the kagome and pyrochlore structures. These phenomena introduce rich energy landscapes and are central to modern condensed-matter and statistical physics.

Néel Order, and Natural and Artificial Frustration

Antiferromagnetic Néel order—in which spins align antiparallel on two interpenetrating sublattices—is realized in many crystalline materials. Classic examples include MnO, NiO, and MnF_2, where neutron scattering has directly confirmed the staggered spin arrangement predicted by simple antiferromagnetic Heisenberg models. In these compounds, the exchange interaction favors alternating spin directions, resulting in a robust ordered phase below a characteristic Néel temperature. When sublattices carry unequal magnetic moments, the same microscopic mechanism yields *ferrimagnetic* order, as in magnetite (Fe_3O_4) and certain rare-earth compounds.

Beyond naturally occurring systems, the physics of frustration has been extensively explored for *natural spin ices* [RHC+99], materials that exhibit the Pauling zero-temperature entropy [Pau35]; we will discuss these in Volume II. The most celebrated class is the family of *spin-ice materials*, such as $Dy_2Ti_2O_7$ and $Ho_2Ti_2O_7$, where the geometry of the pyrochlore lattice imposes local "two-in/two-out" constraints leading to a macroscopically degenerate ground-state manifold. These materials display emergent gauge fields, Coulomb phases, and even quasiparticles behaving as magnetic monopoles, as reviewed in [CMS12].

Artificial spin ices—engineered arrays of nanoscale magnetic islands—provide a highly tunable platform for exploring geometric frustration in real space. As shown in the comprehensive review by Moessner, Nisoli, Moessner, and Schiffer [NMS13], these systems exhibit interactions, collective excitations, and non-equilibrium dynamics that closely parallel the frustrated Ising models discussed above. They are, simplifying, physical toy models for the physics of magnetic systems we described so far. Because the individual magnetic islands can be directly imaged, artificial spin ices offer a rare opportunity to visualize frustration, monopole-like defects, and their propagation in real time.

15.5.1 The Two-Dimensional Ising Model and the Potts Generalization

The two-dimensional Ising model represents a milestone in the history of statistical mechanics, as it provides the simplest example of a system exhibiting a finite-temperature phase transition. Its exact solution, obtained by Lars Onsager in 1944, revealed for the first time the nature of continuous phase transitions and critical phenomena.

Before turning to the two-dimensional Ising Hamiltonian, it is useful to introduce its natural generalization—the *Potts model*—which unifies several discrete spin systems under a common framework.

The q-State Potts Model

The Potts model, proposed by Renfrey B. Potts in 1952 [Pot52], extends the Ising model by allowing each spin variable to take q discrete values,

$$\sigma_i \in \{1, 2, \ldots, q\}. \tag{15.75}$$

The Hamiltonian for a system with nearest-neighbor interactions is

$$H_{\text{Potts}} = -J \sum_{\langle i,j \rangle} \delta_{\sigma_i,\sigma_j} - h \sum_i \delta_{\sigma_i,1}, \tag{15.76}$$

where $\delta_{\sigma_i,\sigma_j}$ is the Kronecker delta, J controls the coupling strength, and h favors one spin orientation (e.g., state 1). For $q = 2$, the Potts model reduces exactly to the Ising model, while larger q values describe systems with multiple equivalent spin orientations, including models of alloys, lattice gases, and percolation.

The Potts model has played a central role in unifying various branches of statistical mechanics: it encompasses the Ising model ($q = 2$), the percolation problem in the limit $q \to 1$, and the random-cluster model introduced by Cornelis Marius Fortuin and Pieter Kasteleyn in 1972 [FK72]. A comprehensive review of its properties and extensions is provided in Wu's classic paper [Wu82].

The transfer-matrix method used for the one-dimensional Ising model generalizes naturally to the Potts case. For zero field ($h = 0$), the transfer matrix is

$$\mathcal{T} = \begin{pmatrix} e^{\beta J} & 1 & \cdots & 1 \\ 1 & e^{\beta J} & \cdots & 1 \\ \vdots & \vdots & \ddots & \vdots \\ 1 & 1 & \cdots & e^{\beta J} \end{pmatrix} = (e^{\beta J} - 1)\delta_{ij} + 1. \tag{15.77}$$

Its eigenvalues are easily obtained:

$$\lambda_1 = e^{\beta J} + q - 1, \qquad \lambda_2 = e^{\beta J} - 1, \tag{15.78}$$

where λ_1 has multiplicity one and λ_2 multiplicity $q - 1$. The corresponding partition function is

$$Z_{q\text{-Potts}}(h = 0) = \lambda_1^N + (q - 1)\lambda_2^N, \tag{15.79}$$

which reduces to the one-dimensional Ising case when $q = 2$.

The Two-Dimensional Ising Hamiltonian

In the two-dimensional Ising model, each spin $\sigma_{x,y} = \pm 1$ sits on a square lattice and interacts with its four nearest neighbors:

$$H_{\text{2D}} = -J \sum_{x=1}^{N} \sum_{y=1}^{N} \big(\sigma_{x,y}\sigma_{x+1,y} + \sigma_{x,y}\sigma_{x,y+1}\big), \tag{15.80}$$

with periodic (toroidal) boundary conditions,

$$\sigma_{N+1,y} = \sigma_{1,y}, \qquad \sigma_{x,N+1} = \sigma_{x,1}. \tag{15.81}$$

The partition function can be expressed as

$$Z = \text{Tr}\,(\mathcal{T}^N) = \sum_{j=1}^{2^N} \lambda_j^N, \tag{15.82}$$

where the transfer matrix $\mathcal{T}$ now acts between entire rows of spins, making its dimension $2^N \times 2^N$. The exponential growth of this matrix dimension explains why the two-dimensional model is far more difficult to solve analytically than its one-dimensional counterpart.

REMARK 15.2 The exact evaluation of the two-dimensional Ising free energy by Onsager (1944) represented a turning point in theoretical physics, demonstrating the power of exact methods and symmetry analysis in statistical mechanics. We will return to Onsager's method and its remarkable consequences, such as the exact critical temperature and spontaneous magnetization, in a later chapter.

Callen's Equation

Beyond approximations such as mean-field or variational theories, one can also derive an *exact* relation for the average magnetization of the Ising model. This relation, known as the *Callen equation* [Cal63], provides an implicit but exact expression for $\langle s_i \rangle$ in terms of local fields and spin correlations. It serves as the formal starting point for many self-consistent and perturbative schemes, including mean-field, Bethe–Peierls, and effective-field theories, which we discuss later.

Consider the classical Ising Hamiltonian

$$H = -\sum_{i<j} J_{ij} s_i s_j - \sum_i h_i s_i, \tag{15.83}$$

with spin variables $s_i = \pm 1$. We wish to compute the mean magnetization at site i:

$$\langle s_i \rangle = \frac{1}{Z} \sum_{\{s\}} s_i\, e^{-\beta H[\{s\}]}, \qquad Z = \sum_{\{s\}} e^{-\beta H[\{s\}]}. \tag{15.84}$$

Separating the sum over s_i from that over the remaining spins $\{s_{k \neq i}\}$, one finds

$$\langle s_i \rangle = \frac{1}{Z} \sum_{\{s_{k\neq i}\}} e^{\beta\left(\sum_{k\neq i,j} J_{kj} s_k s_j + \sum_{k \neq i} h_k s_k\right)} \sum_{s_i = \pm 1} s_i\, e^{\beta s_i\left(\sum_j J_{ij} s_j + h_i\right)}. \tag{15.85}$$

Using the identity $\sum_{s=\pm 1} s\, e^{as} = 2\sinh(a)$, we obtain

$$\langle s_i \rangle = \frac{2}{Z} \sum_{\{s_{k\neq i}\}} e^{\beta(\cdots)} \sinh\big[\beta(\sum_j J_{ij} s_j + h_i)\big]. \tag{15.86}$$

Multiplying and dividing by $2\cosh[\beta(\sum_j J_{ij}s_j + h_i)]$ and recognizing that this normalization reproduces the local partition function, we can rewrite

$$\langle s_i \rangle = \frac{1}{Z}\sum_{\{s\}} e^{-\beta H[\{s\}]} \tanh\big[\beta(\sum_j J_{ij}s_j + h_i)\big] = \big\langle \tanh\big[\beta(\sum_j J_{ij}s_j + h_i)\big]\big\rangle. \tag{15.87}$$

Equation (15.87) is the *Callen identity*—an exact, self-consistent relation connecting the local magnetization to the hyperbolic tangent of the local effective field. Equivalently,

$$\Big\langle s_i - \tanh[\beta(\sum_j J_{ij}s_j + h_i)]\Big\rangle = 0. \tag{15.88}$$

The Callen equation expresses the mean magnetization as the thermal average of a nonlinear function of neighboring spins. The nonlinearity prevents taking the average inside the tanh, which is precisely what mean-field theory does as an approximation:

$$\langle s_i \rangle \approx \tanh\big[\beta(J\sum_j \langle s_j \rangle + h_i)\big]. \tag{15.89}$$

Thus, the Callen relation provides the exact starting point from which mean-field and higher-order corrections can be systematically derived.

For a fully connected (Curie–Weiss) model with $J_{ij} = J/N\delta_{ij}$ and uniform field $h_i = h$, Eq. (15.87) becomes

$$\langle s_i \rangle = \big\langle \tanh[\beta(Jm + h)]\big\rangle, \qquad m = \frac{1}{N}\sum_i s_i. \tag{15.90}$$

Averaging over sites yields

$$\langle m \rangle = \big\langle \tanh[\beta(Jm + h)]\big\rangle. \tag{15.91}$$

Neglecting fluctuations of m—that is, replacing the average inside the tanh by its mean—reduces Eq. (15.91) to the standard mean-field self-consistency equation

$$m = \tanh[\beta(Jm + h)]. \tag{15.92}$$

The importance of the Callen identity is twofold. First, it provides an *exact* relation for the local magnetization, valid for any geometry or coupling matrix J_{ij}, without invoking approximations. Second, it clearly exposes the central approximation made by mean-field and effective-field approaches: the replacement of nonlinear averages by the average of nonlinear functions, i.e., $\langle \tanh(x)\rangle \approx \tanh(\langle x \rangle)$. This step—though uncontrolled—becomes asymptotically valid in systems with weak correlations or a large coordination number.

The TAP Equations and Disordered Mean-Field Theory

In the presence of disorder, the standard mean-field approach can be generalized. For example, in *spin-glass* systems where the couplings J_{ij} are random variables, the mean-field identity acquires corrections due to fluctuations of the local fields. The resulting self-consistency condition, first derived by David Thouless, Paul Anderson, and Robert Palmer [TAP77], is given by the celebrated *TAP equations*:

$$m_i = \tanh\Big[\beta\Big(h_i + \sum_j J_{ij}m_j - \beta(1-q)m_i\Big)\Big], \qquad q = \frac{1}{N}\sum_i m_i^2. \tag{15.93}$$

The additional term $-\beta(1-q)m_i$ is known as the *Onsager reaction term*, and it represents the feedback of spin fluctuations on the local magnetization. The TAP formalism extends mean-field theory to disordered systems by explicitly accounting for these correlations. It forms the conceptual foundation of modern spin-glass theory and has influenced developments in disordered systems, inference problems, and neural networks. A detailed discussion of the TAP equations and their consequences will be presented in Volume II.

16

Ising Model and Phase Transitions

16.1 Low-Temperature Expansion and the Peierls Argument

We now focus on the two-dimensional ferromagnetic Ising model, which provides one of the most fundamental and celebrated examples of a system exhibiting a finite-temperature phase transition. Although the exact solution was obtained only later by Lars Onsager (1944), the existence of a phase transition in two dimensions was first rigorously argued by Rudolf Peierls in 1936 [Pei36]. Here, we introduce the essential ideas of the low-temperature expansion and the Peierls argument.

Definition of the Model

For vanishing external field ($h = 0$), the Hamiltonian of the two-dimensional Ising model is

$$H = -J \sum_{\langle i,j \rangle} \sigma_i \sigma_j, \tag{16.1}$$

DOI: 10.1201/9781003603528-16

where $\sigma_i = \pm 1$ denotes the spin at lattice site i, and $\langle i, j\rangle$ indicates a sum over all nearest-neighbor pairs on a square lattice without double counting. Equivalently, this can be written in graph form as

$$H = -\frac{J}{2}\sum_{i,j} A_{ij}\,\sigma_i\sigma_j, \tag{16.2}$$

where A_{ij} is the adjacency matrix of the lattice.

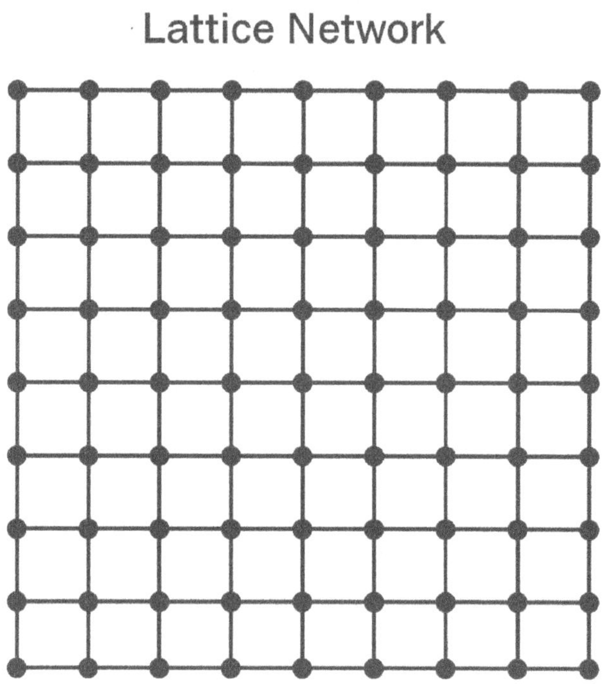

FIGURE 16.1 Schematic of the two-dimensional square-lattice Ising model with nearest-neighbor couplings.

The system exhibits two degenerate ground states: all spins up or all spins down. At finite but low temperature, excitations correspond to the flipping of finite clusters of spins within one of these ordered states.

Low-Temperature Expansion

At $T \to 0$ ($\beta = 1/k_{\mathrm{B}}T \to \infty$), thermal excitations are exponentially suppressed. To analyze the contribution of these excitations to the partition function, we impose periodic (toroidal) boundary conditions to avoid edge effects.

Each spin on the square lattice interacts with four neighbors, so the ground-state energy per spin is $-2J$. Flipping one spin reverses four bonds, yielding an energy cost of $8J$. Since there are N possible spins to flip, the first excited level has degeneracy N and energy $E_1 - E_0 = 8J$.

Flipping two spins leads to two distinct cases: 1. Distant spins: energy cost $16J$, degeneracy $N(N-5)/2$; 2. Adjacent spins: energy cost $12J$, degeneracy $8N$ (two orientations).

These patterns are represented diagrammatically by the number of *frustrated bonds*, bonds joining oppositely aligned spins (shown in red in Fig. 16.2). Each additional flipped spin adds frustrated bonds or modifies existing ones according to whether the flipped spins share an edge.

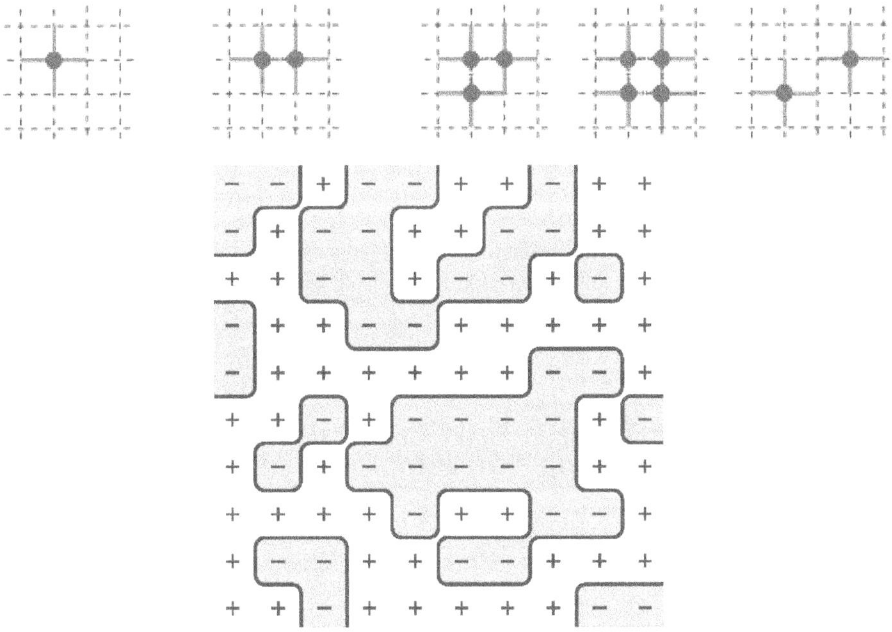

FIGURE 16.2 Low-temperature expansion in terms of frustrated bonds (red). Each closed loop of reversed spins defines a *Peierls droplet*, separating up and down regions. (Source: Wikipedia)

Keeping terms up to order $e^{-20\beta J}$, the partition function can be written as

$$Z = 2e^{2N\beta J}\Big[1 + Ne^{-8\beta J} + 2Ne^{-12\beta J} + \big(\tfrac{N(N-5)}{2} + 7N\big)e^{-16\beta J} + O(e^{-20\beta J})\Big]. \tag{16.3}$$

Introducing $x = e^{-2\beta J}$, and expanding $\ln Z$ up to order x^8,

$$\ln Z = \ln\big(2e^{2N\beta J}\big) + Nx^4 + 2Nx^6 + \tfrac{9Nx^8}{2} + O(x^9), \tag{16.4}$$

we see that $\ln Z$ is proportional to N, since only *connected* excitations contribute to the free energy—analogous to the virial expansion.

The Peierls Argument and Droplet Picture

In the low-temperature phase, excitations correspond to finite clusters of flipped spins embedded in an ordered background. The boundaries of these clusters form closed loops called *Peierls contours* or *droplets*. Let L denote the length of such a boundary; each broken bond contributes energy $2J$, so the total excitation energy scales as $E \propto JL$. Meanwhile, the number of possible closed loops of length L on a square lattice grows combinatorially as $N_L \sim 4^L$.

Thus, the corresponding free energy cost of a droplet is

$$F(L) \approx E - TS = (c_1 - Tc_2)L, \tag{16.5}$$

where $c_1 \propto J$ and $c_2 \propto k_B \ln 4$ are positive constants. At low temperature ($T < T_c$), the energy term dominates and droplets are suppressed, leading to a magnetized phase. At higher temperature ($T > T_c$), the entropy term dominates, and the system becomes disordered. The balance of these two contributions defines the critical temperature,

$$T_c = \frac{c_1}{c_2}, \tag{16.6}$$

where long-range order is lost and the correlation length diverges.

The Peierls argument thus provides an intuitive geometric mechanism for spontaneous symmetry breaking in two dimensions: below T_c, large reversed domains are exponentially unlikely, while above T_c they proliferate, destroying long-range magnetic order. This behavior is absent in one dimension, where the energy of a domain wall is constant rather than extensive in its boundary length.

REMARK 16.1 The Peierls droplet picture was the first rigorous demonstration that a finite-temperature phase transition exists in the two-dimensional Ising model [Pei36]. It predates Onsager's exact solution and remains one of the most elegant and intuitive arguments in statistical mechanics.

16.2 High-Temperature Expansion and Duality in the Two-Dimensional Ising Model

At high temperatures, the Ising model behaves quite differently from its low-temperature ordered phase: the spins are essentially uncorrelated, and thermal fluctuations dominate. The systematic expansion of the partition function in powers of $\tanh(\beta J)$ is known as the van der Waerden expansion, after B. L. van der Waerden [vdW41], and was later refined by Tsutomu Oguchi [Ogu51]. This expansion forms the foundation for the celebrated *Kramers–Wannier duality* [KW41], which relates the high- and low-temperature behaviors of the two-dimensional Ising model.

Van der Waerden Expansion

We start from the Hamiltonian of the two-dimensional Ising model with zero external field,

$$H = -J \sum_{\langle i,j \rangle} \sigma_i \sigma_j, \tag{16.7}$$

where $\sigma_i = \pm 1$. The Boltzmann weight of a single bond can be written as

$$e^{\beta J \sigma_i \sigma_j} = \cosh(\beta J) + \sigma_i \sigma_j \sinh(\beta J) = \cosh(\beta J)\big[1 + \sigma_i \sigma_j \tanh(\beta J)\big]. \tag{16.8}$$

Hence, the partition function becomes

$$\begin{aligned} Z &= \sum_{\{\sigma_i\}} \prod_{\langle i,j \rangle} \cosh(\beta J)\big[1 + \sigma_i \sigma_j \tanh(\beta J)\big] \\ &= \big[\cosh(\beta J)\big]^M \sum_{\{\sigma_i\}} \prod_{\langle i,j \rangle} \big[1 + \sigma_i \sigma_j x\big], \end{aligned} \tag{16.9}$$

where M is the number of bonds on the lattice and $x = \tanh(\beta J)$ is the small expansion parameter for $\beta J \ll 1$.

Expanding the product, we obtain

$$Z = \big[\cosh(\beta J)\big]^M \sum_{\{\sigma_i\}} \Bigg(1 + \sum_{\text{1-bond}} \sigma_i \sigma_j x + \sum_{\text{2-bonds}} (\sigma_i \sigma_j)(\sigma_{i'} \sigma_{j'}) x^2 + \cdots \Bigg). \tag{16.10}$$

Because $\sum_{\sigma_i = \pm 1} \sigma_i = 0$, only those terms survive in which every spin appears an even number of times. These correspond to products of σ_i's forming *closed loops* on the lattice. The first nonzero contribution thus arises from plaquettes (loops of length four).

At zeroth order, $\sum_{\sigma_i} 1 = 2^N$. At fourth order (x^4), the contribution comes from all elementary square plaquettes:

$$\sum_{\sigma_1\sigma_2\sigma_3\sigma_4} \sigma_1\sigma_2\sigma_2\sigma_3\sigma_3\sigma_4\sigma_4\sigma_1 = 2^4. \tag{16.11}$$

Counting all $N/4$ plaquettes on the lattice gives the first correction $2^N Nx^4/4$. Proceeding systematically, one finds

$$Z = 2^N \left[\cosh(\beta J)\right]^M \left(1 + Nx^4 + 2Nx^6 + \left[\tfrac{N(N-5)}{2} + 7N\right]x^8 + \cdots\right). \tag{16.12}$$

The structure of the series mirrors that of the low-temperature expansion, but with the replacement

$$e^{-2\beta J} \longleftrightarrow \tanh(\beta J), \tag{16.13}$$

a correspondence that lies at the heart of the duality discussed below.

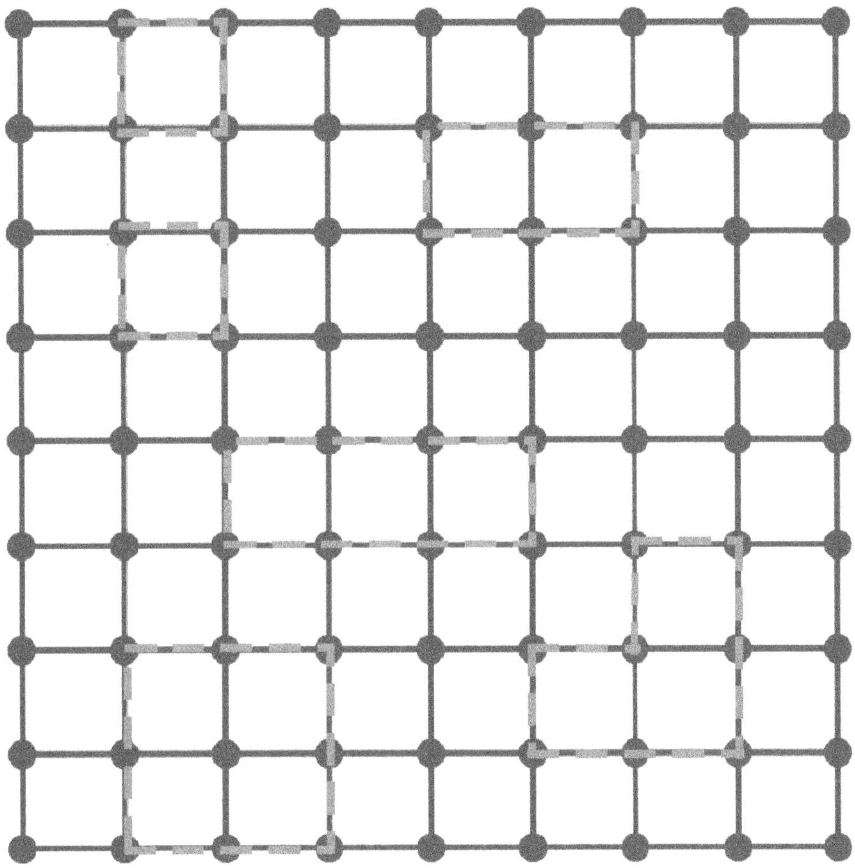

FIGURE 16.3 Representative loop configurations contributing to the high-temperature expansion on a square lattice.

Strong-Weak Coupling Duality

The low-temperature expansion corresponds to the *weak-coupling* regime, expanding around the ordered ground state. In contrast, the high-temperature expansion is a *strong-coupling* expansion, in which the system is disordered. Remarkably, the two regimes are related by a *duality transformation*: the low-temperature diagrams of domain walls are in one-to-one correspondence with the high-temperature closed-loop graphs on the *dual lattice*, as illustrated in Fig. 16.4.

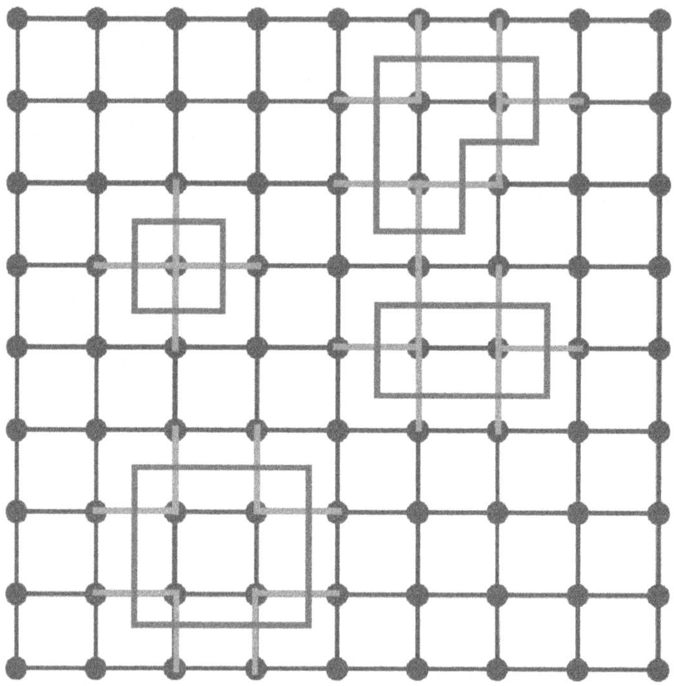

FIGURE 16.4 Correspondence between domain walls in the low-temperature expansion and loop configurations in the high-temperature expansion of the ferromagnetic Ising model on the lattice. This mapping underlies the Kramers–Wannier duality. Spins flipped lead to a cross on the lattice because of broken bonds (red). The plaquette is constructed by dimers/links placed across the broken bond between the dual lattice centers (blue).

Kramers–Wannier Duality and the Critical Temperature

Hendrik Kramers and Gregory Wannier [KW41] proved that the partition function of the two-dimensional Ising model satisfies an exact duality relation connecting couplings βJ and $\tilde{\beta} J$ on the original and dual lattices:

$$\sinh(2\beta J)\,\sinh\big(2\tilde{\beta} J\big) = 1. \tag{16.14}$$

The two temperature regimes are thus inversely related. If a phase transition occurs at temperature T_c, duality implies that it must be self-dual:

$$\tilde{\beta} = \beta_c \quad \Rightarrow \quad \sinh(2\beta_c J) = 1. \tag{16.15}$$

From this, the critical temperature follows as

$$k_{\mathrm{B}} T_c = \frac{2J}{\sinh^{-1}(1)} \approx 2.269\,J. \tag{16.16}$$

This result coincides exactly with the value later derived by Onsager from the full analytic solution of the two-dimensional Ising model's partition function.

REMARK 16.2 The Kramers–Wannier duality establishes a profound symmetry between the high- and low-temperature phases of the Ising model, mapping ordered and disordered configurations onto each other. It also anticipates the broader concept of *strong–weak coupling duality*, a cornerstone of modern theoretical physics.

16.3 Critical Phenomena and Phase Transitions

One of the earliest hints of universality across different gases is the *law of corresponding states.* Consider again the van der Waals equation of state,

$$P = \frac{k_B T}{V - b} - \frac{a}{V^2}, \tag{16.17}$$

and introduce reduced variables with respect to the critical point:

$$\bar{T} = \frac{T}{T_c}, \qquad \bar{V} = \frac{V}{V_c}, \qquad \bar{P} = \frac{P}{P_c}, \tag{16.18}$$

where $k_B T_c = 8a/(27b)$, $V_c = 3b$, and $P_c = a/(27b^2)$. In terms of reduced variables, the equation becomes

$$\bar{P} = \frac{8}{3}\frac{\bar{T}}{\bar{V} - \frac{1}{3}} - \frac{3}{\bar{V}^2}, \tag{16.19}$$

which is independent of the microscopic details of the gas.

Along the coexistence curve, equality of pressure for the liquid and gas branches gives

$$\bar{P} = \frac{8\bar{T}}{3\bar{V}_{\text{liq}} - 1} - \frac{3}{\bar{V}_{\text{liq}}^2} = \frac{8\bar{T}}{3\bar{V}_{\text{gas}} - 1} - \frac{3}{\bar{V}_{\text{gas}}^2}. \tag{16.20}$$

This can be rearranged as

$$\bar{T} = \frac{(3\bar{V}_{\text{liq}} - 1)(3\bar{V}_{\text{gas}} - 1)(\bar{V}_{\text{liq}} + \bar{V}_{\text{gas}})}{8\bar{V}_{\text{gas}}^2\,\bar{V}_{\text{liq}}^2}. \tag{16.21}$$

Close to the critical point set $\bar{V}_{\text{liq}} = 1 + \epsilon/2$ and $\bar{V}_{\text{gas}} = 1 - \epsilon/2$. Expanding to leading order gives

$$\bar{T} \approx 1 - \frac{\epsilon^2}{16}. \tag{16.22}$$

With the reduced temperature

$$t = 1 - \frac{T}{T_c} = 1 - \bar{T}, \tag{16.23}$$

we obtain the mean-field critical law

$$\bar{V}_{\text{gas}} - \bar{V}_{\text{liq}} \propto \epsilon \propto \sqrt{t}. \tag{16.24}$$

Hence the order parameter for the liquid–gas transition (density difference) follows a square-root singularity as $T \to T_c^-$, the same mean-field exponent that appears in magnetic systems.

Historical Note: van der Waals and Corresponding States

The van der Waals equation was introduced in 1873, with the idea that real fluids deviate from ideality due to molecular volume b and attractions a [vdW73a]. Edward Guggenheim

(1945) showed that many fluids obey an approximate law of corresponding states when expressed in reduced variables, collapsing their coexistence curves [Gug45].

Mean-Field Curie–Weiss: Critical Exponents

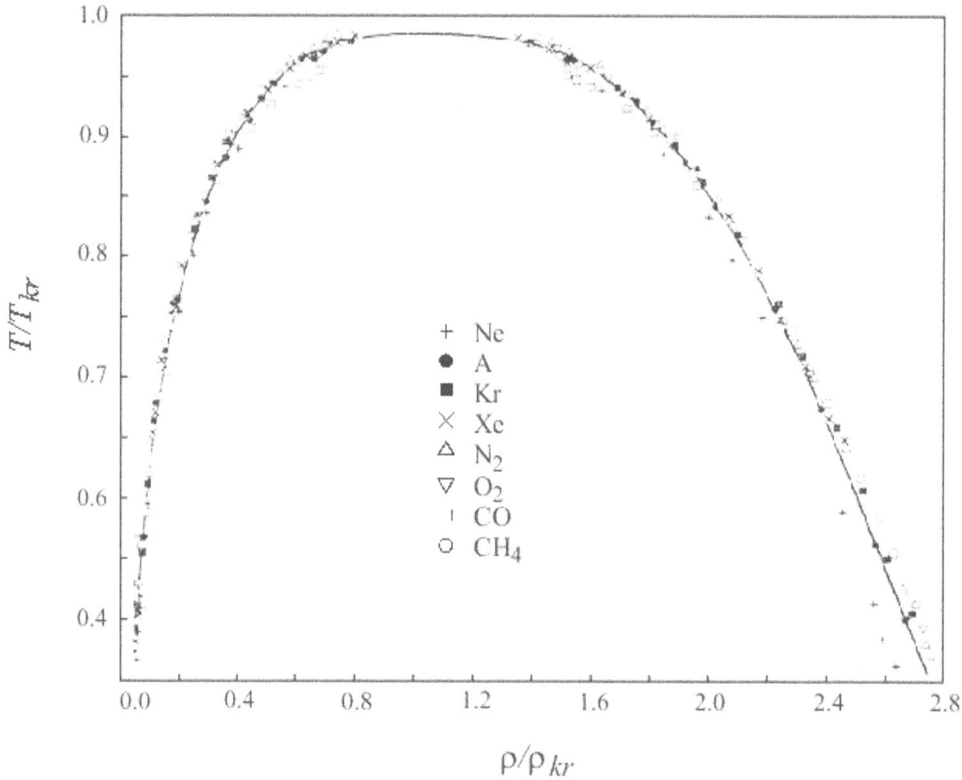

FIGURE 16.5 Coexistence curve for various fluids in reduced variables (Guggenheim plot, 1945). Reproduced from [Gug45].

Around the Curie point T_c, the *Curie–Weiss* mean-field equation

$$M = \tanh\big[\beta(JM + h)\big] \tag{16.25}$$

implies standard mean-field critical behavior. Expanding $\tanh x \simeq x - x^3/3$ gives

$$M \approx \beta JM + \beta h - \frac{\beta^3}{3}(JM + h)^3. \tag{16.26}$$

REMARK 16.3 Critical exponents (definitions). For a second-order transition at $T = T_c$,

$$\begin{aligned}
&\text{Order parameter:} && M(T, h = 0) \sim (T_c - T)^\beta \quad (T \to T_c^-),\\
&\text{Field isotherm:} && M(T_c, h) \sim h^{1/\delta},\\
&\text{Susceptibility:} && \chi(T) = \left.\frac{\partial M}{\partial h}\right|_{h=0} \sim |T - T_c|^{-\gamma}.
\end{aligned} \tag{16.27}$$

These are the conventional signs and powers used throughout this Volume.

From Eq. (16.26) at $h = 0$, the nontrivial solution below T_c is

$$M = \pm \frac{\sqrt{3(\beta J - 1)}}{\beta^{3/2} J^{3/2}} \propto (T_c - T)^{1/2}, \tag{16.28}$$

so $\beta_{\mathrm{MF}} = 1/2$. At $T = T_c$ and small h, the cubic term dominates, yielding $M \propto h^{1/3}$ and $\delta_{\mathrm{MF}} = 3$. Linearizing above T_c gives $\chi \propto (T - T_c)^{-1}$, hence $\gamma_{\mathrm{MF}} = 1$.

Landau Viewpoint

Expanding a mean-field free-energy density as $f(M) = a(T)M^2/2 + bM^4/4 - hM$ with $a(T) \propto T - T_c$, and minimizing reproduces $\beta = 1/2$, $\gamma = 1$, and $\delta = 3$. Deviations in real systems arise from fluctuations beyond mean field; their strength depends on dimensionality and interactions.

Lattice Gas ↔ Curie–Weiss: Similarities and Fluctuations

A simple lattice gas with occupations $\sigma_i \in \{0, 1\}$ maps to an *Ising* model via

$$\sigma_i = 2\sigma_i - 1, \qquad \sigma_i = \frac{1 + \sigma_i}{2}. \tag{16.29}$$

A mean-field lattice-gas Hamiltonian on a graph with adjacency A_{ij} and average coordination d can be written as

$$H = -\frac{\epsilon}{2} c \sum_{i,j} A_{ij}\sigma_i\sigma_j - \mu \sum_i \sigma_i, \qquad c\sum_j A_{ij} = d, \quad c = \frac{d}{N} \text{ if } A_{ij} = 1, \tag{16.30}$$

which, in terms of σ_i, becomes (up to additive constants)

$$H = -\frac{\epsilon d}{2N} \sum_{i,j} \sigma_i\sigma_j - \left(\mu - \frac{d\epsilon}{N}\right) \sum_i \sigma_i. \tag{16.31}$$

Therefore, in Curie–Weiss form,

$$J \leftrightarrow \epsilon, \qquad h \leftrightarrow \mu - \frac{d\epsilon}{N}. \tag{16.32}$$

REMARK 16.4 To avoid confusion, throughout this book we denote the Hamiltonian by $\mathcal{H}$ and the magnetic field by h. Earlier uses of H as both Hamiltonian and field are here harmonized to this convention.

In the lattice gas, the pressure and density are

$$P = \frac{k_{\mathrm{B}}T}{N} \ln Z, \qquad \rho = \frac{\langle n \rangle}{N} = \frac{\partial P}{\partial \mu}, \tag{16.33}$$

and the isothermal compressibility is

$$\kappa_T = \frac{1}{\rho}\frac{\partial \rho}{\partial P}. \tag{16.34}$$

In the Curie–Weiss model,

$$M = \frac{1}{N} \sum_i \langle \sigma_i \rangle = \frac{1}{\beta}\frac{\partial \ln Z}{\partial h}, \qquad \chi = \left.\frac{\partial M}{\partial h}\right|_{h=0}. \tag{16.35}$$

Up to affine rescalings set by the mapping,

$$\rho \leftrightarrow M, \qquad \rho^2 \kappa_T \leftrightarrow \chi. \tag{16.36}$$

Thus liquid–gas and ferromagnetic criticality share the same mean-field exponents and phenomenology.

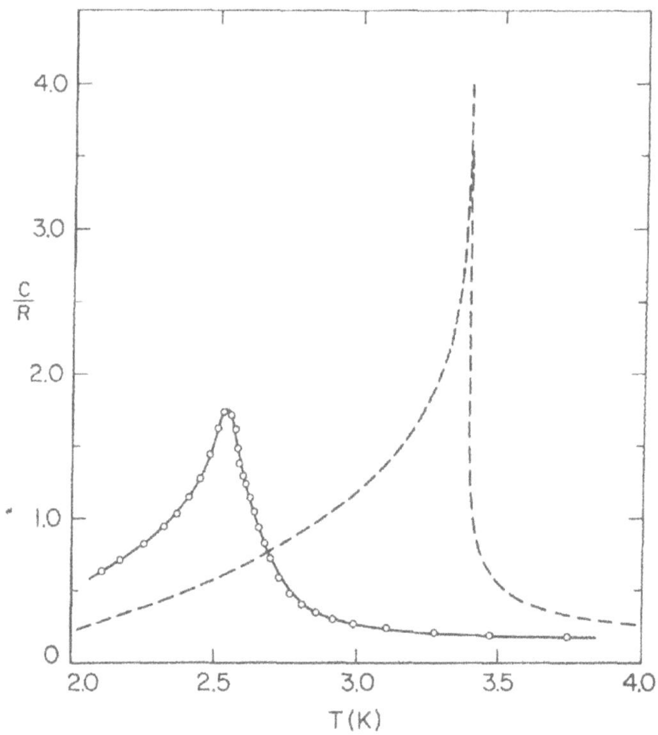

FIGURE 16.6 Magnetic susceptibility $\chi(T)$ for a three-dimensional antiferromagnet, in the presence and in the absence of external magnetic field (dashed). When an external magnetic field is present, the heat capacity divergence disappears. Reprinted from [WMC+71].

Experimental Critical Phenomena

Experimental studies of critical phenomena provide some of the clearest evidence for universality. By the late 1950s–1960s, experiments showed that diverse systems (non-ideal fluids near the liquid–gas critical point and uniaxial ferromagnets near the Curie temperature) exhibit strikingly similar power laws when plotted versus $|T - T_c|$ after appropriate rescaling (data collapse / corresponding states), see, e.g., [Sta71, Gol92].

Remarkably, for many fluids the coexistence (liquid–vapor) curve collapses onto a single master curve (Fig. 16.5), often called the Guggenheim plot [Gug45]. Fig. 16.5 shows that the idea of universality is not far-fetched, but of practical importance, as it allows putting on the same footage, near criticality, materials that would otherwise look very different from each other.

Near a continuous phase transition, macroscopic observables display singularities characterized by critical exponents that depend only on symmetry and dimensionality, not on microscopic details. For instance, Fig. 16.6 shows the heat capacity divergence for an antiferromagnet.

Measurements on simple fluids near the liquid–gas critical point yield representative values

$$\beta \approx 0.33\text{–}0.38, \qquad \gamma \approx 1.2\text{–}1.3, \qquad \delta \approx 4.5\text{–}5,$$

in good agreement with the critical exponents of the three-dimensional Ising universality class [Sta71, Gol92].

For the *Ising model*, the exponents are known with great precision:

Two-dimensional (exact): $\beta = \frac{1}{8}$, $\gamma = \frac{7}{4}$, $\delta = 15$
(Onsager–Yang) [Ons44, Yan52];

Three-dimensional (modern estimates): $\beta \simeq 0.326$, $\gamma \simeq 1.237$, $\delta \simeq 4.789$
(MC/HT series/field theory/bootstrap) [PV02, Has10, KPSDV16].

The close numerical agreement between fluid and three-dimensional Ising critical exponents is a striking confirmation of the concept of *universality classes*: systems that differ microscopically may exhibit identical large-scale behavior near criticality.

Why Universality?

Near criticality, the *correlation length* ξ—the characteristic scale of spatial correlations—diverges as

$$\xi \sim |T - T_c|^{-\nu}. \tag{16.37}$$

When $\xi \to \infty$, the system no longer possesses a natural microscopic length scale: the details of atomic potentials, lattice spacing, or molecular structure become irrelevant. Only a few global features survive, notably:

- the *dimensionality* of space d,
- the *symmetry* of the order parameter,
- and the *range of interactions*.

Systems sharing these ingredients exhibit identical sets of critical exponents and scaling functions—they belong to the same *universality class*.

This observation, originally empirical, received a firm theoretical foundation in the *renormalization-group (RG)* framework developed by Leo Kadanoff and Kenneth Wilson in the late 1960s early 1970s [Kad66, Wil71, WK74b], which we will cover later in this Volume. The key idea is to coarse-grain the system by integrating out short-wavelength degrees of freedom, thereby constructing an effective free energy for long-wavelength modes. Under successive coarse-grainings, the parameters of the theory flow in a multi-dimensional space (the *coupling space*). At the critical point, the system reaches a *fixed point* of this flow, where scale invariance emerges:

$$\vec{g}^\star = \vec{R}(\vec{g}) \tag{16.38}$$

and the eigenvalues of the linearized RG transformation around $\vec{g}^\star$ determine the critical exponents.

The modern classification of universality classes is thus based on the symmetries and dimensionality of the order parameter field $\phi(\vec{r})$, often captured by an effective Landau–Ginzburg Hamiltonian

$$H_{\text{eff}}[\phi] = \int d^d r \left[\frac{1}{2}(\nabla\phi)^2 + \frac{r}{2}\phi^2 + \frac{u}{4!}\phi^4 - h\phi \right], \tag{16.39}$$

where ϕ is scalar for Ising-type systems and vector-valued for multicomponent order parameters.

Examples of Universality Classes:

- *Ising universality class ($n = 1$ scalar order parameter).* Includes uniaxial ferromagnets, liquid–gas transitions, binary fluid mixtures, and lattice gases.
- *XY universality class ($n = 2$).* Describes planar magnets, thin superfluid helium films (^{4}He), and superconductors (phase-only fluctuations).
- *Heisenberg universality class ($n = 3$).* Applies to isotropic three-dimensional ferromagnets such as nickel or iron near T_c.
- *$O(n)$ universality classes (general n).* Cover systems with continuous symmetry of an n-component order parameter. The Ising, XY, and Heisenberg classes correspond to $n = 1, 2, 3$, respectively.
- *Percolation and Potts universality classes.* The q-state Potts model interpolates between discrete-symmetry systems ($q > 2$) and percolation ($q \to 1$).
- *Kosterlitz–Thouless (KT) universality.* Characterizes topological transitions in two-dimensional systems with continuous symmetry, such as two-dimensional superfluids and the XY model.
- *Directed percolation and absorbing-state transitions.* Describe non-equilibrium systems approaching steady states with absorbing configurations.
- *Quantum phase transitions.* At zero temperature, quantum fluctuations replace thermal ones; many are governed by RG fixed points in $(d + z)$ dimensions, where z is the dynamical exponent.

Critical phenomena exhibit scale invariance, and systems differing microscopically but sharing the same symmetry and dimensionality flow under the RG to the same fixed point. Their thermodynamic singularities are thus universal, reflecting the geometry of the fixed point rather than microscopic detail.

16.4 Griffiths Inequalities in Ferromagnetic Systems and Their Consequences

We now turn to a set of powerful exact results concerning ferromagnetic systems, known as the *Griffiths inequalities.* These inequalities, due to Robert B. Griffiths [Gri67], provide rigorous constraints on correlation functions and play a crucial role in understanding monotonicity and phase transitions in Ising-type models.

Statement of the Inequalities

Consider the ferromagnetic Ising model defined on a finite lattice $\Lambda \subset \mathbb{Z}^d$, with spins $\sigma_i \in \{-1, +1\}$ and Hamiltonian

$$H(\sigma) = - \sum_{\{i,j\} \subset \Lambda} J_{ij} \, \sigma_i \sigma_j - \sum_{i \in \Lambda} h_i \, \sigma_i, \tag{16.40}$$

where all couplings are non-negative: $J_{ij} \geq 0$. The corresponding Gibbs measure at inverse temperature β is

$$\mu(\sigma) = \frac{1}{Z} e^{-\beta H(\sigma)}. \tag{16.41}$$

Griffiths I

For any finite subset $A \subset \Lambda$, the expectation of a spin monomial is non-negative:

$$\langle \sigma_A \rangle := \left\langle \prod_{i \in A} \sigma_i \right\rangle \geq 0. \tag{16.42}$$

Griffiths II

For any two finite subsets $A, B \subset \Lambda$, the connected correlation is also non-negative:

$$\langle \sigma_A \sigma_B \rangle \geq \langle \sigma_A \rangle \, \langle \sigma_B \rangle. \tag{16.43}$$

These results follow from the positivity of the Boltzmann weights in ferromagnetic systems and the convexity of the free energy as a function of the external field.

Consequences of the Griffiths Inequalities

The inequalities imply several monotonicity properties:

- *Non-Negativity of Correlations:* $\langle \sigma_i \sigma_j \rangle_c := \langle \sigma_i \sigma_j \rangle - \langle \sigma_i \rangle \langle \sigma_j \rangle \geq 0$.
- *Monotonicity in System Size:* If $\Lambda \subset \Lambda'$, then $\langle \sigma_i \sigma_j \rangle_\Lambda \leq \langle \sigma_i \sigma_j \rangle_{\Lambda'}$.
- *Monotonicity in Boundary Conditions:* Fixing spins on the boundary to $+1$ (equivalent to a positive boundary field) increases all spin expectations: $\langle \sigma_i \rangle_{\text{free}} \leq \langle \sigma_i \rangle_+$.

The magnetic susceptibility in a finite region Λ_L,

$$\chi_L = \sum_{j \in \Lambda_L} \langle \sigma_0 \sigma_j \rangle_c, \tag{16.44}$$

is therefore non-decreasing with L:

$$\chi_L \leq \chi_{L+1} \leq \cdots \leq \chi_\infty = \lim_{L \to \infty} \chi_L. \tag{16.45}$$

A divergence of χ_L as $L \to \infty$ signals the onset of long-range order and a phase transition.

REMARK 16.5 Griffiths inequalities are entirely non-perturbative: they do not rely on approximations or expansions, but follow directly from the positivity and convexity structure of the Gibbs measure. They provide a rigorous foundation for many intuitive statements about ferromagnetic systems, such as "increasing the field increases magnetization" or "larger systems have stronger correlations."

16.5 Approximations for the Ising Model

Bragg–Williams Theory

Historically, one of the earliest mean-field descriptions of ferromagnetism and order–disorder transitions in alloys is the *Bragg–Williams theory.* The model was introduced to capture the competition between ordering tendencies (which lower the internal energy) and thermal agitation (which increases entropy), providing a simple explanation for alloy phase transitions and magnetic order.

Historical Note: The Bragg–Williams Theory

The Bragg–Williams theory was formulated in a series of papers by W. L. Bragg and E. J. Williams in 1934–1935 [BW34, BW35, Wil35]. Their aim was to understand how thermal agitation affects the atomic arrangement in binary alloys such as CuZn and CuAu. They introduced the concept of an *order parameter* to quantify the degree of atomic ordering, long before it became a central idea in statistical mechanics and phase transition theory. For a historical discussion, see also Cahn's review in *The Coming of Materials Science* [Cah01] and Chaikin and Lubensky [CL95].

> "If the metal atoms of either kind were effectively identical, their distribution among the lattice sites would be random. Actually, differences between them tend to cause atoms of one kind to segregate into certain particular sites forming an orderly arrangement, because such an arrangement lowers the potential energy. Thermal agitation has the opposing effect of creating a random arrangement. The actual state of dynamical equilibrium of an alloy is one in which the two processes balance." — **W. L. Bragg and E. J. Williams (1934)**.

The Bragg–Williams approach introduces an important quantity, the *order parameter*, which quantifies the degree of order in a system. In a binary alloy, it measures how strongly one atomic species prefers certain sites over others; in a ferromagnet, it corresponds to the mean magnetization per site. By expressing the free energy as a function of this parameter and minimizing it, one can determine the equilibrium state and locate the transition temperature.

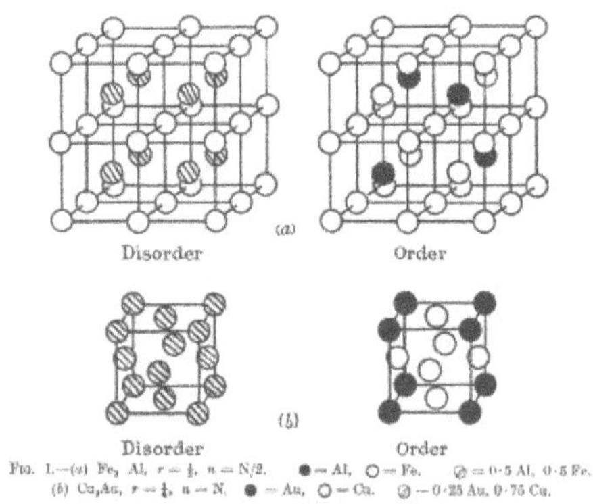

FIGURE 16.7 Illustrations from the original Bragg and Williams papers. Reproduced from [BW34, BW35].

Consider a regular lattice of N sites, each occupied by a magnetic atom with spin $\sigma_i = \pm 1$. In the Bragg–Williams approximation, all spins are assumed to be statistically independent, so that the joint probability distribution factorizes:

$$P(\sigma_1, \ldots, \sigma_N) = \prod_i p_i(\sigma_i), \tag{16.46}$$

where

$$p_i(\uparrow) = \frac{1+m_i}{2}, \qquad p_i(\downarrow) = \frac{1-m_i}{2}, \qquad m_i = \langle \sigma_i \rangle = \bar{m}. \tag{16.47}$$

Energy and Entropy

For a lattice with coordination number z and Hamiltonian

$$\mathcal{H} = -J \sum_{\langle ij \rangle} \sigma_i \sigma_j - h \sum_i \sigma_i, \tag{16.48}$$

the average energy and entropy per spin are

$$E = \langle \mathcal{H} \rangle_P = -\tfrac{1}{2} N z J \bar{m}^2 - N h \bar{m}, \tag{16.49}$$

$$S = -N k_B \left[\tfrac{1+\bar{m}}{2} \ln \frac{1+\bar{m}}{2} + \tfrac{1-\bar{m}}{2} \ln \frac{1-\bar{m}}{2} \right]. \tag{16.50}$$

The free energy is then

$$F = E - TS. \tag{16.51}$$

Minimizing F with respect to $\bar{m}$ yields

$$\frac{dF}{d\bar{m}} = -zJ\bar{m} + \tfrac{1}{2} k_B T \ln \frac{1+\bar{m}}{1-\bar{m}} = 0, \tag{16.52}$$

which gives

$$\bar{m} = \tanh\left[\beta(h + zJ\bar{m})\right]. \tag{16.53}$$

This is the *Bragg–Williams equation of state*, identical in form to the Curie–Weiss mean-field equation, with $\bar{m}$ playing the role of the order parameter.

REMARK 16.6 The Bragg–Williams mean-field theory provides a simple quantitative description of how magnetization (or ordering) varies with temperature. It predicts a continuous transition at

$$T_c = \frac{zJ}{k_B}, \tag{16.54}$$

and an order parameter $\bar{m} \propto (T_c - T)^{1/2}$ near criticality. However, because it neglects correlations between neighboring spins, it overestimates T_c and incorrectly predicts a finite-temperature transition even in one dimension.

This probabilistic formulation of mean-field theory will reappear later when we discuss variational and cluster approximations: it corresponds to assuming a fully factorized probability distribution in the Gibbs free energy functional.

Example: The Curie–Weiss Ferromagnet via Combinatorics

The Curie–Weiss model is the simplest fully-connected Ising ferromagnet. Each spin interacts equally with all others, leading to the mean-field Hamiltonian

$$H = -\frac{J}{2N}\Big(\sum_{i=1}^{N} s_i\Big)^2 - h\sum_{i=1}^{N} s_i, \qquad s_i = \pm 1.$$

The natural order parameter is the total magnetization

$$M = \sum_{i=1}^{N} s_i, \qquad m = M/N.$$

This example illustrates how the full Curie–Weiss mean-field theory emerges purely from *combinatorics* and a *saddle-point approximation*, without invoking variational arguments or heuristic mean-field assumptions.

For a fixed magnetization M, the number of compatible spin configurations is

$$\Omega(M) = \binom{N}{\frac{N+M}{2}},$$

i.e., the number of ways to choose $(N+M)/2$ up-spins. The partition function can then be written exactly as

$$Z = \sum_{M=-N}^{N} \Omega(M)\, \exp\Big[\beta\Big(\frac{J}{2N}M^2 + hM\Big)\Big].$$

For macroscopic systems, the binomial coefficient is approximated using Stirling's formula. Introducing the magnetization per spin $m = M/N$, one finds

$$\Omega(M) \approx \frac{2^N}{\sqrt{2\pi N}}\, \exp[-N\phi(m)]\,,$$

where

$$\phi(m) = \frac{1+m}{2}\ln(1+m) + \frac{1-m}{2}\ln(1-m).$$

The partition function becomes an integral of the form

$$Z \approx 2^N \int_{-1}^{1} \exp\Big\{-N\Big[\phi(m) - \beta\Big(\tfrac{J}{2}m^2 + hm\Big)\Big]\Big\}\, dm.$$

In the thermodynamic limit, the integral is dominated by the minimum of the exponent. Define the Landau free-energy functional

$$f(m) = \phi(m) - \beta\Big(\tfrac{J}{2}m^2 + hm\Big).$$

The saddle point satisfies

$$\frac{df}{dm} = 0.$$

Carrying out the derivative leads to the celebrated Curie–Weiss mean-field equation

$$m = \tanh(\beta J m + \beta h)\,.$$

This is the self-consistency equation for the magnetization, and reproduces all standard mean-field predictions: spontaneous symmetry breaking, the transition at $T_c = J/k_B$, and the magnetic equation of state.

Evaluating the exponent at the saddle point m^* yields the thermodynamic free energy:

$$F \approx -k_B T\Big[\, N \ln 2 \;-\; N\big(\phi(m^*) - \beta(\tfrac{J}{2}(m^*)^2 + hm^*)\big)\Big].$$

All equilibrium properties follow from this expression.

The Curie–Weiss model demonstrates beautifully how a macroscopic order parameter arises purely from the combinatorics of counting microstates. The balance between entropy (encoded in $\phi(m)$) and energy produces spontaneous magnetization below the critical temperature. The entire mean-field theory therefore emerges from basic combinatorial principles and the saddle-point approximation.

The Bethe–Peierls Approximation

The Bragg–Williams approach neglects all correlations between spins. A natural improvement, introduced by Hans Bethe and Rudolf Peierls in 1935–1936 [Bet35, Pei36], is to retain correlations within small clusters while treating different clusters as independent. This *Bethe–Peierls approximation* (or *Bethe cluster method*) forms the basis for later developments such as the Kikuchi and cluster-variational methods.

The system is divided into overlapping clusters consisting of a central spin and its z neighbors. The interactions within each cluster are treated exactly, while the influence of the surrounding lattice is replaced by an effective field $\tilde{h}$. This accounts for short-range correlations neglected in Bragg–Williams theory.

The cluster Hamiltonian is

$$H_{\text{clus}} = -J\sum_{j=1}^{z} \sigma_0\sigma_j - h\sigma_0 - \tilde{h}\sum_{j=1}^{z} \sigma_j, \tag{16.55}$$

where h couples to the central spin σ_0 and $\tilde{h}$ acts on the neighboring spins. The partition function of one cluster is

$$\begin{aligned} Z_{\text{clus}} &= \sum_{\sigma_0=\pm 1}\sum_{\{\sigma_j\}} e^{\beta J\sum_{j=1}^{z}\sigma_0\sigma_j + \beta h\sigma_0 + \beta\tilde{h}\sum_{j=1}^{z}\sigma_j} \\ &= e^{\beta h}\left[2\cosh\big(\beta(J+\tilde{h})\big)\right]^z + e^{-\beta h}\left[2\cosh\big(\beta(J-\tilde{h})\big)\right]^z. \end{aligned} \tag{16.56}$$

The magnetizations of the central and neighboring spins are obtained from

$$\langle\sigma_0\rangle = \frac{1}{\beta}\frac{\partial \ln Z_{\text{clus}}}{\partial h}, \tag{16.57}$$

$$\langle\sigma_j\rangle = \frac{1}{\beta}\frac{\partial \ln Z_{\text{clus}}}{\partial \tilde{h}}. \tag{16.58}$$

For $h = 0$ and translational invariance, we require $\langle\sigma_0\rangle = \langle\sigma_j\rangle$, which yields the self-consistency condition

$$e^{2\beta\tilde{h}} = \frac{\cosh[\beta(J+\tilde{h})]^{z-1}}{\cosh[\beta(J-\tilde{h})]^{z-1}}. \tag{16.59}$$

The critical temperature is determined by the condition for the emergence of a nonzero magnetization:

$$(z-1)\tanh\left(\frac{J}{k_{\mathrm{B}}T_{\mathrm{c}}}\right)=1. \tag{16.60}$$

Unlike the Bragg–Williams result, this formula correctly predicts that a phase transition occurs only for coordination corresponding to $D \geq 2$ (i.e., $z \geq 4$), in agreement with the absence of long-range order in one dimension.

REMARK 16.7 The Bethe–Peierls method thus improves upon the Bragg–Williams approximation by partially restoring correlations. Both can be viewed as particular cases of modern *variational mean-field theories*, where the factorization is imposed at the level of clusters rather than individual spins. In later sections, we will revisit this perspective within the framework of the variational principle and the cluster expansion formalism.

16.6 Variational Principle—Some General Comments

As we have seen, even seemingly simple systems such as the Ising model can be difficult to solve exactly. A powerful approach to treat such interacting problems is to reformulate them as *variational problems*. In essence, this turns a "counting problem" into an *optimization problem*. The central idea is to approximate the free energy of a complicated Hamiltonian by comparing it with a simpler, tractable one whose thermodynamics we can compute exactly.

Historical Note: Origins of the Variational Principle

The variational principle in statistical mechanics has its roots in both thermodynamics and quantum mechanics. In quantum mechanics, it appears as the Rayleigh–Ritz principle for estimating ground-state energies. In statistical mechanics, it was first formulated rigorously by Nikolay Bogoliubov in 1956, building on earlier work by Gibbs, and later generalized by Lars Onsager and others for interacting systems. It provides the theoretical foundation for a wide class of mean-field and cluster approximations used throughout this Volume.

The main idea is to introduce a solvable *trial Hamiltonian* $H_{\text{trial}}(\lambda)$ that depends on a set of parameters λ. This Hamiltonian should approximate the true Hamiltonian H as closely as possible while remaining analytically tractable. The corresponding partition function and free energy are

$$Z_{\text{trial}}(\lambda)=\int e^{-\beta H_{\text{trial}}(\lambda)}\,d\Gamma, \qquad F_{\text{trial}}(\lambda)=-k_{\mathrm{B}}T\ln Z_{\text{trial}}(\lambda), \tag{16.61}$$

where $d\Gamma$ denotes integration over phase space.

We then define the *variational free energy functional*

$$F_{\text{var}}(\lambda)=F_{\text{trial}}(\lambda)+\left\langle H-H_{\text{trial}}(\lambda)\right\rangle_{\text{trial}}, \tag{16.62}$$

where the expectation value is computed with respect to the Gibbs distribution generated by H_{trial}:

$$\left\langle X\right\rangle_{\text{trial}}=\frac{1}{Z_{\text{trial}}(\lambda)}\int X\,e^{-\beta H_{\text{trial}}(\lambda)}\,d\Gamma. \tag{16.63}$$

The Gibbs–Bogoliubov Inequality

The variational method is built upon the *Gibbs–Bogoliubov inequality* (sometimes simply called the Bogoliubov inequality):

$$F \leq F_{\text{var}}(\lambda), \tag{16.64}$$

where F is the true Helmholtz free energy of the system. Thus, F_{var} provides an *upper bound* on the exact free energy. Minimizing F_{var} with respect to the parameters λ yields the best possible approximation within the chosen family of trial Hamiltonians:

$$\frac{\partial F_{\text{var}}}{\partial \lambda} = 0. \tag{16.65}$$

This inequality is a direct consequence of Gibbs's inequality for probability distributions:

$$\sum_i p_i \ln \frac{p_i}{q_i} \geq 0, \tag{16.66}$$

with equality only when $p_i = q_i$. Identifying $p_i \propto e^{-\beta H_i}$ and $q_i \propto e^{-\beta H_{\text{trial},i}}$ immediately yields the Gibbs–Bogoliubov bound.

The variational principle converts the statistical mechanics problem of evaluating a high-dimensional integral (the partition function) into an optimization problem over the parameters λ. If H_{trial} is chosen to capture the dominant interactions of the real system, minimizing F_{var} yields a quantitatively good approximation to F.

For example:

- Choosing H_{trial} as a sum of independent single-spin Hamiltonians gives the *mean-field theory* (e.g., Curie–Weiss or Bragg–Williams).
- Choosing H_{trial} to describe clusters of spins leads to *Bethe–Peierls* or *Kikuchi* cluster approximations.
- Extending H_{trial} to include continuous fields yields the *Landau–Ginzburg* functional, central to modern field-theoretic approaches.

The variational principle balances two contributions to the free energy:

$$F_{\text{var}} = E_{\text{trial}} - TS_{\text{trial}}, \tag{16.67}$$

where the internal energy E_{trial} is computed using the trial Hamiltonian, and S_{trial} is the corresponding entropy. Minimizing F_{var} amounts to finding the most probable macroscopic configuration (or distribution) consistent with thermal equilibrium, given the constraints imposed by H_{trial}.

Mean-field theory and its refinements can thus be understood as specific implementations of this general variational framework, each corresponding to a particular choice of trial Hamiltonian or distribution.

Historical Note: From Bogoliubov to Onsager

The Gibbs–Bogoliubov inequality was first clearly formulated by N. N. Bogoliubov in 1956 [Bog56], though the variational structure it expresses was already implicit in Gibbs's ensemble theory. Onsager later extended similar principles in the treatment of two-dimensional systems and the development of variational methods in thermodynamics. The modern variational mean-field method unifies these ideas into a single framework, forming the backbone of numerous approximate techniques in condensed matter physics and statistical field theory.

Continuous Field Representation of the Ising Model

One might think that models with a discrete state space are peculiar. However, we want to stress that there is an exact mapping between discrete and continuous degrees of freedom with a local potential. To see this, consider the Ising Hamiltonian

$$H[s] = -\frac{1}{2}\sum_{i,j} J_{ij}\,\sigma_i\sigma_j - h\sum_i \sigma_i, \tag{16.68}$$

with $\sigma_i = \pm 1$ and symmetric couplings $J_{ij} = J_{ji}$. The partition function is

$$Z = \sum_{\{\sigma_i\}} \exp\left[\frac{\beta}{2}\sum_{i,j} J_{ij}\sigma_i\sigma_j + \beta h\sum_i \sigma_i\right]. \tag{16.69}$$

Using the Gaussian integral identity,

$$e^{\frac{a^2}{2b}} = \sqrt{\frac{b}{2\pi}}\int_{-\infty}^{\infty} dx\, e^{-\frac{b}{2}x^2+ax}, \tag{16.70}$$

we can decouple the quadratic spin term by introducing a continuous auxiliary field ϕ_i:

$$\exp\left[\frac{\beta}{2}\sum_{i,j} J_{ij}\sigma_i\sigma_j\right] = \mathcal{N}\int \mathcal{D}\phi\, \exp\left[-\frac{1}{2}\sum_{i,j}\phi_i (J^{-1})_{ij}\phi_j + \beta\sum_i \phi_i\sigma_i\right], \tag{16.71}$$

where $\mathcal{N} = [(2\pi)^{N/2}\det J]^{-1/2}$ ensures normalization.

Substituting this identity into Z, we integrate out the spins $\{\sigma_i\}$:

$$\begin{aligned} Z &= \mathcal{N}\int \mathcal{D}\phi\, e^{-\frac{1}{2}\sum_{i,j}\phi_i(J^{-1})_{ij}\phi_j} \prod_i \sum_{\sigma_i=\pm 1} e^{\beta(h+\phi_i)\sigma_i} \\ &= \mathcal{N}\int \mathcal{D}\phi\, \exp\left[-\frac{1}{2}\sum_{i,j}\phi_i(J^{-1})_{ij}\phi_j + \sum_i \ln\bigl(2\cosh[\beta(h+\phi_i)]\bigr)\right]. \end{aligned} \tag{16.72}$$

Equation (16.72) defines a continuous field theory equivalent to the Ising model. The auxiliary field ϕ_i plays the role of a fluctuating local magnetization, and the logarithmic term represents the entropic contribution of the discrete spins.

16.7 The Ising Model on Cayley Trees

Calculating the partition function and expectation values for large interacting systems is, in general, a computationally intractable task. For the Ising model with N spins, the partition function

$$Z = \sum_{\{\sigma_i=\pm 1\}} e^{-\beta H[\{\sigma_i\}]}, \tag{16.73}$$

involves a sum over 2^N configurations. This exponential scaling renders exact enumeration impossible except for very small systems.

However, there exist special geometries for which exact solutions can be obtained. One example is the two-dimensional lattice, which we will discuss later. Another simpler such case is when the interaction network forms a *tree*, i.e., a graph without closed loops. In that case, the absence of cycles allows one to systematically integrate out spins starting from the leaves toward the root of the tree, leading to recursion relations that can be solved exactly. The most famous example is the Ising model on a *Cayley tree* (also known as a Bethe lattice), which exhibits a genuine phase transition despite being locally one-dimensional.

Before treating the Ising case, let us first recall why tree-like structures simplify statistical sums. For continuous Gaussian models, integrals over variables with pairwise interactions of the form $x_i J_{ij} x_j$ can be carried out analytically. Consider the Gaussian partition function

$$Z = \int_{-\infty}^{\infty} \prod_{i=1}^{N} dx_i \, e^{-\frac{1}{2}\sum_{i,j} x_i J_{ij} x_j}, \tag{16.74}$$

where J is a symmetric positive-definite matrix. The result is

$$Z = \frac{(2\pi)^{N/2}}{\sqrt{\det J}}. \tag{16.75}$$

Expectation values can then be obtained by introducing an external field $\vec{h}$:

$$Z(\vec{h}) = \int \prod_i dx_i \, e^{-\frac{1}{2}\sum_{i,j} x_i J_{ij} x_j + \sum_i h_i x_i} = \frac{(2\pi)^{N/2}}{\sqrt{\det J}} e^{\frac{1}{2}\vec{h}^T J^{-1}\vec{h}}, \tag{16.76}$$

so that

$$\langle x_i \rangle = \left. \frac{\partial \ln Z(\vec{h})}{\partial h_i} \right|_{\vec{h}=0} = (J^{-1}\vec{h})_i. \tag{16.77}$$

Thus, once the matrix inverse J^{-1} (or equivalently, the determinant $\det J$) is known, all thermodynamic quantities follow.

In general, computing a determinant for arbitrary J requires $O(N^3)$ operations using LU decomposition—a dramatic improvement over the factorial scaling of direct combinatorial evaluation. Even faster algorithms exist, scaling as $O(N^{2.3})$ [AHU74]. This algebraic simplification illustrates why systems whose interactions form sparse, tree-like networks are much easier to treat analytically: one can effectively eliminate variables recursively without encountering combinatorial explosions due to loops.

The Ising model on a Cayley tree provides an exact, discrete analog of this simplification. In the next subsection, we will use this recursive structure to compute the partition function and magnetization, and show that, unlike the one-dimensional Ising model, the tree geometry indeed supports a finite-temperature phase transition.

This class includes both *Cayley trees* (finite trees with boundary) and *Bethe lattices* (infinite regular trees without boundary). However, the phase transition occurs only on the lattice.

Cayley Trees versus Bethe Lattices

Although the terms "Cayley tree" and "Bethe lattice" are sometimes used interchangeably, they represent conceptually distinct objects:

- A *Cayley tree* is a finite hierarchical structure constructed by starting from a root node and iteratively attaching $q - 1$ new nodes to each non-leaf vertex. It has a well-defined boundary, and the number of boundary sites grows exponentially with the number of layers. As a result, boundary effects dominate the thermodynamics in the finite system.

- A *Bethe lattice*, by contrast, is the *infinite* limit of a Cayley tree in which boundary effects are neglected. It represents a homogeneous infinite graph in which every node has the same coordination number q. In the Bethe lattice, one assumes that bulk properties (e.g., magnetization, correlations) are independent of the choice of root.

In the following, we shall work within the Bethe-lattice picture, which effectively captures the bulk behavior of large trees without explicit boundary corrections. For the Ising model, this provides an analytically solvable case that closely parallels mean-field theory but retains short-range correlations through its local tree structure.

Recursive Solution on the Bethe Lattice

Let us consider the partition function of the Ising model defined on a regular tree with coordination number q:

$$Z = \sum_{\{\sigma\}} e^{\beta J \sum_{\langle i,j \rangle} \sigma_i \sigma_j + \beta h \sum_i \sigma_i}, \tag{16.78}$$

where $\sigma_i = \pm 1$. Defining $K = \beta J$ and $h = \beta \tilde{h}$ for simplicity, the magnetization at a site i is given by

$$m_i = \langle \sigma_i \rangle = \frac{1}{Z} \sum_{\{\sigma\}} \sigma_i e^{K \sum_{\langle i,j \rangle} \sigma_i \sigma_j + h \sum_i \sigma_i}. \tag{16.79}$$

To illustrate the recursive structure, let us recall how the one-dimensional (chain) case can be written as a product of conditional probabilities:

$$P(\sigma_1, \ldots, \sigma_L) = \frac{1}{Z} \prod_{i=1}^{L-1} e^{K\sigma_i \sigma_{i+1} + h\sigma_i} e^{h\sigma_L}. \tag{16.80}$$

From this, one can define the partial partition function (or *message*) of a subtree of size $L-1$ connected to σ_1, denoted $F(\sigma_1)$, such that

$$Z = e^{-h} F(-1) + e^{h} F(1). \tag{16.81}$$

The local magnetization at the root is then

$$m_0 = \frac{e^h F(1) - e^{-h} F(-1)}{e^h F(1) + e^{-h} F(-1)}. \tag{16.82}$$

This reasoning generalizes naturally to trees: each "branch" contributes a partial partition function depending on the spin of the site to which it connects. The key simplification comes from the absence of loops, which allows the partition function to be expressed recursively in terms of such local contributions. In probabilistic terms, this process is equivalent to *marginalization*, where one computes the probability of a spin by summing over all possible configurations of its descendants.

In the next section, we will explicitly derive the recursive relations for the Bethe lattice, showing how they lead to a closed self-consistency equation for the magnetization and how a true phase transition arises in this geometry.

The recursive reasoning used for the one-dimensional Ising model naturally extends to tree-like structures. Because a tree with coordination number q branches into q subtrees at each vertex, one must account for multiple independent subbranches when evaluating marginal probabilities or partial partition functions. This recursive summation process is the foundation of what is known in modern statistical physics and computer science as *belief propagation* or *message passing*.

To make the notation clear, consider a Cayley tree where each node has $q-1$ descendants. We label the spins hierarchically:

- The root spin is denoted τ^0.
- Spins at the first layer are τ_j^1 $(j = 1, \ldots, q)$.
- Spins at the second layer are $\tau_{j_2}^2$, and so on.

A generic spin at level k is then written as $\sigma_{(\tau^0\tau_{j_1}^1\cdots\tau_{j_k}^k)}$. This labeling uniquely identifies each spin and its ancestry within the tree, as illustrated in Fig. 16.8.

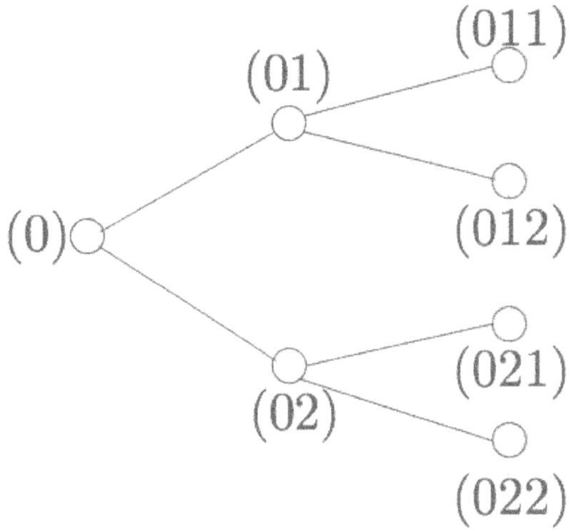

FIGURE 16.8 Tree coordinates for a binary tree. The root spin (0) generates subbranches, each carrying its own subtree. For a binary tree ($q = 3$), only the root has coordination number 2.

Using this notation, the joint probability distribution for all spins can be expressed recursively. At the root node, we write

$$P_{(0)}(\{\sigma\}) = e^{h\sigma_{\tau^0}} \prod_{j_1=1}^{q} e^{K\sigma_{(\tau^0)}\sigma_{(\tau^0\tau_{j_1}^1)}} P_{(0j_1)}\left[(\tau^0) \,|\, (\tau^0\tau_{j_1}^1)\right], \tag{16.83}$$

where $P_{(0j_1)}[(\tau^0) \,|\, (\tau^0\tau_{j_1}^1)]$ represents the contribution from the j_1-th subtree rooted at node $(\tau^0\tau_{j_1}^1)$. Each of these subtrees can in turn be expanded as

$$P_{(0j_1)}\left[(\tau^0) \,|\, (\tau^0\tau_{j_1}^1)\right] = e^{h\sigma_{(\tau^0\tau_{j_1}^1)}} \prod_{j_2=1}^{q} e^{K\sigma_{(0j_1)}\sigma_{(0j_1j_2)}} P_{(0j_1j_2)}\left[(\tau^0\tau_{j_1}^1) \,|\, (\tau^0\tau_{j_1}^1\tau_{j_2}^2)\right]. \tag{16.84}$$

If the tree is finite, the recursion terminates at the leaves:

$$P_{(0j_1\cdots j_k)}\left[(\tau^0\cdots\tau_{j_k}^k) \,|\, (\tau^0\cdots\tau_{j_{k+1}}^k)\right] = e^{h\sigma_{(0j_1\cdots j_k)}}, \tag{16.85}$$

corresponding to a Cayley tree with boundary spins. For an infinite tree (the Bethe lattice), the recursion instead closes through a fixed-point condition, which we will derive shortly.

In a compact notation, one can express the joint probability of the root spin and its branches as

$$P_0(\sigma_0) = e^{h\sigma_0} \prod_{j=1}^{q} P_1(\sigma_0|\sigma^j), \tag{16.86}$$

where σ^j denotes all spins in the j-th subtree attached to σ_0. The partition function can then be written as

$$Z = \sum_{\sigma_0} e^{h\sigma_0} \prod_{j=1}^{q} \sum_{\sigma^j} e^{K\sigma_0\sigma_r^j} P_1(\sigma_0|\sigma^j), \tag{16.87}$$

where σ_r^j is the root spin of the j-th branch. If all branches are statistically equivalent, we define the *partial partition function*

$$g_1(\sigma_0) = \sum_{\sigma^j} e^{K\sigma_0\sigma_r^j} P_1(\sigma_0|\sigma^j), \tag{16.88}$$

so that

$$Z = \sum_{\sigma_0} e^{h\sigma_0} [g_1(\sigma_0)]^q. \tag{16.89}$$

The magnetization at the root is then

$$m_0 = \frac{e^h[g_1(1)]^q - e^{-h}[g_1(-1)]^q}{e^h[g_1(1)]^q + e^{-h}[g_1(-1)]^q}. \tag{16.90}$$

Equation (16.90) closely resembles the solution for the one-dimensional Ising model but includes the q-fold branching structure through $g_1^q(\sigma_0)$. Introducing the ratio

$$x_1^q = \frac{[g_1(-1)]^q}{[g_1(+1)]^q}, \tag{16.91}$$

we can express the magnetization equivalently as

$$m_0 = \frac{e^h - e^{-h}x_1^q}{e^h + e^{-h}x_1^q}. \tag{16.92}$$

The value of x_1 satisfies a recursive fixed-point relation obtained by propagating messages from one generation to the next. This relation ultimately leads to a self-consistent equation for the magnetization and defines the phase diagram of the Ising model on the Bethe lattice.

The recursive formulation presented here, historically known as the Bethe–Peierls method, is mathematically equivalent to the modern *belief-propagation* algorithm used in probabilistic inference and graphical models.

We can now derive a recursion relation for the partial partition functions on successive layers of the Cayley tree. At the second layer, each spin is connected to $q-1$ descendants and to one parent, giving a total coordination number q. The recursive relation for the function $g_1(\sigma_0)$ therefore reads

$$g_1(\sigma_0) = \sum_{\sigma_j=\pm1} e^{K\sigma_0\sigma_j + h\sigma_j} [g_2(\sigma_j)]^{q-1}, \tag{16.93}$$

where $g_2(\sigma_j)$ represents the contribution of the subtree rooted at spin σ_j.

Taking the ratio of Eq. (16.93) for $\sigma_0 = \pm1$, we obtain a recursion relation for the ratio variable $x_n \equiv g_n(-1)/g_n(+1)$:

$$x_1 = \frac{e^{-K+h} + e^{K-h}\,x_2^{q-1}}{e^{K+h} + e^{-K-h}\,x_2^{q-1}}. \tag{16.94}$$

At each subsequent layer, the same argument applies, leading to the general recursion

$$x_n = \frac{e^{-K+h} + e^{K-h}\,x_{n+1}^{q-1}}{e^{K+h} + e^{-K-h}\,x_{n+1}^{q-1}} \equiv f(x_{n+1}). \tag{16.95}$$

By iterating Eq. (16.95), we obtain

$$x_1 = (f \circ f \circ \cdots \circ f)(x_n), \qquad n-1 \text{ times.} \tag{16.96}$$

In the thermodynamic limit (corresponding to an infinite tree or Bethe lattice), the recursion converges to a fixed point x_* satisfying

$$x_* = f(x_*) = \frac{e^{-K+h} + e^{K-h}\, x_*^{q-1}}{e^{K+h} + e^{-K-h}\, x_*^{q-1}}. \tag{16.97}$$

The existence and uniqueness of this limit follow from standard contraction-mapping theorems such as the Browder–Kirk fixed-point theorem.

The quantity x_*, which depends on temperature ($K = \beta J$) and field h, encodes all macroscopic observables of the Bethe lattice. In particular, the magnetization of the central (root) spin is given by

$$m_0 = \frac{e^h - e^{-h}\, x_*^q}{e^h + e^{-h}\, x_*^q}. \tag{16.98}$$

In the thermodynamic limit, this magnetization corresponds to that of the bulk.

The fixed-point Eq. (16.97) can have one or three real solutions depending on the temperature. At high temperature, a single stable solution ($m_0 = 0$) exists, corresponding to the paramagnetic phase. Below a critical coupling K_c, two symmetric nonzero solutions ($m_0 = \pm m$) emerge, marking the onset of spontaneous symmetry breaking, e.g., the emergence of macroscopically observable states in which the underlying $\mathcal{Z}_2$ symmetry is broken. This is what we will discuss at length in the next chapter.

A detailed discussion of the exact solution of the Ising model on the Bethe lattice can be found in Baxter's *Exactly Solved Models in Statistical Mechanics* [Bax82].

Kittel's Molecular Zipper: A Statistical Model of Boiling

The techniques we have developed so far are sufficient to understand how "boiling" looks like in a discrete model such as the Ising model. Charles Kittel's "molecular zipper" model [Kit69] of a DNA strand provides an elegant example of how a simple statistical system can display a sharp transition resembling a boiling point. The model consists of a chain of N links, each of which may be *closed* with energy 0 or *open* with energy ϵ. Crucially, the unzipping can only proceed from one end: the s-th link may open only if all links $1, \ldots, s-1$ are already open. The final link is constrained to remain closed, ensuring the chain remains intact.

When a link opens, the two strands can rotate freely and occupy one of G distinct orientations; this leads to a degeneracy factor of G for every open link. The energy cost and the configurational multiplicity compete thermally, and the balance between these two ingredients drives the behavior of the model.

One can enumerate all configurations by the number of open links s, running from 0 to $N-1$. Each such state has energy $E_s = s\epsilon$ and degeneracy G^s, so the partition function becomes a geometric sum. The result is a compact closed-form expression, from which thermodynamic quantities such as the average number of open links, free energy, and entropy follow directly. These derivations will be treated carefully as a worked exercise in Volume III; here we highlight only the physical conclusions.

The dimensionless combination

$$x \equiv G e^{-\beta\epsilon}$$

naturally appears in the solution. When $x < 1$, the system behaves smoothly: only a fraction of links are open, and thermodynamic quantities vary regularly with temperature. However, when $x \to 1$, the system approaches a singular point. This defines a critical temperature

$$T_H = \frac{\epsilon}{k_B \ln G},$$

where several thermodynamic quantities (including the average number of open links and the entropy) diverge as $T \to T_H^-$.

This behavior is analogous to a *boiling point.* As the system approaches T_H, adding more energy does not significantly increase the temperature. Instead, the extra energy overwhelmingly contributes to opening additional links. This mirrors how, in a boiling liquid, added heat goes into converting liquid to vapor rather than raising the temperature.

Exponential Growth of States and Hagedorn Behavior

The number of states with energy $E = s\epsilon$ is

$$\rho(E) = G^s = \exp\left(\frac{E}{k_B T_H}\right).$$

Thus the density of states grows exponentially with energy. This feature is the hallmark of *Hagedorn behavior*, originally identified in the context of hadronic matter by Rolf Hagedorn [Hag65]. In such systems, T_H acts as a limiting temperature: further heating merely produces new degrees of freedom (in QCD, new hadrons), rather than increasing thermal motion.

The zipper model provides a minimal example of this phenomenon. As $T \to T_H$, the system can absorb large amounts of energy by opening more and more links, producing an explosive growth of accessible configurations. The entropy diverges for the same reason: many more microstates become available at nearly no energetic cost.

Thus Kittel's molecular zipper illustrates a deep concept: when a system's density of states grows exponentially, it exhibits a limiting temperature and phase-transition–like behavior. In the zipper, the phase transition corresponds to a sudden unzipping cascade; in a fluid, it corresponds to boiling; in high-energy physics, it is associated with the Hagedorn transition of hadrons.

The full derivation of these results—partition function, average open links, free energy, entropy, and critical behavior—will be presented as a detailed exercise in Volume 3 of this series.

17

Landau Theory of Phase Transitions

17.1 Introduction and Physical Motivation

In the previous chapters, we discussed mean-field approaches such as the Curie–Weiss and Bragg–Williams theories, which provide a microscopic justification for macroscopic order parameters and spontaneous symmetry breaking. However, these approaches were based on specific microscopic Hamiltonians. Lev Davidovich Landau introduced a far more general and elegant framework: one that does not depend on microscopic details but on symmetry principles and analyticity of the free energy near the transition.

The essential concept is that near a continuous phase transition, a single macroscopic quantity—the *order parameter*—captures the onset of order. For a ferromagnet, this is the magnetization m; for a fluid, the density difference between liquid and vapor; for a superconductor, the complex Cooper-pair field.

Historical Note: Landau's Theory of Phase Transitions

In 1937, Lev D. Landau proposed a unifying phenomenological theory of continuous (second-order) phase transitions [Lan37]. He postulated that the free energy of a system near a critical point can be expressed as an analytic expansion in an *order parameter*, constrained only by the symmetries of the system. This idea represented a paradigm shift:

DOI: 10.1201/9781003603528-17

instead of deriving macroscopic behavior from microscopic interactions, one could classify and predict phase transitions based solely on symmetry and dimensionality. Landau's theory later became the foundation of the modern *Landau–Ginzburg* and *renormalization group* approaches.

17.2 Landau Expansion of the Free Energy

Landau's central postulate is that the free energy can be expressed as an analytic function of the order parameter m:

$$F = F_0 + N\left(hm + \frac{a(T)}{2}m^2 + \frac{b(T)}{4}m^4 + \cdots\right), \tag{17.1}$$

where h represents an external field conjugate to m (e.g., magnetic field), and $a(T)$, $b(T)$ are smooth, temperature-dependent coefficients.

This expansion is valid provided that m is small near T_c and that the free energy is symmetric under the transformation $m \to -m$ when $h = 0$. Only even powers of m are then allowed. Odd terms appear when symmetry is explicitly broken by a field h.

The equilibrium state minimizes the free energy:

$$\frac{\partial F}{\partial m} = 0 \quad \Rightarrow \quad 0 = a(T)m + b(T)m^3 + h. \tag{17.2}$$

Spontaneous Symmetry Breaking

At zero external field ($h = 0$), Eq. (17.2) yields

$$m = 0, \qquad m = \pm\sqrt{\frac{-a(T)}{b(T)}}. \tag{17.3}$$

For physical stability, we require $b(T) > 0$ so that $F(m) \to +\infty$ as $|m| \to \infty$. If $a(T) > 0$, the only minimum is at $m = 0$ (disordered or paramagnetic phase). If $a(T) < 0$, two equivalent minima appear at $\pm m_0$, representing spontaneous symmetry breaking. Assuming that $a(T)$ changes sign linearly near T_c,

$$a(T) = a_0(T - T_c), \tag{17.4}$$

we find that for $T < T_c$,

$$m_0(T) = \pm\sqrt{\frac{a_0(T_c - T)}{b(T)}} \quad \propto \quad (T_c - T)^{1/2}. \tag{17.5}$$

Thus, the critical exponent for the order parameter is $\beta = 1/2$, in agreement with mean-field theory.

Symmetry and Universality

Landau's construction is the first explicit realization of an *effective field theory.* By including all analytic terms consistent with the symmetries of the order parameter, it describes the universal behavior of a wide class of systems—independent of their microscopic details. The specific form of $a(T)$ and $b(T)$ depends on the system, but the scaling

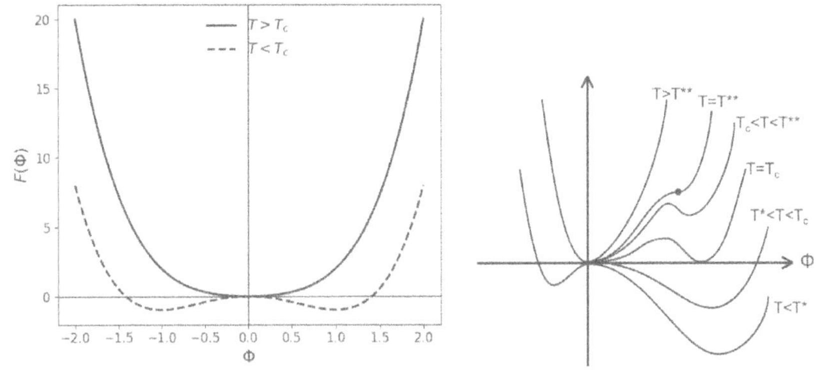

FIGURE 17.1 Landau free energy near a second-order (left, $h = 0$) and a first-order (right, $h \neq 0$) phase transition.

exponents and qualitative phase diagram do not.

17.3 Critical Exponents from Landau Theory

From Eq. (17.1), the equilibrium free energy density at $h = 0$ is

$$f(m) = \frac{a(T)}{2}m^2 + \frac{b(T)}{4}m^4. \tag{17.6}$$

Minimizing $f(m)$ gives

$$f(T) = \begin{cases} 0, & T > T_c, \\ -\dfrac{a(T)^2}{4b(T)}, & T < T_c. \end{cases} \tag{17.7}$$

The specific heat C_V is obtained from $C_V = -T\partial^2 f/\partial T^2$. Using $a(T) \approx a_0(T - T_c)$, we find

$$C_V = \begin{cases} 0, & T > T_c, \\ \dfrac{k_B a_0^2 T_c}{2b(T_c)}, & T < T_c, \end{cases} \tag{17.8}$$

showing a finite discontinuity at T_c, corresponding to $\alpha = 0$.

The inverse susceptibility follows from Eq. (17.2):

$$\chi^{-1} = \left.\frac{\partial h}{\partial m}\right|_{h=0} = a(T) + 3b(T)m^2. \tag{17.9}$$

Evaluating above and below T_c gives

$$\chi^{-1} = \begin{cases} a(T), & T > T_c, \\ -2a(T), & T < T_c, \end{cases} \tag{17.10}$$

so that $\gamma = 1$, and the universal ratio of amplitudes is $|\chi^{-1}(T > T_c)/\chi^{-1}(T < T_c)| = 2$.

Landau theory yields the mean-field critical exponents

$$\alpha = 0, \quad \beta = \frac{1}{2}, \quad \gamma = 1, \quad \delta = 3. \tag{17.11}$$

These satisfy the standard scaling relations (e.g., $\alpha + 2\beta + \gamma = 2$).

Limitations of Landau Theory

While Landau theory captures the qualitative features of continuous phase transitions, it neglects fluctuations of the order parameter. As a result, it fails quantitatively near critical points where correlations diverge. The fluctuation corrections are systematically included in the *Landau–Ginzburg* and *renormalization group* theories discussed below.

17.4 Landau–Ginzburg Theory

Landau's phenomenological approach considers only the spatially averaged order parameter. To describe spatial variations and fluctuations, Lev Ginzburg and Vitaly Landau extended the theory by promoting the order parameter m to a continuous field $\phi(\mathbf{r})$:

$$\mathcal{F}[\phi] = \int d^d r \left[\frac{a(T)}{2}\phi^2 + \frac{b(T)}{4}\phi^4 + \frac{\kappa}{2}|\nabla\phi|^2 - h\,\phi \right]. \tag{17.12}$$

This *Landau–Ginzburg free energy functional* forms the starting point for field-theoretic and renormalization-group analyses of critical phenomena.

The gradient term $(\nabla\phi)^2$ penalizes spatial inhomogeneities and introduces a characteristic correlation length

$$\xi = \sqrt{\frac{\kappa}{a(T)}}, \tag{17.13}$$

which diverges as $T \to T_c$. This divergence underlies universality: on large scales, the microscopic details of the system become irrelevant.

The Landau–Ginzburg functional is a direct continuum generalization of the Curie–Weiss and Bragg–Williams models. It can also be derived systematically from the lattice Ising model using coarse-graining, leading to effective Hamiltonians with kernels analogous to lattice Laplacians or Green functions, as discussed below.

In lattice systems, one often encounters translationally invariant matrices A_{ij} describing nearest-neighbor interactions. For a d-dimensional hypercubic lattice with spacing a, we can represent these matrices as Fourier integrals:

$$A_{km} = \left(\frac{a}{2\pi}\right)^d \int_{-\frac{\pi}{a}}^{\frac{\pi}{a}} d\vec{q}\, A(\vec{q})\, e^{i(\vec{r}_k - \vec{r}_m)\cdot\vec{q}}, \tag{17.14}$$

where $\vec{r}_i$ is the lattice vector of site i, and $A(\vec{q})$ is defined on the first Brillouin zone $B = [-\pi/a, \pi/a]^d$.

For nearest-neighbor couplings,

$$A(\vec{q}) = \alpha + 2\beta \sum_{\nu=1}^{d} \cos(q_\nu a). \tag{17.15}$$

The inverse matrix, or *lattice Green function*, is then

$$(A^{-1})_{km} = \left(\frac{a}{2\pi}\right)^d \int_B \frac{e^{i(\vec{r}_k - \vec{r}_m)\cdot\vec{q}}}{A(\vec{q})}\, d\vec{q}. \tag{17.16}$$

In the long-wavelength (continuum) limit $a \to 0$,

$$A(\vec{q}) \approx \alpha + 2\beta - \beta a^2 \|\vec{q}\|^2 + O(q^4), \tag{17.17}$$

so that $A(\vec{q})^{-1}$ yields the propagator

$$G(\vec{q}) \sim \frac{1}{r + \kappa q^2}, \tag{17.18}$$

the hallmark of the Landau–Ginzburg continuum theory.

REMARK 17.1 The lattice-to-continuum mapping illustrates that the gradient term in Eq. (17.12) arises from expanding the discrete Laplacian at small momenta. Thus, the Landau–Ginzburg model can be viewed as the smooth, long-wavelength limit of microscopic spin models.

17.5 From Lattice to Continuum: The Landau–Ginzburg Functional

Having introduced the phenomenological Landau theory and its expansion in powers of the order parameter, we now wish to connect it explicitly to microscopic lattice models such as the Ising ferromagnet. This derivation reveals how the Landau–Ginzburg free energy arises as the *continuum limit* of a lattice theory, and clarifies the role of the correlation length and of the dimension d.

The goal of this section is to start from a microscopic lattice model and derive a coarse-grained continuum description. In doing so, we will identify the correlation length ξ, study its divergence near criticality, and understand how the large-d limit justifies mean-field theory. This construction also bridges the discrete Ising model with the continuous Landau–Ginzburg functional introduced earlier.

We recall the expression for the inverse matrix A^{-1} representing lattice correlations,

$$(A^{-1})_{km} = \left(\frac{a}{2\pi}\right)^d \int_{-\frac{\pi}{a}}^{\frac{\pi}{a}} d\vec{p}\, \frac{e^{i(\vec{r}_k - \vec{r}_m)\cdot\vec{p}}}{\alpha + 2\beta \sum_{\nu=1}^{d} \cos(p_\nu a)}. \tag{17.19}$$

This quantity defines the *lattice Green function*, whose decay as a function of distance determines the correlation length of the model.

To analyze its asymptotic behavior, we use the identity

$$\frac{1}{x} = \int_0^\infty dq\, e^{-qx}, \tag{17.20}$$

yielding

$$(A^{-1})_{km} = \left(\frac{a}{2\pi}\right)^d \int_{-\frac{\pi}{a}}^{\frac{\pi}{a}} d\vec{p} \int_0^\infty dq\, e^{-q(\alpha + 2\beta \sum_{\nu=1}^{d} \cos p_\nu a)} e^{i(\vec{r}_k - \vec{r}_m)\cdot\vec{p}}. \tag{17.21}$$

For a fixed Euclidean distance between lattice points, we can choose coordinates such that $(\vec{r}_k - \vec{r}_m)_j = \delta_{1j}(k - m)a$. Using the integral representation of the modified Bessel function of the first kind,

$$\frac{1}{2\pi} \int_{-\pi}^{\pi} dp\, e^{r \cos p + inp} = I_n(r), \tag{17.22}$$

we find

$$(A^{-1})_{k-m} = \left(\frac{a}{2\pi}\right)^d \int_0^\infty dq\, e^{-q\alpha} I_0^{d-1}(2\beta q)\, I_{k-m}(2\beta q). \tag{17.23}$$

Large-d Limit

Let us consider the limit $d \to \infty$ with $\beta = 2d\tilde{\beta}$. For large dimension, the modified Bessel functions behave as $I_n(x \ll 1) \simeq x^n/n!$, giving

$$(A^{-1})_{k-m} \approx \left(\frac{\tilde{\beta}}{d}\right)^{k-m}. \tag{17.24}$$

Thus, correlations become short-ranged as d increases, and mean-field theory becomes exact in the limit $d \to \infty$. This is the origin of the statement that mean-field theories correspond to infinite-dimensional or fully connected limits.

For $|k - m| \gg 1$, one can use the following general theorem: if $f(p)$ is a periodic analytic function with its nearest singularity at $p = ip_{\text{sing}}$, then

$$\int_{-\pi}^{\pi} dp\, f(p) e^{ixp} \sim e^{-xp_{\text{sing}}}, \quad (x \gg 1). \tag{17.25}$$

In our case, the singularity satisfies

$$\alpha + 2\beta \cosh(p_{\text{sing}} a) + 2\beta(d-1) = 0, \tag{17.26}$$

from which we identify the inverse correlation length

$$\xi^{-1} = p_{\text{sing}} = \operatorname{arccosh}\left[\frac{\alpha - 2\beta(d-1)}{\beta}\right]. \tag{17.27}$$

As $\alpha - 2\beta(d-1) \to 0$, the correlation length diverges, $\xi \to \infty$, corresponding to criticality. Near this point, correlation functions exhibit power-law decay, and long-range order emerges.

Physical Meaning

The large-d expansion demonstrates that correlations become weaker as the coordination number grows, justifying the success of mean-field and Landau approximations. Conversely, in low dimensions, fluctuations are strong, the correlation length grows rapidly, and mean-field theory fails. The divergence of ξ signals the emergence of critical behavior.

17.6 From the Ising Lattice to a Field Theory

We now proceed to derive the Landau–Ginzburg functional directly from the Ising model. We begin with the generic partition function for an $O(1)$ spin system,

$$Z = \sum_{\{\sigma_i = \pm 1\}} \exp\left[-\beta\left(J\sum_{ij}(D_{ij} - A_{ij})\sigma_i\sigma_j - \sum_i h_i\sigma_i\right)\right], \tag{17.28}$$

where A_{ij} represents the adjacency matrix of the lattice, and D_{ij} is the diagonal matrix of vertex degrees, such that $L_{ij} = D_{ij} - A_{ij}$ is the lattice Laplacian.

To reformulate this as an integral, we replace discrete spins by continuous variables via

$$\sum_{\{\sigma_i = \pm 1\}} \to \lim_{\gamma \to \infty} \int_{-\infty}^{\infty} d\sigma_i\, e^{-\gamma(\sigma_i^2 - 1)^2}, \tag{17.29}$$

which enforces the Ising constraint $\sigma_i^2 = 1$ in the limit $\gamma \to \infty$.

The partition function becomes

$$Z = \lim_{\gamma\to\infty} \int_{-\infty}^{\infty} \prod_i d\sigma_i \, \exp\Big[-\gamma \sum_i (\sigma_i^4 - 2\sigma_i^2 + 1) - \beta\big(J \sum_{ij} L_{ij}\sigma_i\sigma_j - \sum_i h_i\sigma_i\big)\Big]. \quad (17.30)$$

On a lattice of spacing a, we introduce the coarse-grained field $m(\vec{x}_i) \equiv \sigma_i$ and approximate $L_{ij}/a^2 \to -\nabla^2$. The sum over lattice sites becomes an integral, leading to

$$Z \simeq \lim_{\gamma\to\infty} \int_{-\infty}^{\infty} \mathcal{D}m(\vec{x}) \, \exp\Big[- \int d^d x \, \big(\gamma m^4(\vec{x}) - \gamma m^2(\vec{x}) - \beta J a^2 |\nabla m(\vec{x})|^2 + \beta h(\vec{x}) m(\vec{x})\big)\Big]. \quad (17.31)$$

For finite γ, we arrive at the general form

$$Z[h] = \int \mathcal{D}m(\vec{x}) \, \exp\Big\{ - \int d^d x \, \Big[\frac{r}{2} m^2(\vec{x}) + \frac{K}{2} |\nabla m(\vec{x})|^2 + \frac{g}{4!} m^4(\vec{x}) - h(\vec{x}) m(\vec{x}) \Big]\Big\}. \quad (17.32)$$

This is the *Landau–Ginzburg field theory* for an $O(1)$ scalar order parameter $m(\vec{x})$.

Equation (17.32) defines a statistical field theory whose "action" plays the role of the free energy functional. The parameters r, K, and g correspond to the microscopic couplings ($r \sim a(T - T_c)$, $K \sim Ja^2$, $g \sim \gamma$). The field $m(\vec{x})$ represents the coarse-grained magnetization, and the integral $\int \mathcal{D}m(\vec{x})$ is the sum over all possible configurations weighted by their Boltzmann factor.

REMARK 17.2 The path integral formulation in Eq. (17.32) is primarily formal: its rigorous meaning is defined perturbatively, typically as a power series in the coupling g. Nevertheless, it serves as the natural starting point for perturbation theory, diagrammatic expansions, and the renormalization group, which will be developed later in this Volume.

17.7 Fluctuations Around Mean Field and the Continuum Limit

Up to this point, we have derived the Landau–Ginzburg functional from coarse-graining the lattice Hamiltonian directly. There is, however, another elegant route to the same result—one that emphasizes the role of *fluctuations* around the mean-field solution. This approach connects the Ising model to an equivalent *continuous field theory* via a Gaussian (Hubbard–Stratonovich) transformation.

REMARK 17.3 The goal is to rewrite the partition function of the Ising model as an integral over an auxiliary continuous field ϕ, representing local fluctuations of the mean magnetization. This transformation introduces an effective "action" for ϕ, whose saddle point corresponds to the mean-field equation, while fluctuations around it encode corrections beyond mean field.

Starting from the ferromagnetic Ising model ($J > 0$):

$$Z = \sum_{\{\sigma_i=\pm1\}} \exp\Big[\beta\Big(J \sum_{ij} (D_{ij} - A_{ij})\sigma_i\sigma_j + \sum_i h_i\sigma_i \Big)\Big], \quad (17.33)$$

we introduce the positive-definite matrix $\tilde{L} = (\beta L)^{-1} = (\beta(D - A))^{-1}$ and apply the Gaussian identity called Hubbard–Stratonovich transformation:

$$e^{\frac{1}{2}\vec{s}\tilde{L}\vec{s} + \vec{h}\vec{s}} = \sqrt{\det(\tilde{L})} \int \prod_i \frac{d\phi_i}{\sqrt{2\pi}} \, e^{-\frac{1}{2}\vec{\phi}\tilde{L}\vec{\phi} + (\vec{\phi} + \beta\vec{h})\vec{s}}. \quad (17.34)$$

Assuming L is invertible (up to an arbitrary constant diagonal shift), one finds

$$\det\big(\tilde{L}\big) = \beta^{-N/2} |\det J|^{-1/2}. \tag{17.35}$$

Substituting into the partition function and summing over the spins gives

$$Z = 2^N \beta^N \sqrt{\det\big(\tilde{L}\big)} \int \prod_i \frac{d\phi_i}{\sqrt{2\pi}} e^{-\frac{\beta}{2}\vec{\phi}L^{-1}\vec{\phi}} \prod_i \cosh[\beta(\phi_i + h_i)]. \tag{17.36}$$

Equivalently,

$$Z = \beta^N \sqrt{\det\big(\tilde{L}\big)} \int \prod_i \frac{d\phi_i}{\sqrt{2\pi}} e^{-\frac{\beta}{2}\vec{\phi}L^{-1}\vec{\phi} + \sum_i \log[\cosh(\beta\phi_i + \beta h_i)]}. \tag{17.37}$$

The exponent defines an effective energy functional:

$$H[\phi] = \frac{1}{2} \sum_{ij} \phi_i J_{ij} \phi_j - \frac{1}{\beta} \sum_i \log\big[2\cosh\big(\beta(\phi_i + h_i)\big)\big]. \tag{17.38}$$

To interpret ϕ_i, we differentiate $\log Z$ with respect to the external field:

$$m_i = \langle \sigma_i \rangle = \frac{1}{\beta}\frac{\partial \log Z}{\partial h_i} = \sum_j L_{ij}^{-1}(\phi_j - h_j), \tag{17.39}$$

so that

$$h_i + \sum_j L_{ij} m_j = \phi_i. \tag{17.40}$$

Thus ϕ_i represents the *effective local field* acting on site i, averaged over neighboring spins. In matrix form:

$$\vec{\phi} - \vec{h} = L\vec{m}. \tag{17.41}$$

The partition function can now be rewritten in terms of the magnetization field $\vec{m}$:

$$Z = \int \prod_i \frac{dm_i}{\sqrt{2\pi}} \exp\left[-\frac{\beta}{2}\vec{m} \cdot Ł \cdot \vec{m} + \sum_i \log(\cosh)\big(\beta[h_i + \sum_j J_{ij} m_j]\big)\right]. \tag{17.42}$$

Expanding $\log\cosh x$ for small β (high-temperature expansion),

$$\log\cosh[\beta(h_i + \sum_j L_{ij} m_j)] \approx \frac{\beta^2}{2}(h_i + \sum_j L_{ij} m_j)^2 - \frac{\beta^4}{12}(h_i + \sum_j L_{ij} m_j)^4 + \cdots, \tag{17.43}$$

leads to an effective quartic theory.

If A is a lattice matrix, any function of it, such as $J = a_1 I + a_2 A$, defines another translationally invariant lattice operator. In the continuum limit, we replace lattice sums by integrals and discrete fields by continuous ones:

$$\sum_i \phi_i \longrightarrow \frac{1}{\sqrt{a^d}} \int d^d x\, \phi(\vec{x}), \qquad \phi_i \longrightarrow \sqrt{a^d}\, \phi(\vec{x}_i). \tag{17.44}$$

For small wavevectors, the lattice Laplacian satisfies

$$L(\vec{q}) \approx a_0 + b_0 \|\vec{q}\|^2, \tag{17.45}$$

which implies in real space

$$L \rightarrow a_0 + b_0 \nabla^2. \tag{17.46}$$

Collecting all contributions from the expansion, the partition function takes the schematic form

$$Z = \int \mathcal{D}m(\vec{x})\, e^{-\int d^dx\, \left[K_0 m^2 + K_1(\nabla m)^2 + K_2 m^4 + K_3 m^4(\nabla m)^2 + K_4 m^6 + \cdots\right]}, \tag{17.47}$$

where the coefficients K_i are smooth functions of temperature, encoding microscopic parameters.

The corresponding effective Hamiltonian reads

$$-\beta H = \int d^dx \left[K_0 m^2 + K_1(\nabla m)^2 + K_2 m^4 + K_3 m^4(\nabla m)^2 + K_4 m^6 + \cdots\right]. \tag{17.48}$$

For convenience, one defines $K_0 = t/2$, $K_1 = 1$, and $K_2 = u/4$, leaving higher-order coefficients implicit.

REMARK 17.4 Near the critical temperature T_c, only the lowest-order, symmetry-allowed terms are relevant. Equation (17.47) therefore reduces to the standard Landau–Ginzburg Hamiltonian,

$$H[m] = \int d^dx \left[\frac{t}{2}m^2 + \frac{1}{2}(\nabla m)^2 + \frac{u}{4}m^4 - hm\right], \tag{17.49}$$

which governs the universal critical behavior of the Ising universality class.

Symmetries and Physical Meaning

The Landau–Ginzburg theory is a *phenomenological field theory* constrained by symmetry and locality:

- The field $m(\vec{x})$ represents the coarse-grained local order parameter. For Ising-like systems, it is a scalar field invariant under the $\mathbb{Z}_2$ symmetry $m(\vec{x}) \to -m(\vec{x})$.
- The energy functional must be *local*, involving only $m(\vec{x})$ and its spatial derivatives.
- Rotational and translational invariance in space require invariance under

$$\vec{x} \to \vec{x} + \vec{x}_0, \qquad \vec{x} \to R\vec{x}, \tag{17.50}$$

where R is any orthogonal transformation.

Finally, the mean-field approximation corresponds to minimizing the functional under the assumption of a uniform order parameter:

$$Z \approx e^{-\beta\mathcal{H}[\bar{m}^*]}, \qquad \bar{m}^* = \arg\min_m \int d^dx\, [t\, m^2 + (\nabla m)^2 + u\, m^4 + \cdots]. \tag{17.51}$$

Setting $m(\vec{x}) = \bar{m}$ and $\nabla\bar{m} = 0$ recovers the Landau mean-field equations of state discussed earlier.

REMARK 17.5 The Hubbard–Stratonovich approach provides a systematic path from the discrete Ising model to the continuous Landau–Ginzburg field theory. The resulting functional encodes both the mean-field behavior (through its stationary points) and the structure of fluctuations (through the field integral). In the next sections, we will explore how to analyze these fluctuations and determine the limits of validity of mean-field theory through the *Ginzburg criterion.*

17.8 Mean-Field Solutions and Spatial Fluctuations

In the Landau–Ginzburg theory, the mean-field approximation corresponds to neglecting spatial variations of the order parameter. This assumption reproduces the results of the Curie–Weiss and Bragg–Williams theories discussed earlier, but now derived directly from the continuum free-energy functional.

Starting from the Landau–Ginzburg Hamiltonian

$$-\beta H = \int d^d x \left(t\, m^2(\vec{x}) + u\, m^4(\vec{x}) + c\,|\nabla m(\vec{x})|^2 + \cdots \right), \tag{17.52}$$

we neglect spatial gradients by setting $\nabla^2 m(\vec{x}) = 0$, corresponding to a spatially uniform magnetization. The mean-field equation of state then reads

$$2t\, m + 4u\, m^3 = 0, \tag{17.53}$$

or equivalently,

$$m\,[\,t + 2um^2\,] = 0. \tag{17.54}$$

The solutions are

$$m = \begin{cases} 0, & t > 0, \\ \pm\sqrt{\dfrac{-t}{2u}}, & t < 0, \end{cases} \tag{17.55}$$

which are precisely the same as those obtained in the Curie–Weiss model once we identify

$$t = \frac{a(T)}{2}, \qquad u = \frac{b(T)}{4} > 0. \tag{17.56}$$

Since $t \propto (T - T_c)$, we recover the mean-field scaling of the magnetization:

$$m \propto (T_c - T)^{1/2}. \tag{17.57}$$

REMARK 17.6 In the Curie–Weiss and Bragg–Williams models, the free energy was expressed as

$$f(m) = \tfrac{a(T)}{2} m^2 + \tfrac{b(T)}{4} m^4 - hm. \tag{17.58}$$

Minimizing this free energy yields the same equation of state as Eq. (17.55). Thus, the Landau–Ginzburg mean-field approximation generalizes these lattice mean-field theories to a continuum field description, in which $m(\vec{x})$ represents the coarse-grained local magnetization.

Domain Walls and Solitons

When spatial variations are included, the equilibrium equation becomes

$$c\,\nabla^2 m(\vec{x}) = 2t\, m(\vec{x}) + 4u\, m^3(\vec{x}), \tag{17.59}$$

where $c \equiv K_1 > 0$ is the stiffness (gradient) coefficient. In one spatial dimension this reduces to

$$c\, m''(x) = 2t\, m(x) + 4u\, m^3(x). \tag{17.60}$$

This is a nonlinear differential equation whose constant solutions reproduce the mean-field minima in Eq. (17.55). However, it also admits spatially varying solutions corresponding to *domain walls* (or *kinks*), where the order parameter smoothly interpolates between the two minima $m = \pm m_0$.

A standard solitonic solution is

$$m(x) = m_0 \tanh\left(\frac{x}{\xi}\right), \tag{17.61}$$

where

$$m_0 = \sqrt{\frac{-t}{2u}}, \qquad \xi = \sqrt{\frac{c}{-t}}, \tag{17.62}$$

and ξ is the *domain-wall width*, equal to the correlation length near the critical point.

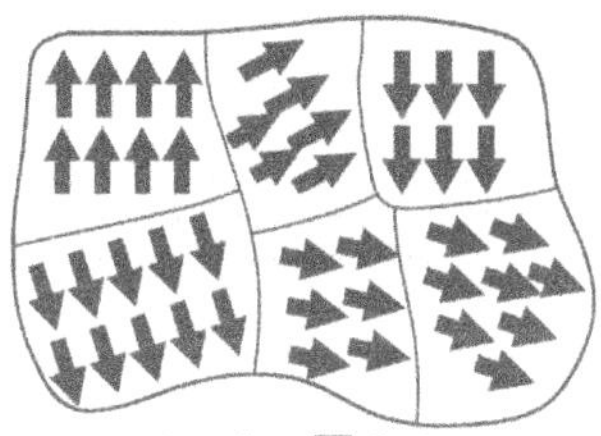

FIGURE 17.2 The field ϕ (or m) represents an effective local magnetization, varying smoothly across a domain wall between $m = \pm m_0$. (Source: Wikipedia)

The domain-wall (or kink) solution Eq. (17.61) represents the transition region between two locally ordered domains of opposite magnetization. The wall has a finite energy per unit area proportional to $\sqrt{c|t|^3/u}$ and a width ξ that diverges as $t \to 0^-$. Such solutions illustrate how the Landau–Ginzburg model naturally captures the existence of interfaces and topological solitons, including domain walls, vortices, and skyrmions in appropriate symmetry classes [GL50, Abr57, Sky62, BP75].

In higher dimensions, the same equation admits richer solitonic configurations, such as domain walls, vortices, and skyrmions, depending on the symmetry of the order parameter.

Dimensional Analysis of the Field Theory

Before proceeding to fluctuation corrections, it is useful to examine the dimensional structure of the Landau–Ginzburg Hamiltonian. The action

$$-\beta H = \int d^d x \left(K_0 m^2 + K_1(\nabla m)^2 + K_2 m^4 + \cdots\right) \tag{17.63}$$

is dimensionless, and the measure $d^d x$ carries the dimension $[L]^d$. Each term in the integrand must therefore have dimension $[L]^{-d}$.

We recall that $[\nabla] = [L]^{-1}$. From the gradient term we obtain

$$[K_1(\nabla m)^2] = [L]^{-d} \quad \Rightarrow \quad [m] = [L]^{1-\frac{d}{2}}. \tag{17.64}$$

The other parameters then have dimensions:

$$[K_0 m^2] = [L]^{-d} \quad \Rightarrow \quad [K_0] = [L]^{-2}, \tag{17.65}$$

$$[K_2 m^4] = [L]^{-d} \quad \Rightarrow \quad [K_2] = [L]^{d-4}. \tag{17.66}$$

Adopting the field-theoretic convention $[L] = -1$ (i.e., expressing everything in momentum dimensions), we find

$$
\begin{aligned}
[m] &= \frac{d}{2} - 1, \\
[K_0] &= 2, \\
[K_2] &= 4 - d.
\end{aligned}
\tag{17.67}
$$

Upper Critical Dimension

The quartic coupling K_2 becomes dimensionless at $d_c = 4$. This dimension is known as the *upper critical dimension.* For $d > 4$, the coupling K_2 is irrelevant in the renormalization-group sense, and the system's behavior is correctly described by mean-field theory. For $d < 4$, fluctuations become increasingly important and modify the critical exponents from their mean-field values.

REMARK 17.7 This analysis shows that the Curie–Weiss and Bragg–Williams mean-field theories can be viewed as the $d \geq 4$ limit of the Landau–Ginzburg field theory, where spatial fluctuations are negligible. In lower dimensions, as we shall see in the next sections, fluctuations dominate and require a renormalized treatment.

17.9 Gaussian Theory: From Real to Momentum Space

The Gaussian (or *free*) theory corresponds to the Landau–Ginzburg functional truncated at quadratic order in the field. It represents the simplest approximation capturing the essential features of correlations and fluctuations near the critical point.

Real-Space Formulation

We start from the Gaussian partition function in real space:

$$
Z = \int \mathcal{D}m(\vec{x}) \, \exp\left[- \int d^d x \, \left(K_0 \, m^2(\vec{x}) + K_1 \, |\nabla m(\vec{x})|^2 - h(\vec{x}) \, m(\vec{x}) \right) \right]. \tag{17.68}
$$

Here $m(\vec{x})$ is the coarse-grained order-parameter field, $h(\vec{x})$ is the external field, and the constants K_0 and K_1 control the "mass" and stiffness of the field, respectively.

In operator form, this can be written as

$$
Z = \int \mathcal{D}m(\vec{x}) \, \exp\left[- \int d^d x \, d^d y \, \left(m(\vec{x}) \, \mathcal{K}(\vec{x}, \vec{y}) \, m(\vec{y}) - h(\vec{x}) \, m(\vec{x}) \right) \right], \tag{17.69}
$$

where

$$
\mathcal{K}(\vec{x}, \vec{y}) = \left[K_0 - K_1 \nabla^2 \right] \delta(\vec{x} - \vec{y}). \tag{17.70}
$$

Correlation functions are obtained as functional derivatives of $\log Z[h]$:

$$
\langle m(\vec{x}_1) \cdots m(\vec{x}_n) \rangle_c = \frac{\delta^n \log Z[h]}{\delta h(\vec{x}_1) \cdots \delta h(\vec{x}_n)}. \tag{17.71}
$$

The two-point function

$$
G(\vec{x}_1, \vec{x}_2) = \langle m(\vec{x}_1) m(\vec{x}_2) \rangle \tag{17.72}
$$

is the central quantity of interest, as it describes correlations of the order parameter and can be probed directly in scattering experiments.

Fourier Representation

To evaluate $G(\vec{x}_1, \vec{x}_2)$, we diagonalize the quadratic Hamiltonian by Fourier transforming the field:

$$\tilde{m}(\vec{q}) = \int \frac{d^d x}{\sqrt{(2\pi)^d}}\, e^{i\vec{q}\cdot\vec{x}}\, m(\vec{x}), \tag{17.73}$$

$$m(\vec{x}) = \int \frac{d^d q}{\sqrt{(2\pi)^d}}\, e^{-i\vec{q}\cdot\vec{x}}\, \tilde{m}(\vec{q}). \tag{17.74}$$

The Parseval identity gives

$$\int d^d x\, |m(\vec{x})|^2 = \int d^d q\, |\tilde{m}(\vec{q})|^2. \tag{17.75}$$

Using $\nabla \to i\vec{q}$ in Fourier space and integrating by parts, Eq. (17.68) becomes

$$Z = \int \mathcal{D}\tilde{m}(\vec{q})\, \exp\left[-\int d^d q \left((K_0 + K_1 q^2)\, |\tilde{m}(\vec{q})|^2 - \tilde{h}^*(\vec{q})\, \tilde{m}(\vec{q})\right)\right]. \tag{17.76}$$

We see that each Fourier mode $\tilde{m}(\vec{q})$ is independent and Gaussian distributed, with variance proportional to $(K_0 + K_1 q^2)^{-1}$. Setting $K_1 = 1$ and $K_0 = t$ for simplicity, we identify the *Gaussian propagator* or *Green function* in momentum space:

$$G(\vec{q}) = \frac{1}{q^2 + t}. \tag{17.77}$$

REMARK 17.8 The parameter t acts as an effective "mass term" proportional to $T - T_c$, and controls the correlation length $\xi \sim t^{-1/2}$. At criticality ($t = 0$), the correlation length diverges and the system becomes scale invariant, leading to power-law correlations $G(r) \sim r^{-(d-2)}$.

Two-Point Function

In this momentum-space representation, correlations between modes are diagonal:

$$\langle \tilde{m}(\vec{q})\tilde{m}(\vec{q}\,')\rangle_{h=0} = (2\pi)^d\, \delta^d(\vec{q} + \vec{q}\,')\, G(\vec{q}), \tag{17.78}$$

where $G(\vec{q})$ is given by Eq. (17.77). Transforming back to real space yields the Ornstein–Zernike correlation function:

$$G(r) = \int \frac{d^d q}{(2\pi)^d}\, \frac{e^{i\vec{q}\cdot\vec{r}}}{q^2 + \xi^{-2}} \sim \frac{e^{-r/\xi}}{r^{(d-2)/2}}, \tag{17.79}$$

which decays exponentially at large r with correlation length ξ.

REMARK 17.9 The Gaussian theory thus provides a consistent microscopic-to-macroscopic link:

$$\text{field correlations} \longleftrightarrow \text{scattering intensity } S(q), \tag{17.80}$$

and quantitatively predicts the critical broadening of diffraction peaks observed experimentally near second-order phase transitions.

Experimental Connection: Scattering and Correlations

The Fourier transform of the two-point correlation function,

$$S(\vec{q}) = \int d^d r \, e^{-i\vec{q}\cdot\vec{r}} \, G(\vec{r}) = \langle |\tilde{m}(\vec{q})|^2 \rangle, \tag{17.81}$$

is known as the *static structure factor*. It quantifies the intensity of fluctuations in the order parameter at wavevector $\vec{q}$, and it is the central quantity measured in scattering experiments.

Experimental Setup

In small-angle *neutron* or *X-ray scattering* experiments, a monochromatic beam of neutrons or photons with incident wavevector $\vec{k}_{\text{in}}$ and energy E_{in} strikes the sample. After interaction with the sample, particles are scattered with outgoing wavevector $\vec{k}_{\text{out}}$ and energy E_{out}. The momentum transfer is defined as

$$\vec{q} = \vec{k}_{\text{in}} - \vec{k}_{\text{out}}, \tag{17.82}$$

and the differential scattering cross-section is proportional to the structure factor:

$$\frac{d\sigma}{d\Omega} \propto S(\vec{q}). \tag{17.83}$$

For neutrons, the scattering amplitude is sensitive to the nuclear potential or to the magnetic moment of atoms (depending on polarization), while for X-rays it arises from the electron density.

Relation to Correlations

The structure factor provides a direct probe of spatial correlations in the system. If $G(\vec{r}) = \langle m(\vec{x}) m(\vec{x} + \vec{r}) \rangle$ denotes the pair correlation of the local order parameter, then

$$S(\vec{q}) = \int d^d r \, e^{-i\vec{q}\cdot\vec{r}} \, G(\vec{r}) \tag{17.84}$$

is its Fourier transform. This implies that large-scale correlations (large ξ) produce a narrow peak in $S(\vec{q})$ near $\vec{q} = 0$, while short-range correlations yield a broad, featureless response.

Ornstein–Zernike Form

Within the Gaussian (free-field) theory, the two-point correlation function is

$$G(\vec{r}) = \int \frac{d^d q}{(2\pi)^d} \frac{e^{i\vec{q}\cdot\vec{r}}}{q^2 + \xi^{-2}} \sim \frac{e^{-r/\xi}}{r^{(d-2)/2}}, \tag{17.85}$$

whose Fourier transform gives

$$S(\vec{q}) \propto \frac{1}{q^2 + \xi^{-2}}. \tag{17.86}$$

This is the celebrated *Ornstein–Zernike form* [OZ14], which successfully describes the shape of the scattering peak observed near continuous phase transitions.

Experimental Signatures

As the temperature approaches the critical point:

- The correlation length ξ diverges as $\xi \sim |T - T_c|^{-\nu}$, where $\nu \approx 0.63$ for the three-dimensional Ising universality class.
- The peak in $S(\vec{q})$ at $q = 0$ sharpens and increases in height as ξ^2, while its width decreases as ξ^{-1}.
- Neutron and X-ray experiments directly observe this phenomenon as *critical opalescence*—the enhanced scattering of light or neutrons due to large-scale density or magnetization fluctuations.

Quantitatively, the experimental scattering intensity is written as

$$I(\vec{q}, T) \propto \frac{A(T)}{q^2 + \xi(T)^{-2}}, \tag{17.87}$$

where $A(T)$ is a temperature-dependent amplitude. Measurements of $I(q, T)$ near T_c provide precise determinations of the critical exponents ν (from ξ) and γ (from $A(T) \propto \chi \sim |T - T_c|^{-\gamma}$).

Examples

- In magnetic systems (e.g., iron or nickel), *polarized neutron scattering* probes spin–spin correlations and confirms that $S(\vec{q})$ follows the Ornstein–Zernike form near the Curie temperature [Lov84].
- In binary fluid mixtures and liquid–gas systems, *X-ray and light scattering* experiments display identical critical broadening, verifying universality [CL95, Sta71].
- The phenomenon of *critical opalescence* in fluids near the liquid–gas critical point is the optical manifestation of the divergence of $S(\vec{q} \to 0)$ [AM76].

Further Reading

Detailed discussions of scattering theory and its connection to critical phenomena can be found in several classic sources. A comprehensive treatment of the underlying principles and experimental methods is presented by Lovesey in *Theory of Neutron Scattering from Condensed Matter* [Lov84], which remains a standard reference for the analysis of both elastic and inelastic neutron scattering in condensed matter systems. A clear and pedagogical exposition of the relation between the structure factor, correlation functions, and critical scattering is given by Chaikin and Lubensky in *Principles of Condensed Matter Physics* [CL95], particularly in Chapter 4, where the Ornstein–Zernike form of the structure factor is derived from the Gaussian field theory. The broader theoretical framework of critical phenomena, scaling, and universality is developed in Stanley's classic monograph *Introduction to Phase Transitions and Critical Phenomena* [Sta71], which situates scattering experiments within the modern renormalization-group perspective. For an introduction to diffraction and the definition of the structure factor from a solid-state physics viewpoint, see Ashcroft and Mermin's *Solid State Physics* [AM76], Section 2.4. Finally, the theoretical origin of the Ornstein–Zernike form itself traces back to the pioneering paper by Ornstein and Zernike (1914) [OZ14], who first related density fluctuations near the critical point to the phenomenon of critical opalescence observed in fluids.

18

Landau–Ginzburg and Fluctuations

18.1 From Mean Field to Fluctuations: The Landau–Ginzburg Picture

In the previous chapter, we developed the Landau theory of phase transitions, emphasizing how macroscopic order parameters and symmetry considerations determine the general form of the free energy near criticality. There, spatial variations of the order parameter were neglected, corresponding to the *mean-field approximation*. However, as we have seen, close to the critical temperature T_c the correlation length ξ diverges, and fluctuations occur over arbitrarily large scales. In this regime, the assumption of spatial uniformity breaks down, and a local field description becomes necessary.

The *Landau–Ginzburg theory* extends the mean-field framework by promoting the order parameter to a spatially varying field $m(\vec{x})$ and introducing gradient terms that account for the cost of inhomogeneities. In this formulation, both local values and spatial correlations of the field become dynamical variables, allowing one to describe not only uniform phases but also domains, interfaces, and critical fluctuations.

To analyze these fluctuations systematically, it is convenient to express the theory in momentum space. Each Fourier mode $\tilde{m}(\vec{q})$ represents a collective excitation of wavelength $\lambda = 2\pi/|\vec{q}|$, and their statistical weights determine the system's response and correlation functions. This approach naturally introduces a *momentum cutoff* $\Lambda \sim 2\pi/a$ set by the microscopic lattice spacing a, beyond which the continuum description ceases to be valid.

DOI: 10.1201/9781003603528-18

From Coarse-Graining to Field Theory

The Landau–Ginzburg functional can be interpreted as the continuum limit of Kadanoff's block-spin coarse-graining procedure: the local magnetization field $m(\vec{x})$ plays the role of the "block variable" averaged over a region of size comparable to the correlation length. Fluctuations of this field encode the physics that survives after microscopic degrees of freedom have been integrated out, bridging the lattice description and the renormalization-group viewpoint developed later.

In the following sections, we reformulate the Landau–Ginzburg theory in Fourier space and study how its Gaussian approximation captures the essential features of correlations and scattering phenomena near criticality.

18.2 Some Comments on Fourier Transforms

In the previous chapter, we replaced the discrete lattice theory by a continuum approximation. This simplification should, however, be treated with care. In reality, the lattice spacing a sets a natural upper limit to the wavenumbers that can occur in the system, since wavelengths shorter than a have no physical meaning. Hence, the continuum Fourier transform should properly be replaced by a Fourier *series* defined on a finite lattice.

For a cubic system of volume $V = L^D$ with periodic boundary conditions, we have

$$m(\vec{x}) = \frac{1}{V} \sum_{\vec{q}_n} e^{i\vec{q}_n \cdot \vec{x}}\, \tilde{m}(\vec{q}_n), \qquad \vec{q}_n = \frac{2\pi}{L}\,\vec{n}, \tag{18.1}$$

where $\vec{n}$ is a vector of integers. In the thermodynamic limit, the discrete sum can be approximated by a continuous integral with a momentum cutoff $\Lambda = 2\pi/a$:

$$m(\vec{x}) \approx \int^{\Lambda} \frac{d^D q}{(2\pi)^D}\, e^{i\vec{q}\cdot\vec{x}}\, \tilde{m}(\vec{q}). \tag{18.2}$$

The cutoff Λ serves as a reminder that the continuum theory is valid only for momenta $|\vec{q}| < \Lambda$, corresponding to length scales much larger than the lattice spacing.

Measurable Quantities

Scattering experiments probe spatial correlations of the order parameter. The probability for scattering at momentum transfer $\vec{q}$ is proportional to the Fourier transform of the correlation function,

$$G(\vec{q}) = \langle |\tilde{m}(\vec{q})|^2 \rangle \propto \frac{1}{q^2 + \xi^{-2}}. \tag{18.3}$$

The real-space correlation function is the inverse transform:

$$G(\vec{r}) = \langle m(\vec{x}) m(\vec{x} + \vec{r}) \rangle.$$

At the critical point ($\xi \to \infty$), the correlations become scale invariant, and $G(\vec{r})$ assumes a power-law form,

$$G(\vec{r}) \propto \frac{1}{r^{D-2-\eta}}, \tag{18.4}$$

where η is the *anomalous dimension* characterizing critical fluctuations. Under a scale transformation $\vec{r} \to \lambda\vec{r}$, the correlation function obeys

$$G(\lambda r) = \lambda^s\, G(r), \tag{18.5}$$

for some scaling exponent s. At the critical point, this fractal-like behavior manifests in simulations and experiments as shown in Fig. 18.1.

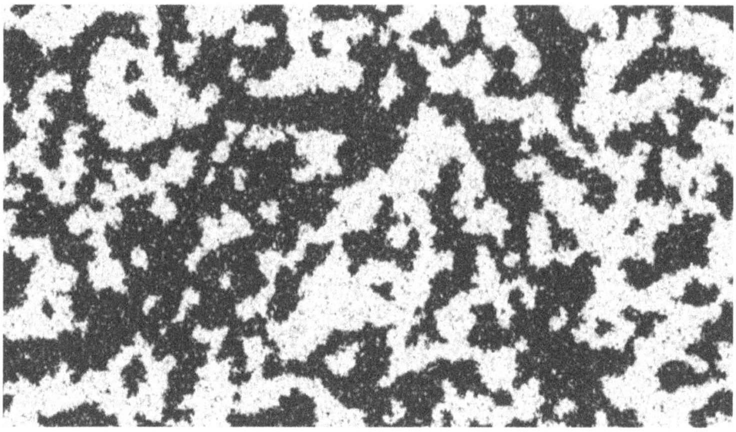

FIGURE 18.1 Typical spin configuration of the two-dimensional Ising model at the critical temperature. Black and white regions represent domains of opposite magnetization, exhibiting scale-invariant (fractal) structure. Obtained from numerical simulations.

18.3 Gaussian Theory

For noninteracting or weakly coupled modes, the Landau–Ginzburg partition function reduces to a Gaussian integral. In Fourier representation,

$$Z = \int \mathcal{D}\tilde{m}(\vec{q}) \, \exp\left[- \int d^D q \, \left((K_0 + K_1 q^2) \, |\tilde{m}(\vec{q})|^2 - \tilde{h}^*(\vec{q}) \, \tilde{m}(\vec{q}) \right) \right], \tag{18.6}$$

where $K_0 = t$ and K_1 have been chosen such that the modes are diagonal and decoupled. The two-point correlation function in momentum space is then

$$\langle \tilde{m}(\vec{q}) \tilde{m}(\vec{q}\,') \rangle_{h=0} = (2\pi)^D \, \delta^D(\vec{q} + \vec{q}\,') \, G(\vec{q}), \qquad G(\vec{q}) = \frac{1}{q^2 + t}. \tag{18.7}$$

This function $G(\vec{q})$ is the *Green function* of the Gaussian theory. Each mode contributes independently, and the variance of $\tilde{m}(\vec{q})$ satisfies the equipartition relation $\langle |\tilde{m}(\vec{q})|^2 \rangle \propto k_B T$.

To obtain the real-space correlation function $G(\vec{x})$, we compute the Fourier transform of $G(\vec{q})$:

$$G(\vec{x}) = \int \frac{d^D q}{(2\pi)^D} \, \frac{e^{i\vec{q}\cdot\vec{x}}}{q^2 + t} = \int \frac{d^D q}{(2\pi)^D} \int_0^\infty ds \, e^{-s(q^2+t)} \, e^{i\vec{q}\cdot\vec{x}}. \tag{18.8}$$

Shifting the integration variable $\vec{q} \to \vec{q} + i\vec{x}/(2s)$ and performing the Gaussian integral gives

$$G(\vec{x}) = \frac{1}{(4\pi)^{D/2}} \int_0^\infty ds \, s^{-D/2} \, e^{-st - \frac{x^2}{4s}}. \tag{18.9}$$

Using the integral representation of the modified Bessel function I_ν,

$$\int_0^\infty dx \, x^{\nu-1} e^{-\gamma x - \beta/x} = 2 \left(\frac{\beta}{\gamma} \right)^{\nu/2} K_\nu(2\sqrt{\beta\gamma}),$$

we obtain

$$G(\vec{x}) = \frac{1}{(2\pi)^{D/2}} \left(\frac{\sqrt{t}}{r}\right)^{\frac{D}{2}-1} K_{\frac{D}{2}-1}(\sqrt{t}\, r), \tag{18.10}$$

where $r = |\vec{x}|$ and K_ν is the modified Bessel function of the second kind.

For large r, the Bessel function has the asymptotic form

$$K_\nu(z) \simeq \sqrt{\frac{\pi}{2z}}\, e^{-z}, \qquad (z \gg 1),$$

so that

$$G(\vec{x}) \sim \frac{e^{-r/\xi}}{r^{(D-1)/2}}, \qquad \xi = \frac{1}{\sqrt{t}}. \tag{18.11}$$

The quantity ξ is the *correlation length*, which diverges near the critical temperature as

$$\xi \propto |T - T_c|^{-\nu}, \qquad \text{with} \quad \nu_{\text{MF}} = \frac{1}{2}.$$

This exponential decay of correlations away from the critical point is a defining feature of disordered phases.

At criticality ($t = 0$), $\xi \to \infty$, and Eq. (18.10) reduce to the power-law form

$$G(\vec{x}) = \int \frac{d^D q}{(2\pi)^D} \frac{e^{i\vec{q}\cdot\vec{x}}}{q^2} \sim \frac{1}{r^{D-2}}, \tag{18.12}$$

which reflects scale invariance and the absence of a characteristic length scale at the phase transition.

REMARK 18.1 The Gaussian (free-field) theory yields a universal description of correlations in the disordered phase:

$$G(\vec{r}) \sim \frac{e^{-r/\xi}}{r^{(D-1)/2}}, \qquad G(\vec{q}) = \frac{1}{q^2 + \xi^{-2}}.$$

The exponential decay defines the correlation length ξ, which diverges at T_c. At the critical point, $G(\vec{r})$ follows a power law $r^{-(D-2)}$, corresponding to the *Ornstein–Zernike form* observed in scattering experiments. This result forms the foundation for the analysis of fluctuations and critical scattering discussed in later sections.

Thus, the overall behavior of the correlation function is summarized as:

$$G(x) \approx \begin{cases} \dfrac{e^{-x/\xi}}{x^{(D-1)/2}}, & x > \xi, \\ \dfrac{1}{x^{D-2}}, & x < \xi, \end{cases} \tag{18.13}$$

showing that the correlation length ξ diverges at the critical point.

The phenomenon of the divergence of the correlation length is also referred to as the emergence of *long-range order*. Using correlations as the key descriptor of second-order phase transitions was first proposed by Leonard Ornstein and Frits Zernike [OZ14].

We now introduce two new critical exponents by assuming that the correlation function behaves as

$$G(|\vec{x}|) \approx \frac{e^{-|\vec{x}|/\xi}}{|\vec{x}|^{D-2+\eta}}, \tag{18.14}$$

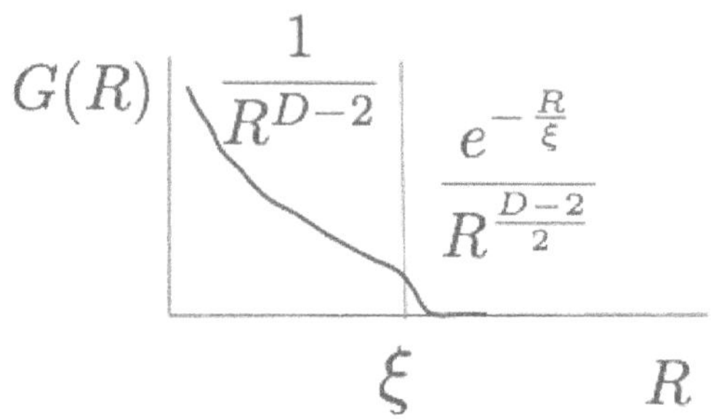

FIGURE 18.2 Qualitative behavior of the correlation function above and below the correlation length ξ.

with the correlation length diverging near the critical point as

$$\xi \approx t^{-\nu}. \tag{18.15}$$

We have thus introduced the exponents η and ν, which complement the previously defined exponents α, β, and γ. These critical exponents are not independent and satisfy several scaling relations, which we shall discuss later. For the mean-field case, $\nu = \frac{1}{2}$ and $\eta = 0$.

FIGURE 18.3 Critical opalescence near the liquid–gas critical point. Large-scale fluctuations scatter light, producing a milky appearance. Taken from [SBC$^+$00], from R. A. Ferrell.

Experimental measurements of critical exponents for various systems are summarized in Tab. 18.1. Although mean-field theory provides qualitatively correct trends, its predictions deviate significantly from the experimentally observed values, particularly in low-dimensional systems where fluctuations are important.

TABLE 18.1 Experimental values of the critical exponents for several systems, compared with the mean-field predictions.

System	Material	α	β	γ	ν
Ferromagnet ($n = 3$)	Fe, Ni	-0.1	0.40	1.3	0.70
Superfluid ($n = 2$)	^{4}He	0	0.32	unknown	0.67
Liquid–Gas ($n = 1$)	CO_2, N_2	0.1	0.30	1.24	0.63
Superconductor ($n = 2$)	Cu	0	$\frac{1}{2}$	1	$\frac{1}{2}$
Mean-Field Theory	—	0	$\frac{1}{2}$	1	$\frac{1}{2}$

These data raise two fundamental questions:

(i) Where do the scaling laws originate?
(ii) Why does mean-field theory fail to predict the correct critical exponents?

In principle, one might also ask why mean-field theory nevertheless works so well for superconductors and other systems with long-range interactions. As we shall see, the first question was answered through the development of the *scaling hypothesis* and the *renormalization group*, pioneered by Leo Kadanoff, Michael Fisher, and Kenneth Wilson, who recognized that near criticality, physical systems must exhibit *self-similarity*. The second question is addressed by analyzing the role of fluctuations, which we now turn to.

18.4 Role of Fluctuations

Let us now analyze how fluctuations contribute around the saddle point in a ferromagnet. We decompose the order parameter field as

$$m(\vec{x}) = \bar{m} + \delta m(\vec{x}), \tag{18.16}$$

where $\bar{m}$ is the uniform mean-field solution, and $\delta m(\vec{x})$ represents fluctuations about it. Since $\nabla \bar{m} = 0$, we have $\nabla m = \nabla \delta m(\vec{x})$. Expanding in powers of δm, we obtain

$$\begin{aligned} m^2 &= \bar{m}^2 + 2\bar{m}\,\delta m(\vec{x}) + [\delta m(\vec{x})]^2, \\ (\nabla m)^2 &= (\nabla \delta m)^2, \\ m^4 &= \bar{m}^4 + 4\bar{m}^3\,\delta m(\vec{x}) + 6\bar{m}^2[\delta m(\vec{x})]^2 + 4\bar{m}[\delta m(\vec{x})]^3 + [\delta m(\vec{x})]^4. \end{aligned} \tag{18.17}$$

The Hamiltonian becomes

$$\begin{aligned} \beta H &= \int d^D x \left[K_0 \bar{m}^2 + K_2 \bar{m}^4 + (K_0 + 6K_2\bar{m}^2)(\delta m)^2 + K_1(\nabla \delta m)^2\right] + \mathcal{O}(\delta m^3, \delta m^4) \\ &\approx V(K_0\bar{m}^2 + K_2\bar{m}^4) + \int d^D x \left[(K_0 + 6K_2\bar{m}^2)(\delta m)^2 + K_1(\nabla \delta m)^2\right]. \end{aligned} \tag{18.18}$$

We see that the quadratic fluctuations are Gaussian, with an effective "mass term"

$$K_0 + 6K_2\bar{m}^2 = \begin{cases} K_0, & K_0 > 0, \\ -2K_0, & K_0 < 0. \end{cases} \tag{18.19}$$

Hence, as in the correlation function, fluctuations become large when $K_0 \to 0$, i.e., near the critical point.

The contribution of these Gaussian fluctuations to the free energy can be computed exactly using the identity $\log\det A = \operatorname{Tr}\log A$, giving

$$f = K_0\bar{m}^2 + K_2\bar{m}^4 + \frac{1}{2}\int \frac{d^D q}{(2\pi)^D}\log\bigl(K_1 q^2 + K_0 + 6K_2\bar{m}^2\bigr). \tag{18.20}$$

The fluctuation correction to the specific heat is then

$$C_{\text{Ising}} \propto -\frac{\partial^2 f}{\partial t^2} = \begin{cases} C^{\text{MFT}}(K_0 < 0) + \frac{1}{2}\int \frac{d^D q}{(2\pi)^D}\frac{1}{(K_1 q^2 + K_0)^2}, & K_0 > 0, \\ C^{\text{MFT}}(K_0 > 0) + \frac{1}{2}\int \frac{d^D q}{(2\pi)^D}\frac{1}{(K_1 q^2 - 2K_0)^2}, & K_0 < 0. \end{cases} \tag{18.21}$$

Since the system has a finite lattice spacing a, we must introduce a cutoff $\Lambda \propto 1/a$. The integrals take the general form

$$\int^{\Lambda} dq \, \frac{q^{D-1}}{(q^2 + c\xi^{-2})^2} \approx \int^{\Lambda} dq \, \frac{q^{D-1}}{q^4}. \tag{18.22}$$

For $D > 4$, this integral diverges, but the cutoff renders it finite, scaling as a^{4-D}. For $D < 4$, the integral converges and, after rescaling q by ξ^{-1}, behaves as ξ^{4-D}. Hence,

$$C_{\text{sing}} \propto \begin{cases} a^{4-D}, & D > 4, \\ \xi^{4-D}, & D < 4. \end{cases} \tag{18.23}$$

Near criticality, $\xi \propto t^{-\nu}$, so for $D < 4$ the fluctuation corrections diverge strongly as $t \to 0$, indicating that the mean-field approximation fails in low dimensions. For $D < 2$, the integral is nevertheless convergent, consistent with the absence of spontaneous symmetry breaking for local order parameters below two dimensions.

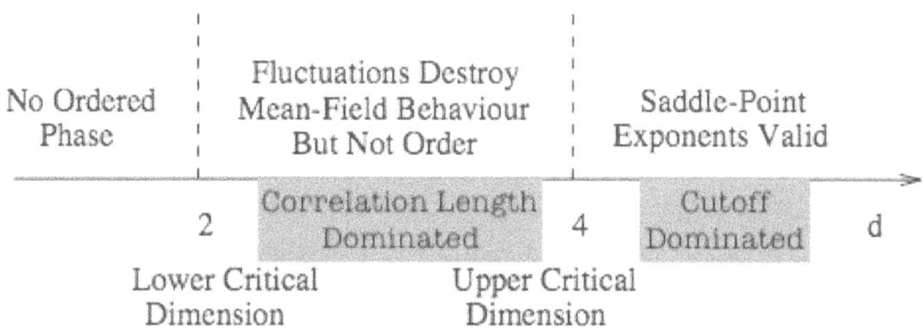

FIGURE 18.4 Upper and lower critical dimensions. Mean-field theory is valid for $D \geq 4$ (fluctuations negligible) and fails for $D < 4$, where fluctuations dominate. No long-range order occurs for $D \leq 2$.

The dimension $D = 4$ is called the *upper critical dimension*, above which mean-field theory is exact. The dimension $D = 2$ is the *lower critical dimension*, below which long-range order and symmetry breaking cannot occur in systems with short-range interactions and continuous symmetry.

We now have a criterion to decide when fluctuations can be neglected in favor of the mean-field description. Let us assume that the correlation length behaves as $\xi \approx \xi_0 t^{-\nu}$ near the critical temperature. If the fluctuations within a correlation volume are small compared to the mean-field variation of the specific heat, then the mean-field approximation is valid.

From the previous analysis, we saw that the singular part of the heat capacity scales as ξ^{4-D}. Therefore, fluctuations are small if the ratio between the fluctuation contribution and the discontinuity of the heat capacity δC is negligible. This comparison can be expressed as

$$\frac{\xi^{-D}}{a^{-D}} \ll \frac{\delta C}{k_{\mathrm{B}}}, \tag{18.24}$$

where a is the microscopic cutoff (typically the lattice spacing). If the inequality is reversed, $\xi^{-D}/a^{-D} \gg \delta C/k_{\mathrm{B}}$, then fluctuations dominate near the transition, and mean-field theory fails.

Combining these observations gives the *Ginzburg criterion.* Introducing the proportionality $\xi \sim t^{-\nu}$, we can write

$$t^{-(D-4)\nu} \gg \frac{\xi_0^D\, \delta C}{a^D\, k_{\mathrm{B}}}, \tag{18.25}$$

and hence define the reduced temperature t_G below which fluctuations become important:

$$t_G = \frac{T_G - T_c}{T_c} \sim \left(\frac{\xi_0^D\, \delta C}{a^D\, k_{\mathrm{B}}}\right)^{\frac{2}{4-D}}, \tag{18.26}$$

where we used $\nu = \frac{1}{2}$ from mean-field theory.

REMARK 18.2 For most systems, t_G is not very small: fluctuations dominate in an experimentally accessible region near T_c. However, for superconductors, the Ginzburg number is extraordinarily small, $t_G \sim 10^{-12}$, meaning that one would have to tune the temperature within one part in 10^{12} of T_c to observe fluctuation effects. This explains why mean-field predictions (such as those of the Ginzburg–Landau theory of superconductivity) agree so well with experiments: the fluctuation-dominated region is practically unobservable.

Some Comments on Continuous Symmetries and Symmetry Breaking

The case of continuous symmetries introduces new and subtle phenomena, especially in low dimensions. Let us consider a vector field $\vec{m}(\vec{x})$ with n components, $m_\alpha(\vec{x})$, representing for instance a magnetization that can freely rotate in an n-dimensional internal space. This corresponds to an $O(n)$ *model*, whose Landau–Ginzburg Hamiltonian reads

$$\beta H = \int d^D x \left[K_0 \sum_\alpha m_\alpha^2(\vec{x}) + K_1 \sum_\alpha \vec{\nabla} m_\alpha(\vec{x}) \cdot \vec{\nabla} m_\alpha(\vec{x}) + \cdots\right]. \tag{18.27}$$

The energy is invariant under global rotations in the internal indices:

$$\vec{m} \to O\,\vec{m}, \qquad O \in O(n),$$

meaning that all directions in spin space are equivalent. The Fourier-space correlation function is therefore diagonal:

$$\langle \tilde{m}_\alpha(\vec{q}) \tilde{m}_\beta(\vec{q}\,') \rangle = \frac{\delta_{\alpha\beta}\, \delta(\vec{q} - \vec{q}\,')}{K_1 q^2 + K_0}. \tag{18.28}$$

Mermin–Wagner Theorem

When the symmetry is continuous, the fluctuations of the orientation of $\vec{m}$ can be long-ranged and destroy long-range order in low dimensions. The *Mermin–Wagner theorem* [MW66] states that for systems with short-range interactions and continuous symmetries, no spontaneous symmetry breaking can occur in dimensions $D \leq 2$. This means, for example, that there is no true ferromagnetic order in the two-dimensional XY or Heisenberg models at any finite temperature.

Discrete Versus Continuous Symmetries

The Ising model, which breaks a discrete $\mathbb{Z}_2$ symmetry, does undergo a phase transition in two dimensions because its order parameter does not involve continuous angular degrees of freedom. By contrast, in the two-dimensional XY model, where spins have an orientation in the plane, there is no conventional symmetry breaking. Instead, there occurs a *Kosterlitz–Thouless transition* [KT73]: pairs of vortices and antivortices become bound below a critical temperature, leading to a topological phase transition with algebraic (rather than exponential) decay of correlations.

Goldstone Theorem

For $D > 2$, when a continuous symmetry is spontaneously broken, small fluctuations of the order parameter in directions orthogonal to the chosen symmetry-breaking direction correspond to *massless modes* known as *Goldstone modes*. The *Goldstone theorem* guarantees the existence of such modes whenever a continuous global symmetry is spontaneously broken.

18.5 Scaling Relations

We now summarize the six critical exponents introduced so far:

$$C_{\text{Ising}} \sim |t|^{-\alpha}, \qquad M \sim |t|^{\beta}, \qquad \chi \sim |t|^{-\gamma},$$
$$M \sim h^{1/\delta}, \qquad \xi \sim |t|^{-\nu}, \qquad G(r) \sim \frac{e^{-r/\xi}}{r^{D-2+\eta}}. \tag{18.29}$$

These exponents are not independent: experimental and theoretical results reveal that they satisfy four fundamental *scaling relations*:

$$\gamma = \nu(2-\eta) \qquad \textit{(Fisher identity)}, \tag{18.30}$$
$$\alpha + 2\beta + \gamma = 2 \qquad \textit{(Rushbrooke identity)}, \tag{18.31}$$
$$\gamma = \beta(\delta - 1) \qquad \textit{(Widom identity)}, \tag{18.32}$$
$$\nu D = 2 - \alpha \qquad \textit{(Josephson identity)}. \tag{18.33}$$

As a consequence, only two of the six critical exponents are independent: all others can be expressed in terms of them.

REMARK 18.3 The fact that these relations hold across a wide variety of physical systems is one of the most striking manifestations of *universality*. Systems as diverse as magnets, liquid–gas transitions, and binary fluids share identical sets of exponents because their large-scale physics depends only on a few general features—dimensionality D, the symmetry of the order parameter, and the range of interactions—not on microscopic details. Each set of critical exponents defines a distinct *universality class*. Although many combinations are mathematically possible, only a few correspond to stable physical fixed points of the renormalization group, which is why nature realizes a limited number of universality classes despite the vast range of microscopic systems.

18.6 Scaling Hypothesis and Widom Scaling Form

The scaling relations between the critical exponents can be derived by assuming that the *singular part of the free energy* obeys a scaling form near the critical point. We recall that the thermodynamic state of the system is controlled by two independent variables: the reduced temperature $t = (T - T_c)/T_c$ and the external magnetic field h.

Scaling Assumption

As the system approaches the phase transition, the correlation length ξ sets the characteristic scale over which spins are correlated. The external field effectively acts on correlated regions of size ξ, suggesting that its influence is "renormalized" by this length scale. We thus define an effective field

$$h_{\text{eff}} = h\,\xi^{y},$$

where y is a scaling exponent to be determined. Since $\xi \sim |t|^{-\nu}$, we expect the singular part of the free energy to depend on the ratio of these quantities in the form

$$F_{\text{sing}}(t,h) = |t|^{2-\alpha}\, g_f\left(\frac{h}{|t|^{\Delta}}\right), \tag{18.34}$$

where α and $\Delta = y\nu$ are critical exponents, and $g_f(x)$ is a universal scaling function that is regular for finite x and homogeneous under scaling transformations.

$$g_f(x) = \begin{cases} g_f(0) = \text{const}, & x \to 0, \\ x^p, & x \gg 1. \end{cases} \tag{18.35}$$

Equation (18.34) is known as the *Widom scaling form* for the free energy.

Consequences of the Scaling Hypothesis

The energy density is obtained from $E_{\text{sing}} = \partial_t F_{\text{sing}}$, giving

$$E_{\text{sing}} \propto |t|^{1-\alpha}\left[(2-\alpha)\, g_f\left(\frac{h}{|t|^{\Delta}}\right) - \frac{h}{|t|^{\Delta}}\, g_f'\left(\frac{h}{|t|^{\Delta}}\right)\right]. \tag{18.36}$$

The function in brackets is itself homogeneous, and thus the specific heat at zero field behaves as

$$C_{\text{sing}} = \left.\frac{\partial E_{\text{sing}}}{\partial T}\right|_{h=0} \propto |t|^{-\alpha}. \tag{18.37}$$

The magnetization follows from

$$m(t,h) = \frac{\partial F_{\text{sing}}}{\partial h} = |t|^{2-\alpha-\Delta}\, g_f'\left(\frac{h}{|t|^{\Delta}}\right), \tag{18.38}$$

and behaves as

$$m(t,h) \sim \begin{cases} |t|^{2-\alpha-\Delta}, & h \to 0, \\ |t|^{2-\alpha-\Delta}\left(\dfrac{h}{|t|^{\Delta}}\right)^{p}, & t \to 0. \end{cases}$$

Derivation of Scaling Relations

The critical exponents β and δ are defined by the asymptotic forms

$$m(t=0,h) \sim h^{1/\delta}, \qquad m(t,h=0) \sim |t|^{\beta}.$$

Comparing with Eq. (18.38), we find:

$$\beta = 2 - \alpha - \Delta, \tag{18.39}$$

$$\delta = \frac{\Delta}{2-\alpha-\Delta} = \frac{1}{p}. \tag{18.40}$$

The magnetic susceptibility is

$$\chi = \left.\frac{\partial m}{\partial h}\right|_{h=0} \sim |t|^{2-\alpha-2\Delta} \equiv |t|^{-\gamma}, \tag{18.41}$$

giving

$$\gamma = 2\Delta + \alpha - 2. \tag{18.42}$$

Substituting Eqs. (18.38)–(18.39) yields the following identity:

$$\alpha + 2\beta + \gamma = \alpha + 2(2-\alpha-\Delta) + (2\Delta + \alpha - 2) = 2, \tag{18.43}$$

which is precisely the *Rushbrooke identity*:

$$\alpha + 2\beta + \gamma = 2. \tag{18.44}$$

For the liquid–gas transition, experiments confirm this remarkably well, giving $\alpha_{\mathrm{LG}} + 2\beta_{\mathrm{LG}} + \gamma_{\mathrm{LG}} \approx 2.01$.

Similarly, from the definition of δ we obtain

$$\delta - 1 = \frac{\Delta}{2-\alpha-\Delta} - 1 = \frac{2\Delta+\alpha-2}{2-\alpha-\Delta} = \frac{\gamma}{\beta}, \tag{18.45}$$

which gives the *Widom identity*:

$$\gamma = \beta(\delta - 1). \tag{18.46}$$

Thus, two of the four scaling relations can be derived directly from the simple assumption of a homogeneous scaling form for the free energy.

Scaling of the Correlation Length

The correlation length also obeys a similar scaling hypothesis:

$$\xi(t,h) = |t|^{-\nu}\, g_{\xi}\left(\frac{h}{|t|^{\Delta}}\right), \tag{18.47}$$

where $g_{\xi}(x)$ is a universal scaling function. Because the free energy is extensive, its singular part must scale with the number of correlation volumes $(L/\xi)^D$. Decomposing the free energy into regular and singular parts,

$$f = \left(\frac{L}{a}\right)^{D} f_{\mathrm{reg}}(t,h) + \left(\frac{L}{\xi}\right)^{D} f_{\mathrm{sing}}(t,h), \tag{18.48}$$

we infer that

$$f_{\mathrm{sing}} \propto \xi^{-D} \sim |t|^{D\nu} \propto |t|^{2-\alpha}. \tag{18.49}$$

This leads directly to the *Josephson identity*:

$$D\nu = 2 - \alpha. \tag{18.50}$$

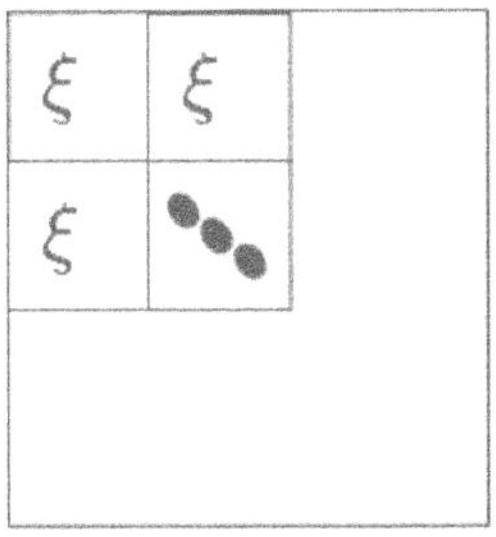

FIGURE 18.5 The total system divided into blocks of the correlation length.

The Scaling Hypothesis

The scaling hypothesis assumes that near the critical point, the singular part of the free energy is a homogeneous function of t and h. From this assumption follow the scaling forms

$$F_{\text{sing}}(t,h) = |t|^{2-\alpha}\, g_f\left(\frac{h}{|t|^{\Delta}}\right), \qquad \xi(t,h) = |t|^{-\nu}\, g_\xi\left(\frac{h}{|t|^{\Delta}}\right),$$

and the fundamental scaling relations:

$$\begin{aligned} \gamma &= \beta(\delta-1) && \text{(Widom identity)},\\ \alpha + 2\beta + \gamma &= 2 && \text{(Rushbrooke identity)},\\ D\nu &= 2-\alpha && \text{(Josephson identity)}. \end{aligned}$$

These relations confirm that all thermodynamic singularities near the critical point are governed by only a few independent exponents. Their validity across disparate systems illustrates the universality of critical phenomena and the power of scaling ideas, later formalized by the renormalization group.

Finally, looking at the correlation function at the critical point, we have

$$\chi(t,h=0) \propto \int^{\xi} d^D x \langle m(x) m(0)\rangle \propto \int^{\xi} d^D x \frac{1}{x^{D-2+\eta}} \propto \xi^{2-\eta}, \tag{18.51}$$

and we see that

$$\xi^{2-\eta} \propto t^{(2-\eta)\nu} \propto t^{-\gamma}. \tag{18.52}$$

Hence, we get the last scaling relation, the *Fisher identity*:

$$(2-\eta)\nu = \gamma. \tag{18.53}$$

The scaling assumption recovers all the observed hyperscaling relationships between the exponents.

18.7 Renormalization and the Scaling Hypothesis

The concept of *scaling* lies at the heart of modern critical phenomena. It was Leo P. Kadanoff (1966) [Kad66] who first suggested that the physics of the critical point could be understood as the manifestation of *scale invariance*. He proposed that the singular behavior of physical quantities near criticality arises from the fact that the system "looks the same" at different length scales.

The key question is then the following:

Is it possible to define a transformation that maps a statistical model into itself by integrating out some of its microscopic degrees of freedom, while reabsorbing their effect into new, renormalized coupling constants?

If such a transformation exists, one can look for values of the coupling constants that remain unchanged under it. These are the *fixed points* of the transformation, and they correspond to *scale-invariant theories*—that is, the physical description remains unchanged under coarse-graining. At criticality, the system is expected to sit precisely at such a fixed point.

To illustrate this idea concretely, let us examine the simplest example where it can be worked out exactly: the one-dimensional Ising model.

Block-Spin Renormalization in One Dimension

Consider the one-dimensional Ising model with nearest-neighbor interactions:

$$Z = \sum_{\{\sigma_i\}} e^{K \sum_i s_i s_{i+1}}, \tag{18.54}$$

where $K = \beta J$. We focus on three consecutive spins s_1, s_2, and s_3, and we imagine "integrating out" the central spin s_2 (see Fig. 18.6). This is the simplest form of a *block-spin transformation.*

We write

$$Z = \sum_{s_1, s_2, s_3 = \pm 1} e^{K(s_1 s_2 + s_2 s_3)} = \sum_{s_1, s_3 = \pm 1} 2\cosh[K(s_1 + s_3)]. \tag{18.55}$$

Our goal is to rewrite this expression in the same functional form as the original Hamiltonian, but with an effective coupling K' and an additive constant K'_0:

$$2\cosh[K(s_1 + s_3)] = e^{K' s_1 s_3 + K'_0}. \tag{18.56}$$

This equality must hold for all combinations of $s_1, s_3 = \pm 1$. From the two possible cases ($s_1 s_3 = \pm 1$) we obtain:

$$b = -K' + K'_0 = \log[2\cosh(0)] = \log 2, \tag{18.57}$$

$$a = K' + K'_0 = \log[2\cosh(2K)]. \tag{18.58}$$

Solving for K' and K'_0 gives

$$K' = \frac{1}{2} \log[\cosh(2K)], \tag{18.59}$$

$$K'_0 = \frac{1}{2} \log[2(1 + \cosh(2K))] = \log[2\cosh K]. \tag{18.60}$$

We thus obtain a new Hamiltonian of the same form but with a renormalized coupling:

$$Z = e^{K'_0} \sum_{\{s_1, s_3\}} e^{K' s_1 s_3}. \tag{18.61}$$

For a chain of $2N$ spins, the same argument gives

$$Z = e^{N K'_0} \sum_{\{s_i\}} e^{K' \sum_{i=1}^{N} s_i s_{i+1}}. \tag{18.62}$$

FIGURE 18.6 Block-spin renormalization: we remove only white spins.

The term NK_0' represents an additive contribution to the free energy. After coarse-graining, the number of spins is halved. To preserve the overall normalization of the free energy, we rescale the system size by doubling the number of spins, keeping the same total length:

$$Z = \sum_{\{s_i\}} e^{K' \sum_{i=1}^{2N} s_i s_{i+1}}, \tag{18.63}$$

so that the renormalized coupling constant is

$$K' = \mathcal{R}(K) = \frac{1}{2} \log[\cosh(2K)]. \tag{18.64}$$

Repeated application of this transformation defines the *renormalization group (RG) flow*:

$$K_{n+1} = \mathcal{R}(K_n), \qquad K_n = \mathcal{R}^n(K_1). \tag{18.65}$$

The RG transformation $\mathcal{R}$ acts on the space of coupling constants. The spins have effectively been "integrated out," and the system's behavior is now captured by the evolution of its couplings under coarse-graining.

Fixed Points and Scale Invariance

A fixed point K^* satisfies

$$K^* = \mathcal{R}(K^*), \tag{18.66}$$

meaning that the system remains unchanged under coarse-graining. The one-dimensional Ising model has only two fixed points:

$$K = 0, \qquad K = \infty, \tag{18.67}$$

corresponding respectively to infinite temperature (disordered phase) and zero temperature (perfect order). Hence, no nontrivial finite-temperature fixed point exists in one dimension, consistent with the absence of a phase transition.

Although the RG transformation is not a true group (since coarse-graining is not invertible), it forms a *semi-group*, describing a continuous flow in the space of couplings. For historical reasons, this flow is nevertheless called the *renormalization group*.

Behavior of the Correlation Length

The spin–spin correlation function in one dimension decays exponentially:

$$\langle \sigma_0 \sigma_r \rangle \approx e^{-r/\xi(K)}, \qquad \xi(K) = \frac{1}{|\log \tanh K|}. \tag{18.68}$$

Under the transformation $K' = \mathcal{R}(K)$, one finds

$$\log\left[\tanh\left(\log\sqrt{\cosh(2K)}\right)\right] = \log[\tanh^2 K] = 2\log[\tanh K], \tag{18.69}$$

which implies

$$\xi(\mathcal{R}(K)) = \frac{\xi(K)}{2}. \tag{18.70}$$

Iterating the transformation n times,

$$\xi(\mathcal{R}^n(K)) = \frac{\xi(K)}{2^n}, \tag{18.71}$$

or equivalently,

$$2^n \xi(\mathcal{R}^n(K)) = \xi(K). \tag{18.72}$$

Thus, $\xi(K) \to \infty$ as $K \to \infty$ (the ordered phase) and $\xi(K) \to 0$ as $K \to 0$ (the disordered phase), in perfect agreement with the exact solution of the one-dimensional Ising model.

The block-spin renormalization approach, while illustrative, fails for the two-dimensional Ising model when applied to larger clusters (e.g., four spins), because it generates additional interaction terms not present in the original Hamiltonian. Nevertheless, this simple example contained the essential insight: by iteratively coarse-graining and rescaling, one can describe the flow of coupling constants in an abstract space of theories. This idea was later generalized by Kenneth Wilson to the continuous Landau–Ginzburg model in momentum space, leading to the modern formulation of the *renormalization group*.

19

Renormalization Group

19.1 Partition Function under Rescaling

In the previous chapters, we introduced the scaling hypothesis and saw how Kadanoff's block-spin idea provides a conceptual bridge between microscopic models and macroscopic critical behavior. We also saw how Widom's scaling form and the resulting relations among critical exponents emerge from the assumption of scale invariance near the critical point. We are now ready to formalize these ideas into a mathematical framework: the *renormalization group* (RG).

The RG provides a systematic way to study how the parameters of a system evolve when we *change the length scale* at which we describe it. This is the quantitative realization of Kadanoff's insight: if we "zoom out" and integrate out the microscopic fluctuations within a block of spins, the system can be rewritten in terms of new, effective coupling constants. If this transformation leaves the form of the Hamiltonian unchanged, the system is said to be *scale invariant*, a property characteristic of critical points.

1. Integrate out short-wavelength (microscopic) degrees of freedom to obtain an *effective* partition function at a larger scale.
2. Rescale the coordinates and fields to restore the system to its original size, obtaining a new set of coupling constants.

This process defines a transformation in the space of coupling parameters:

$$K' = \mathcal{R}(K), \tag{19.1}$$

DOI: 10.1201/9781003603528-19

known as the *renormalization transformation.* The operation of rescaling and integrating out fluctuations is equivalent to "looking at the system under a different magnification." This approach is often called *real-space renormalization group* when applied to lattice models.

Near the phase transition, the free energy can be separated into a regular and a singular part:

$$f(h,t) = f_0(h,t) + f_{\text{sing}}(h,t), \tag{19.2}$$

where $t = (T - T_c)/T_c$ is the reduced temperature. Close to the critical point, the singular part of the free energy is expected to be scale invariant, taking the form

$$f_{\text{sing}}(h,t) \sim t^{2-\alpha}\, g_{\text{sing}}\left(\frac{h}{t^{\Delta}}\right), \tag{19.3}$$

as we saw in Widom's scaling form.

Scaling under a Change of Length Scale

Let us now assume that we rescale the system by a factor $b > 1$, corresponding to a coarse-graining of the microscopic degrees of freedom:

$$L \to bL.$$

Under this transformation, the free energy density scales as

$$f_{\text{sing}}(h,t) \to b^{-D}\, f_{\text{sing}}(t',h'), \tag{19.4}$$

where t and h themselves transform as

$$t' = b^{D_t}\, t, \tag{19.5}$$
$$h' = b^{D_h}\, h. \tag{19.6}$$

The exponents D_t and D_h are the *scaling dimensions* of t and h, respectively, and are directly related to the critical exponents.

Choosing the rescaling factor such that $b^{D_t}\, t = 1$ (i.e., scaling to the correlation length), we find

$$f_{\text{sing}}(h,t) = t^{D/D_t}\, g_f\left(\frac{h}{t^{D_h/D_t}}\right). \tag{19.7}$$

By comparison with Widom's scaling form, we can identify:

$$2 - \alpha = \frac{D}{D_t}, \tag{19.8}$$

$$\Delta = \frac{D_h}{D_t}. \tag{19.9}$$

Thus, the critical exponents that describe the singular behavior of thermodynamic quantities can be traced back to how the parameters of the theory rescale under changes of length.

Alternatively, we may express this idea in terms of the correlation length ξ, which rescales linearly:

$$\xi' = \frac{\xi}{b}. \tag{19.10}$$

In the RG framework, this behavior will emerge naturally from the flow of the coupling constants.

19.2 Renormalization Group

If we repeat the coarse-graining procedure multiple times, we generate a sequence of renormalized coupling constants:

$$K_n = \underbrace{\mathcal{R} \circ \mathcal{R} \circ \cdots \circ \mathcal{R}}_{n \text{ times}}(K_0), \tag{19.11}$$

defining a discrete *dynamical system* in the space of couplings. If the system is scale invariant, there must exist values of the couplings that remain unchanged under this transformation:

$$K^* = \mathcal{R}(K^*), \tag{19.12}$$

which are the *fixed points* of the renormalization group. At these points, the system appears identical under rescaling, and the fixed point thus describes the physics of the phase transition.

The RG transformation (19.11) acts on the space of all possible Hamiltonians. This viewpoint—that renormalization is a transformation on the *space of theories* rather than on a single model—was one of Wilson's key conceptual breakthroughs.

Linearization Around a Fixed Point

Let us consider two coupling parameters, for instance $K = \beta J$ (interaction strength) and $H = \beta h$ (magnetic field). We may then define the RG transformation as

$$\begin{pmatrix} K_n \\ H_n \end{pmatrix} = \mathcal{R} \begin{pmatrix} K_{n-1} \\ H_{n-1} \end{pmatrix} = \begin{pmatrix} \mathcal{R}_K(K_{n-1}, H_{n-1}) \\ \mathcal{R}_H(K_{n-1}, H_{n-1}) \end{pmatrix}. \tag{19.13}$$

Close to a fixed point (K^*, H^*), we expand the transformation to first order:

$$\begin{pmatrix} \delta K_n \\ \delta H_n \end{pmatrix} \approx W \begin{pmatrix} \delta K_{n-1} \\ \delta H_{n-1} \end{pmatrix}, \tag{19.14}$$

where $\delta K = K - K^*$, $\delta H = H - H^*$, and W is the Jacobian matrix

$$W = \begin{pmatrix} \partial_K \mathcal{R}_K & \partial_H \mathcal{R}_K \\ \partial_K \mathcal{R}_H & \partial_H \mathcal{R}_H \end{pmatrix}_{(K^*,H^*)}. \tag{19.15}$$

Diagonalizing W gives independent scaling directions corresponding to its eigenvectors $\vec{v}_i$ and eigenvalues λ_i. After k iterations,

$$W^k \vec{v}_i = \lambda_i^k \vec{v}_i, \tag{19.16}$$

so that each coupling grows, remains constant, or decreases depending on whether $\lambda_i > 1$, $\lambda_i = 1$, or $\lambda_i < 1$, respectively.

- *Relevant direction:* the coupling grows near the fixed point ($\lambda > 1$);
- *Marginal direction:* the coupling remains unchanged ($\lambda = 1$);
- *Irrelevant direction:* the coupling decreases near the fixed point ($\lambda < 1$).

The physical meaning is clear: relevant couplings become increasingly important as we approach the critical point, while irrelevant ones fade away. In terms of the rescaling factor b, the eigenvalues can be expressed as

$$\lambda_i = b^{y_i}, \tag{19.17}$$

where y_i are the *RG scaling exponents*. For the two primary couplings t and h, this gives

$$\begin{aligned} t' &= b^{y_t}\, t, && (19.18)\\ h' &= b^{y_h}\, h. && (19.19) \end{aligned}$$

Comparing with Eqs. (19.8) and (19.9), we identify

$$\begin{aligned} y_t &= D_t, && (19.20)\\ y_h &= D_h, && (19.21) \end{aligned}$$

and thus the exponents controlling the scaling of thermodynamic quantities, such as α and Δ, are directly related to the eigenvalues of the linearized RG transformation:

$$2-\alpha = \frac{D}{y_t}. \tag{19.22}$$

The renormalization group therefore provides a geometrical interpretation of critical exponents: they correspond to the scaling eigenvalues of the RG flow linearized around its fixed points.

19.3 Hyperscaling and Critical Exponents

We have previously introduced the hyperscaling relation,

$$\underbrace{D\nu = 2-\alpha}_{\text{Hyperscaling relation}} \qquad \text{with} \quad \underbrace{2-\alpha = \frac{D}{D_t}}_{\text{RG scaling}}, \tag{19.23}$$

from which we immediately obtain the critical exponent

$$\nu = \frac{1}{D_t}. \tag{19.24}$$

For the exponent δ, which characterizes the scaling of the magnetization with the external field at $t = 0$ via $m \sim h^{1/\delta}$, we recall the relationship derived earlier:

$$\delta = \frac{\Delta}{2-\alpha-\Delta} = \frac{\frac{D_h}{D_t}}{\frac{D}{D_t}-\frac{D_h}{D_t}} = \frac{D_h}{D-D_h}. \tag{19.25}$$

Similarly, the other exponents can be expressed in terms of the scaling dimensions D_t and D_h:

$$\alpha = 2-\frac{D}{D_t}, \qquad \beta = \frac{D-D_h}{D_t}, \qquad \gamma = \frac{2D_h-D}{D_t}, \qquad \eta = 2+D-2D_h. \tag{19.26}$$

These relations provide a complete "recipe" for obtaining all critical exponents once the scaling dimensions D_t and D_h are known.

19.3.1 Continuous Renormalization Group

So far, the renormalization group (RG) transformation has been described as a discrete process, in which we successively integrate out short-distance degrees of freedom. To move toward a differential formulation, we consider an infinitesimal transformation characterized by a scale factor

$$b = 1 + dl.$$

Since $b^y \approx 1 + y\,dl$ for small dl, the infinitesimal change of the coupling vector $\vec{K}$ can be written as

$$\vec{K}(1+dl) - \vec{K}(1) \approx dl\,(W - I)\,\vec{K}(1), \tag{19.27}$$

where W is the linearized RG operator and I the identity. This motivates the definition of the infinitesimal RG flow equations:

$$\frac{d\vec{K}}{dl} = \tilde{\mathcal{R}}(\vec{K}), \tag{19.28}$$

where $l = \log b$ represents the logarithm of the rescaling factor. Equivalently, one can write the continuous version of the RG equations as

$$\frac{dK}{d\log b} = \tilde{\mathcal{R}}(K), \tag{19.29}$$

whose stationary points correspond to the fixed points of the discrete transformation. This differential form is the starting point for the modern, field-theoretic formulation of the renormalization group introduced by Kenneth G. Wilson.

Historical Landmarks of the Renormalization Group

The differential form of the RG equations, and the notion of flow in coupling-constant space, emerged only after a sequence of conceptual breakthroughs.

- *Fisher's scaling theory (1965–1974).* Michael E. Fisher developed the theory of scaling and scaling functions, establishing relationships among critical exponents and clarifying the structure of singularities near critical points [Fis67].
- *Kadanoff's block-spin picture (1966).* Leo P. Kadanoff introduced the idea of scale transformations in real space, proposing that near criticality one should "block" spins together and rescale the system, leading to the concept of universality classes [Kad66].
- *Wilson's formulation of the RG (1971–1975).* Kenneth G. Wilson placed the RG on firm mathematical footing, introducing momentum-shell integration, fixed points, critical exponents, and the ϵ-expansion [WF72, Wil75, Wil71, WK74a]. His framework produced the modern understanding of universality and scaling.
- *Wegner (1972): corrections to scaling.* Franz Wegner identified subleading contributions near RG fixed points, now known as "Wegner corrections," crucial for accurate quantitative predictions [Weg72].
- *Polchinski (1984): RG as an EFT flow.* Joseph Polchinski reformulated RG as a flow equation for the effective action, greatly influencing modern effective field theory and functional RG approaches [Pol84]. Building on Wilson's framework, Polchinski's flow equation was generalized into a fully nonperturbative form by Wetterich, who introduced the exact evolution equation for the effective average action [Wet93]. Shortly thereafter, Morris developed independently alternative exact RG formulations and approximation schemes that remain influential in modern field theory and condensed-matter applications [Mor94]. Closely related in spirit are the Callan–Symanzik equations [Cal70, Sym70], which describe how correlation functions evolve with scale and provide a bridge between perturbative renormalization and Wilsonian effective field theory.

Together, these works established the renormalization group as the central conceptual and computational framework for understanding phase transitions, universality, and scale dependence in both statistical physics and quantum field theory.

Suggested Reading

For further study of renormalization-group techniques, several comprehensive and pedagogical treatments are highly recommended. Kardar's Volume II [Kar07] provides a modern, elegant field-theoretic introduction, while Binney *et al.* [BDFN92] offer a clear statistical–mechanics perspective. Goldenfeld's text [Gol92] presents the RG with an emphasis on universality and scaling, and Zinn-Justin's monograph [ZJ02] remains the authoritative field-theoretic reference. For functional RG methods, introductory reviews such as Berges *et al.* [BTW02] complement the original works of Wetterich and Morris.

19.4 Wilson's Renormalization Group Approach

Wilson reformulated the RG in terms of *momentum-space coarse-graining* applied to the Landau–Ginzburg field theory. The idea is to integrate out the high-momentum (short-wavelength) modes of the order parameter field, thereby obtaining an effective theory valid at longer length scales.

We start from the partition function of the Gaussian (free) Landau–Ginzburg model:

$$Z = \int \mathcal{D}\tilde{m}(\vec{q})\,\exp\left[-\int^{\Lambda} d^D q\,(K_0 + K_1 q^2)\,|\tilde{m}(\vec{q})|^2\right], \tag{19.30}$$

where $\Lambda \sim 2\pi/a$ is the momentum cutoff corresponding to the lattice spacing a. The field $\tilde{m}(\vec{q})$ represents the Fourier transform of the coarse-grained order parameter $m(\vec{x})$.

Integrating Out Fast Modes

In analogy with Kadanoff's block-spin procedure, we divide the Fourier modes into "slow" and "fast" components. Modes with $|\vec{q}| < \Lambda/b$ represent long-wavelength fluctuations, while those with $\Lambda/b < |\vec{q}| < \Lambda$ correspond to short-wavelength fluctuations:

$$F = [\Lambda/b, \Lambda]. \tag{19.31}$$

We decompose the field as

$$\tilde{m}(\vec{q}) = \tilde{m}_<(\vec{q}) + \tilde{m}_>(\vec{q}), \tag{19.32}$$

where $\tilde{m}_<$ and $\tilde{m}_>$ contain momenta below and above Λ/b, respectively. The path-integral measure factorizes accordingly:

$$\mathcal{D}\tilde{m}(\vec{q}) = \mathcal{D}\tilde{m}_<(\vec{q})\,\mathcal{D}\tilde{m}_>(\vec{q}). \tag{19.33}$$

The partition function can then be written as

$$Z = \int \mathcal{D}\tilde{m}_<\,\mathcal{D}\tilde{m}_>\,\exp[-\beta H(\tilde{m}_<, \tilde{m}_>)]. \tag{19.34}$$

Because the Gaussian Hamiltonian is quadratic and diagonal in momentum space,

$$-\beta H = \int d^D q\,(K_0 + K_1 q^2)\,|\tilde{m}(\vec{q})|^2 + h\,\tilde{m}(q=0), \tag{19.35}$$

the partition function factorizes into independent contributions from the two regions of momentum space:

$$Z = \underbrace{\int \mathcal{D}\tilde{m}_> \, e^{-\beta H(\tilde{m}_>)}}_{\text{constant factor}} \int \mathcal{D}\tilde{m}_< \, e^{-\beta H(\tilde{m}_<)}. \tag{19.36}$$

Integrating out the high-momentum modes simply produces a multiplicative constant independent of $\tilde{m}_<$:

$$Z = \overbrace{\frac{1}{\prod_{q>\Lambda/b}\sqrt{\frac{(K_0+K_1q^2)}{V}}}}^{\text{constant (field independent)}} \int \mathcal{D}\tilde{m}_< \, e^{-\beta H(\tilde{m}_<)}. \tag{19.37}$$

Thus, for the Gaussian theory, the coarse-grained Hamiltonian retains the same functional form.

Rescaling

To restore the cutoff to its original value Λ, we perform a rescaling of momenta and coordinates:

$$\vec{q}\,' = b\,\vec{q}, \qquad \vec{x}\,' = \frac{\vec{x}}{b}.$$

After this transformation, the partition function becomes

$$Z = \int \mathcal{D}\tilde{m}_< \exp\left[-\int^{\Lambda} d^D q\, b^{-D}\left(K_0 + b^{-2}K_1 q^2\right)|\tilde{m}_<(\vec{q})|^2 + h\,\tilde{m}_<(q=0)\right]. \tag{19.38}$$

Finally, to maintain canonical normalization of the quadratic term, we rescale the field as

$$\tilde{m}_< = Z_m\,\tilde{m},$$

where the renormalization factor Z_m ensures invariance of the gradient term:

$$b^{-D-2}Z_m^2 = 1 \quad \Rightarrow \quad Z_m = b^{1+D/2}. \tag{19.39}$$

The renormalized couplings then transform as

$$K_0' = b^{-D}Z_m^2 K_0 = b^2 K_0, \tag{19.40}$$
$$K_1' = b^{-D-2}Z_m^2 K_1 = K_1, \tag{19.41}$$
$$K_r' = b^{-D-r}Z_m^2 K_r = b^{-r+2}K_r, \tag{19.42}$$
$$h' = Z_m\,h = b^{1+D/2}h. \tag{19.43}$$

We identify $K_0 = t$, the reduced temperature parameter, so that

$$t' = b^2 t, \qquad h' = b^{1+D/2}h.$$

Hence, the scaling dimensions of t and h are

$$D_t = 2, \qquad D_h = 1 + \frac{D}{2}.$$

From these, the critical exponents follow:

$$\nu = \frac{1}{D_t} = \frac{1}{2}, \qquad \delta = \frac{D_h}{D - D_h} = \frac{1+\frac{D}{2}}{D-1-\frac{D}{2}}. \tag{19.44}$$

For $D = 4$, this gives $\delta = 3$, in agreement with mean-field theory; for $D = 3$, we find $\delta = 5$, consistent with the Gaussian (noninteracting) field theory prediction. The exponent η also takes its mean-field value $\eta = 0$.

Finally, note that for $r > 2$, all higher-derivative couplings K_r are *irrelevant*, since their scaling dimension $-r+2$ is negative. This means that such terms vanish under successive coarse-graining, and only the lowest-order relevant terms (t, h, and the gradient term) determine the universal behavior near criticality.

Let us now summarize the operational steps that led to the formulation of the *Renormalization Group* (RG) procedure:

1. Start from the path-integral representation of the partition function, written in terms of the Fourier-transformed field in momentum space.
2. Split the field into two parts: low-momentum and high-momentum components, separated by a scale $\tilde{k} = \Lambda/b$.
3. Integrate over all momenta above $\tilde{k}$—these correspond to the "fast" modes.
4. Rewrite the energy (or Hamiltonian) in a form identical to the original one but expressed in terms of a renormalized field $Z_m \, \tilde{m}$.
5. Impose the invariance condition $K_1 = 1$ to fix the field renormalization factor Z_m.
6. Absorb the scale dependence b into the coupling constants to obtain the RG transformation for the couplings.

For the Gaussian (free-field) theory, these steps are straightforward. However, the Gaussian theory alone is insufficient to describe real critical phenomena, since it yields only the mean-field exponents. To obtain corrections beyond this approximation, we must include interactions—specifically, the quartic term in the Landau–Ginzburg Hamiltonian—and study their effect perturbatively.

19.5 Field Dimensions

Before moving to perturbation theory, it is useful to examine the dimensional analysis of the couplings entering the field theory. We start from the general Landau–Ginzburg Hamiltonian:

$$-\beta H = \int d^D x \left[K_0 \, m^2(\vec{x}) + K_1 \, (\nabla m(\vec{x}))^2 + K_2 \, m^4(\vec{x}) + \cdots \right]. \tag{19.45}$$

The left-hand side is dimensionless, so each term in the integrand must have dimensions of $[L]^{-D}$. Since $d^D x$ carries dimension $[L]^D$ and derivatives contribute powers of $[L]^{-1}$, we can determine the scaling dimensions of each coupling.

If we impose $[K_1] = 1$ (i.e., the gradient term is dimensionless and normalized), the integrand in the Hamiltonian must have overall dimension $[L]^{-D}$. Since ∇ carries dimension $[\nabla] = [L]^{-1}$, we have

$$[(\nabla m)^2] = [L]^{-D} \quad \Rightarrow \quad [L]^{-2} \, [m]^2 = [L]^{-D} \quad \Rightarrow \quad [m] = [L]^{-\frac{D-2}{2}} = [L]^{-\frac{D}{2}+1}. \tag{19.46}$$

From this, we deduce the dimensions of the couplings:

$$\begin{aligned} [m] &= [L]^{-\frac{D}{2}+1}, \\ [K_0] &= [L]^{-2}, \\ [K_1] &= [L]^0 = 1, \\ [K_2] &= [L]^{D-4}, \\ [g_{d,n}] &= [L]^{\, d-n+D\left(\frac{n}{2}-1\right)}, \end{aligned}$$

where $g_{d,n}$ denotes a general coupling multiplying an operator of the form $\nabla^d m^n$.
With the momentum-dimension convention $[L] = -1$ (equivalently $[q] = +1$), the engineering dimensions are

$$[m] = \frac{D}{2} - 1,$$
$$[K_0] = 2,$$
$$[K_2] = 4 - D,$$
$$[g_{d,n}] = n - d - D\Big(\frac{n}{2} - 1\Big).$$

These relations match the scaling factors obtained from the RG transformation. They allow us to determine, purely from dimensional analysis, which couplings are *relevant* and which are *irrelevant* at the critical point.

A coupling is *irrelevant* if its scaling dimension is negative:

$$n - d - D\Big(\frac{n}{2} - 1\Big) < 0, \tag{19.47}$$

which can also be written as

$$D > \frac{2(n-d)}{n-2}. \tag{19.48}$$

These quantities are known as *operator dimensions.* They correspond to the eigenvalues of the linearized RG flow at the fixed point and classify the possible perturbations of the theory into relevant, marginal, or irrelevant directions.

19.6 Perturbative Partition Function and Feynman Diagrams

To understand how fluctuations modify the mean-field critical exponents, we now introduce perturbation theory around the Gaussian approximation.

The interacting Landau–Ginzburg partition function can be written as:

$$Z = \int \mathcal{D}m(\vec{x})\, e^{-\beta H[m]} = \int \mathcal{D}m(\vec{x})\, \exp\Big[-\int d^D x\, \big(t\, m^2(\vec{x}) + (\nabla m(\vec{x}))^2 + u\, m^4(\vec{x}) + \cdots\big)\Big], \tag{19.49}$$

where $t = K_0$ and $u = K_2$ are the coupling constants for the quadratic and quartic terms, respectively.

Applying the Fourier transform

$$m(\vec{x}) = \int \frac{d^D q}{\sqrt{(2\pi)^D}} e^{i\vec{q}\cdot\vec{x}}\, \tilde{m}(\vec{q}), \tag{19.50}$$

the quartic term becomes

$$\begin{aligned}\int d^D x\, u\, m^4(\vec{x}) &= u \int d^D x \int \frac{d^D q_1\, d^D q_2\, d^D q_3\, d^D q_4}{(2\pi)^{2D}} e^{i(\vec{q}_1+\vec{q}_2+\vec{q}_3+\vec{q}_4)\cdot\vec{x}} \tilde{m}(\vec{q}_1)\tilde{m}(\vec{q}_2)\tilde{m}(\vec{q}_3)\tilde{m}(\vec{q}_4) \\ &= u \int \frac{d^D q_1\, d^D q_2\, d^D q_3\, d^D q_4}{(2\pi)^{3D}} (2\pi)^D\, \delta(\vec{q}_1+\vec{q}_2+\vec{q}_3+\vec{q}_4)\tilde{m}(\vec{q}_1)\tilde{m}(\vec{q}_2)\tilde{m}(\vec{q}_3)\tilde{m}(\vec{q}_4).\end{aligned} \tag{19.51}$$

Let us denote this quartic interaction term compactly as $u\,\tilde{V}$.

Assuming that u is small, we can expand the exponential of the interaction term as a power series:

$$Z = \int \mathcal{D}\tilde{m}(\vec{q})\, e^{-\int d^D q\,(K_0+q^2)\,|\tilde{m}(\vec{q})|^2} \left(1 + u\tilde{V} + \frac{u^2}{2!}\tilde{V}^2 + \cdots\right)$$
$$= \sum_{k=0}^{\infty} \frac{u^k}{k!} Z_k, \tag{19.52}$$

where Z_k contains the k-th order term in the perturbative expansion:

$$Z_k = \int \mathcal{D}\tilde{m}(\vec{q})\, e^{-\int d^D q\,(K_0+q^2)\,|\tilde{m}(\vec{q})|^2}\, \tilde{V}^k.$$

For brevity, we introduce the shorthand notation:

$$\int \frac{d^D q_1\, d^D q_2\, d^D q_3\, d^D q_4}{(2\pi)^{2D}} \equiv \int_q . \tag{19.53}$$

This expansion forms the basis of the *diagrammatic perturbation theory*, where each term Z_k can be represented as a sum of Feynman diagrams corresponding to Gaussian integrals with different topologies. In the next section, we will introduce the Feynman rules and discuss how these diagrams contribute to physical quantities such as the correlation function and the renormalized couplings.

19.7 Gaussian Averages, Wick's Theorem, and First Corrections

At zeroth order in u (i.e., Z_0), we recover the Gaussian theory. At first order in u, we have

$$Z_1 = \int_q\!\int \mathcal{D}\tilde{m}\, \exp\!\left[\int d^D q\,(K_0+q^2)\,|\tilde{m}(q)|^2\right] \tilde{V}$$
$$= \int_q\!\int \mathcal{D}\tilde{m}\, \exp\!\left[\int d^D q\,(K_0+q^2)\,|\tilde{m}(q)|^2\right] (2\pi)^D \delta(\vec{q}_1+\vec{q}_2+\vec{q}_3+\vec{q}_4)\, \tilde{m}(\vec{q}_1)\tilde{m}(\vec{q}_2)\tilde{m}(\vec{q}_3)\tilde{m}(\vec{q}_4), \tag{19.54}$$

so that

$$Z_1 = \int_q (2\pi)^D \delta(\vec{q}_1+\vec{q}_2+\vec{q}_3+\vec{q}_4) \left\langle \tilde{m}(\vec{q}_1)\tilde{m}(\vec{q}_2)\tilde{m}(\vec{q}_3)\tilde{m}(\vec{q}_4)\right\rangle_0, \tag{19.55}$$

with the Gaussian average

$$\langle\cdot\rangle_0 \equiv \frac{1}{Z_0}\int \mathcal{D}\tilde{m}(q)\, \exp\!\left[\int d^D q\,(t+q^2)\,|\tilde{m}(q)|^2\right] \cdot \tag{19.56}$$

(we have written $t \equiv K_0$ for brevity). In compact form,

$$Z = \left\langle e^{u\tilde{V}}\right\rangle_0, \qquad Z_0 = \langle 1\rangle_0. \tag{19.57}$$

For a finite system (discrete momenta), the two-point function is

$$\langle \tilde{m}(q)\,\tilde{m}(q')\rangle_0 = \frac{(2\pi)^D\, V\, \delta_{q',-q}}{2\,(t+q^2)}. \tag{19.58}$$

In the thermodynamic limit $V \to \infty$ (continuum momenta),

$$C(q,q') \equiv \langle \tilde{m}(q)\,\tilde{m}(q')\rangle_0 = \frac{(2\pi)^D\, \delta(\vec{q}' + \vec{q})}{2\,(t+q^2)}. \tag{19.59}$$

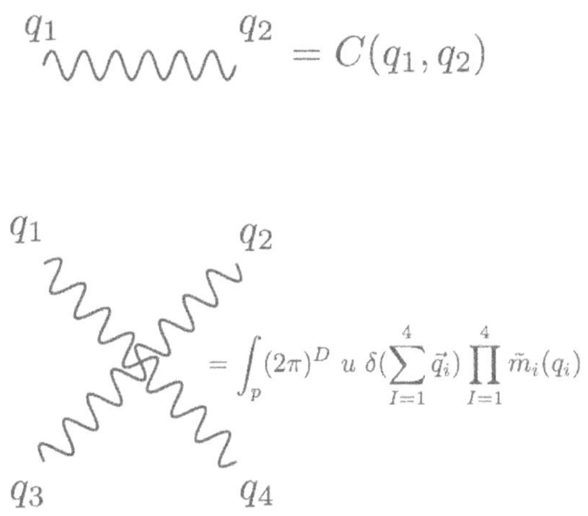

FIGURE 19.1 Diagrammatic representation of the perturbative expansion in u. Solid lines denote Gaussian propagators; four-legged vertices denote the $u\,m^4$ interaction with momentum conservation.

Wick-Isserlis Theorem

For Gaussian fields,

$$\Big\langle \prod_{i=1}^{\ell} \tilde{m}(q_i) \Big\rangle_0 = \begin{cases} 0, & \ell \text{ odd}, \\ \text{sum over all pairwise contractions}, & \ell \text{ even}, \end{cases} \tag{19.60}$$

so, for instance,

$$\begin{aligned} \langle \tilde{m}(q_1)\tilde{m}(q_2)\tilde{m}(q_3)\tilde{m}(q_4)\rangle_0 = (2\pi)^{2D} \bigg[& \frac{\delta(\vec{q}_1 + \vec{q}_2)\,\delta(\vec{q}_3 + \vec{q}_4)}{2(t + q_1^2)\,2(t + q_3^2)} + \frac{\delta(\vec{q}_1 + \vec{q}_4)\,\delta(\vec{q}_2 + \vec{q}_3)}{2(t + q_1^2)\,2(t + q_2^2)} \\ & + \frac{\delta(\vec{q}_1 + \vec{q}_3)\,\delta(\vec{q}_2 + \vec{q}_4)}{2(t + q_1^2)\,2(t + q_2^2)} \bigg]. \end{aligned} \tag{19.61}$$

Diagrammatically, each contraction corresponds to a propagator line; the overall $(2\pi)^D\delta(\sum \vec{q}_i)$ at a vertex enforces momentum conservation.

$O(n)$ Generalization

For an $O(n)$ order parameter $\vec{m} = (m_1, \ldots, m_n)$, the Gaussian propagator becomes

$$C_{\alpha\beta}(q, q') = \langle \tilde{m}_\alpha(q)\,\tilde{m}_\beta(q')\rangle_0 = \frac{(2\pi)^D\, V\, \delta_{q',-q}\,\delta_{\alpha\beta}}{2\,(t + q^2)} \quad \text{(discrete)}, \tag{19.62}$$

and the quartic interaction must respect $O(n)$ symmetry:

$$m^4(x) \longrightarrow \Big(\sum_{\alpha=1}^{n} m_\alpha(x)^2\Big)^2. \tag{19.63}$$

At the diagrammatic level, the rules are identical except for the additional Kronecker $\delta_{\alpha\beta}$ carried by each propagator and appropriate index contractions at the vertices.

First-Order Correction to the Two-Point Function

Consider the perturbative expansion of the two-point correlator:

$$\langle \tilde{m}(q_a)\tilde{m}(q_b)\rangle = \left\langle \tilde{m}(q_a)\tilde{m}(q_b) \left(1 - u \int_q (2\pi)^D \delta(\textstyle\sum_{i=1}^4 \vec{q}_i)\, \tilde{m}(q_1)\tilde{m}(q_2)\tilde{m}(q_3)\tilde{m}(q_4) + \cdots \right)\right\rangle_0. \tag{19.64}$$

At zeroth order in u, we obtain the Gaussian result. At first order, Wick's theorem produces contractions of the six fields. Combinatorially, one finds a connected contribution (the "tadpole"/"bubble" attached to the external line) and a disconnected part proportional to $\langle \tilde{m}(q_a)\tilde{m}(q_b)\rangle_0 \langle \tilde{m}_1 \cdots \tilde{m}_4\rangle_0$. Organizing these pieces, we can write

$$\langle \tilde{m}_a \tilde{m}_b\rangle = \langle \tilde{m}_a \tilde{m}_b\rangle_0 \; - \; 12\, u\, \langle \tilde{m}_a \tilde{m}_b \tilde{m}_1 \cdots \tilde{m}_4\rangle_{0,\mathrm{conn}} \; - \; 3\, u\, \langle \tilde{m}_a \tilde{m}_b\rangle_0\, \langle \tilde{m}_1 \cdots \tilde{m}_4\rangle_0 \; + \; O(u^2), \tag{19.65}$$

where the numerical coefficients come from the number of distinct contractions leading to each topology.

Physical correlators are normalized by the full partition function:

$$\langle \tilde{m}_a \tilde{m}_b\rangle_{\mathrm{phys}} = \frac{\langle \tilde{m}_a \tilde{m}_b\rangle}{Z} = \frac{\langle \tilde{m}_a \tilde{m}_b\rangle_0 - 12\, u\, \langle \tilde{m}_a \tilde{m}_b \tilde{m}_1 \cdots \tilde{m}_4\rangle_{0,\mathrm{conn}} - 3\, u\, \langle \tilde{m}_a \tilde{m}_b\rangle_0 \langle \tilde{m}_1 \cdots \tilde{m}_4\rangle_0 + \cdots}{1 - 3\, u\, \langle \tilde{m}_1 \cdots \tilde{m}_4\rangle_0 + \cdots}. \tag{19.66}$$

Expanding the denominator to $O(u)$ and keeping terms consistently to first order yields

$$\begin{aligned}\langle \tilde{m}_a \tilde{m}_b\rangle_{\mathrm{phys}} &\approx (\langle \tilde{m}_a \tilde{m}_b\rangle_0 - 12\, u\, \langle \tilde{m}_a \tilde{m}_b \tilde{m}_1 \cdots \tilde{m}_4\rangle_{0,\mathrm{conn}} - 3\, u\, \langle \tilde{m}_a \tilde{m}_b\rangle_0 \langle \tilde{m}_1 \cdots \tilde{m}_4\rangle_0) \\ &\quad (1 + 3\, u\, \langle \tilde{m}_1 \cdots \tilde{m}_4\rangle_0) \\ &= \langle \tilde{m}_a \tilde{m}_b\rangle_0 - 12\, u\, \langle \tilde{m}_a \tilde{m}_b \tilde{m}_1 \cdots \tilde{m}_4\rangle_{0,\mathrm{conn}} \; + \; O(u^2).\end{aligned} \tag{19.67}$$

Thus, *only the connected diagrams* contribute to properly normalized correlation functions. This general fact—correlators are generated by $\log Z$—will be proved in the next chapter; diagrammatically, it is the statement that disconnected vacuum bubbles cancel between numerator and denominator, leaving exactly the sum of connected graphs attached to the external legs.

20

Critical Exponents

20.1 From Mean Field to Field Theory: Why Go Beyond

In the previous chapter, we developed the perturbative formulation of the Landau–Ginzburg theory and introduced Feynman diagrams as a systematic way to compute corrections to the Gaussian (mean-field) results. Mean-field theory provides a qualitatively correct picture of phase transitions but fails quantitatively: it neglects fluctuations, leading to incorrect values of the critical exponents, especially in low dimensions ($D < 4$). To obtain accurate exponents, one must incorporate fluctuations perturbatively and study how couplings evolve with scale.

The field-theoretic renormalization group (RG) provides precisely such a framework. It connects the behavior of correlation functions at different length scales to the flow of coupling constants, allowing one to compute universal quantities such as the critical exponents α, β, γ, ν, and η beyond mean-field theory.

In this chapter, we first revisit the definitions of Fourier transforms and propagators, which play a central role in field-theoretic perturbation theory. We then show how connected diagrams—the physically meaningful ones—are generated by the logarithm of the partition function. These results will serve as the foundation for computing fluctuation corrections to mean-field exponents in the following sections.

DOI: 10.1201/9781003603528-20

20.2 Fourier Transforms, Green Functions, and Covariance Revisited

Before proceeding, let us clarify the precise normalization conventions for Fourier transforms and propagators, as different definitions lead to slightly different prefactors in correlation functions.

We have defined the symmetric Fourier transform and its inverse as

$$\begin{aligned} m(\vec{x}) &= \int \frac{d^D q}{\sqrt{(2\pi)^D}}\, e^{-i\vec{q}\cdot\vec{x}}\, \tilde{m}(\vec{q}), \\ \tilde{m}(\vec{q}) &= \int \frac{d^D x}{\sqrt{(2\pi)^D}}\, e^{i\vec{q}\cdot\vec{x}}\, m(\vec{x}). \end{aligned} \tag{20.1}$$

Alternatively, one can use the more common convention employed in many textbooks:

$$\begin{aligned} m(\vec{x}) &= \int \frac{d^D q}{(2\pi)^D}\, e^{-i\vec{q}\cdot\vec{x}}\, \tilde{m}(\vec{q}), \\ \tilde{m}(\vec{q}) &= \int d^D x\, e^{i\vec{q}\cdot\vec{x}}\, m(\vec{x}), \end{aligned} \tag{20.2}$$

which modifies the normalization of the propagator.

Further Reading

A clear and pedagogical introduction to the field-theoretic renormalization group and its connection to critical exponents can be found in M. Kardar, *Statistical Physics of Fields* (Cambridge University Press, 2007) [Kar07]. These works provide an intuitive derivation of the diagrammatic rules, the meaning of connected diagrams, and the computation of exponents using perturbative RG techniques.

The covariance in momentum space is defined by

$$C(q_1, q_2) = \frac{\displaystyle\int \mathcal{D}\tilde{m}(q)\, \tilde{m}(q_1)\tilde{m}(-q_2)\, e^{-\beta H[\tilde{m}]}}{\displaystyle\int \mathcal{D}\tilde{m}(q)\, e^{-\beta H[\tilde{m}]}}. \tag{20.3}$$

For the quadratic Hamiltonian,

$$H[\tilde{m}] = \int \frac{d^D q}{a}\, (K_0 + K_1 q^2)\, \tilde{m}(q)\tilde{m}(-q), \tag{20.4}$$

we include a normalization constant a to track the effect of Fourier conventions.

To illustrate, consider adding an external field $h(q)$ and performing the Gaussian integral:

$$Z = \int_{-\infty}^{\infty} d^m x\, e^{-\frac{1}{a}\vec{x}^T G\vec{x} + \vec{y}\cdot\vec{x}} = \frac{\pi^{m/2}}{\sqrt{\det G}}\, e^{\frac{a}{4}\vec{y}^T G^{-1}\vec{y}}. \tag{20.5}$$

Differentiating with respect to the source components gives

$$\langle x_i x_j \rangle_{h=0} = \frac{a}{2}\, G^{-1}_{ij}, \tag{20.6}$$

so the covariance reads

$$C_{ij} = \frac{a}{2}\, G^{-1}_{ij}. \tag{20.7}$$

If we set $a = 2$, we recover the familiar relation $\langle x_i x_j \rangle = G_{ij}^{-1}$. In Fourier space, where G is diagonal,

$$G^{-1}(q, q') = \frac{\delta(q - q')}{K_0 + K_1 q^2}. \tag{20.8}$$

In our symmetric convention, where $a = \sqrt{(2\pi)^D}$, the propagator becomes

$$C(q, q') = \frac{\sqrt{(2\pi)^D}\,\delta(q - q')}{2\,(K_0 + K_1 q^2)}. \tag{20.9}$$

In most textbooks using the convention (20.2), one has $a = 2(2\pi)^D$, giving instead

$$C(q, q') = \frac{(2\pi)^D\,\delta(q - q')}{K_0 + K_1 q^2}, \tag{20.10}$$

which is the form most commonly used in field theory and condensed matter texts.

20.3 Connected Diagrams and Physical Correlators

We now turn to an important conceptual result: only *connected diagrams* contribute to physical expectation values. Formally, any observable that can be written as

$$\langle \mathcal{O}(\tilde{m}) \rangle_{\text{phys}} = \frac{\langle \mathcal{O}(\tilde{m}) \rangle}{\langle 1 \rangle} \tag{20.11}$$

will receive contributions only from diagrams that are connected to the external points of $\mathcal{O}$.

To see this, let us expand perturbatively:

$$\langle \mathcal{O} \rangle_0 = \int \mathcal{D}\tilde{m}\, e^{-\int d^D q\,(K_0 + K_1 q^2)\,|\tilde{m}(q)|^2}\, \mathcal{O}, \tag{20.12}$$

and

$$\langle \mathcal{O}(\tilde{m}) \rangle_{\text{phys}} = \frac{\langle \mathcal{O}(\tilde{m})\, e^{-u\tilde{V}} \rangle_0}{\langle e^{-u\tilde{V}} \rangle_0} \approx \frac{\langle \mathcal{O}(\tilde{m})\, \sum_{k=0}^{\infty} \frac{(-u)^k}{k!} \tilde{V}^k \rangle_0}{\langle \sum_{k=0}^{\infty} \frac{(-u)^k}{k!} \tilde{V}^k \rangle_0}. \tag{20.13}$$

Expanding both numerator and denominator, one finds

$$\langle \mathcal{O}(\tilde{m}) \rangle_{\text{phys}} = \sum_{k=0}^{\infty} \frac{(-u)^k}{k!}\, \langle \tilde{V}^k\, \mathcal{O} \rangle_{\text{conn}}, \tag{20.14}$$

where $\langle \cdot \rangle_{\text{conn}}$ denotes the expectation value restricted to connected diagrams.

Generating Functionals and Connected Diagrams

Physical n-point correlation functions are generated by functional derivatives of the partition function with sources:

$$Z[J] = \int \mathcal{D}m\, e^{-\beta H[m] - \int J(q)\,\tilde{m}(q)}, \tag{20.15}$$

and

$$\langle \tilde{m}(q_1) \cdots \tilde{m}(q_n) \rangle = \frac{1}{Z} \prod_{i=1}^{n} \frac{\delta}{\delta J(q_i)}\, Z[J] \Big|_{J=0}. \tag{20.16}$$

The key observation is that $\log Z$ generates only connected diagrams. To prove this, we can use the *replica trick*, or equivalently the cumulant expansion.

Consider, for simplicity, the real-space form of the partition function:

$$Z = \int \mathcal{D}\phi(x)\, e^{-\int d^D x\, [\, K_0 \phi^2(x) + K_1 (\nabla\phi)^2 + V(\phi)\,]}. \tag{20.17}$$

The same argument holds in momentum space, where the propagator is diagonal and the interaction $V(\phi)$ generates the same diagrammatic structure.

We now introduce n replicas of the field:

$$Z^n = \int \mathcal{D}\phi_\alpha(x)\, e^{-\sum_{\alpha=1}^{n} \int d^D x\, [\, K_0 \phi_\alpha^2 + K_1 (\nabla\phi_\alpha)^2 + V(\phi_\alpha)\,]}. \tag{20.18}$$

Propagators are diagonal in the replica index:

$$\langle \phi_\alpha\, \phi_\beta \rangle \propto \delta_{\alpha\beta}. \tag{20.19}$$

When computing expectation values involving $V(\phi_\alpha)$, each connected diagram carries one overall replica index, while disconnected diagrams contribute higher powers of n. After summing over replica indices,

$$Z^n = \sum_{k=0}^{\infty} n^k \,(k\text{-connected diagrams}). \tag{20.20}$$

Taking the limit $n \to 0$ yields

$$\log Z = \lim_{n\to 0} \frac{Z^n - 1}{n} = (1\text{-connected diagrams}) + \lim_{n\to 0} \sum_{k=2}^{\infty} n^{k-1} = \text{connected diagrams only.} \tag{20.21}$$

Hence, $\log Z$ generates exclusively connected diagrams. This is precisely the field-theoretic analog of the cumulant expansion in statistics: $\log\langle e^U \rangle_0$ collects only connected cumulants of the random variable U. We will make use of this result in the next chapter when computing fluctuation corrections and the renormalization of the coupling constants.

Probabilistic Approaches to the Renormalization Group

Long before the renormalization group (RG) became a central tool in high-energy and condensed-matter physics, Giovanni Jona-Lasinio emphasized a fundamentally *probabilistic* interpretation of RG transformations. In his formulation [JL75], RG flows arise from the successive coarse-graining of probability measures, much like the convergence theorems in probability theory governing sums of random variables. Under coarse-graining, probability distributions approach fixed points characterized by universal scaling behavior, providing a statistical foundation for critical exponents and universality classes. This perspective highlights the deep connection between RG, central-limit-type phenomena, and large-deviation principles. In Volume II we will introduce this probabilistic RG framework in detail for a toy model, showing how coarse-graining of Gibbs measures leads naturally to scaling, fixed points, and universality in statistical systems.

20.4 Perturbative Renormalization Group

We now turn to the central goal of this section: the computation of fluctuation corrections to the correlation function and, from it, the critical exponents. As discussed earlier, the mean-field approximation predicts the Gaussian propagator

$$F_0(q) = \frac{1}{2(K_0 + K_1 q^2)},$$

which leads to $\nu = \frac{1}{2}$ and $\eta = 0$. To go beyond this approximation, we include the first perturbative corrections in the quartic coupling u. These corrections will reveal how the apparent divergence of the perturbative expansion can be reabsorbed into a renormalization of the parameters, laying the foundation for Wilson's renormalization group.

Let us recall that the susceptibility is given by

$$\chi(t) = \frac{\partial M}{\partial h} = \frac{\partial^2 \log Z}{\partial h^2}, \tag{20.22}$$

which, in momentum space, is proportional to the correlation function at zero momentum:

$$\chi(t) = F(q = 0). \tag{20.23}$$

Because translational invariance implies momentum conservation, the two-point function takes the form

$$\langle \tilde{m}(q)\tilde{m}(q')\rangle = \sqrt{(2\pi)^D}\, \delta(q - q')\, F(q), \tag{20.24}$$

where $F(q)$ is the (connected) propagator or correlation function.

Diagrammatically, $F(q)$ receives corrections from the one-loop ("tadpole") and two-loop ("sunset") diagrams shown in Fig. 20.1. The first-order diagram simply renormalizes the mass term (shifts the critical temperature), while higher-order diagrams generate nontrivial momentum dependence and anomalous scaling.

FIGURE 20.1 Leading diagrams contributing to the correlation function: (left) the tree-level propagator; (right) the one-loop correction (tadpole).

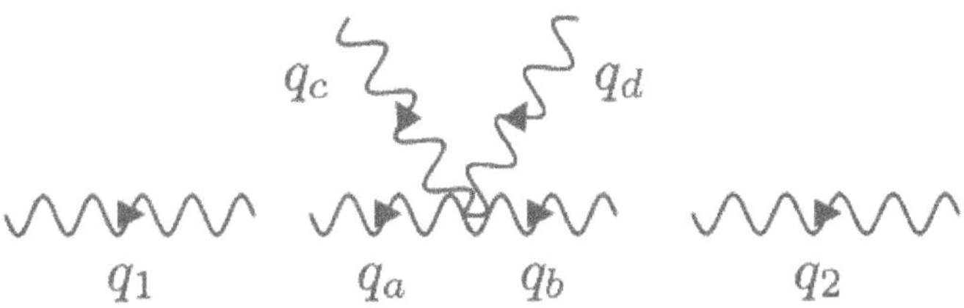

FIGURE 20.2 Parametrization of momenta at the one-loop vertex.

First-Order Correction to the Propagator

We write the full propagator as a perturbative expansion:

$$F(q) = \sum_{k=0}^{\infty} u^k\, F_k(q), \tag{20.25}$$

where $F_0(q)$ is the Gaussian result. At first order in u, the correction is obtained from the connected part of the six-point function:

$$\langle\tilde{m}(q_1)\tilde{m}(q_2)\rangle_{\rm conn} = \langle\tilde{m}(q_1)\tilde{m}(q_2)\rangle_0 \tag{20.26}$$

$$+ 12u\int_q (2\pi)^D\delta(q_a+q_b+q_c+q_d)\,\langle\tilde{m}(q_1)\tilde{m}(q_2)\tilde{m}(q_a)\tilde{m}(q_b)\tilde{m}(q_c)\tilde{m}(q_d)\rangle_{0,\rm conn}. \tag{20.27}$$

The multiplicity factor 12 accounts for the distinct Wick contractions leading to the same topology.

Evaluating the connected term gives the one-loop correction (the tadpole diagram). Carrying out the delta-function integrations yields

$$\langle\tilde{m}(q_1)\tilde{m}(q_2)\rangle_{\rm conn} = 12u\,\delta(q_1-q_2)\int^\Lambda \frac{d^Dq_c}{(2\pi)^D}\frac{1}{4\,[K_0+K_1q_1^2]^2\,[K_0+K_1q_c^2]}, \tag{20.28}$$

where Λ is the ultraviolet cutoff (set by the inverse lattice spacing).

Thus, to first order,

$$F(q) = \frac{1}{2(K_0+K_1q^2)}\left[1+12u\,\frac{1}{2(K_0+K_1q^2)}\int^\Lambda \frac{d^Dq_c}{(2\pi)^D}\frac{1}{2(K_0+K_1q_c^2)}\right]. \tag{20.29}$$

The loop integral is easily evaluated using spherical coordinates:

$$\int^\Lambda d^Dq\,\frac{1}{a+bq^2} = S_D\int_0^\Lambda dq\,\frac{q^{D-1}}{a+bq^2} \simeq \frac{S_D\,\Lambda^{D-2}}{b(D-2)}, \tag{20.30}$$

where $S_D = 2\pi^{D/2}/\Gamma(D/2)$ is the surface area of the unit sphere in D dimensions.

This result shows two key points:

- For $D<4$, the integral diverges with the cutoff Λ; perturbation theory alone breaks down.
- The first-order term shifts the position of the critical point by renormalizing K_0.

Indeed, recall that in mean-field theory the critical point corresponds to $K_0=0$. With the correction included, the susceptibility becomes

$$\begin{aligned}\chi^{-1}(K_0) &= F(q=0)^{-1}\\ &= 2(K_0+K_1q^2)\left[1-12u\,\frac{1}{2(K_0+K_1q^2)}\int^\Lambda \frac{d^Dq_c}{(2\pi)^D}\frac{1}{2(K_0+K_1q_c^2)}\right]^{-1}\\ &\approx 2(K_0+K_1q^2)\left[1+12u\,\frac{1}{2(K_0+K_1q^2)}\int^\Lambda \frac{d^Dq_c}{(2\pi)^D}\frac{1}{2(K_0+K_1q_c^2)}\right].\end{aligned} \tag{20.31}$$

Expanding to first order in u, we obtain

$$\chi^{-1}(K_0) = 2K_0 + 12u\int^\Lambda \frac{d^Dq_c}{(2\pi)^D}\frac{1}{2(K_0+K_1q_c^2)}. \tag{20.32}$$

Hence $\chi^{-1}(K_0)$ no longer vanishes at $K_0=0$; the true critical point is shifted to a negative value $K_0=K_{0c}$ determined by the divergent integral. Physically, the fluctuations lower the critical temperature compared to the mean-field prediction.

REMARK 20.1 The divergence encountered here signals the failure of naive perturbation theory. However, the key observation is that this divergence can be systematically removed by *renormalization* of the parameters—that is, by redefining K_0 and u as scale-dependent ("running") couplings. This leads naturally to the momentum-space renormalization group developed by Wilson, in which high-momentum (short-distance) modes are integrated out step by step, producing finite effective parameters at each scale.

In the next section, we will explicitly implement Wilson's idea by splitting the Fourier modes of the field into "fast" and "slow" components and integrating out the former. This procedure will give rise to the RG flow equations that describe how the couplings evolve with scale and determine the universal critical exponents.

Splitting of Fast and Slow Modes

In Wilson's formulation of the renormalization group, we separate the field into "slow" and "fast" modes according to momentum scale:

$$\tilde{m}(q) = \tilde{m}_<(q) + \tilde{m}_>(q), \tag{20.33}$$

where $\tilde{m}_<(q)$ contains momenta $|\vec{q}| < \Lambda/b$ and $\tilde{m}_>(q)$ contains momenta in the shell $\Lambda/b < |\vec{q}| < \Lambda$. For convenience, we denote $\tilde{m}_>(q) \equiv \tilde{\sigma}(q)$ and $\tilde{m}_<(q) \equiv \tilde{m}(q)$.

The interaction term is not diagonal in this decomposition. Expanding the quartic term, we obtain

$$\left(\int_0^{\Lambda/b} \frac{d^D q}{\sqrt{(2\pi)^D}}\, e^{i\vec{q}\cdot\vec{x}}\, m(q) + \int_{\Lambda/b}^{\Lambda} \frac{d^D q}{\sqrt{(2\pi)^D}}\, e^{i\vec{q}\cdot\vec{x}}\, \sigma(q)\right)^4. \tag{20.34}$$

Let p and k denote momenta of the slow and fast modes, respectively. After expansion, we have

$$\int_{k,p} \Big[\tilde{m}^4(p) + 4\,\tilde{\sigma}^3(k)\,\tilde{m}(p) + 6\,\tilde{\sigma}^2(k)\,\tilde{m}^2(p) + 4\,\tilde{\sigma}(k)\,\tilde{m}^3(p) + \tilde{\sigma}^4(k)\Big]. \tag{20.35}$$

The partition function can now be written as

$$\begin{aligned} Z = \int \mathcal{D}\tilde{m}\,\mathcal{D}\tilde{\sigma}\, \exp\Big[&- \beta H_0[\tilde{m}] - u\int_p \tilde{m}^4(p) - \beta H_0[\tilde{\sigma}] \\ &- u\int_{k,p} \big(4\,\tilde{\sigma}^3(k)\tilde{m}(p) + 6\,\tilde{\sigma}^2(k)\tilde{m}^2(p) + 4\,\tilde{\sigma}(k)\tilde{m}^3(p) + \tilde{\sigma}^4(k)\big)\Big], \end{aligned} \tag{20.36}$$

where H_0 denotes the Gaussian (quadratic) part of the Hamiltonian.

The fast field $\tilde{\sigma}$ enters quadratically in H_0 and can thus be integrated out as a Gaussian. We define its free partition function as

$$Z_0^{\tilde{\sigma}} = \int \mathcal{D}\tilde{\sigma}\, e^{-\beta H_0[\tilde{\sigma}]}. \tag{20.37}$$

Dividing and multiplying by $Z_0^{\tilde{\sigma}}$, we rewrite

$$Z = \int \mathcal{D}\tilde{m}\, e^{-\beta H[\tilde{m}]}\, Z_0^{\tilde{\sigma}} \int \mathcal{D}\tilde{\sigma}\, \frac{\exp\Big[-u\int_{k,p}\big(4\,\tilde{\sigma}^3(k)\tilde{m}(p) + 6\,\tilde{\sigma}^2(k)\tilde{m}^2(p) + 4\,\tilde{\sigma}(k)\tilde{m}^3(p) + \tilde{\sigma}^4(k)\big)\Big]}{Z_0^{\tilde{\sigma}}}. \tag{20.38}$$

We define the Gaussian average over the fast modes as

$$\left\langle e^{-u\tilde{V}}\right\rangle_{\tilde{\sigma}} = \int \mathcal{D}\tilde{\sigma}\, \frac{e^{-u\int_{k,p}\left(4\,\tilde{\sigma}^3(k)\tilde{m}(p)+6\,\tilde{\sigma}^2(k)\tilde{m}^2(p)+4\,\tilde{\sigma}(k)\tilde{m}^3(p)+\tilde{\sigma}^4(k)\right)}}{Z_0^{\tilde{\sigma}}}. \tag{20.39}$$

Then the full partition function becomes

$$Z = Z_0^{\tilde{\sigma}} \int \mathcal{D}\tilde{m}\, \exp\Big[-\beta H[\tilde{m}] + \log\left\langle e^{-u\tilde{V}}\right\rangle_{\tilde{\sigma}}\Big]. \tag{20.40}$$

The logarithm produces precisely the sum of *connected* diagrams generated by averaging over the fast field $\tilde{\sigma}$.

At this stage, the strategy is clear: expand $\log\langle e^{-u\tilde{V}}\rangle_{\tilde{\sigma}}$ in powers of u, keeping only those terms that reproduce the functional form of $\beta H[\tilde{m}]$. These corrections renormalize the coefficients of the quadratic and quartic terms in the effective Hamiltonian for the slow field.

To first order in u, only one diagram contributes.

- Diagram (a) in Fig. 20.3 vanishes because it contains a single power of $\tilde{\sigma}$ (an odd term).
- Diagram (c) also vanishes since it contains three powers of $\tilde{\sigma}$.
- Diagram (d) is nonzero but involves only $\tilde{\sigma}$ and thus does not contribute to $\beta H[\tilde{m}]$.
- Diagram (b) in Fig. 20.3 contributes: it has two powers of $\tilde{m}$ and gives a correction to the quadratic term.

We now note that the propagator for the $\tilde{\sigma}$ field is identical to that of the $\tilde{m}$ field; only the domain of integration differs. The one-loop diagram contributing to the quadratic term of the effective Hamiltonian can be written as

$$-6u\int_0^{\Lambda/b} \frac{d^Dp_1\,d^Dp_2}{\sqrt{(2\pi)^D}} \int_{\Lambda/b}^{\Lambda} d^Dk_1\,(2\pi)^D\delta(k_1+k_2+p_1-p_2)\,\frac{\sqrt{(2\pi)^D}\,\delta(k_1+k_2)}{2(K_0+K_1k_1^2)}\,\tilde{m}(p_1)\tilde{m}(p_2), \tag{20.41}$$

which simplifies to

$$-6u\int_0^{\Lambda/b} \frac{d^Dp_1}{\sqrt{(2\pi)^D}}\,|\tilde{m}(p_1)|^2 \int_{\Lambda/b}^{\Lambda} \frac{d^Dk_1}{2(K_0+K_1k_1^2)}. \tag{20.42}$$

Hence, the effective Hamiltonian for the slow field $\tilde{m}$ becomes

$$\begin{aligned}\beta H[\tilde{m}] = &\int_0^{\Lambda/b} \frac{d^Dp}{\sqrt{(2\pi)^D}}|\tilde{m}(p)|^2\left[K_0 + 6u\int_{\Lambda/b}^{\Lambda}\frac{d^Dk_1}{2(K_0+K_1k_1^2)} + K_1p^2\right]\\ &+ u\int_0^{\Lambda/b}\frac{d^Dp_1\cdots d^Dp_4}{\sqrt{(2\pi)^{4D}}}(2\pi)^D\delta(p_1+p_2+p_3+p_4)\,\tilde{m}(p_1)\tilde{m}(p_2)\tilde{m}(p_3)\tilde{m}(p_4).\end{aligned} \tag{20.43}$$

This Hamiltonian has the same functional form as the original one but with renormalized coefficients. We now rescale momenta and fields:

$$\begin{aligned}\tilde{m} &\to Z_m\,\tilde{m},\\ p &\to b^{-1}p,\end{aligned}$$

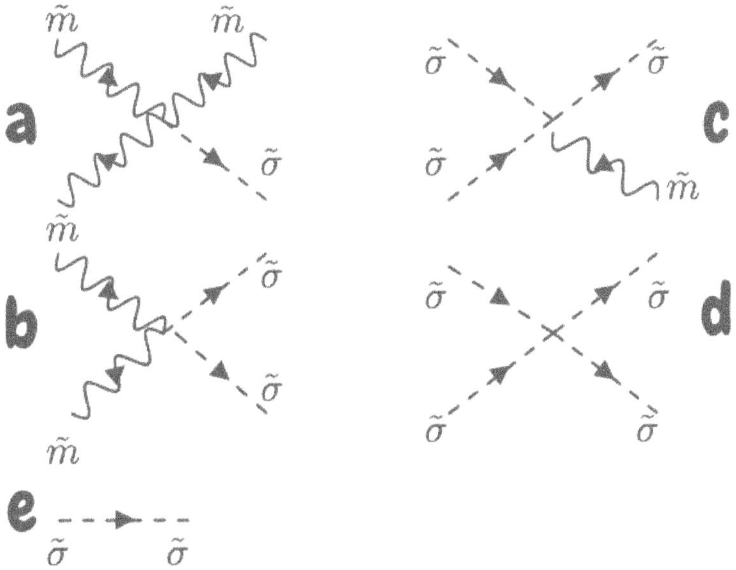

FIGURE 20.3 Top: All interaction terms and the new propagator which appear in the effective model from σ. Bottom: The only diagram that contributes at the first order.

and express the Hamiltonian again with a momentum cutoff Λ:

$$\beta H[\tilde{m}] = \int_0^\Lambda \frac{d^D p}{\sqrt{(2\pi)^D}} |\tilde{m}(p)|^2 \left[b^{-D} Z_m^2 K_0 + 6u\, b^{-D} Z_m^2 \int_{\Lambda/b}^{\Lambda} \frac{d^D k_1}{2(K_0 + K_1 k_1^2)} + K_1 b^{-D-2} p^2 \right]$$
$$+ u\, b^{-3D} Z_m^4 \int_0^\Lambda \frac{d^D p_1 \cdots d^D p_4}{\sqrt{(2\pi)^{4D}}} (2\pi)^D \delta(p_1 + p_2 + p_3 + p_4)\, \tilde{m}(p_1)\tilde{m}(p_2)\tilde{m}(p_3)\tilde{m}(p_4). \tag{20.44}$$

The factor b^{-3D} arises because one power of the momentum integral is removed by the δ-function enforcing momentum conservation at the vertex.

From this expression we identify the renormalized couplings (restoring t for K_0):

$$t' = b^{-D} Z_m^2\, t + 6u\, Z_m^2 b^{-D} \int_{\Lambda/b}^{\Lambda} \frac{d^D k_1}{2(K_0 + K_1 k_1^2)}, \tag{20.45}$$

$$K_1' = Z_m^2 b^{-D-2}\, K_1, \tag{20.46}$$

$$u' = b^{-3D} Z_m^4\, u. \tag{20.47}$$

Imposing the normalization condition $K_1' = K_1 = 1$, we fix

$$Z_m = b^{\frac{D}{2}+1}. \tag{20.48}$$

Substituting this result into Eqs. (20.45)–(20.47), we obtain

$$t' = b^2 \left[t + 6u \int_{\Lambda/b}^{\Lambda} \frac{d^D k_1}{2(K_0 + K_1 k_1^2)} \right], \tag{20.49}$$

$$u' = b^{4-D} u. \tag{20.50}$$

The integral in Eq. (20.49) behaves as a constant times Λ^{D-2}, so we may write

$$t' = b^2 \left[t + 6u\, C\, \Lambda^{D-2} \right], \tag{20.51}$$

$$u' = b^{4-D} u, \tag{20.52}$$

where C is a dimension-dependent constant proportional to the solid angle S_D.

For infinitesimal transformations $b = 1 + \tau$, we have $d \log b = d\tau$, and the flow equations become

$$\frac{dt}{d\tau} = 2 \left[t + 6u\, C\, \Lambda^{D-2} \right], \tag{20.53}$$

$$\frac{du}{d\tau} = (4 - D)\, u. \tag{20.54}$$

At next order in u, the renormalization of the quartic term introduces corrections proportional to u^2, giving

$$\frac{dt}{d\tau} = 2 \left[t + 6u\, C\, \Lambda^{D-2} - 6\, C\, \Lambda^{D-4} u\, t \right], \tag{20.55}$$

$$\frac{du}{d\tau} = (4 - D) u - 36\, C\, \Lambda^{D-4} u^2. \tag{20.56}$$

To express these equations in dimensionless form, we define the scaled quantities

$$\tilde{\Lambda}^2 = C\, \Lambda^2,$$

$$\tilde{u} = \frac{u}{\tilde{\Lambda}^{4-D}},$$

$$\tilde{t} = \frac{t}{\tilde{\Lambda}^2},$$

and introduce $\epsilon = 4 - D$. Equations (20.55)–(20.56) then become

$$\frac{d\tilde{t}}{d\tau} = 2\tilde{t} - 12\, \tilde{t}\tilde{u} + 12\, \tilde{u}, \tag{20.57}$$

$$\frac{d\tilde{u}}{d\tau} = \epsilon\, \tilde{u} - 36\, \tilde{u}^2. \tag{20.58}$$

The fixed points of these equations are readily obtained:

$$\tilde{t}^* = 0, \qquad \tilde{u}^* = 0, \quad \text{(Gaussian fixed point, mean-field behavior)}, \tag{20.59}$$

$$\tilde{t}^* = -\frac{\epsilon}{6 + \epsilon}, \qquad \tilde{u}^* = \frac{\epsilon}{36}, \quad \text{(nontrivial Wilson–Fisher fixed point)}. \tag{20.60}$$

As $\epsilon \to 0$ (i.e., $D \to 4$), the nontrivial fixed point merges with the Gaussian one, confirming that mean-field theory is exact above the upper critical dimension.

Fixed Points and Linearization

A few comments are in order:

- The nontrivial fixed point $\tilde{u}^* = \epsilon/36$ corresponds to a negative $\tilde{t}^*$.
- The two fixed points (Gaussian and Wilson–Fisher) are close to each other when $\epsilon = 4 - D$ is small; hence, ϵ is the natural expansion parameter.
- The critical exponents correspond to the eigenvalues of the linearized RG flow at the fixed points.

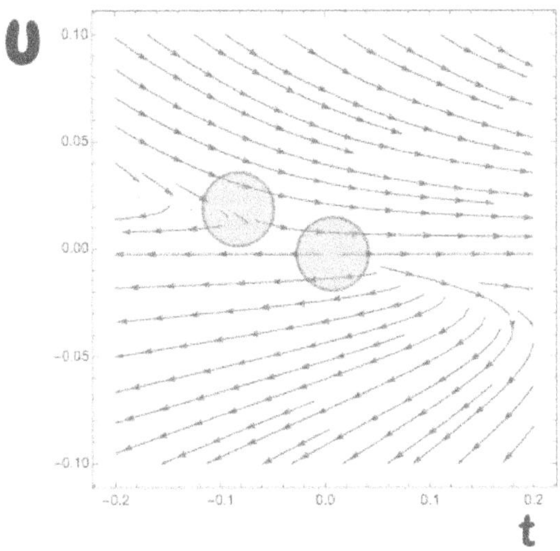

FIGURE 20.4 Flow of the RG equations showing the Gaussian and Wilson–Fisher fixed points.

To determine the critical exponents, we linearize the RG flow equations near a fixed point. Close to $(\tilde{t}^*, \tilde{u}^*)$, we define small perturbations $\delta\tilde{t} = \tilde{t} - \tilde{t}^*$ and $\delta\tilde{u} = \tilde{u} - \tilde{u}^*$, and write

$$\frac{d}{d\tau}\begin{pmatrix}\delta\tilde{t}\\ \delta\tilde{u}\end{pmatrix} = T\begin{pmatrix}\delta\tilde{t}\\ \delta\tilde{u}\end{pmatrix}, \tag{20.61}$$

where the stability matrix T is

$$T = \begin{pmatrix}2(1-6y^*) & 12(1-x^*)\\ 0 & \epsilon - 72y^*\end{pmatrix}. \tag{20.62}$$

The eigenvalues and eigenvectors of T determine the scaling directions and the corresponding exponents.

For the *Gaussian fixed point* $(\tilde{t}^*, \tilde{u}^*) = (0, 0)$:

$$\begin{aligned} \vec{v}_{1,g} &= \delta\tilde{t} + 6\,\delta\tilde{u}, & \lambda_1 &= 2,\\ \vec{v}_{2,g} &= \delta\tilde{t} + 6\,\delta\tilde{u}, & \lambda_2 &= \epsilon. \end{aligned} \tag{20.63}$$

For the *Wilson–Fisher fixed point* $(\tilde{t}^*, \tilde{u}^*) = (-\epsilon/(6+\epsilon), \epsilon/36)$:

$$\begin{aligned} \vec{v}_{1,\mathrm{WF}} &= \delta\tilde{t} + 6\,\delta\tilde{u}, & \lambda_1 &= 2 - \frac{\epsilon}{3},\\ \vec{v}_{2,\mathrm{WF}} &= \delta\tilde{t} + (6-\epsilon)\,\delta\tilde{u}, & \lambda_2 &= -\epsilon. \end{aligned} \tag{20.64}$$

The relevant direction corresponds to the positive eigenvalue λ_1, and the irrelevant one to $\lambda_2 < 0$. The correlation-length critical exponent ν is the inverse of the relevant eigenvalue:

$$\nu = \frac{1}{D_t} = \frac{1}{2 - \frac{\epsilon}{3}} \approx 0.599. \tag{20.65}$$

For the three-dimensional Ising model ($D = 3$, $\epsilon = 1$), this gives $\nu \approx 0.63$, in excellent agreement with experimental and numerical results: $\nu_{\text{Ising}} \approx 0.6299$ for magnets and $\nu_{\text{LG}} \approx 0.62$ for the liquid–gas critical point. The ϵ-expansion thus improves significantly upon mean-field predictions.

Critical Surfaces and Relevant Directions

The geometry of the RG flow provides an intuitive picture of criticality. Consider the set of all points in parameter space that flow into a given fixed point under RG transformations. This set is known as the *basin of attraction* or *critical surface* of the fixed point. All points on the critical surface share the same large-scale (universal) physics and correspond to critical states.

For the Ising universality class, there are two relevant couplings: the reduced temperature t and the external field h. Thus, the renormalization-group flow is characterized by two relevant directions; all other directions are irrelevant. If we set $h = 0$, only one relevant direction remains, corresponding to temperature deviations from T_c. In a continuous RG framework, irrelevant couplings correspond to negative eigenvalues of the stability matrix at the fixed point.

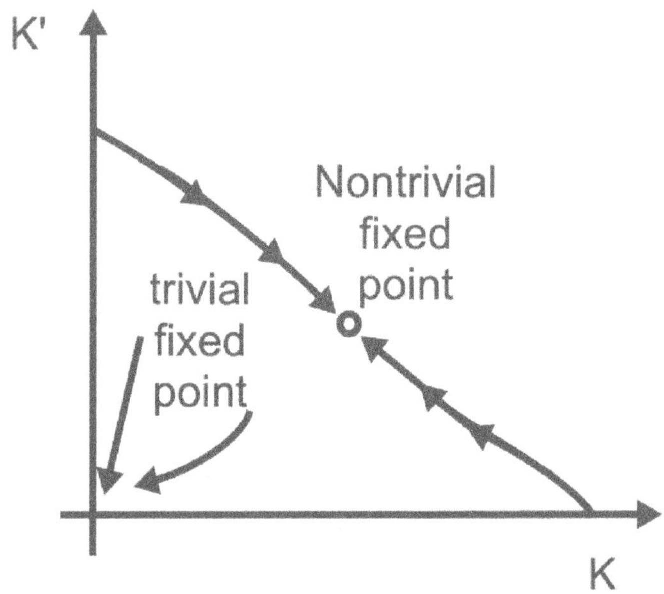

FIGURE 20.5 Schematic RG flow near a fixed point. For $h = 0$, the physical fixed point has a single relevant direction. Its basin of attraction (the critical surface) has co-dimension one, meaning that a small change in t moves the system across the phase transition.

Because the critical surface has dimension $N - 1$ (or co-dimension 1) in an N-dimensional coupling space, trajectories starting at large scales ($K_i \to 0$) will necessarily cross it as one tunes the control parameter, such as temperature. Physically, this means that as we raise the temperature from $T = 0$, the system must pass through T_c and undergo a phase transition. This interpretation connects the geometry of RG flows to the physical structure of phase diagrams.

Experimental Realizations of Universality Classes

A powerful validation of universality comes from experiments across fluids, magnets, superconductors, and non-equilibrium systems. Here we highlight representative experimental platforms for each universality class discussed above.

The liquid–vapor transition of simple fluids exhibits critical exponents consistent with the three-dimensional Ising universality class, beautifully confirmed in light-scattering experiments on xenon by Michael Moldover *et al.* [MSM76]. Binary liquid mixtures—such as methanol–cyclohexane—show identical scaling near demixing transitions, matching the same fixed point. Uniaxial ferromagnets such as $CoCl_2 \cdot 2H_2O$ and $FeCl_2$ also display three-dimensional Ising critical exponents as measured in neutron scattering.

Thin films of ^{4}He provide the clearest experimental realization of the Berezinskii–Kosterlitz–Thouless (BKT) transition: the universal jump in the superfluid density was observed in the classic experiments of David Bishop and John Reppy [BR78]. Ultrathin superconducting films (Al, Nb) display vortex–antivortex unbinding and BKT-like resistive transitions. Cold atomic gases confined to two-dimensional traps have directly imaged the proliferation of vortices at the transition.

Isotropic three-dimensional ferromagnets such as Ni, EuO, and $RbMnF_3$ exhibit the critical exponents predicted for the three-dimensional Heisenberg universality class. Precision neutron-scattering experiments on $RbMnF_3$ by Robert Birgeneau, H. J. Guggenheim, and coworkers [BG70] demonstrated the expected power-law correlations near T_c.

Liquid crystals near the nematic–isotropic transition behave as systems with vector order parameters and are well described by $O(3)$-like fluctuations. Dynamic light-scattering and birefringence measurements have mapped these critical exponents, in agreement with the theoretical $O(n)$ classification.

The q-state Potts model is realized experimentally in multicomponent alloys where different atomic species correspond to Potts states. Random resistor networks and gelation processes provide accurate realizations of percolation criticality. Percolation exponents were strikingly verified in the classic experiments on carbon–metal mixtures by de Gennes and coworkers [dGP93].

Beyond helium films, the BKT universality class appears in Josephson junction arrays and superconducting wire networks: experiments by David Resnick *et al.* [RGB$^+$81] measured the characteristic exponential divergence of the correlation length as predicted by KT theory.

Instead, experiments on turbulence in liquid crystals (the DSM1 $\to$ DSM2 transition) by Kazumasa Takeuchi and Masaki Sano [TS10] provided the first quantitative experimental confirmation of directed percolation critical exponents in a natural system.

Quantum phase transitions, which we discuss in Volume II, have been observed in many materials. A prototypical example is the quantum paramagnet-to-antiferromagnet transition in $TlCuCl_3$, where neutron scattering revealed scaling in $(d+z)$ dimensions consistent with quantum criticality [RNM+08]. Cold atoms in optical lattices also realize quantum critical points, such as the superfluid–Mott insulator transition.

These experiments, spanning over fifty years and many different materials, establish the empirical foundation of universality: systems with distinct microscopic physics nevertheless share the same critical behavior when governed by the same symmetry and dimensionality.

20.5 Kadanoff's Picture and the Physical Meaning of the Renormalization Group

Having developed the formalism of the Wilsonian renormalization group, it is useful to return to the original and remarkably intuitive picture proposed by Leo Kadanoff in 1966 [Kad66]. Kadanoff's key insight was that the complex microscopic degrees of freedom in a many-body system could be replaced by effective, coarse-grained "block" variables without changing the large-scale physics. This conceptual step transformed the understanding of critical phenomena: the details of atomic interactions become irrelevant at long wavelengths, and the system's behavior is governed by scale invariance.

Kadanoff's argument begins with the observation that near a critical point the correlation length ξ diverges. Microscopic details at scales much smaller than ξ cannot affect the macroscopic behavior, since the system "looks the same" on scales smaller than ξ. One can therefore imagine grouping spins into blocks of linear size b, each containing roughly b^D microscopic spins. Each block spin represents the average magnetization of its block. Interactions between blocks define a new Hamiltonian with renormalized couplings $(t', h', u', \ldots)$ that depend on b. If the system is at criticality, it should be invariant under this transformation: the block-spin rescaling should reproduce the same form of the Hamiltonian, differing only by the scale of observation.

Formally, the Kadanoff transformation acts as a mapping

$$(t, h, u, \ldots) \longrightarrow (t', h', u', \ldots),$$

where t is the reduced temperature, h the external field, and u the interaction strength. The couplings evolve under successive coarse-graining steps, defining trajectories in the space of all possible Hamiltonians. Fixed points of this transformation correspond to scale-invariant theories, and the directions of the flow near those fixed points determine the relevant, marginal, and irrelevant operators.

This heuristic picture already contains the essential ingredients of Wilson's renormalization group:

- *Mode elimination:* averaging over short-distance (high-momentum) degrees of freedom, or equivalently, integrating out fast modes of the field.
- *Rescaling:* restoring the system to its original size by redefining lengths and fields.
- *Flow in coupling space:* tracking how the effective parameters change with scale.

Wilson's contribution was to formalize this intuition using the language of field theory and to derive quantitative flow equations for the couplings, such as

$$\frac{d\tilde{t}}{d\tau} = 2\tilde{t} - 12\tilde{t}\tilde{u} + 12\tilde{u}, \qquad \frac{d\tilde{u}}{d\tau} = \epsilon\tilde{u} - 36\tilde{u}^2,$$

whose fixed points describe the universal behavior of systems near criticality. In this way, the renormalization group bridges the conceptual picture of Kadanoff's real-space scaling and the computational power of Wilson's momentum-space formalism.

Kadanoff and Wilson: Two Views of the Same Idea

- *Kadanoff's view (1966):* a conceptual, real-space argument based on block spins and scale invariance; emphasizes the redundancy of microscopic details and the origin of universality.
- *Wilson's view (1971–1974):* a quantitative, field-theoretic implementation us-

ing coarse-graining in momentum space; provides explicit flow equations, fixed points, and critical exponents.

Together, they constitute one of the most profound unifications in theoretical physics: microscopic models, regardless of their details, can be classified into a small number of *universality classes*, each governed by a scale-invariant fixed point of the renormalization group.

Thus, the renormalization group can be understood as both a *flow* in the space of Hamiltonians and as a *physical process* of coarse-graining and rescaling that captures the essence of critical phenomena. It embodies the principle that physics at different length scales is connected, and that the universal behavior near phase transitions is a direct consequence of this scale invariance.

One of the most powerful modern interpretations of the renormalization group is its *information-theoretic and variational formulation.* In this view, the RG can be regarded as a systematic reduction of variables that minimizes redundant information while preserving the essential features governing macroscopic behavior. This perspective emphasizes that coarse-graining corresponds to finding an *optimal encoding* of microscopic configurations that retains maximal information about long-wavelength observables, while minimizing the free energy functional.

This interpretation makes explicit the connection between the renormalization group and *information theory*: coarse-graining can be seen as minimizing the information content (or relative entropy) between the microscopic and macroscopic descriptions. Each RG step produces an effective Hamiltonian with fewer degrees of freedom but equivalent thermodynamic predictions for the long-wavelength observables.

References

1. Andrei A. Abrikosov. On the magnetic properties of superconductors of the second type. *Sov. Phys. JETP*, 5:1174–1182, 1957. Zh. Eksp. Teor. Fiz. 32, 1442–1452 (1957).
2. George Arnold and Frank Cahill. *The Magician's Own Book, or, the Whole Art of Conjuring.* Dick & Fitzgerald, 1862. Reprinted by Forgotten Books.
3. Michael H. Anderson, Jason R. Ensher, Mark R. Matthews, Carl Edwin Wieman, and Eric Allen Cornell. Observation of bose-einstein condensation in a dilute atomic vapor. *Science*, 269(5221):198–201, July 1995.
4. Alexei A. Abrikosov, Vitaly L. Ginzburg, and Anthony J. Leggett. *Nobel Lectures in Physics 2003.* World Scientific, 2004.
5. A. V. Aho, J. E. Hopcroft, and J. D. Ullman. *The Design and Analysis of Computer Algorithms.* Addison-Wesley, Reading, MA, 1974.
6. Jason Alicea. New directions in the pursuit of majorana fermions in solid state systems. *Reports on Progress in Physics*, 75:076501, 2012.
7. Neil W. Ashcroft and Nathaniel D. Mermin. *Solid State Physics.* Holt, Rinehart and Winston, New York, 1976. Section 2.4.
8. Hans C. Andersen. Molecular dynamics simulations at constant pressure and/or temperature. *The Journal of Chemical Physics*, 72(4):2384–2393, 1980.
9. Vladimir I. Arnold. *Mathematical Methods of Classical Mechanics*, volume 60 of *Graduate Texts in Mathematics.* Springer, New York, 2nd edition, 1989.
10. Daniel Arovas, John R. Schrieffer, and Frank Wilczek. Fractional statistics and the quantum hall effect. *Physical Review Letters*, 53:722–723, 1984.
11. Michael P. Allen and Dominic J. Tildesley. *Computer Simulation of Liquids.* Oxford University Press, Oxford, 2nd edition, 2017.
12. Peter Atkins. *The Laws of Thermodynamics: A Very Short Introduction.* Oxford University Press, Oxford, 2010. A concise and accessible overview of the fundamental laws of thermodynamics.
13. Olivier Avenel and Elie Varoquaux. Observation of singly quantized dissipation events obeying the Josephson frequency relation in the critical flow of superfluid ^{4}He through an aperture. *Physical Review Letters*, 55(24):2704–2707, 1985.
14. Berni J. Alder and Thomas E. Wainwright. Phase transition for a hard sphere system. *The Journal of Chemical Physics*, 27(5):1208–1209, 1957.
15. Roger Balian. Francois Massieu and the thermodynamic potentials. *Comptes Rendus Physique*, 18(9–10):520–528, 2017. A modern analysis of Massieu's contributions and their relation to modern thermodynamics.
16. Rodney J. Baxter. *Exactly Solved Models in Statistical Mechanics.* Academic Press, London, 1982.
17. Ludovic Berthier and Giulio Biroli. Theoretical perspective on the glass transition and amorphous materials. *Reviews of Modern Physics*, 83(2):587–645, 2011.
18. Valentin Blickle and Clemens Bechinger. Realization of a micrometre-sized stochastic heat engine. *Nature Physics*, 8:143–146, 2012.
19. John Bardeen, Leon N. Cooper, and John Robert Schrieffer. Theory of superconductivity. *Physical Review*, 108:1175–1204, 1957.
20. James J. Binney, Nigel J. Dowrick, Andrew J. Fisher, and Mark E. J. Newman. *The Theory of Critical Phenomena: An Introduction to the Renormalization Group.* Oxford University Press, 1992.
21. Ehrhard Behrends. *Introduction to Markov Chains.* Vieweg+Teubner Verlag, 2000.

22. Jacob D. Bekenstein. Black holes and entropy. *Physical Review D*, 7:2333–2346, 1973.
23. Hans A. Bethe. Statistical theory of superlattices. *Proceedings of the Royal Society A*, 150:552–575, 1935.
24. Max Born and Herbert S. Green. A general kinetic theory of liquids. *Proceedings of the Royal Society A*, 188(1012):10–18, 1946.
25. Robert J. Birgeneau and H. J. Guggenheim. Critical behavior of mnf2. *Physical Review B*, 1:2211–2230, 1970.
26. George David Birkhoff. Proof of the ergodic theorem. *Proceedings of the National Academy of Sciences of the USA*, 17(12):656–660, 1931.
27. Theodor H. Berlin and Mark Kac. The spherical model of a ferromagnet. *Physical Review*, 86(6):821–835, 1952.
28. Etienne P. Bernard and Werner Krauth. Two-step melting in two dimensions: First-order liquid–hexatic transition. *Physical Review Letters*, 107(15):155704, 2011.
29. Theodore L. Brown, H. Eugene LeMay, Bruce E. Bursten, Catherine Murphy, Patrick Woodward, and Matthew E. Stoltzfus. *Chemistry: The Central Science*. Pearson, Upper Saddle River, NJ, 14th edition, 2018. Comprehensive general chemistry text; see Chapters 5 and 19 on thermochemistry and Gibbs free energy.
30. Stephen Blundell. *Magnetism in Condensed Matter*. Oxford University Press, Oxford, 2001.
31. Cedric Bernardin and Stefano Olla. Fourier's law for a microscopic model of heat conduction. *Journal of Statistical Physics*, 121(1–2):271–289, 2005.
32. Herman Boerhaave. *Elementa Chemiae*. Apud Isaacum Severinum, Lugduni Batavorum, Leiden, 1732. Available online at `https://wellcomecollection.org/works/jums6t8q`.
33. Nikolay N. Bogoliubov. Kinetic equations. *Journal of Physics USSR*, 10:265–274, 1946.
34. Nikolay N. Bogoliubov. A variational principle in the theory of the superfluid state. *Soviet Physics Doklady*, 1:23–26, 1956.
35. Niels Bohr. Studies on the electron theory of metals. In L. Rosenfeld and J. Rud Nielsen, editors, *Early Works (1905–1911), Niels Bohr Collected Works, Vol. 1*, pages 163–393. Elsevier, 1972. Original 1911 doctoral dissertation.
36. Ludwig Boltzmann. Weitere studien ber das wrmegleichgewicht unter gasmoleklen. *Sitzungsberichte der Kaiserlichen Akademie der Wissenschaften*, 66:275–370, 1872.
37. Ludwig Boltzmann. On the relation between the second law of thermodynamics and probability calculations. *Sitzungsberichte der Kaiserlichen Akademie der Wissenschaften*, 52:171–190, 1895. Originally: *Über die Beziehung zwischen dem zweiten Hauptsatze der mechanischen Wärmetheorie und der Wahrscheinlichkeitsrechnung*.
38. Ludwig Boltzmann. *Vorlesungen über Gastheorie*. J.A. Barth, Leipzig, 1896.
39. Ludwig Boltzmann. *Lectures on Gas Theory*. Dover Publications, New York, 1964. English translation of *Vorlesungen über Gastheorie* (1896–1898).
40. Bela Bollobas. *Modern Graph Theory*, volume 184 of *Graduate Texts in Mathematics*. Springer, New York, 1998.
41. Max Born. Zur quantenmechanik der stovorgnge. *Zeitschrift für Physik*, 37:863–867, 1926.
42. Rufus Bowen. Markov partitions for axiom a diffeomorphisms. *American Journal of Mathematics*, 92(3):725–747, 1970.
43. Robert Boyle. *A Defence of the Doctrine Touching the Spring and Weight of the Air, and Its Effects*. Thomas Robinson, Oxford, 1662. Includes the formulation of Boyle's law, describing the inverse relationship between the pressure and volume of a gas at constant temperature.

44. Richard M. Bozorth. *Ferromagnetism.* IEEE Press, New York, 1993. Originally published by D. Van Nostrand, 1951.
45. Alexander A. Belavin and Alexander M. Polyakov. Metastable states of two-dimensional isotropic ferromagnets. *JETP Lett.*, 22:245–248, 1975.
46. Heinz-Peter Breuer and Francesco Petruccione. *The Theory of Open Quantum Systems.* Oxford University Press, Oxford, 2002.
47. Herman J. C. Berendsen, Jan P. M. Postma, Wilfred F. van Gunsteren, Antonio Di Nola, and Johan R. Haak. Molecular dynamics with coupling to an external bath. *The Journal of Chemical Physics*, 81(8):3684–3690, 1984.
48. David J. Bishop and John D. Reppy. Study of the superfluid transition in two-dimensional helium films. *Physical Review Letters*, 40:1727–1730, 1978.
49. Percy W. Bridgman. The electron theory of metals in the light of new experimental data. *Physical Review*, 19(2):114–134, February 1922.
50. James H. Brinkworth. On the temperature variation of the specific heats of hydrogen and nitrogen. *Proceedings of the Royal Society of London. Series A, Containing Papers of a Mathematical and Physical Character*, 126(801):204–212, January 1930.
51. Stephen G. Brush. *The Kind of Motion We Call Heat.* North-Holland, Amsterdam, 1976.
52. Juergen Berges, Nikolaos Tetradis, and Christof Wetterich. Non-perturbative renormalization flow in quantum field theory and statistical physics. *Physics Reports*, 363:223–386, 2002.
53. William L. Bragg and Evan J. Williams. The effect of thermal agitation on atomic arrangement in alloys. *Proceedings of the Royal Society A*, 145:699–730, 1934.
54. William L. Bragg and Evan J. Williams. The effect of thermal agitation on atomic arrangement in alloys. ii. *Proceedings of the Royal Society A*, 151:540–566, 1935.
55. Robert W. Cahn. *The Coming of Materials Science.* Pergamon Press, 2001.
56. Herbert B. Callen. *Thermodynamics: An Introduction to the Physical Theories of Equilibrium Thermostatics and Irreversible Thermodynamics.* John Wiley & Sons, New York, 1960. Original edition introducing Callen's postulates and the modern axiomatic formulation of thermodynamics.
57. Herbert B. Callen. A note on the statistical mechanics of ferromagnetism. *Physics Letters*, 4(6):161–162, 1963.
58. Curtis G. Callan. Broken scale invariance in scalar field theory. *Physical Review D*, 2:1541–1547, 1970.
59. Herbert B. Callen. *Thermodynamics and an Introduction to Thermostatistics.* John Wiley & Sons, New York, 2nd edition, 1985.
60. Predrag Cvitanović, Roberto Artuso, Ronnie Mainieri, Gnter Tanner, and Gregor Vattay. *Chaos: Classical and Quantum.* Niels Bohr Institute, 2016. Open-access online monograph: ChaosBook.org.
61. Michele Campisi. *Lectures on the Mechanical Foundations of Thermodynamics.* SpringerBriefs in Physics. Springer, 2024. Chapter 1 discusses the heat theorem and the existence of temperature as an integrating factor.
62. Kenneth L. Caneva. *Robert Mayer and the Conservation of Energy.* Princeton University Press, Princeton, NJ, 1974. Comprehensive account of Mayer's contribution to the principle of energy conservation.
63. Constantin Carathéodory. Untersuchungen ber die grundlagen der thermodynamik. *Mathematische Annalen*, 67:355–386, 1909. English translation by D. H. Delphenich (2009), available at `http://neo-classical-physics.info/uploads/3/0/6/5/3065888/caratheodory_-_thermodynamics.pdf`.
64. Francesco Caravelli. *Spectral Methods in Complex Systems.* Independently published, USA, 2025. Also available at arXiv:2509.05793.

65. Sydney Chapman and Thomas G. Cowling. *The Mathematical Theory of Non-Uniform Gases.* Cambridge University Press, Cambridge, 3rd edition, 1970.
66. Carlo Cercignani. *Ludwig Boltzmann: The Man Who Trusted Atoms.* Oxford University Press, Oxford, 1998.
67. Subrahmanyan Chandrasekhar. *An Introduction to the Study of Stellar Structure.* The University of Chicago Press, Chicago, 1939.
68. Steven Chu, Leo Hollberg, John E. Bjorkholm, Alex Cable, and Arthur Ashkin. Three-dimensional viscous confinement and cooling of atoms by resonance radiation pressure. *Physical Review Letters*, 55(1):48–51, July 1985.
69. Carlo Cercignani and Gilberto M. Kremer. *The Relativistic Boltzmann Equation: Theory and Applications.* Birkhuser, Basel, 2002.
70. Paul M. Chaikin and Tom C. Lubensky. *Principles of Condensed Matter Physics.* Cambridge University Press, Cambridge, 1995.
71. Rudolf Clausius. On different forms of the fundamental equations of the mechanical theory of heat and their convenient application. *Annalen der Physik*, 125:353–400, 1865. Original statement of the heat theorem and introduction of entropy.
72. Alfred Clebsch. Ber die bildung der acht-undzwanzig doppeltangenten des allgemeinen curve dritten grades. *Journal für die reine und angewandte Mathematik*, 65:359–376, 1866.
73. Claudio Castelnovo, Roderich Moessner, and Shivaji L. Sondhi. Spin ice, fractionalization, and topological order. *Reviews of Modern Physics*, 85:253–298, 2012.
74. Daniel Collin, Fernando Ritort, Christopher Jarzynski, Steven B. Smith, Ignacio Tinoco, and Carlos Bustamante. Verification of the crooks fluctuation theorem and recovery of RNA folding free energies. *Nature*, 437:231–234, 2005.
75. Gavin E. Crooks. Entropy production fluctuation theorem and the nonequilibrium work relation for free energy differences. *Physical Review E*, 60(3):2721–2726, 1999.
76. Radu Coldea and David A. Tennant and Elena M. Wheeler and et al., Quantum Criticality in an Ising Chain: Experimental Evidence for Emergent E8 Symmetry, *Science*, 327:177–180, 2010.
77. Pierre Curie. Proprits magntiques des corps diffrentes tempratures. *Annales de Chimie et de Physique*, 5:289–405, 1895.
78. Francesco Caravelli and Marc D. Vuffray. Boundary-induced classical generalized Gibbs ensemble with angular momentum. *arXiv preprint*, arXiv:2407.08833, 2024.
79. John Dalton. *A New System of Chemical Philosophy.* R. Bickerstaff, Manchester, 1808. First edition; introduced the atomic theory and early atomic weight measurements.
80. Thierry Dauxois. Fermi, Pasta, Ulam, and a mysterious lady. *Physics Today*, 61(1):55–57, 2008.
81. Peter Debye. Zur theorie der spezifischen wrmen. *Annalen der Physik*, 39:789–839, 1912. English translation in *The Old Quantum Theory*, ed. D. ter Haar, Pergamon Press, 1967.
82. Peter Debye. Zur theorie der spezifischen wrmen. *Annalen der Physik*, 386:1154–1160, 1926.
83. Julius M. Deutsch. Quantum statistical mechanics in a closed system. *Physical Review A*, 43(4):2046–2049, 1991.
84. Pierre-Gilles de Gennes and Jacques Prost. *The Physics of Liquid Crystals.* Oxford University Press, 2nd edition, 1993.
85. Sybrandus R. de Groot, Willem A. van Leeuwen, and Cornelis G. van Weert. *Relativistic Kinetic Theory: Principles and Applications.* North-Holland, Amsterdam, 1980.
86. Paul A. M. Dirac. Quantum mechanics of many-electron systems. *Proceedings of the Royal Society A*, 123(792):714–733, 1929.

87. Pierre-Louis Dulong and Alexis-Thérèse Petit. Recherches sur quelques points importants de la thorie de la chaleur. *Annales de Chimie et de Physique*, 10:395–413, 1819.
88. Théophile de Donder and Paul Van Rysselberghe. *Affinity*. Stanford University Press, Stanford, 1936. Introduces the concept of chemical affinity and the degree of advancement of reaction.
89. Igor M. Dubrovskii. The role of angular momentum conservation law in statistical mechanics. *Condensed Matter Physics*, 11(4):585–596, 2008.
90. Igor M. Dubrovskyi. Statistical mechanics that takes into account angular momentum conservation law. In *Thermodynamics—Interaction Studies—Solids, Liquids and Gases*. InTech, 2011.
91. Igor E. Dzyaloshinskii. A thermodynamic theory of "weak" ferromagnetism of antiferromagnetics. *Journal of Physics and Chemistry of Solids*, 4(4):241–255, 1958.
92. Charles A. Eckert. Molecular thermodynamics of chemical reactions. *Industrial and Engineering Chemistry*, 59(9):20–32, 1967. Discussion of molecular-level thermodynamics of chemical reactions.
93. Alan R. Edmonds. *Angular Momentum in Quantum Mechanics*. Princeton University Press, Princeton, 1957.
94. Albert Einstein. Zur theorie der opaleszenz von homogenen flssigkeiten und flssigkeitsgemischen in der nhe des kritischen zustandes. *Annalen der Physik*, 33:1275–1298, 1910. Introduces the entropy-based theory of thermodynamic fluctuations and the Einstein fluctuation formula.
95. Harold J. T. Ellingham. Reducibility of oxides and sulphides in metallurgical processes. *Journal of the Society of Chemical Industry (London)*, 63:125–160, 1944. Introduced the Ellingham diagram, a graphical method for representing Gibbs free energy of formation as a function of temperature.
96. Kerry A. Emanuel. An air–sea interaction theory for tropical cyclones, part i. *Journal of the Atmospheric Sciences*, 42:1062–1071, 1986. Introduced the thermodynamic theory of hurricanes as heat engines.
97. Kerry A. Emanuel. Tropical cyclones. *Annual Review of Earth and Planetary Sciences*, 31:75–104, 2003. Comprehensive review of tropical cyclone thermodynamics and dynamics.
98. Kerry A. Emanuel. Hurricanes: Tempests in a greenhouse. *Physics Today*, 59(8):74–75, 2006. Discusses hurricanes in the context of climate change and thermodynamic limits.
99. Robert Emden. *Gaskugeln: Anwendungen der Mechanischen Wärmetheorie auf Kosmologische und Meteorologische Probleme*. B. G. Teubner, Leipzig, 1907. Applied polytropic processes to the thermodynamic structure of stars, leading to the Lane–Emden equation.
100. Arnold Eucken. Die spezifischen wrmen der gase als funktion der temperatur. *Annalen der Physik*, 343(4):185–221, 1912.
101. Enrico Fermi. *Thermodynamics*. Prentice-Hall, New York, 1936. Reprinted by Dover Publications, New York, 1956. A concise and clear introduction to classical thermodynamics.
102. Alan Fersht. *Structure and Mechanism in Protein Science*. W. H. Freeman, New York, 1999.
103. Ralph Fowler and Edward A. Guggenheim. Statistical thermodynamics. In *Statistical Thermodynamics*. Cambridge University Press, London, 1939.
104. Paul M. Fishbane, Stephen G. Gasiorowicz, and Stephen T. Thornton. *Physics for Scientists and Engineers*. Prentice Hall, Upper Saddle River, NJ, 3rd edition, 2005.

105. Michael E. Fisher. The theory of critical point singularities. *Reports on Progress in Physics*, 30:615–730, 1967.
106. Gottfried Falk and Henning Jung. Axiomatic formulation of classical thermodynamics. *Zeitschrift für Physik*, 157(1):141–162, 1959.
107. Cornelis M. Fortuin and Pieter W. Kasteleyn. On the random-cluster model. i. introduction and relation to other models. *Physica*, 57(4):536–564, 1972.
108. Reinhard Folk. The survival of Ernst Ising and the struggle to solve his model. *arXiv preprint arXiv:2203.13849*, 2022.
109. Enrico Fermi, John Pasta, and Stanislaw Ulam. Studies of nonlinear problems. *Los Alamos Scientific Laboratory Report LA-1940*, 1955. Reprinted in *Collected Papers of Enrico Fermi*, University of Chicago Press, 1965.
110. Enrico Fermi, John Pasta, Stanislaw Ulam, and Mary Tsingou. Studies of nonlinear problems. Technical Report LA-1940, Los Alamos Scientific Laboratory, 1955. Reprinted in *Collected Papers of Enrico Fermi*, Vol. II, University of Chicago Press (1965).
111. Fabio Franchini. The tonks-girardeau gas as a luttinger liquid. *Physics Reports*, 524(3):1–91, 2013.
112. Daan Frenkel and Berend Smit. *Understanding Molecular Simulation: From Algorithms to Applications.* Academic Press, San Diego, 2nd edition, 2002.
113. Giovanni Gentile. Osservazioni sopra le statistiche intermedie. *Il Nuovo Cimento*, 17:493–497, 1940.
114. Iulia Georgescu. 25 years of bec. *Nature Reviews Physics*, 2(8):396–396, July 2020.
115. Josiah W. Gibbs. A method of geometrical representation of the thermodynamic properties of substances by means of surfaces. *Transactions of the Connecticut Academy of Arts and Sciences*, 2:382–404, 1873. Introduces the concept of free energy and geometrical thermodynamics.
116. Josiah W. Gibbs. On the equilibrium of heterogeneous substances. *Transactions of the Connecticut Academy of Arts and Sciences*, 3:108–248, 343–524, 1876. Part I (1876), Part II (1878); foundational work establishing chemical thermodynamics.
117. Josiah W. Gibbs. On the equilibrium of heterogeneous substances. *Transactions of the Connecticut Academy of Arts and Sciences*, 3:108–248, 343–524, 1876–1878. Landmark paper introducing the concepts of chemical potential, phase equilibrium, and the method of intercepts.
118. Josiah W. Gibbs. *Elementary Principles in Statistical Mechanics.* Yale University Press, New Haven, 1902. Defines entropy as a thermodynamic potential and discusses integrability of $\delta Q/T$.
119. Carmen J. Giunta. Understanding the mole: The historical and conceptual development of a chemistry standard. *Journal of Chemical Education*, 92(10):1593–1598, 2015.
120. Vittorio Gorini, Andrzej Marek Kossakowski, and E. C. George Sudarshan. Completely positive dynamical semigroups of n-level systems. *Journal of Mathematical Physics*, 17(5):821–825, 1976.
121. Joseph Gay-Lussac. Mmoire sur la dilatation et la contraction des gaz et des vapeurs. *Annales de Chimie*, 63:137–175, 1807. Classic paper reporting the free expansion of gases and related observations.
122. Vitaly L. Ginzburg and Lev D. Landau. On the theory of superconductivity. *Zh. Eksp. Teor. Fiz.*, 20:1064–1082, 1950. English translation in *Collected Papers of L. D. Landau*, ed. D. ter Haar (Pergamon, 1965).
123. William F. Giauque and Duncan P. MacDougall. The production of temperatures below one degree absolute by adiabatic demagnetization of gadolinium sulfate octahydrate. *Physical Review*, 43:768–781, 1933.

124. Nigel Goldenfeld. *Lectures on Phase Transitions and the Renormalization Group.* Addison-Wesley, 1992.
125. David L. Goodstein. *States of Matter.* Dover Publications, New York, 1975.
126. Paul Gordan. Ueber die darstellung der symmetrischen functionen durch linere formen. *Journal für die reine und angewandte Mathematik*, 69:323–354, 1868.
127. R. B. Griffiths. Correlations in Ising ferromagnets. I. *Journal of Mathematical Physics*, 8(3):478–483, 1967.
128. William F. Giauque and M. Ashley Stout. The entropy of water and the third law of thermodynamics. The heat capacity of ice from 15 to 273 K. *Journal of the American Chemical Society*, 58(7):1144–1150, 1936. First experimental determination of residual entropy in ice, confirming the Third Law of Thermodynamics.
129. Robert Giles and Ian Sneddon. *Mathematical Foundations of Thermodynamics.* Pergamon Press, Oxford, 1964.
130. Edward A. Guggenheim. The principle of corresponding states. *Journal of Chemical Physics*, 13(7):253–261, 1945.
131. Rolf Hagedorn. Statistical thermodynamics of strong interactions at high-energies. *Supplemento al Nuovo Cimento*, 3:147–186, 1965.
132. James M. Haile. *Molecular Dynamics Simulation: Elementary Methods.* John Wiley & Sons, New York, 1997.
133. Frederick D. M. Haldane. Fractional statistics in arbitrary dimensions: A generalization of the Pauli principle. *Physical Review Letters*, 67:937–940, 1991.
134. Martin Hasenbusch. Finite size scaling study of lattice models in the three-dimensional Ising universality class. *Physical Review B*, 82:174433, 2010.
135. Stephen W. Hawking. Particle creation by black holes. *Communications in Mathematical Physics*, 43:199–220, 1975.
136. Paul Hertz. On the mechanical foundations of thermodynamics. *Annalen der Physik*, 338(1):225–274, 1910. English translation of *Über die mechanischen Grundlagen der Thermodynamik.*
137. Terrell L. Hill. *An Introduction to Statistical Thermodynamics.* Dover Publications, New York, 1960.
138. Zoran Hadzibabic, Peter Kruger, Marc Cheneau, Baptiste Battelier, and Jean Dalibard. Berezinskii-Kosterlitz-Thouless crossover in a trapped atomic gas. *Nature*, 441:1118–1121, 2006.
139. Bertrand I. Halperin and David R. Nelson. Theory of two-dimensional melting. *Physical Review Letters*, 41(2):121–124, 1978.
140. Bertrand I. Halperin and David R. Nelson. Resistive transition in superconducting films. *Journal of Low Temperature Physics*, 36:599–616, 1979.
141. William G. Hoover. Canonical dynamics: Equilibrium phase-space distributions. *Physical Review A*, 31(3):1695–1697, 1985.
142. James E. House. *Principles of Chemical Kinetics.* Academic Press, Burlington, MA, 2nd edition, 2007.
143. Kerson Huang. *Statistical Mechanics.* John Wiley & Sons, New York, 2nd edition, 1987.
144. Harold Edwin Hall and William Francis Vinen. The rotation of liquid Helium ii. I. Experiments on the propagation of second sound in uniformly rotating Helium ii. *Proceedings of the Royal Society A: Mathematical and Physical Sciences*, 238(1213):215–234, 1956.
145. Jan Ingenhousz. *Expériences et Observations sur Divers Objets de Physique.* Paris, 1785. Includes early experiments on thermal conductivity using metal rods coated with wax.
146. Noboru Innami. Geometry of geodesics for convex billiards and circular billiards. *Nihonkai Mathematical Journal*, 13:73–120, 2002.

147. Ernst Ising. Beitrag zur theorie des ferromagnetismus. *Zeitschrift für Physik*, 31:253–258, 1925.
148. Ted Jacobson. Thermodynamics of spacetime: The Einstein equation of state. *Physical Review Letters*, 75:1260–1263, 1995.
149. Gregg Jaeger. The Ehrenfest classification of phase transitions: Introduction and evolution. *Archive for History of Exact Sciences*, 53(1):51–81, May 1998.
150. Christopher Jarzynski. Nonequilibrium equality for free energy differences. *Physical Review Letters*, 78(14):2690–2693, 1997.
151. Edwin T. Jaynes. Information theory and statistical mechanics. *Physical Review*, 106(4):620–630, 1957. Introduces the maximum entropy principle, reinterpreting Gibbs's ensembles in terms of information theory.
152. Edwin T. Jaynes. *The Gibbs Paradox*, pages 1–21. Springer Netherlands, 1992.
153. Donald W. Jepsen. Dynamics of a simple many-body system of hard rods. *Journal of Mathematical Physics*, 6(3):405–413, 1965.
154. Giovanni Jona-Lasinio. The renormalization group: A probabilistic view. *Il Nuovo Cimento B*, 26(1):99–119, 1975.
155. James Prescott Joule. On the mechanical equivalent of heat. *Philosophical Transactions of the Royal Society of London*, 140:61–82, 1843. Joule's foundational paper quantifying the mechanical equivalent of heat.
156. Ferencz Juettner. Das maxwellsche gesetz der geschwindigkeitsverteilung in der relativtheorie. *Annalen der Physik*, 339(5):856–882, 1911.
157. Erich Joos, H. Dieter Zeh, Claus Kiefer, Domenico Giulini, Jrgen Kupsch, and Ion-Olimpiu Stamatescu. *Decoherence and the Appearance of a Classical World in Quantum Theory.* Springer, 2nd edition, 2003.
158. Leo P. Kadanoff. Scaling laws for Ising models near t_c. *Physics*, 2(6):263–272, 1966.
159. Peter Kapitza. Viscosity of liquid Helium below the λ-point. *Nature*, 141:74–74, 1938.
160. Mehran Kardar. *Statistical Physics of Fields*, volume 2. Cambridge University Press, 2007.
161. Aleksandr I. Khinchin. *Mathematical Foundations of Statistical Mechanics.* Dover Publications, New York, 1949. Translated from the Russian edition (1943).
162. John G. Kirkwood. The statistical mechanical theory of transport processes I. general theory. *Journal of Chemical Physics*, 14(3):180–201, 1946.
163. Charles Kittel. *Phase Transitions and Critical Phenomena.* Academic Press, 1969.
164. Alexei Y. Kitaev. Unpaired majorana fermions in quantum wires. *Physics-Uspekhi*, 44:131–136, 2001.
165. Charles Kittel. *Introduction to Solid State Physics.* John Wiley & Sons, Hoboken, NJ, 8th edition, 2005.
166. Alexei Kitaev. Anyons in an exactly solved model and beyond. *Annals of Physics*, 321:2–111, 2006.
167. Charles Kittel and Herbert Kroemer. *Thermal Physics.* W. H. Freeman, New York, 2nd edition, 1980.
168. Edward H. Knight. *The Practical Dictionary of Mechanics: Supplementary Volume.* Cassell & Co., London, 1884. Supplementary volume.
169. John M. Kosterlitz. The critical properties of the two-dimensional XY model. *Journal of Physics C: Solid State Physics*, 7(6):1046–1060, 1974.
170. Filip Kos, David Poland, David Simmons-Duffin, and Alessandro Vichi. Precision islands in the Ising and o(n) models. *Journal of High Energy Physics*, 2016(8):36, 2016.
171. Werner Krauth. Statistical mechanics: Crystallization in two dimensions. *Nature Physics*, 8:154–155, 2012.
172. Vadim Kaloshin and Alfonso Sorrentino. On the integrability of Birkhoff billiards. *Philosophical Transactions of the Royal Society A*, 376(2131):20170419, 2018.

173. Jan Klaers, Julian Schmitt, Frank Vewinger, and Martin Weitz. Bose–Einstein condensation of photons in an optical microcavity. *Nature*, 468:545–548, 2010.
174. John M. Kosterlitz and David J. Thouless. Ordering, metastability and phase transitions in two-dimensional systems. *Journal of Physics C: Solid State Physics*, 6:1181–1203, 1973.
175. Edwin C. Kemble and John H. van Vleck. On the theory of the temperature variation of the specific heat of Hydrogen. *Physical Review*, 21(6):653–661, June 1923.
176. Hendrik A. Kramers and Gregory H. Wannier. Statistics of the two-dimensional ferromagnet. part I. *Physical Review*, 60(3):252–262, 1941.
177. Keith J. Laidler. *Chemical Kinetics.* Pearson, Upper Saddle River, NJ, 3rd edition, 1997.
178. Lev D. Landau. On the theory of phase transitions. *Zh. Eksp. Teor. Fiz.*, 7:19–32, 1937. English translation in *Collected Papers of L. D. Landau*, Pergamon Press, 1965.
179. Oscar E. Lanford. Time evolution of large classical systems. *Dynamical Systems, Theory and Applications*, pages 1–111, 1975.
180. Robert B. Laughlin. Anomalous quantum hall effect: An incompressible quantum fluid with fractionally charged excitations. *Physical Review Letters*, 50:1395–1398, 1983.
181. Antoine-Laurent Lavoisier. *Traité Élémentaire de Chimie.* Chez Cuchet, Paris, 1789. First edition; includes the formulation of the law of conservation of mass and the modern chemical nomenclature.
182. Melvin Lax. Relation between canonical and microcanonical ensembles. *Physical Review*, 97(6):1419–1420, March 1955.
183. Jan Liphardt, Sophie Dumont, Steven B. Smith, Ignacio Tinoco, and Carlos Bustamante. Equilibrium information from nonequilibrium measurements in an experimental test of the Jarzynski equality. *Science*, 296:1832–1835, 2002.
184. Anthony James Leggett. *Quantum Liquids: Bose Condensation and Cooper Pairing in Condensed Matter.* Oxford University Press, 2006.
185. Andrew Lenard. Momentum distribution in the ground state of the one-dimensional system of impenetrable bosons. *Journal of Mathematical Physics*, 5(7):930–943, 1964.
186. Goran Lindblad. On the generators of quantum dynamical semigroups. *Communications in Mathematical Physics*, 48:119–130, 1976.
187. Otto Lummer and Ernst Kurlbaum. Die strahlung des schwarzen krpers und des hohlraums. *Verhandlungen der Deutschen Physikalischen Gesellschaft*, 17:106–111, 1898.
188. Paul E. Lammert, Xianglin Ke, Jie Li, Cristiano Nisoli, David M. Garand, Vincent H. Crespi, and Peter Schiffer. Direct entropy determination and application to artificial spin ice. *Nature Physics*, 6(10):786–789, 2010.
189. Antoine-Laurent Lavoisier and Pierre-Simon Laplace. Mémoires sur la chaleur. *Mémoires de l'Académie Royale des Sciences*, 1780. Presented in 1780, published in 1783.
190. Elliott H. Lieb and Werner Liniger. Exact analysis of an interacting Bose gas. I. The general solution and the ground state. *Physical Review*, 130(4):1605–1616, 1963.
191. Lev D. Landau and E. M. Lifshitz. *Statistical Physics, Part 1*, volume 5 of *Course of Theoretical Physics.* Pergamon Press, 3rd edition, 1980.
192. Stefano Luzzatto, Ian Melbourne, and Frederic Paccaut. The Lorenz attractor is mixing. *Communications in Mathematical Physics*, 260:393–401, 2005.
193. Andrea J. Liu and Sidney R. Nagel. Nonlinear dynamics: Jamming is not just cool any more. *Nature*, 396:21–22, 1998.
194. David M. Lee, Douglas D. Osheroff, and Robert C. Richardson. *Nobel Lectures in Physics 1996.* World Scientific, 1997.

195. Josef Loschmidt. Ber den zustand des wrmegleichgewichtes eines systems von krpern mit rcksicht auf die schwerkraft. *Sitzungsberichte der Kaiserlichen Akademie der Wissenschaften*, 73:128–142, 1876.
196. Osmo Vin Lounasmaa. *Experimental Principles and Methods Below 1 K.* Academic Press, 1974.
197. Stephen W. Lovesey. *Theory of Neutron Scattering from Condensed Matter*, volume 1. Clarendon Press, Oxford, 1984. Elastic Scattering.
198. Binhua Lin and Stuart A. Rice. Direct measurements of constrained Brownian motion of an individual particle. *Physical Review E*, 72(5):051606, 2005.
199. Elliot H. Lieb and Jakob Yngvason. The physics and mathematics of the second law of thermodynamics. *Physics Reports*, 310(1):1–96, 1999.
200. Ernst Mach. *Principien der Wärmelehre.* J. A. Barth, Leipzig, 1896.
201. Edme Mariotte. *Essai de la Nature de l'Air.* Jean Cusson, Paris, 1676. Independent formulation of the pressure–volume relationship for gases, later known as Boyle–Mariotte's law.
202. Thomas V. Marcella. Entropy production and the second law of thermodynamics: An introduction to the second law analysis. *American Journal of Physics*, 60(10):888–895, 1992. An accessible pedagogical treatment of entropy production and second law analysis.
203. Francois Massieu. Sur les fonctions caractristiques des divers fluides. *Comptes Rendus Hebdomadaires des Séances de l'Académie des Sciences*, 69:858–862, 1869. Introduces the concept of characteristic functions, now known as Massieu potentials.
204. Stephen F. Mason. *A History of the Sciences.* Science Library, Collier Books, New York, 1970.
205. James Clerk Maxwell. Illustrations of the dynamical theory of gases. *Philosophical Magazine*, 19:19–32, 149–164, 1860.
206. James Clerk Maxwell. On the dynamical theory of gases. *Philosophical Transactions of the Royal Society of London*, 157:49–88, 1867.
207. Victor Munoz and William A. Eaton. A simple model for calculating the kinetics of protein folding from three-dimensional structures. *Proceedings of the National Academy of Sciences*, 96:11311–11316, 1999.
208. Anastassia M. Makarieva, Victor G. Gorshkov, and Bai-Lian Li. On the validity of representing hurricanes as Carnot heat engines. *Atmospheric Chemistry and Physics*, 8:17423–17437, 2008. Examines limitations of the Carnot analogy in hurricane thermodynamics.
209. Glenn J. Martyna, Michael L. Klein, and Mark Tuckerman. Nosé–hoover chains: The canonical ensemble via continuous dynamics. *The Journal of Chemical Physics*, 97(4):2635–2643, 1992.
210. Tru Moriya. Anisotropic superexchange interaction and weak ferromagnetism. *Physical Review*, 120(1):91–98, 1960.
211. Tim R. Morris. The exact renormalization group and approximate solutions. *International Journal of Modern Physics A*, 9(14):2411–2450, 1994.
212. Nicholas Metropolis, Arianna W. Rosenbluth, Marshall N. Rosenbluth, Augusta H. Teller, and Edward Teller. Equation of state calculations by fast computing machines. *The Journal of Chemical Physics*, 21(6):1087–1092, 1953.
213. Pirooz Mohazzabi, John A. Smith, and Adriano Citati. Thermodynamics of hurricanes. *Journal of Applied Mathematics and Physics*, 10:1198–1211, 2022. Revisits the thermodynamic description of hurricanes as heat engines.
214. Michael R. Moldover, Jack W. Schmidt, and Charles W. Meyer. Refractive index of Xenon near the critical point. *Physical Review A*, 14:179–188, 1976.

215. Nathaniel David Mermin and Herbert Wagner. Absence of ferromagnetism or antiferromagnetism in one- or two-dimensional isotropic Heisenberg models. *Physical Review Letters*, 17:1133–1136, 1966.
216. National Museum of American History. Apparatus used by Jan Ingenhousz to study thermal conductivity. `https://americanhistory.si.edu/collections/search/object/nmah_1749`, 2025. Smithsonian Institution. Accessed November 10, 2025.
217. Walther Nernst. *Nobel Lectures in Chemistry, 1921–1941*. Elsevier Publishing Company, Amsterdam, 1966. Includes Nernst's original discussion of the heat theorem and its implications for the Third Law of Thermodynamics.
218. Mark E. J. Newman. *Networks: An Introduction*. Oxford University Press, Oxford, 2010.
219. Cristiano Nisoli, Roderich Moessner, and Peter Schiffer. Colloquium: Artificial spin ice: Designing and imaging magnetic frustration. *Reviews of Modern Physics*, 85:1473–1490, 2013.
220. Shuichi Nosé. A unified formulation of the constant temperature molecular dynamics methods. *The Journal of Chemical Physics*, 81(1):511–519, 1984.
221. Chetan Nayak, Steven H. Simon, Ady Stern, Michael Freedman, and Sankar Das Sarma. Non-abelian anyons and topological quantum computation. *Reviews of Modern Physics*, 80:1083–1159, 2008.
222. T. Oguchi. Theory of spin systems. *Journal of the Physical Society of Japan*, 6:31–38, 1951.
223. Lars Onsager. Reciprocal relations in irreversible processes. I. *Physical Review*, 37(4):405–426, 1931.
224. Lars Onsager. Reciprocal relations in irreversible processes. II. *Physical Review*, 38(12):2265–2279, 1931.
225. Lars Onsager. Crystal statistics. I. A two-dimensional model with an order–disorder transition. *Physical Review*, 65:117–149, 1944.
226. Douglas D. Osheroff, Robert C. Richardson, and David M. Lee. Evidence for a new phase of solid He-3. *Physical Review Letters*, 28:885–888, 1972.
227. Wilhelm Ostwald. *Grundriß der allgemeinen Chemie*. Wilhelm Engelmann, Leipzig, 1890.
228. Edward Ott. *Chaos in Dynamical Systems*. Cambridge University Press, Cambridge, 2nd edition, 2002.
229. Leonard Salomon Ornstein and Frits Zernike. Accidental deviations of density and opalescence at the critical point of a single substance. *Proc. Acad. Sci. Amsterdam*, 17:793–806, 1914.
230. Thanu Padmanabhan. Thermodynamical aspects of gravity: New insights. *Reports on Progress in Physics*, 73:046901, 2010.
231. Giorgio Parisi. *The Theory of Simple Glasses: Exact Solutions in Infinite Dimensions*. Cambridge University Press, 2020.
232. Wolfgang Pauli. Ber das h-theorem vom anwachsen der entropie vom standpunkt der neuen quantenmechanik. In *Probleme der modernen Physik: Arnold Sommerfeld Festschrift*. S. Hirzel, Leipzig, 1928.
233. Linus Pauling. The structure and entropy of ice and of other crystals with some randomness of atomic arrangement. *Journal of the American Chemical Society*, 57(12):2680–2684, 1935.
234. Rama K. Pathria and Paul D. Beale. *Statistical Mechanics*. Elsevier, Oxford, 3rd edition, 2011.
235. Ilya Prigogine and Robert Defay. *Chemical Thermodynamics*. Longmans, Green and Co., London, 1961. Classic monograph on chemical equilibrium, reaction energetics, and irreversible processes.

236. Rudolf Peierls. On Ising's model of ferromagnetism. *Mathematical Proceedings of the Cambridge Philosophical Society*, 32(3):477–481, 1936.
237. Jerome K. Percus. An exact result for the one-dimensional hard-rod fluid in a field. *Journal of Statistical Physics*, 15(6):505–511, 1976.
238. William D. Phillips. Laser cooling and trapping of neutral atoms. *Reviews of Modern Physics*, 70(3):721–741, 1998.
239. Jos Pinto Peixoto and M. Ali Kettani. The control of the water cycle. *Scientific American*, 228(4):46–62, 1973. Describes the thermodynamic concept of a "rainmaker" chimney based on adiabatic cooling and atmospheric convection.
240. Max Planck. *Vorlesungen über Thermodynamik*. Veitel Heine, Berlin, 1897.
241. Joseph Polchinski. Renormalization and effective Lagrangians. *Nuclear Physics B*, 231:269–295, 1984.
242. Isaak Yakovlevich Pomeranchuk. On the theory of liquid He-3. *ZhETF*, 20:919–926, 1950. English trans.: Soviet Phys. JETP.
243. Renfrey B. Potts. Some generalized order-disorder transformations. *Mathematical Proceedings of the Cambridge Philosophical Society*, 48:106–109, 1952.
244. Ilya Prigogine. *Étude thermodynamique des phénomènes irréversibles*. Desoer, Lige, 1947.
245. Ilya Prigogine. *Introduction to the Thermodynamics of Irreversible Processes*. Interscience / Wiley, New York, 3rd edition, 1968. Foundational monograph developing the modern theory of entropy production.
246. David Poland and Harold A. Scheraga. *Theory of Helix-Coil Transitions in Biopolymers*. Academic Press, 1970.
247. Andrea Pelissetto and Ettore Vicari. Critical phenomena and renormalization-group theory. *Physics Reports*, 368(6):549–727, 2002.
248. David J. Resnick, John C. Garland, J. T. Boyd, Siegfried Shoemaker, and Robert S. Newrock. Kosterlitz–Thouless transition in proximity-coupled superconducting arrays. *Physical Review Letters*, 47:1542–1545, 1981.
249. Arthur P. Ramirez, Atsushi Hayashi, Robert J. Cava, R. Siddharthan, and B. Sriram Shastry. Zero-point entropy in "spin ice". *Nature*, 399:333–335, 1999. Experimental measurement of zero-point entropy in $Dy_2Ti_2O_7$ and $Ho_2Ti_2O_7$.
250. Malvin A. Ruderman and Charles Kittel. Indirect exchange coupling of nuclear magnetic moments by conduction electrons. *Physical Review*, 96(1):99–102, 1954.
251. Ch. Regg, B. Normand, M. Matsumoto, *et al.* Quantum magnets under pressure: Controlling elementary excitations in $tlcucl_3$. *Physical Review Letters*, 100:205701, 2008.
252. Duane H. D. Roller. *The Early Development of the Concepts of Temperature and Heat*. Harvard University Press, Cambridge, MA, 1950.
253. David Ruelle. *Thermodynamic Formalism*. Addison-Wesley, Reading, MA, 1978. Reprinted by Cambridge University Press, 2004.
254. David Ruelle. *Chaotic Evolution and Strange Attractors*. Cambridge University Press, Cambridge, 1989.
255. Ernest Rutherford. The scattering of α and β particles by matter and the structure of the atom. *Philosophical Magazine*, 21:669–688, 1911.
256. Wayne M. Saslow. A history of thermodynamics: The missing manual. *Entropy*, 22(1):77, 2020.
257. H. Eugene Stanley, Sergey V. Buldyrev, Murat Canpolat, Osamu Mishima, Mehdi R. Sadr-Lahijany, Alessandro Scala, and Francis W. Starr. The puzzling behavior of water at very low temperature. *Physical Chemistry Chemical Physics*, 2(8):1551–1558, 2000.

258. Elad Schneidman, Michael J. Berry, Ronen Segev, and William Bialek. Weak pairwise correlations imply strongly correlated network states in a neural population. *Nature*, 440:1007–1012, 2006.
259. Douglas Self. Ammonia motors, 2024. Museum of RetroTech. Accessed March 5, 2024. A historical overview of ammonia-based engines and their thermodynamic principles.
260. Otto Stern and Walther Gerlach. Der experimentelle nachweis der richtungsquantelung im magnetfeld. *Zeitschrift für Physik*, 9:349–352, 1922.
261. Claude E. Shannon. A mathematical theory of communication. *Bell System Technical Journal*, 27(3):379–423, 1948. Reprinted in *Claude E. Shannon: Collected Papers*, IEEE Press, 1993.
262. Francis E. Simon. The third law of thermodynamics. *Reports on Progress in Physics*, 1(1):1–31, 1931. A comprehensive review of the Third Law and low-temperature thermodynamics.
263. Yakov G. Sinai. On the notion of entropy of a dynamical system. *Doklady Akademii Nauk SSSR*, 124:768–771, 1963.
264. Yakov G. Sinai. Dynamical systems with elastic reflections. Ergodic properties of dispersing billiards. *Russian Mathematical Surveys*, 25(2):137–189, 1970.
265. Dan M. Stamper-Kurn and Masahito Ueda. Spinor Bose gases: Symmetries, magnetism, and quantum dynamics. *Reviews of Modern Physics*, 85:1191–1244, 2013.
266. Tony H. R. Skyrme. A unified field theory of mesons and baryons. *Nucl. Phys.*, 31:556–569, 1962.
267. Stephen Smale. Differentiable dynamical systems. *Bulletin of the American Mathematical Society*, 73(6):747–817, 1967.
268. Jun John Sakurai and Jim Napolitano. *Modern Quantum Mechanics*. Addison-Wesley, San Francisco, 2nd edition, 1994.
269. Igor M. Sokolov. Statistical mechanics of entropic forces: Disassembling a toy. *European Journal of Physics*, 31(6):1353–1367, 2010.
270. Herbert Spohn. Entropy production for quantum dynamical semigroups. *Journal of Mathematical Physics*, 19(5):1227–1230, 1978.
271. Peter Schiffer and Arthur J. Ramirez. Frustration and disorder in magnetic materials. *Nature Physics*, 3(9):547–553, 2007. Review of experimental studies of frustration and entropy in spin ices and related systems.
272. Mark Srednicki. Chaos and quantum thermalization. *Physical Review E*, 50(2):888–901, 1994.
273. H. Eugene Stanley. *Introduction to Phase Transitions and Critical Phenomena*. Oxford University Press, Oxford, 1971.
274. Otto Stern. Ber das wesen der Maxwell–Boltzmannschen geschwindigkeitsverteilung bei atomstrahlen. *Zeitschrift für Physik*, 52:110–127, 1919. Atomic beam velocity distribution experiment.
275. Miles V. Sullivan. Drinking bird, July 1946. Filed September 23, 1945.
276. Kurt Symanzik. Small distance behaviour in field theory and power counting. *Communications in Mathematical Physics*, 18:227–246, 1970.
277. John L. Synge. *The Relativistic Gas*. North-Holland, Amsterdam, 1957.
278. David J. Thouless, Paul W. Anderson, and Robert G. Palmer. Solution of "solvable model of a spin glass." *Philosophical Magazine*, 35(3):593–601, 1977.
279. Richard C. Tolman and Paul C. Fine. On the irreversible production of entropy. *Reviews of Modern Physics*, 20(1):51–77, 1948. A classic discussion of entropy production and the work–entropy relationship in irreversible thermodynamics.
280. Benjamin Thompson. An experimental enquiry concerning the source of the heat which is excited by friction. *Philosophical Transactions of the Royal Society of London*,

88:80–102, 1798. Count Rumford's classic paper on heat generation by friction, challenging the caloric theory.
281. Laszlo Tisza. *Generalized Thermodynamics*. MIT Press, Cambridge, MA, 1966. Includes chapter "The Evolution of the Concepts of Thermodynamics."
282. Richard Chace Tolman. *Relativity, Thermodynamics and Cosmology*. Oxford University Press, Oxford, 1934.
283. Richard C. Tolman. *The Principles of Statistical Mechanics*. Oxford University Press, Oxford, 1938. Reprinted by Dover Publications (1979).
284. Lewi Tonks. The complete equation of state of one-, two- and three-dimensional gases of hard elastic spheres. *Physical Review*, 50(10):955–963, 1936.
285. Kazumasa A. Takeuchi and Masaki Sano. Universal fluctuations in growing interfaces of turbulent liquid crystals. *Physical Review Letters*, 104:230601, 2010.
286. Daniel C. Tsui, Horst L. Stormer, and Arthur C. Gossard. Two-dimensional magnetotransport in the extreme quantum limit. *Physical Review Letters*, 48:1559–1562, 1982.
287. Mary Tsingou. Fermi-Pasta-Ulam numerical experiment documentation. Los Alamos National Laboratory internal report LA-1940, 1955. Co-author of the Fermi-Pasta-Ulam-Tsingou experiment; see also D. D. Holm and N. J. Zabusky, *Phys. Today* 57, 31 (2004).
288. George E. Uhlenbeck and George W. Ford. *Lectures in Statistical Mechanics*. American Mathematical Society, Providence, RI, 1963.
289. Johannes D. van der Waals. *On the Continuity of the Gaseous and Liquid States*. 1873a. Ph.D. thesis, University of Leiden (English translation title).
290. Johannes D. van der Waals. *Over de Continuiteit van den Gas- en Vloeistoftoestand (On the Continuity of the Gaseous and Liquid States)*. PhD thesis, Universiteit Leiden, Leiden, 1873b. Reprinted in English in *Physics Today*, 1988.
291. Johannes D. van der Waals. Nobel lecture: The equation of state for gases and liquids. *Nobel Lectures, Physics 1901–1921*, 1910.
292. Bartel L. van der Waerden. Die lange reichweite der regelmigen atomanordnung in mischkristallen. *Zeitschrift für Physik*, 118:473–488, 1941.
293. Hermann von Helmholtz. *Principien der Thermodynamik*, volume 22. Annalen der Physik und Chemie, 1884. Original formulation of the Helmholtz theorem on mechanical foundations of thermodynamics.
294. Martin van Hecke. Jamming of soft particles: Geometry, mechanics, scaling and isostaticity. *Journal of Physics: Condensed Matter*, 22(3):033101, 2010.
295. Cédric Villani. *A Review of Mathematical Topics in Collisional Kinetic Theory*, volume 1 of *Handbook of Mathematical Fluid Dynamics*. Elsevier, 2002.
296. Cédric Villani. *Hypocoercivity*. American Mathematical Society, Providence, RI, 2010.
297. Hendrika J. van Leeuwen. Problmes de la thorie lectronique du magntisme. *Journal de Physique et le Radium*, 2(12):361–377, 1921.
298. John H. Van Vleck. *The Theory of Electric and Magnetic Susceptibilities*. Oxford University Press, Oxford, 1932.
299. John von Neumann. Wahrscheinlichkeitstheoretischer aufbau der quantenmechanik. *Nachrichten von der Gesellschaft der Wissenschaften zu Göttingen*, pages 245–272, 1927.
300. John von Neumann. Beweis des ergodensatzes und des h-theorems in der neuen mechanik. *Zeitschrift für Physik*, 57:30–70, 1929.
301. John von Neumann. Proof of the quasi-ergodic hypothesis. *Proceedings of the National Academy of Sciences of the USA*, 18(1):70–82, 1932.

302. Gregory H. Wannier. Antiferromagnetism. The triangular Ising net. *Physical Review*, 79:357–364, 1950. Classic paper introducing geometric frustration and residual entropy in spin systems.
303. Qi-Huo Wei, Clemens Bechinger, and Paul Leiderer. Single-file diffusion of colloids in one-dimensional channels. *Science*, 287(5453):625–627, 2000.
304. Franz J. Wegner. Corrections to scaling laws. *Physical Review B*, 5:4529–4536, 1972.
305. Pierre Weiss. L'hypothse du champ molculaire et la proprit ferromagntique. *Journal de Physique Théorique et Appliquée*, 6(1):661–690, 1907. Introduced the concept of the molecular field and the mean-field theory of ferromagnetism.
306. Xiao-Gang Wen. *Quantum Field Theory of Many-Body Systems.* Oxford University Press, Oxford, 2004.
307. Christof Wetterich. Exact evolution equation for the effective potential. *Physics Letters B*, 301(1):90–94, 1993.
308. Kenneth G. Wilson and Michael E. Fisher. Critical exponents in 3.99 dimensions. *Physical Review Letters*, 28:240–243, 1972.
309. John C. Wheatley. Superfluidity in He-3. *Reviews of Modern Physics*, 47:415–470, 1975.
310. Evan J. Williams. The effect of thermal agitation on atomic arrangement in alloys. III. *Proceedings of the Royal Society A*, 152:231–252, 1935.
311. Kenneth G. Wilson. Renormalization group and critical phenomena. I. renormalization group and the Kadanoff scaling picture. *Physical Review B*, 4:3174–3183, 1971.
312. Kenneth G. Wilson. The renormalization group: Critical phenomena and the Kondo problem. *Reviews of Modern Physics*, 47(4):773–840, 1975.
313. Frank Wilczek. Quantum mechanics of fractional-spin particles. *Physical Review Letters*, 49:957–959, 1982.
314. Kenneth G. Wilson and J. Kogut. The renormalization group and the ϵ expansion. *Physics Reports*, 12(2):75–200, 1974a.
315. Kenneth G. Wilson and John B. Kogut. The renormalization group and the ϵ expansion. *Physics Reports*, 12(2):75–199, 1974b.
316. Jonathan C. Wright, Henry W. Moos, John H. Colwell, Bobby W. Mangum, and Donald D. Thornton. Dypo$_4$: A three-dimensional ising antiferromagnet. *Physical Review B*, 3:843–849, 1971.
317. Gang Wang, Eric M. Sevick, Emil Mittag, Daniel J. Searles, and Denis J. Evans. Experimental demonstration of violations of the second law of thermodynamics for small systems and short time scales. *Physical Review Letters*, 89:050601, 2002.
318. Fa-Yueh Wu. The Potts model. *Reviews of Modern Physics*, 54(1):235–268, 1982.
319. X. Sunney Xie, Huan-Xiang Lu, *et al. Single-Molecule Biophysics.* Springer, New York, 2018. Includes dwell-time analysis and single-molecule kinetics.
320. Chen-Ning Yang. The spontaneous magnetization of a two-dimensional Ising model. *Physical Review*, 85:808–816, 1952.
321. A. Peter Young. Melting and the vector Coulomb gas in two dimensions. *Physical Review B*, 19(4):1855–1866, 1979.
322. Will Yeadon and Mark Quinn. Thermodynamics education for energy transformation: A stirling engine experiment. *Physics Education*, 56(5):055033, 2021.
323. Jacques Yvon. La thorie statistique des fluides et l'quation d'tat. *Actualités Scientifiques et Industrielles*, 203:1–88, 1935.
324. Mark W. Zemansky. *Heat and Thermodynamics: An Intermediate Textbook.* McGraw-Hill, New York, 7th edition, 1997. A classic intermediate text providing clear discussions of the laws of thermodynamics and entropy.
325. Gustav Zeuner. *Technical Thermodynamics. Volume I: Fundamental Laws of Thermodynamics, Theory of Gases.* D. Van Nostrand Company, New York, 1906. Introduced the concept of polytropic transformations as a generalization of adiabatic processes.

326. Jean Zinn-Justin. *Quantum Field Theory and Critical Phenomena.* Oxford University Press, 4th edition, 2002.
327. Wojciech H. Zurek. Decoherence, Einselection, and the quantum origins of the classical. *Reviews of Modern Physics*, 75(3):715–775, 2003.

Indices

Index by keywords

For Product Safety Concerns and Information please contact our EU
representative GPSR@taylorandfrancis.com
Taylor & Francis Verlag GmbH, Kaufingerstraße 24, 80331 München, Germany

www.ingramcontent.com/pod-product-compliance
Ingram Content Group UK Ltd.
Pitfield, Milton Keynes, MK11 3LW, UK
UKHW052217190926
475702UK00006B/1065
* 9 7 8 1 0 3 2 9 8 2 0 4 5 *